ECOSYSTEM FUNCTION IN SAVANNAS

MEASUREMENT AND MODELING
AT LANDSCAPE TO GLOBAL SCALES

ECOSYSTEM FUNCTION IN SAVANNAS

MEASUREMENT AND MODELING AT LANDSCAPE TO GLOBAL SCALES

EDITED BY

MICHAEL J. HILL
NIALL P. HANAN

CRC Press
Taylor & Francis Group
Boca Raton London New York

CRC Press is an imprint of the
Taylor & Francis Group, an **informa** business

CRC Press
Taylor & Francis Group
6000 Broken Sound Parkway NW, Suite 300
Boca Raton, FL 33487-2742

First issued in paperback 2019

© 2011 by Taylor and Francis Group, LLC
CRC Press is an imprint of Taylor & Francis Group, an Informa business

No claim to original U.S. Government works

ISBN-13: 978-1-4398-0470-4 (hbk)
ISBN-13: 978-0-367-86455-2 (pbk)

Library of Congress Cataloging-in-Publication Data

Ecosystem function in Savannas : measurement and modeling at landscape to global
  scales / edited by Michael J. Hill, Niall P. Hanan.
       p. cm.
  Includes bibliographical references and index.
  ISBN 978-1-4398-0470-4 (hardcover : alk. paper)
    1. Savanna ecology--Mathematical models. 2. Savanna ecology--Statistical methods.
  3. Savanna plants--Phenology. 4. Savanna plants--Geographical distribution. I. Hill,
  Michael J., 1953- II. Hanan, Niall P. III. Title.

QH541.5.P7E27 2010
577.4'8015118--dc22   2010036947

Visit the Taylor & Francis Web site at
http://www.taylorandfrancis.com

and the CRC Press Web site at
http://www.crcpress.com

Contents

Section I   Savannas: Biogeographical and Ecological Perspectives

Section II   Carbon, Water and Trace Gas Fluxes in Global Savannas

Section III Remote Sensing of Biophysical and Biochemical Characteristics in Savannas

Section IV Perspectives on Patch- to Landscape-Scale Savanna Processes and Modeling

Section V Regional Carbon Dynamics in Savanna Systems: Remote Sensing and Modeling Applications

Section VI Continental and Global Scale Modeling of Savannas

Section VII   Understanding Savannas as Coupled Human–Natural Systems

Section VIII   Synthesis

List of Figures

Acknowledgments

The editors would like to thank all the contributors for their responsiveness and good grace in dealing with editorial instructions and the review process. We would also like to thank those contributing authors who attended the special session on measurement and modeling in savannas organized by the editors at the AGU Fall Meeting in San Francisco in 2007. The response to this session was a catalyst for the development of this book.

Acknowledgements

Introduction

This book seeks to be the key reference work on the quantitative spatiotemporal dynamics of global savannas spanning biophysics, biochemistry, measurement, and modeling. The book is explicitly targeted to the measurement and modeling of the dynamics of the natural savanna systems and their interaction with anthropogenic forces. It seeks ultimately to build a synthesis across scales; measurement and modeling methods; natural and human processes, including information on measurement of carbon and water exchange in savannas; remote sensing of surface properties; spatiotemporal measurement and modeling of tree–grass systems; and impacts of disturbance from grazing, fire, and other human actions. Hence, conversion of savanna to agriculture or forestry is examined for only two large South American savannas where these conversions are highly significant and from the viewpoint of the resulting changes in ecosystem function in terms of carbon fluxes. Either directly, or by brief review of the literature, contributed chapters cover virtually all of the measurements, methods, remote sensing approaches, and model types applied to savanna systems. The individual chapters in the book represent not only the diversity of approaches among disciplines but also the different geographical perspectives among savanna specialists in the major temperate and tropical savanna regions. In the final chapter, we forge a synthesis among these diverse perspectives, scales, methods, sensors, and modeling approaches that defines gaps, connections, and potential synergies that can only be realized through greater collaboration and more definite focus on savannas by the global science community.

Although we fully understand that savannas cannot be compartmentalized between their natural biophysical and anthropogenic functions and interactions, this book is focused on capturing the current state of knowledge within the biophysical realm. In part, we still lack a sufficiently detailed understanding of the biophysical and ecosystem function in natural savanna systems, and this places a limitation on the development of a truly integrated analysis across the coupled human–environment domain. We try to partly forge this bridge by describing the coupled human–savanna systems and discussing the current implementation and future possibilities for linked biophysical and sociological modeling. However, we recognize that this book largely comes from a biophysical science paradigm and that, therefore, there is a place for another volume that seeks to treat in detail, the methods and approaches that might be needed to form a truly integrated, cross-paradigm approach to measuring, modeling, and assisting decision making in the global populated coupled human–savanna system.

The book is constructed in eight sections. The first section provides a spatially explicit description of the major tropical and subtropical regions within the global savanna biome in terms of the broad biogeographical properties, temporal dynamics, and disturbance levels. It introduces the reader to the global distribution of savannas, both at the large scale with spatially explicit mapping and in smaller subregional patchworks that are currently not easily mapped. It then examines the current understanding of tree–grass interactions (the fundamental property of savanna systems as they lie in the continuum between grasslands and dry forests). The second section examines carbon, water, energy, and trace gas fluxes and their response to climate variability and anthropogenic disturbances for a selection of the major global savanna regions. The global flux network is substantial, but savannas tend to be under-represented at present. The chapters cover the Australian tropical savanna, the southern African savanna in Kruger National Park, the savanna gradient from the cerrado to the Amazon forest, and oak savannas in Texas and California.

The third section examines quantitative surface properties of savannas that can be retrieved using remote sensing, can characterize the tree and grass layers in terms of tree demography and vegetative cover, and can provide inputs to models. These include structure and biomass description with active sensors; fractional cover, biochemistry, and canopy description with optical and hyperspectral sensors; canopy cover and land clearing with multispectral time series data; and fire and burn scar detection with a variety of sensors. The fourth section introduces the reader to the diversity of conceptual and numerical approaches currently used to explore savanna dynamics at patch-landscape scales; the chapters, in turn, examine environmental factors, spatially explicit modeling, modeling of fires, and combination of grid modeling including climatological and disturbance factors with upscaling approaches to represent aggregate effects at regional scale.

The fifth section focuses on the use of remote sensing at regional-continental scales in a variety of ways including land surface–climate linkages in Africa, basic land cover change linked to process change by ground measurements, land surface models that explicitly target forage dynamics for cattle production and drought assessment, and model-data assimilation with a remote sensing–specific model incorporating ground and remote sensing observations. The sixth section describes model applications at continental-global scales, including several well-established biogeochemical and dynamic global vegetation models and contrasts results and patterns using these models across the global savanna regions.

The seventh section provides a review of the characteristics of land use and human societies that have coevolved with savannas in the different regions. It then examines how biophysical analysis and modeling can be combined with socioeconomic analysis to give a systems level synthesis of global savannas as coupled human–natural systems. The eighth and last section draws together the major strands of work and intersecting themes from the book to explore how savanna modeling and measurement approaches might be better unified and integrated.

Savannas worldwide are fascinating ecosystems that support unique communities of pastoral and agropastoral people alongside wild and domestic herbivores. The book addresses some of the discontinuities in the treatment of savannas by the scientific community and documents a range of measurements, methods, technologies, applications, and modeling approaches side by side. This confluence of information allows synergies, connections, integrative opportunities, and complementarities among approaches and data sources to emerge for the reader and the savanna research community. Hopefully, this will help to indicate where measurement and modeling could be used to harmonize among scales and across disciplinary boundaries. In particular, we hope that this book will help to bridge across the markedly different perspectives on savannas by which ecologists, biogeochemists, remote sensors, geographers, anthropologists, and modelers approach their science.

Michael J. Hill
Niall P. Hanan

Editors

Michael J. Hill received his PhD degree from the University of Sydney, Australia, in 1985. He spent 12 years in the CSIRO Division of Animal Production and then 6 years in the Bureau of Rural Sciences in the Department of Agriculture, Fisheries, and Forestry of the Australian Government, where he carried out research in and contributed to the management of the Cooperative Research Centre for Greenhouse Accounting. In 2006, he became a professor of earth systems science in the Department of Earth Systems Science and Policy at the University of North Dakota. He has a background in grassland agronomy, but he has been working with spatial information and remote sensing of land systems for the past 17 years. Dr. Hill has published widely on agronomy, ecology, biogeography and production of grasslands, and radar, multispectral, and hyperspectral remote sensing of grasslands, and more recently he has been involved in the development of scenario analysis models for assessment of carbon dynamics in Australian rangeland and savanna systems. His current interests are in the use of MODIS land product data in model-data assimilation, application of quantitative information from hyperspectral and multiangle imaging to vegetation description, multicriteria and decision frameworks for coupled human–environment systems, and methods and approaches to application of spatial data for land use management.

Niall P. Hanan received his PhD from the University of London in 1990. Since 1998, he has been a research scientist in the Natural Resource Ecology Laboratory at Colorado State University. His research interests include savanna ecology function and dynamics; biosphere–atmosphere interactions and global change; ecophysiology of carbon, water, and energy exchange between plants and atmosphere; and radiative and aerodynamic transfer in vegetation canopies. His current research centers on the ecology of semiarid grassland and savanna systems in Africa and North America. Ongoing projects include research into the ecological determinants and dynamics of tree–grass interactions and coexistence in African savannas; the measurement of biosphere–atmosphere exchange of water and carbon and the implications for competitive interactions and productivity in savannas; impact of land use change in semiarid systems on long-term carbon and water dynamics; and the greenhouse gas implications of cattle and intensive cattle production systems. In 2007, he spent a year as a U.S. Department of State Fulbright Scholar at the University of Bamako and the Institut Polytechnique Rural in Mali.

Lead Authors

Almut Arneth is an associate professor at the Department for Physical Geography and Ecosystem Analysis at Lund University (almut.arneth@nateko.lu.se). Her research interests are biosphere–atmosphere interactions and the role of vegetation dynamics and terrestrial biogeochemical cycles in the climate system. A specific objective is to investigate these interactively, considering processes that link ecosystem greenhouse gas emissions and emissions of atmospherically reactive substances, combining field observations and modeling.

Gregory P. Asner is a professor in the Department of Global Ecology, Carnegie Institution and at the Department of Earth System Science, Stanford University (gpa@stanford.edu). His research aims at understanding connections between ecosystems, resource use, and climate change. His remote sensing work focuses on the use of new technologies for studies of ecosystem structure, function, and biodiversity.

Dennis Baldocchi is a professor of biometeorology at the University of California, Berkeley (baldocchi@berkeley.edu). His research interests include measuring and modeling trace gas exchange between ecosystems and the atmosphere and understanding their linkage with climatic, physiological, and ecological drivers. He pursues this research with a set of studies focusing on oak savanna, annual grasslands, and peatlands in California and through international coordination of the global FLUXNET project.

Damian J. Barrett is a professorial research fellow with a joint appointment in the Centre for Water in the Minerals Industry (CWiMI) and Centre for Mined Land Rehabilitation (CMLR) in the Sustainable Minerals Institute at the University of Queensland (d.barrett@smi.uq.edu.au). His research interests include biogeochemical modeling, environmental biophysics, ecosystem functioning, biodiversity, and sustainability.

Gabriela Bucini is a researcher in the Natural Resources and Environment Laboratory at Colorado State University (gbucini@nrel.colostate.edu). She studies African savanna ecosystems with a focus on woody cover. She is interested in woody cover mapping methods and upscaling effects. Her research has explored woody cover interactions with climate, soil, herbivores, fire, and humans from regional to continental scales.

Mercedes M. C. Bustamante (mercedes@unb.br) is a professor of the Department of Ecology at the Universidade de Brasília and is working on

Ecosystem Ecology (Biogeochemistry). Her research interests include the effects of global environmental changes and land uses changes on the functioning of savannas of Central Brazil, especially C and N cycling.

John O. Carter is a principal scientist with the Queensland Government (John.Carter@climatechange.qld.gov.au). His research interests are applying the techniques of multiple constraints to model parameterization and model-data fusion for model calibration and using models to assess the response of natural resource systems to climate variability and management.

Tim Danaher is currently the manager of the Remote Sensing Unit within the NSW Department of Environment and Climate Change and is responsible for developing systems to map and monitor vegetation across NSW using Landsat imagery and high-resolution SPOT 5 imagery (Tim.Danaher@ environment.nsw.gov.au). He was previously a principal scientist and team leader for the Queensland Statewide Landcover and Trees Study (SLATS). He has worked on radiometric calibration and correction techniques that enable the better use of Landsat satellite imagery for spatial and time-series mapping of vegetation changes, has developed indices for mapping of woody vegetation cover, and has developed methods for mapping vegetation cover change.

Andrés Etter is a professor in the Department of Ecology and Territory of the School of Environmental and Rural Studies from Javeriana University (Bogotá-COLOMBIA) (aetter@javeriana.edu.co). He has 20 years of experience in landscape ecological mapping and research in the Colombian Amazon forests, Orinoco savannas, and Andean Montane forest regions integrating biophysical, socioeconomic, and historic data, using remote sensing and GIS. His current research deals with the modeling and understanding of land use change processes aimed at better informed and more dynamic conservation planning processes.

Kathleen A. Galvin is a professor in the Department of Anthropology and a senior research scientist at the Natural Resource Ecology Laboratory, Colorado State University (Kathleen.Galvin@colostate.edu). Her research interests include human adaptability and vulnerability, human dimensions of global environmental change, arid and semiarid tropical ecosystems, pastoralism, household economics and household agent–based modeling, diet, nutrition, and food security in pastoral populations.

Niall P. Hanan is a research scientist at the Natural Resource Ecology Laboratory, Colorado State University, Fort Collins (niall.hanan@colostate.edu). His research interests include ecological determinants and dynamics of tree–grass interactions and coexistence in African savannas; impact and feedbacks associated with land use change; and long-term carbon and water dynamics in semiarid and savanna systems.

Michael J. Hill is a professor in Earth system science and policy at the University of North Dakota (hillmj@aero.und.edu). His research interests include remote sensing, biogeochemical processes and land use change in savanna and grassland ecosystems, and analysis of coupled human–environment systems using spatial multicriteria analysis.

Lindsay B. Hutley is an associate professor at Charles Darwin University, Darwin, Australia (Lindsay.hutley@cdu.edu.au). He has research interests in tropical savanna biogeography with a focus on carbon and water cycles and how fire, grazing, weed invasion, and climate change influence these processes.

Florian Jeltsch is a professor of plant ecology and conservation biology at the University of Potsdam, Germany (jeltsch@rz.uni-potsdam.de). He has a background in spatial modeling with a focus on understanding and predicting the dynamics of populations, communities, and landscapes in a variable environment. A major research topic is the effect of land use and climate change on arid and semiarid savanna systems.

Frank van Langevelde is an assistant professor in resource ecology at Wageningen University in the Netherlands (frank.vanlangevelde@wur.nl). His current research interest focuses on understanding herbivore–plant interactions in a spatial context. It is grouped around three themes that examine (1) the effects of limiting resources, competition with grasses, and defoliation (grazing and fire) on savanna tree seedlings from different environmental conditions; (2) the spatial dynamics of tree–grass interactions and species coexistence in savanna systems; and (3) how competition and facilitation shape communities of herbivores and plants.

Adam Liedloff is a research scientist with CSIRO based in Darwin, Australia (adam.liedloff@csiro.au). His research involves developing models of landscape function and his interests include fire ecology, tree and grass dynamics, ecohydrology, and carbon accounting in the tropical savannas.

Marcy E. Litvak is an assistant professor in biology at the University of New Mexico (mlitvak@unm.edu). Her research interests are focused on quantifying biotic and abiotic controls over ecosystem–atmosphere exchange of carbon, water, and energy and examining biogeochemical processes across ecological gradients.

Richard M. Lucas is a professor at Aberystwyth University in Wales, United Kingdom (rml@aber.ac.uk), where he is involved primarily in the integration of single data and time-series radar, optical (hyperspectral), and LiDAR data for retrieving the biomass, structure, and speces or community composition of ecosystems ranging from subtropical woodlands to tropical forests and mangroves.

Christopher Potter is a senior research scientist at the Biospheric Science Branch at NASA Ames Research Center, Moffett Field, California (Chris. Potter@nasa.gov). His research has involved the development of the CASA global model of ecosystem exchange and its application to understanding the dynamics of a range of global ecosystems including the Amazonia and the Boreal zone.

David P. Roy is a professor at the Geographic Information Science Center of Excellence, South Dakota State University, South Dakota (david.roy@sdstate.edu). His research interests include the development of remote sensing and advanced computing methods to integrate or fuse satellite sensor data and to map and characterize terrestrial change, in particular the occurrence and spatial extent of vegetation fires; the causes and consequences of land cover and land use change; and fire–climate–vegetation interactions. He is also interested in development of methodologies to facilitate the transfer of remote sensing products into the user domain, particularly in developing countries.

Matthew D. Turner is a professor in geography at the University of Wisconsin, Madison (mturner2@wisc.edu). His research interests include political ecology, nutrient cycling, and grazing ecology in tropical savanna and steppe environments; GIS for integrated socioecological analysis; and political decentralization and common property management.

George Louis Vourlitis is a professor in biological sciences at California State University in San Marcos, California (georgev@csusm.edu). His research interests lie in quantifying the rates of mass (CO_2, H_2O vapor, and other trace gases) and energy exchange of terrestrial ecosystems and how these surface–atmosphere exchanges are affected by, and feedback on, global change.

Kerstin Wiegand is a professor of ecosystem modeling at the University of Göttingen, Germany (Kerstin.Wiegand@uni-goettingen.de). Her research interests revolve around spatial ecology, biocomplexity, biodiversity, plant population and community ecology, and nature conservation, with a methodological emphasis on spatial simulation models and spatial statistics.

Christopher A. Williams is an assistant professor in the Graduate School of Geography at Clark University in Massachusetts (CWilliams@clarku.edu). His research addresses how the Earth's biosphere responds to natural and human perturbations and combines field, laboratory, and remote sensing data with process-based modeling aimed at understanding how terrestrial biophysical and biogeochemical processes are influenced by hydroclimatic variability and disturbance.

Contributors

Sally Archibald
Council for Scientific and Industrial
    Research
Pretoria, South Africa

John Armston
Department of Environment and
    Resource Management
Queensland Climate Change
    Centre of Excellence
Queensland, Australia

Almut Arneth
Department of Physical Geography
    and Ecosystem Analysis
Lund University
Lund, Sweden

Gregory P. Asner
Department of Global Ecology
Carnegie Institution of Washington
Washington, D.C.

and

Department of Earth System
    Science
Stanford University
Stanford, California

Dennis Baldocchi
Berkeley Atmospheric Science
    Center
University of California-Berkeley
Berkeley, California

Eduardo R. M. Barbosa
Resource Ecology Group
Wageningen University
Wageningen, the Netherlands

Damian J. Barrett
Centre for Water in the Minerals
    Industry

and

Centre for Mined Land
    Rehabilitation
Sustainable Minerals Institute
University of Queensland
Queensland, Australia

John Battles
Berkeley Atmospheric Science
    Center
University of California-Berkeley
Berkeley, California

Jason Beringer
School of Geography and
    Environmental Science
Monash University
Victoria, Australia

Steven de Bie
Resource Ecology Group
Wageningen University
Wageningen, the Netherlands

Niels Blaum
Plant Ecology and Conservation
    Biology
University of Potsdam
Potsdam, Germany

Randall B. Boone
Natural Resource Ecology
    Laboratory
Colorado State University
Fort Collins, Colorado

Luigi Boschetti
Department of Geography
University of Maryland
College Park, Maryland

Nicolas Boulain
Department of Environmental
   Sciences
University of Technology Sydney
Sydney, New South Wales,
   Australia

Dorine Bruget
Department of Environment and
   Resource Management
Queensland Climate Change
   Centre of Excellence
Queensland, Australia

Gabriela Bucini
Natural Resource Ecology
   Laboratory
Colorado State University
Fort Collins, Colorado

Peter Bunting
Institute of Geography and
   Earth Sciences
The University of Wales
   Aberystwyth
Aberystwyth, United Kingdom

Mercedes M. C. Bustamante
Departmento de Ecologia
Universidade de Brasília
Distrito Federal, Brazil

Joao M. B. Carreiras
Departamento de Ciências
   Naturais
Instituto de Investigação
   Científica Tropical (IICT)
Lisboa, Portugal

John O. Carter
Department of Environment and
   Resource Management
Queensland Climate Change
   Centre of Excellence
Queensland, Australia

Qi Chen
Berkeley Atmospheric Science
   Center
University of California-Berkeley
Berkeley, California

Xingyuan Chen
Berkeley Atmospheric Science
   Center
University of California-Berkeley
Berkeley, California

Daniel Clewley
Institute of Geography and Earth
   Sciences
The University of Wales
   Aberystwyth
Aberystwyth, United Kingdom

Lisa Collett
Department of Environment and
   Resource Management
Queensland Government
Brisbane, Australia

Garry D. Cook
CSIRO Sustainable Ecosystems
Darwin, Northern Territory,
   Australia

Tim Danaher
Department of Environment and
   Climate Change
Information Sciences Branch
Alstonville, New South Wales,
   Australia

Ken Day
Department of Environment and
    Resource Management
Queensland Climate Change
    Centre of Excellence
Queensland, Australia

Andrés Etter
Departamento de Ecología y
    Territorio
Universidad Javeriana
Bogotá D.C., Colombia

L. G. Ferreira
Instituto de Estudos Sócio-
    Ambientais
Universidade Federal de Goiás
Goiás, Brazil

Neil Flood
Department of Environment and
    Resource Management
Queensland Climate Change
    Centre of Excellence
Queensland, Australia

Kathleen A. Galvin
Department of Anthropology and
    Natural Resource Ecology
    Laboratory
Colorado State University
Fort Collins, Colorado

Louis Giglio
Science Systems and
    Applications, Inc.
NASA Goddard Space Flight
    Centre
Greenbelt, Maryland

Sam Gillingham
Department of Environment and
    Resource Management
Queensland Government
Brisbane, Australia

Niall P. Hanan
Natural Resource Ecology
    Laboratory
Colorado State University
Fort Collins, Colorado

Robert Hassett
Department of Environment and
    Resource Management
Queensland Climate Change
    Centre of Excellence
Queensland, Australia

James L. Heilman
Department of Soil and Crop
    Sciences
Texas A&M University
College Station, Texas

Beverley Henry
Department of Environment and
    Resource Management
Queensland Climate Change
    Centre of Excellence
Queensland, Australia

Steven I. Higgins
Institut für Physische
    Geographie
Johann Wolfgang Goethe-
    Universität Frankfurt
    am Main
Frankfurt am Main, Germany

Michael J. Hill
Department of Earth System
    Science and Policy
University of North Dakota
Grand Forks, North Dakota

Lindsay B. Hutley
School of Science and Primary
    Industries
Charles Darwin University
Northern Territory, Australia

Florian Jeltsch
Department of Plant Ecology and
 Conservation Biology
University of Potsdam
Potsdam, Germany

Robin Kelly
Natural Resource Ecology Lab
Colorado State University
Fort Collins, Colorado

Joanna Kitchen
Department of Environment and
 Resource Management
Queensland Government
Brisbane, Australia

Frank van Langevelde
Resource Ecology Group
Wageningen University
Wageningen, the Netherlands

Alex C. Lee
Geomatic and Spatial Systems
Defence Imagery and Geospatial
 Organization
Australian Capital Territory,
 Australia

Michael A. Lefsky
Department of Forest, Rangeland
 and Watershed Stewardship
Colorado State University
Fort Collins, Colorado

Caroline E. R. Lehmann
School for Environmental
 Research
Charles Darwin University
Northern Territory, Australia

Veiko Lehsten
Department of Physical Geography
 and Ecosystem Analysis
Lund University
Lund, Sweden

Shaun R. Levick
Department of Global Ecology
Carnegie Institution
Stanford, California

Adam Liedloff
CSIRO Sustainable Ecosystems
Darwin, Northern Territory,
 Australia

Marcy E. Litvak
Department of Biology
University of New Mexico
Albuquerque, New Mexico

Richard M. Lucas
Institute of Geography and Earth
 Sciences
The University of Wales
 Aberystwyth
Aberystwyth, United Kingdom

Siyan Ma
Berkeley Atmospheric Science Centre
University of California-Berkeley
Berkeley, California

Greg McKeon
Department of Environment and
 Resource Management
Queensland Climate Change
 Centre of Excellence
Queensland, Australia

Katrin M. Meyer
Department of Ecosystem
 Modelling
Georg-August-University of
 Göttingen
Göttingen, Germany

Gretchen Miller
Berkeley Atmospheric Science Center
University of California-Berkeley
Berkeley, California

Mahta Moghaddam
Department of Electrical
  Engineering and Computer
  Science
University of Michigan
Ann Arbor, Michigan

William J. Parton
Natural Resource Ecology Lab
Colorado State University
Fort Collins, Colorado

Christopher Potter
NASA Ames Research Center
Moffett Field, California

Herbert H. T. Prins
Resource Ecology Group
Wageningen University
Wageningen, the Netherlands

Robin S. Reid
Warner College of Natural
  Resource Sciences
Colorado State University
Fort Collins, Colorado

Humberto Ribeiro da Rocha
Departamento de Ciências
  Atmosféricas
Universidade de São Paulo
  Rua do Matão
São Paulo, Brazil

Miguel O. Román
Terrestrial Information Systems
  Branch
NASA Goddard Space Flight
  Centre
Greenbelt, Maryland

Milton H. Romero
Department of Geography
University of Leicester
Leicester, United Kingdom

Eva Rossmanith
Department of Plant Ecology and
  Conservation Biology
University of Potsdam
Potsdam, Germany

David P. Roy
Geographic Information Science
  Center of Excellence
South Dakota State University
Brookings, South Dakota

Youngryel Ryu
Berkeley Atmospheric Science Center
University of California-Berkeley
Berkeley, California

Sassan S. Saatchi
Jet Propulsion Laboratory
California Institute of Technology
Pasadena, California

Armando Sarmiento
Departamento de Ecología y Territorio
Universidad Javeriana
Bogotá D.C., Colombia

Peter Scarth
Department of Environment and
  Resource Management
Queensland Climate Change
  Centre of Excellence
Queensland, Australia

Crystal B. Schaaf
Department of Geography
Boston University
Boston, Massachusetts

Robert J. Scholes
Council for Scientific and
  Industrial Research,
  Natural Resources and
  Environment
Pretoria, South Africa

Susan Schwinning
Department of Biology
Texas State University
San Marcos, Texas

Paul Siqueira
Department of Electrical and
   Computer Engineering
University of Massachusetts
Amherst, Massachusetts

Izak P. J. Smit
South African National Parks
Scientific Services
Kruger Park, South Africa

Allan Spessa
Department of Meteorology
University of Reading
Reading, United Kingdom

Grant Stone
Department of Environment and
   Resource Management
Queensland Climate Change
   Centre of Excellence
Queensland, Australia

Kirsten Thonicke
Potsdam Institute for Climate
   Impact Research
Potsdam, Germany

Britta Tietjen
Potsdam Institute for Climate
   Impact Research
Potsdam, Germany

Kyle Tomlinson
Resource Ecology Group
Wageningen University
Wageningen, the Netherlands

Matthew D. Turner
Department of Geography
University of Wisconsin
Madison, Wisconsin

Karin M. Viergever
School of GeoSciences
   (Geography)
The University of Edinburgh
Scotland, United Kingdom

George Louis Vourlitis
Department of Biology
California State University
San Marcos, California

David Ward
School of Biological and
   Conservation Sciences
University of KwaZulu-Natal
Scottsville, South Africa

Rebecca Wenk
Berkeley Atmospheric Science
   Center
University of California-Berkeley
Berkeley, California

Kerstin Wiegand
Department of Ecosystem
   Modelling
Georg-August-University of
   Göttingen
Göttingen, Germany

Christopher A. Williams
Graduate School of
   Geography
Clark University
Worcester, Massachusetts

Iain Woodhouse
School of GeoSciences
   (Geography)
The University of Edinburgh
Scotland, United Kingdom

Jingfeng Xiao
Berkeley Atmospheric Science
   Center
University of California-Berkeley
Berkeley, California

Reviewers

Emma Archer, Gary Bastin, Peter Baxter, Nicolas Boulain, Christian Brannstrom, Daniel Brown, Mercedes Bustamante, Kelly Caylor, Chris Chilcott, Pablo Cipriotti, Derek Eamus, Rod Fensham, Jelle Ferwerda, Jacques Gignoux, Niall P. Hanan, Alex Held, Geoff Henebry, Michael Hill, Ricardo Holdo, Francesco Holecz, Eric Lambin, Caroline Lehman, Veiko Lehsten, Renaud Mathieu, Tim McVicar, Lutz Merbold, Katrin Meyer, Andrew Millington, Chris Neill, Asko Noormets, Bill Parton, Lara Prihodko, Stephen Prince, Elaine Prins, John Raison, Corina Riginos, Michael Roderick, Stephen Roxburgh, Brad Rundquist, Joe Scanlon, Bill Sea, Jonathon Seaquist, Tony Swemmer, Larry Tieszen, Anna Treydte, Matthew Turner, Yingping Wang, Chris Williams, Dick Williams, and Matt Williams.

Section I

Savannas: Biogeographical and Ecological Perspectives

1

Biogeography and Dynamics of Global Tropical and Subtropical Savannas: A Spatiotemporal View

Michael J. Hill, Miguel O. Román, and Crystal B. Schaaf

CONTENTS

Introduction

The global savanna biome is a complex entity. Savannas occupy the continuum between forest and grassland steppe with a variable mixture of trees and grassland. A range of definitions is provided by Mistry (2000), some of which are climatic and some of which are based on vegetation. In our opinion, the best definition is provided by Dansereau (1957): "a mixed physiognomy of grasses and woody plants in any geographical area." However, this mixture may encompass a very wide set of ecosystems: McPherson (1997) considers that savannas of various types covered much of North America, including many systems in which the woody plants are shrubs and grass cover is extremely sparse. Most of temperate and tropical Australia was originally covered in savanna (Moore, 1975a; Harrington et al., 1984a). At the other end of the scale, dry woodlands (Miles et al., 2006) inhabit the "gray zone" between woody savanna and woodland or forest.

In terms of an association of plant functional types, the definition is easy: Any mixture of trees and grass extended from the point where tree density and canopy cover precludes grass growth, to the point where trees are no longer present. However, savannas are really manifested at multiple spatial scales, and they are defined by variable classifications of the continuum between ecosystems dominated by trees and grasses. At the biome level, they include classical open woodland with grassland as well as complicated patch arrangements of grassland, woodland, and forest. Hence the definition of Dansereau (1957) works well if one adds "naturally occurring," "at landscape or regional scale," because this accommodates the patch complex and some definition of trees, shrubs, and grass density that distinguishes savannas from arid shrublands. For our purposes, there are two ways of conceptualizing these systems: (1) as tree–grass systems from an ecosystem function viewpoint and (2) as savannas from a region—global scale land cover viewpoint.

As a result of this complexity, the spatial identification of the global extent of savannas is one of the most difficult tasks. One estimate suggests that savannas cover 33 million km^2 (Ramankutty and Foley, 1999), but every estimate depends on the definition of thresholds for tree cover, density, and grass cover beyond which other physiognomic classes such as woodland or arid shrubland become more appropriate. The drivers for their distribution have been described by the balance between plant available moisture and plant available nutrients (well described by Mistry, 2000). However, more recent analysis speculates that savannas are the product of feedback between fire, forest loss, C_4 grass expansion, and low atmospheric CO_2 (Beerling and Osborne, 2006). The coevolution of large and diverse herds of grazing ungulates (e.g., Du Toit and Cumming, 1999) and the influence of the indirect interactive effects of herbivory and fire (e.g., van Langevelde et al., 2003) are also very important factors.

In this introductory chapter, our aim is to describe the global savanna biome using the remote sensing and geospatial data sets that have been developed in the last 10 years. This analysis is, therefore, restricted to available geospatial boundaries and classifications, and it does not attempt to map or describe all tree–grass systems or small remnants or fragments of savanna within other major biomes. The ecology, grassland dynamics, fire patterns, and some human dimensions of the major savanna regions of the world have been described in detail by various authors: global savannas (Mistry, 2000; Furley, 2004; Pennington et al., 2006); the Brazilian cerrado (Oliveira and Marquis, 2002); North American savanna (McPherson, 1997); Neotropical savannas (Sarmiento, 1984; Pennington et al., 2006); Australian tropical savanna (Shaw and Norman, 1975; Mott and Tothill, 1984; Orr and Holmes, 1984; Tothill and Gillies, 1992; Dyer et al., 2001); and the African savannas (Sinclair et al., 2008 and preceding volumes; Bassett and Crummery, 2003; Fairhead and Leach,

1996). The book edited by Mistry (2000) provides an excellent overview of every major savanna region. Readers are directed to these excellent volumes for a descriptive understanding of the ecology and fine-scale dynamics of savannas. However, a spatiotemporal description at the global scale has not been attempted. This spatially and temporally explicit description is a fundamental requirement to introduce the detailed treatment of measurement and modeling of these tree–grass systems in the subsequent sections and chapters in this book. Hence, this chapter is largely descriptive: synthesis is provided within various chapters and in the final section of the book.

The complexity, and consequent difficulty with mapping, of the global savanna biome demands that we establish a very clear basis for this description rooted in the available data sets, their spatial scales, assumptions, and known uncertainties. Here we will examine the following globally significant factors that can be described by geospatial data and multitemporal remote sensing: climate; land cover; fractional cover of plant functional types; vegetation and biodiversity; domesticated livestock density; human rural population density; and human appropriation of net primary productivity. We omit fire and soil factors from explicit spatial or temporal analysis. Fire is the focus of Roy et al. in Chapter 12. Global soils data are not amenable to spatially explicit analysis due to aggregation error. Instead, we examine the pattern of soil order decomposition within savanna ecoregions (see Palm et al., 2007 for an effective global assessment of tropical soils).

Concept and Assumptions

Our analysis starts from the premise that global Geographic Information System (GIS) data sets and time series of coarse resolution remote sensing products can be used to give an improved overall description of the global savanna biome. The analysis concentrates on the tropical and subtropical savannas. The dry forests of southeast Asia and temperate savannas in Australia (Moore, 1975b; Rossiter and Ozanne, 1975; Harrington et al., 1984b; Lodge et al., 1984) and North America (McPherson, 1997) are included in Figure 1.1 (although in the latter case, only the oak woodlands in California and the central forest–grass transition zone, and not McPherson's much broader definition), but they are not included in subsequent analysis. However, the Edwards Plateau juniper-oak woodlands of Texas (Chapter 7) and the Californian oak woodlands (Chapter 8) are explicitly dealt with in the next section. The spatiotemporal data used in this analysis are described in Table 1.1. The analysis is structured as follows:

1. *Spatial definition of global tropical and subtropical savannas:* a biogeographic and ecological boundary for the analysis from the World

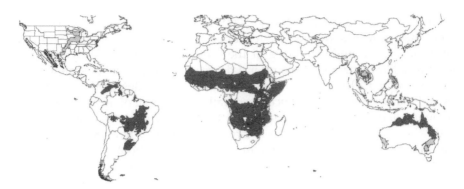

FIGURE 1.1
Global savanna ecoregions. The dry forests of southeast Asia, temperate savannas in south eastern and western Australia, the forest–grassland transition zone in the central North America (both largely altered), and the Californian oak savannas shown here in a lighter shade of gray are excluded from subsequent analysis.

Wildlife Fund (WWF) Terrestrial Ecoregions of the World (Olson et al., 2001). It should be noted that there may be disagreements among biogeographers and ecologists about ecoregion designations as savannas or some other vegetation types, but we have based our analysis strictly on the WWF Ecoregion classification.

2. *Definition of land cover within global savannas:* a broad definition of proportion of the ecoregions occupied by savanna and non-savanna land cover based on the Moderate Resolution Imaging Spectroradiometer (MODIS) land cover product (Friedl et al., 2002; Friedl et al., 2010). This is contrasted with the vegetation structure–based GLOBCOVER product developed from MERIS (MEdium Resolution Imaging Spectrometer) data (Defourny et al., 2006; Bicheron et al., 2008).

3. *Climate and soils of global savannas:* based on a revision of the Köppen–Geiger climate classification (Peel et al., 2007), a global map of Soil Orders (Reich, 2005), and a recent assessment of world soils in the context of ecosystem services (Palm et al., 2007).

4. *Vegetation and biodiversity:* vascular plant species richness mapping (Kier at al., 2005; Kreft and Jetz, 2007) and tree cover percentage from remote sensing (Hansen et al., 2002, 2003).

5. *Fire:* seasonal distribution and regional intensity of fires from literature (Bond et al., 2005; Carmona-Moreno et al., 2005; Giglio et al., 2005; Mouillot and Field, 2005; Riano et al., 2007).

6. *Domesticated livestock:* cattle densities from the FAO-Oxford Gridded Livestock of the World (Robinson et al., 2007). Cattle density is used as an indicator for all livestock distribution. However, cattle densities are not indicative of overall livestock density and, hence, grazing or browsing pressure. Although there are virtually no other

TABLE 1.1

Global Spatial Data Layers Used in this Analysis

Data	Description and Application
WWF Terrestrial Ecoregions of the World (Olson et al., 2001)	Global subdivision based on biomes, biogeographic realms, floristic and zoogeographic provinces, animal and plant groups, biotic provinces and vegetation types, and detailed regional data. Geographical Boundary—only savanna or woodland/grassland ecoregions accepted
Collection 5500 m MODIS MOD12Q1 2009 IGBP land cover classification (Friedl et al., 2010)	Land cover class filter for time series profiles—savanna, woody savanna, open shrubland, grassland, cropland, and cropland/natural
Collection 51 km MOD43B4 Nadir BRDF adjusted reflectance (Schaaf et al., 2002)	Time series of reflectances and calculated mean and spatial standard deviation of EVI for each land cover class in each ecoregion
Revised global Köppen–Geiger Climates and WMO point database used for revision (Peel et al., 2007)	Variability broad climate types associated with savanna ecoregions and seasonal rainfall from WMO point database for each ecoregion. Association of point data for seasonal rainfall with cattle density also analyzed
Collection 4500 m MOD44B Vegetation Continuous Fields (Hansen et al., 2003)	Percentage cover of trees, herbaceous vegetation, and bare ground. Tree cover variability by ecoregion subdivided by land cover class
Global gridded livestock database (Robinson et al., 2007)	Global grids of animal density (animals/km^2) for cattle (original database kindly provided by William Wint, Oxford University)
Global gridded rural population density (0.0083° resolution) (Salvatore et al., 2005)	Rural population densities (persons/km^2) within each savanna ecoregion for South America and Africa (downloaded from http://www.fao.org/geonetwork/srv/en/main.home)
Global gridded population density (2.5 arc minute resolution) CIESIN 2005	Population density (persons/km^2) within each savanna ecoregion in northern Australia
Global gridded percent human appropriation of net primary productivity (0.25 gridcell) (Imhoff et al., 2004a, b)	Mean and standard deviation of HANPP (Gg C/0.25 deg) for each savanna ecoregion

domesticated livestock in the Australian tropical savannas, very large populations of sheep and goats are present in the savanna regions of East, West, and the drier parts of Southern Africa; and there are significant populations of sheep and camelids in South American savannas outside the cerrado. Readers are referred to the FAO database for a detailed documentation of global densities of all types of domesticated livestock.

7. *Human footprint and use:* based on human rural population densities from the FAO Poverty Mapping project (Salvatore et al., 2005) and

human appropriation of net primary production (HANPP: Imhoff et al., 2004a, b; Boone et al., 2007; Krausmann et al., 2007). There is a more recent estimate of the global HANPP (Haberl et al., 2007); for comparison the overall relative percentage appropriations were 20% (Imhoff et al., 2004a) and 24% (Haberl et al., 2007). However, we had only the 2004 data available as a GIS layer.

8. *Temporal profiles of land surface photosynthetic activity:* based on time series profiles of the enhanced vegetation index (EVI) (Huete et al., 2002) calculated from the MODIS Nadir BRDF (Bidirectional Reflectance Distribution Function)—Adjusted Reflectance (NBAR) product for each savanna land cover type within each savanna ecoregion. The EVI performs well in high aerosol environments such as savannas, wherein biomass burning is a major feature. The EVI is also used to drive global productivity models (see Potter, Chapter 22).

The analysis in this chapter specifically excludes treatment of land cover change involving major agriculturalization. This volume is concerned with the tree–grass systems. The presence of this class is documented where major land cover change and cropping have occurred. However, in all other respects, the focus is on the tree–grass systems and their characterization. The results are presented with full knowledge of the approximations and averaging involved in such global scale, coarse resolution data analysis. However, the results provide an overarching view of what we identify as tropical and subtropical savannas.

Spatial Definition of Global Savannas: Ecoregions and Land Cover

Global tropical and subtropical savannas are identified by all ecoregions within the WWF Terrestrial Ecoregions of the World (Olson et al., 2001) denoted as savannas or grassy woodlands lying within 25° of the equator (dark gray areas, Figure 1.1). The ecoregions were developed on the basis of endemism, species assemblages, and geological history (Olson et al., 2001) and, hence, represent a definition that is largely independent of the other data we use. The map shows their global distribution. We made a specific decision to exclude the dry tropical forest in southeast Asia; however, there is some overlap in definition between these dry forests and the broad savanna biome (Miles et al., 2006). We also made a decision to include the Uruguyan savanna, because it forms a continuum with the cerrado, even though many would consider it a grassland (William Hoffman, personal communication). In this analysis, the ecoregions are subdivided into five easily recognizable groups on three continental land masses (Figure 1.2a):

(a) Australasia—Arnhem Land (AL), Brigalow (Br), Cape York (CY), Carpenteria (Ca), Einasleigh Uplands (EU), Kimberly (K), Trans Fly (TF: New Guinea), and Victoria Plains (VP); (b) South America—Beni Moxos (BM), Campos Rupestres (CR), Cerrado (Ce), Guianian savanna (Ga), Llanos del Orinocco (L), Pantanal (U), and Uruguayan (U) savanna; (c) West Africa— East Sudanian savanna (ES1: western portion), East Sudanian savanna (ES2: eastern portion), Guinean forest–savanna mosaic (Ge), Inner Niger Delta flooded savanna (INDf), Lake Chad flooded savanna (LCf), Northern Congolian forest–savanna mosaic (NC), Sahelian Acacia savanna (SA), and West Sudanian savanna (WS); (d) East Africa—Northern Acacia– Commiphora bushland and thicket (NAC), Southern Acacia–Commiphora bushland and thicket (SACB), Somali Acacia–Commiphora bushland and thicket (SoA), and Victoria Basin forest–savanna mosaic (VB); and (e) Southern Africa—Angolan Miombo woodlands (AM), Angolan Mopane woodlands (Amo), Central Zambezian Miombo woodlands (CZM), Eastern Miombo woodlands (EM), Southern Miombo Woodlands (SM), Southern Congolian forest–savanna mosaic (SC), West Congolian forest–savanna mosaic (WC), Kalahari Acacia–Baikiaea woodlands (KAB), Southern African Bushveld (Sab), Zambezian Baikiaea woodlands (ZB), Zambezian and Mopane woodlands (Zmo), and Zambezian coast flooded savanna (ZCf). The vegetation, ecology, and dynamics of these five savanna groups are described in detail in Mistry (2000). The flooded savannas, which include the Llanos del Orinocco, Pantanal, and parts of the Beni Moxos in South America; and the Lake Chad, Inner Niger Delta ecoregions, and Okavango Delta in Africa, represent a special case where hydrologic and edaphic conditions interact to determine vegetation structure.

The remote sensing-based land cover classes within these savanna ecoregions are quite diverse and subject to the effects of human disturbance, the spatial and spectral resolution of the remote sensing data used, and the land cover classification schemes applied. There are five global land cover products: IGBP DISCover (Loveland et al., 2000); UMD land cover (Hansen et al., 2000); Global Land Cover 2000 (Bartholme and Belward, 2005); the MODIS land cover (Friedl et al., 2002); and the GLOBCover product (Defourny et al., 2006; Bicheron et al., 2008). However, they are not easily compared (Herold et al., 2008), because land cover is defined on different bases, particularly for those landscapes characterized by mixtures of trees, shrubs, and grasses. In order to remain internally consistent, we have chosen to undertake our analysis using products from only the MODIS sensor. The MODIS land cover data set (Table 1.2a) has the necessary biome-centric savanna, woody savanna, open shrubland, and grassland classes needed for our purpose (Figure 1.2b). However, we do draw a contrast between the MODIS land cover and the land cover classification provided by GLOBCOVER (Table 1.2b). There are very significant differences in height definitions distinguishing between trees and shrubs and for canopy cover percentages in relation to land cover types.

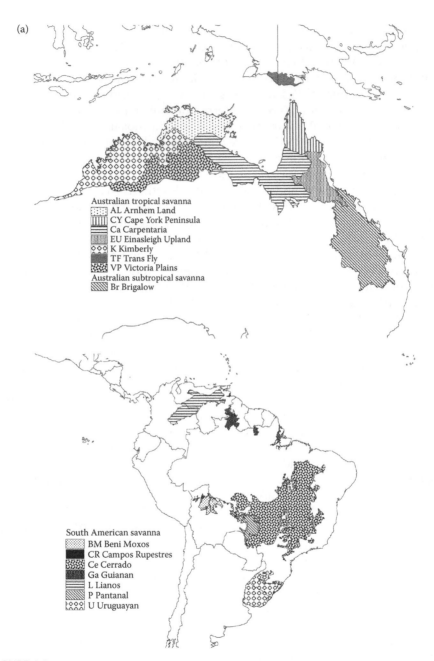

FIGURE 1.2
(See color insert following page 320.) (a) WWF ecoregions for Australian, South American, and African tropical and subtropical savannas. (Adapted from Olson, D. M. et al. 2001. *Bioscience* 51, 933–938.) (b) MODIS IGBP land cover maps for Australian, South American, and African tropical and subtropical savannas. (Adapted from Friedl, M. A. et al., 2010. *Remote Sensing of Environment* 114, 168–182.)

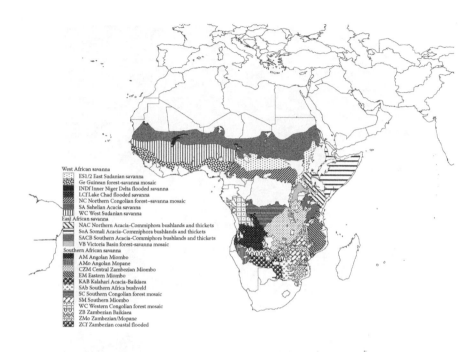

West African savanna
ES1/2 East Sudanian savanna
Ge Guinean forest–savanna mosaic
INDf Inner Niger Delta flooded savanna
LCf Lake Chad flooded savanna
NC Northern Congolian forest–savanna mosaic
SA Sahelian Acacia savanna
WC West Sudanian savanna
East African savanna
NAC Northern Acacia-Commiphora bushlands and thickets
SoA Somali Acacia-Commiphora bushlands and thickets
SACB Southern Acacia-Commiphora bushlands and thickets
VB Victoria Basin forest-savanna mosaic
Southern African savanna
AM Angolan Miombo
AMo Angolan Mopane
CZM Central Zambezian Miombo
EM Eastern Miombo
KAB Kalahari Acacia-Baikiaea
SAb Southern Africa bushveld
SC Southern Congolian forest mosaic
SM Southern Miombo
WC Western Congolian forest mosaic
ZB Zambezian Baikiaea
ZMo Zambezian/Mopane
ZCf Zambezian coastal flooded

(b)

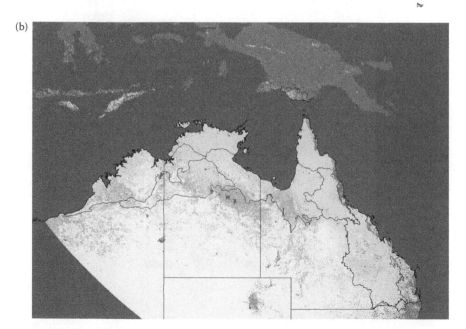

FIGURE 1.2 Continued.

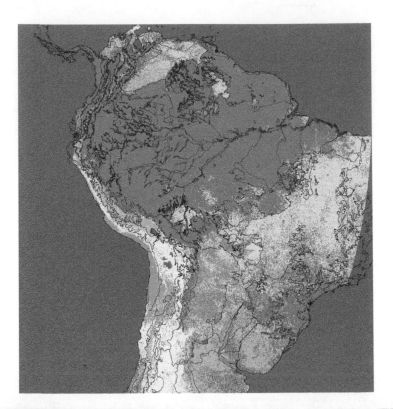

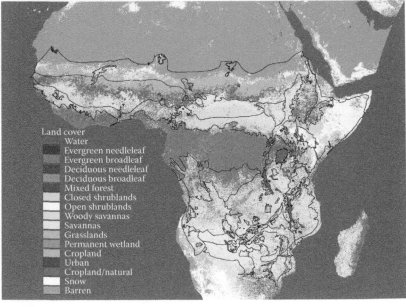

Land cover
Water
Evergreen needleleaf
Evergreen broadleaf
Deciduous needleleaf
Deciduous broadleaf
Mixed forest
Closed shrublands
Open shrublands
Woody savannas
Savannas
Grasslands
Permanent wetland
Cropland
Urban
Cropland/natural
Snow
Barren

FIGURE 1.2   Continued.

TABLE 1.2

(a) MODIS IGBP Land Cover Classification: Characterization of Savanna Zone Classes. (b) Generalized ESA GLOBCOVER Classes for Savanna Ecoregions (Excluding Bare, Artificial, and Water)

	Characterization
(a) IGBP Land Cover Unit	
Open shrubland	Lands with woody vegetation less than 2 m tall and with shrub canopy cover between 10% and 60%. The shrub foliage can be either evergreen or deciduous.
Woody savanna	Lands with herbaceous and other understory systems and with forest canopy cover between 30% and 60%. The forest cover height exceeds 2 m.
Savanna	Lands with herbaceous and other understory systems and with forest canopy cover between 10% and 30%. The forest cover height exceeds 2 m.
Grassland	Lands with herbaceous types of cover. Tree and shrub cover is less than 10%.
Cropland	Lands covered with temporary crops followed by harvest and a bare soil period (e.g., single and multiple cropping systems). Note that perennial woody crops will be classified as the appropriate forest or shrub land cover type.
Cropland/natural	Lands with a mosaic of croplands, forests, shrublands, and grasslands in which no one component comprises more than 60% of the landscape.
(b) GLOBCOVER	
Croplands	Rainfed and post-flooding or irrigated croplands including tree crops
Mosaic cropland	Crop-dominant (50–70%) and natural vegetation-dominant (50–70%) mosaics
Tree, grass, and shrub mosaics	Forest shrubland (50–70%)/grassland or grassland (50–70%)/forest shrubland
Closed/open shrubland	Shrubland (<5 m) either 15–40% or >40% cover
Closed forest	>40% cover of trees (>5 m)
Open forest	15–40% cover of trees (>5 m)
Grassland	Closed (>40%) or open (15–40%)
Flooded lands— permanent or temporary	Includes flooded forests, savannas, and grasslands and, therefore, major parts of African and South American savanna ecoregions
Sparse vegetation	Cover <15%

Source: Data from (a) Friedl, M. A. et al. 2002. *Remote Sensing of Environment* 83, 287–302; (b) Bicheron, P. et al. 2008. GLOBCOVER. *Products description and validation report.* Medias France/POSTEL, Toulouse, France, 47pp.

The tropical and subtropical savanna ecoregions show wide variation in percentage area associated with MODIS savanna land cover classes and other associated MODIS land cover classes (Figure 1.3a). Open shrublands and grasslands form significant components of the drier Australian and East and West African ecoregions, and croplands and cropland/natural mosaics form a significant component of South American and East and West African savanna ecoregions. By contrast, in the GLOBCOVER classification, savanna and woody savanna are variously described by closed or open shrubland (very significant for Australian ecoregions), grassland, forest, shrubland mosaic, or open forest (15–40% cover) (Table 1.2b; Figure 1.3b). The full GLOBCOVER product provides more detail in terms of structural classes and class descriptions than are summarized in Table 1.2b or shown in Figure 1.3b. It is evident that the MODIS biome-based classification and the GLOBCOVER structural classification are, in fact, complementary for the savanna ecoregions; land cover classes from MODIS benefit from spectral information in the short wave infrared bands not available on MERIS, whereas land cover classes from GLOBCOVER benefit from higher spectral resolution among visible and near infrared bands available on MERIS. However, for this savanna biome analysis, the MODIS product provides us with the most appropriate definition and analysis filter.

Vegetation and Biodiversity

The extent and variability of tree cover provides one quantitative diagnostic for savanna ecoregions (Figures 1.3c and 1.4). In some African ecoregions, tree density and ecosystem diversity are intimately linked with the utilization patterns of large herbivores (Asner et al., 2009). The Vegetation Continuous Fields (VCF) product from the MODIS sensor (Table 1.1) provides a quantitative global measurement of the fractional cover of trees, herbaceous plants, and bare ground (Hansen et al., 2002). Here, VCF tree cover percentages are compared only for the savanna and woody savanna land cover classes within each ecoregion (Figure 1.3b). The results show that apart from the Trans Fly savanna, which is heavily forested, Lake Chad and Inner Niger flooded savannas, and the Kalahari Baikiaea woodlands and Angolan mopane woodlands, which are very sparsely treed; tree cover ranges between 8% and 25% for the savanna class and up to 40% for the wooded savanna class. The maps of percentage tree cover (Figure 1.4) clearly show the patterns of woody savanna surrounding the equatorial forest in Africa. However, VCF data have been shown to have limitations at site scale when compared with field data (Bucini and Hanan, 2007).

Many global savanna ecoregions are important centers of biodiversity. Global scale mapping of biodiversity is an emerging science with the

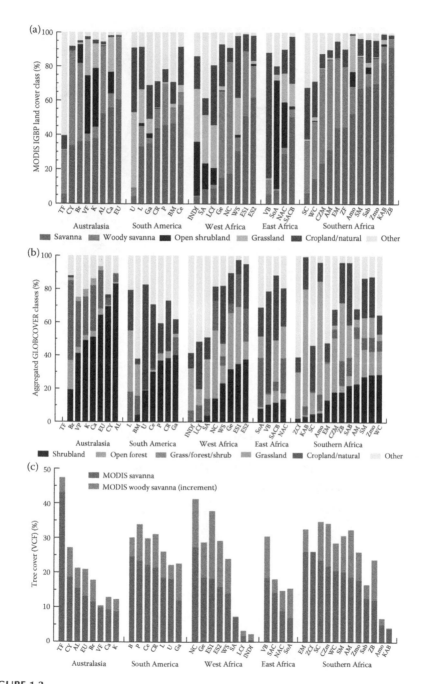

FIGURE 1.3
(See color insert following page 320.) Proportion of savanna ecoregions occupied by (a) MODIS IGBP land cover classes; (b) aggregated classes from the ESA GLOBCOVER; and (c) percentage tree cover from MOD44 VCF data for MODIS IGBP savanna and woody savanna (full bar height) land cover classes.

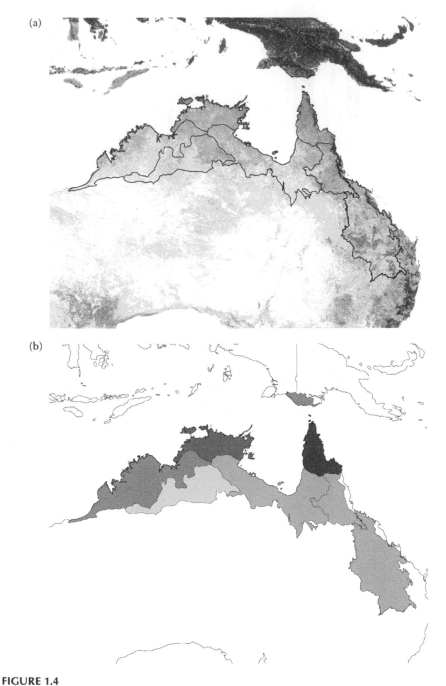

FIGURE 1.4
Percentage tree cover (a, c, and e) and species richness (b, d, and f) (species number per ecoregion) for the savanna ecoregions in Australia, South America, and Africa. (Adapted from Kier, G. et al. 2005. *Journal of Biogeography* 32, 1107–1116.)

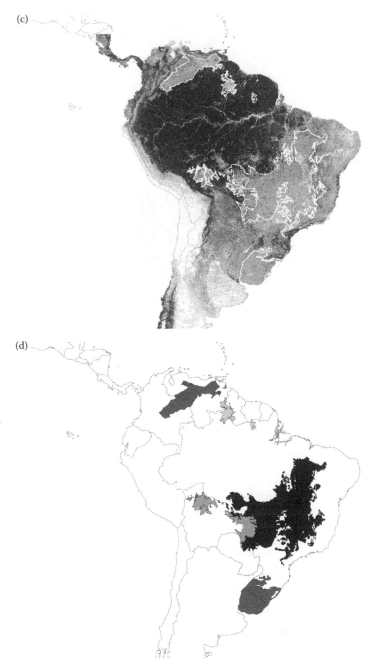

FIGURE 1.4 Continued.

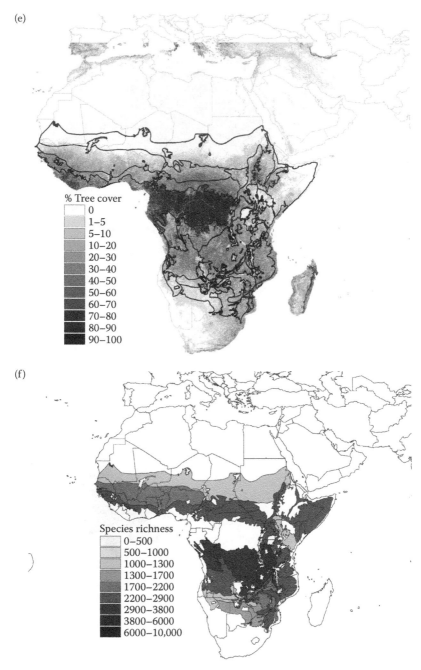

FIGURE 1.4 Continued.

combination of inventory methods, digitizing of museum and herbaria specimen collections, and global coverage of spatial data at relatively fine resolution (Ferrier et al., 2004; Mutke and Barthlott, 2005; Barthlott et al., 2005). Measuring and mapping endemism significantly helps with definition of biodiversity hotspots (Kier and Barthlott, 2001). Maps of species richness of vascular plants (Figure 4; Kier et al., 2005) show most savanna ecoregions having in excess of 1000 species. Among specific ecoregions, the Cerrado (6500), the Cape York savanna (3100), and the Central Zambezian miombo woodlands (3800) have the highest species richness in the savanna biome within the three continental land masses (Kier et al., 2005).

Climate and Soils of Global Savannas

A comprehensive analysis of climate for the global savannas could occupy a book all by itself. However, it is instructive to examine some broad characteristics. The Köppen–Geiger Climate Classification has been recently updated (Peel et al., 2007), and both the WMO (World Meteorological Organization) point database and GIS map layers have been made freely available. An analysis of the intersection between savanna ecoregions and the Köppen–Geiger climate map shows that savanna ecoregions are predominantly within the following climate classes: tropical savanna (Aw); arid steppe hot (B classes); and temperate, dry winter, hot or warm summer (Cwa, Cwb) (Figure 1.5a; B and C subtypes aggregated). The Guianian, Beni-Moxos, and Trans Fly savannas lie partially within the tropical forest, rainforest class (Af), whereas tropical savanna ecoregions furtherest from the equator within Australia and Africa, substantially occupy the arid, desert, and hot summer climate class (BWh). The various miombo woodland ecoregions are associated with the temperate, dry winter, hot summer (Cwa), and warm summer (Cwb) climate classes. The Uruguyan savanana lies within the temperate, without dry season, hot summer class (Cfa) and is located at similar latitude to the Australian temperate savannas.

The point database from the Köppen–Geiger classification (Peel et al., 2007) was intersected with the savanna ecoregions (998 individual sites), and the averages and standard deviations were extracted for the 6-month periods (October to March and April to September) that capture the gross seasonal differences associated with latitudinal location. This analysis contains some imperfections, because point coverage is much poorer for the South American ecoregions than for either Africa or Australia (e.g., only one point is available for the Pantanal, and two each for the Beni Moxos and Guianian savannas). However, in general, these data are instructive in highlighting absolute and relative levels of dry and wet season rainfall (remembering that the wet

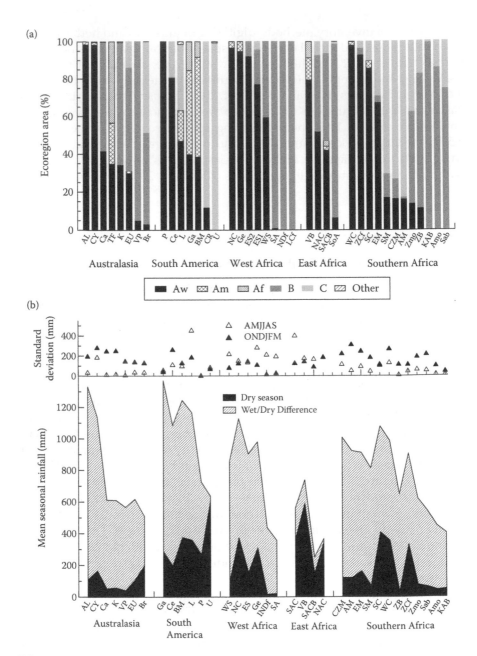

FIGURE 1.5
Pattern of seasonal rainfall for global savanna ecoregions based on meterological data used by Peel et al., 2007 to revise the Köppen–Geiger climate classification. (a) Proportion of savanna ecoregions falling into updated Köppen–Geiger climate zones. Temperate and arid subzones aggregated. (b) Mean seasonal rainfall (based on 6-month aggregates) for savanna ecoregions grouped by broad regions and ordered from highest to lowest seasonal difference.

seasons are generally reversed between the southern and northern hemispheres) (Figure 1.5b). In particular, they highlight

1. The relative aridity of many of the inland Australian savanna ecoregions, which have significant areas of open shrubland and grassland (Figure 1.3a)
2. The relative wetness of the major South American savanna ecoregions
3. The arid nature of the sub-Saharan Sahelian savanna ecoregions contrasting with the good wet and dry season rainfall in other West African savanna ecoregions, thus surely providing potential for major agriculturalization similar to the cerrado
4. The similar rainfall between "wet" and "dry" seasons in East Africa—the result of the bimodal rainfall patterns influenced by both Atlantic and Indian ocean oscillations
5. The relatively higher wet season rainfall associated with miombo woodland ecoregions and the parallels in seasonal patterns and land cover classes (Figure 1.3a) between the more arid southern African ecoregions (Sab, Amo, and Kab) and inland Australian ecoregions

The data also highlight the spatial variability across these ecoregions with spatial standard deviations in mean wet season rainfall predominantly ranging between 100 and 300 mm per season.

Since tree density is such a key property of savannas, we used the climate point database to examine the relationship between seasonal rainfall and percentage tree cover. For each ecoregion, percentage tree cover was extracted from the VCF data at each climate point within the savanna land cover class only, and the mean tree cover was plotted against mean wet season and dry season rainfall (Figure 1.6). There is large variation in tree cover as expected (and spatial standard deviations were relatively large), but clear response curves to both wet season and dry season rainfall are observed. Although there are many variables and factors embedded in such broad global relationships, it would be expected that, within savannas, tree density would increase with rainfall up to the level at which extended periods of inundation or extreme wetting and drying regimes may impact tree survival such as in seasonally flooded savannas.

The soils associated with global savannas are summarized here by relating the global ecoregions to the Global Soil Suborder Map (Reich, 2005) aggregated to order level and, by reference, to a recent assessment of global soils that align the core soil properties of texture, mineralogy, and organic matter with ecosystem services (Palm et al., 2007). Readers are referred to in-depth assessments of global soil carbon stocks (e.g., Eswaran et al., 1993; Lal, 2004; Maquere et al., 2008). Analysis of soil orders by savanna ecoregions

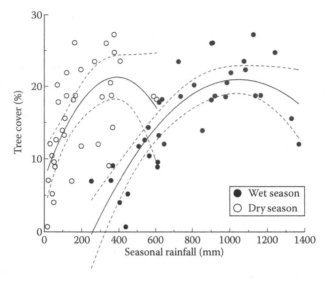

FIGURE 1.6
Relationship between mean seasonal rainfall in wet and dry seasons and average percentage tree cover from MOD44 VCF in the MODIS savanna land cover class for savanna ecoregions. The dashed lines show the confidence intervals for the fit.

(Table 1.3a) confirms the high spatial variability of soils across the savanna biome. Entisols, the most extensive soils globally (Palm et al., 2007), are high in weatherable minerals and common across all savannas except North America. Alfisols are common across all savannas and are threatened by nutrient depletion (Table 1.3b). Aridisols are common in East Africa and Australia; these savanna zones have strong rainfall gradients toward major deserts. The Oxisols and Ultisols, with high acidity, low nutrient content, and significant iron and aluminum (Palm et al., 2007), are most common in South American and Southern African savanna ecoregions. The Vertisols, fertile black cracking clays, are significant in Australasia, East, and West Africa and are commonly associated with grasslands, both seasonal and flooded (Palm et al., 2007). There are significant areas of shifting sand in the Kalahari desert of South Africa and the Sahelian area of West Africa bordering the Sahara desert. Inceptisols are minor components except in East Africa, and Mollisols are black soils dominant in the prairies and the grassland-forest transition zone that constitutes the oak savannas in North America. In general, much of the savanna biome is subject to soil moisture constraints, but the better soils are more susceptible to erosion and nutrient loss resulting from land use change, and the typical tropical soils suffer from major constraints associated with high acidity and leaching rates. Table 1.3b represents an overall assessment of attributes and constraints from Palm et al. (2007), but it includes grasslands and shrublands, and so is not exclusive to the savanna ecoregions we have defined here.

TABLE 1.3a

Soil Types of the Global Savanna Ecoregions Aggregated Over the Major Continental Groups

Soil Order	Area[a] 000 km²	Shifting Sand	Oxisols	Vertisols	Aridisols	Ultisols	Mollisols	Alfisols	Inceptisols	Entisols	Other
Australasia	1700	2.6	2.9	12.9	17.7	7.9	0.6	17.5	0.7	35.8	1.5
South America	3065	0.1	34.1	2.0	0.0	19.6	3.4	14.8	4.8	21.0	0.2
West Africa	7028	9.6	10.0	6.8	5.3	6.9	0.0	22.3	2.1	36.2	0.7
East Africa	2679	1.9	7.5	8.3	32.9	6.0	0.0	7.3	10.4	24.8	0.8
Southern Africa	5150	11	23.8	1.6	3.1	14.0	0.0	15.9	6.0	30.4	0.7
North America	554	0.0	0.0	0.2	1.3	2.7	56.3	32.4	3.8	3.0	0.3
All savannas	20,176	5.0	16.3	5.1	7.7	10.7	2.6	17.8	4.4	29.9	Trace

Note: North American Savannas are Californian and Mid-Continental Oak Woodlands. Australian temperate savannas excluded.

[a] Area calculation based on Albers Azimuthal Equal Area projection (continental).

TABLE 1.3b

Major Soil Attributes and Constraints for Tropical and Subtropical
Grasslands, Savanna, and Shrublands

Attributes and Constraints	Area (%)	10^6 ha
Seasonal soil moisture stress	63.6	1244
Low nutrient capital reserves	37.1	726
Aridic	24.4	477
Aluminum toxic	21.6	423
High leaching potential	15.8	310
High soil erosion risk	12.1	236
High P fixation	8	157
Waterlogged	7.2	141
Calcareous	7	137
Cracking clays	6.3	123

Source: After Palm, C. et al. 2007. *Annual Review of Environment and Resources*
32, 99–129.

Fire

It has been argued that fire is an underappreciated factor in determination of
the global vegetation structure (Bond et al., 2005) and that climate–fire–C_4
grassland feedbacks are responsible for the global savanna biome distribution
(Beerling and Osborne, 2006). The former authors found in a simulation study
that closed forests doubled in a fire-free world, primarily at the expense of C_4
plants that dominate the understorey of tropical savannas. Analysis of global
fire patterns for the twentieth century via synthesis of post-1980 data and historical reconstruction (Mouillot and Field, 2005) found that burned areas of
savanna and dry forests in Africa, South Asia, South America, and Australia
increased from 430 Mha year^{-1} in 1900–1910 to 520.1 Mha year^{-1} in 1999–2000
and that savannas were responsible for between 81 (1900–1910) and 93% (1960–
1970) of global burned area. However, data reliability for savannas was assessed
as very poor before 1980. An analysis of global spatial patterns of burned area
from NOAA AVHRR 8 km Pathfinder data derived a four-class (SON, JJA,
MAM, and DJF) map of global fire activity, a graphical representation of latitudinal distribution, and maps of global fire probability for the four periods
(Carmona-Moreno et al., 2005). Peak seasonal fire activity was associated with
September to November for the Australian, Southern African savanna, and
southern cerrado; June to August for the northern cerrado and miombo woodlands; March to May for the northern areas of West Africa and the Llanos; and
December to February for the rest of West Africa and the northern Llanos
(Carmona-Moreno et al., 2005). In a further analysis of this Pathfinder data,
global monthly fire patterns were classified using principal component analysis

(Riano et al., 2007). The savannas were predominantly associated with a burned area peak of 5% in the July to October (Southern Hemisphere pattern) period and 10% in the November to February (Northern Hemisphere pattern) period (Riano et al., 2007). These analyses all emphasize the importance of savannas as the major biome for global fire activity.

Rural Population, Livestock, and Human Appropriation of Net Primary Productivity

Assessment of the human impact in savannas is complex, and it is addressed in greater detail in Chapters 25 and 26. At the scale of this analysis, we seek only to show the distribution of some major factors, rural human population and cattle density, and to summarize the assessments of HANPP (Imhoff et al., 2004b). The maps of rural population density (Salvatore et al., 2005) and cattle density (Robinson et al., 2007) reflect a complex suite of climatic and ecological factors and human history (Figure 1.7). They also reflect a strong contrast between Africa on one hand and Australia and South America on the other.

A very broad analysis of what could be termed *pastoral pressure* can be carried out by contrasting the human rural population density with the cattle density in the savanna ecoregion. There are high levels of spatial variability in cattle and rural population density within ecoregions, with substantial areas of some ecoregions having no cattle (Figure 1.8). However, carrying out the analysis only for areas with greater-than-zero cattle density versus using the whole ecoregion made little difference to the pattern of response. The two-way response space between cattle and population shows distinct contrasts for the three continental land masses (Figure 1.8). The South American savannas have a relatively high cattle population density and low human rural population density. The Australian savannas are very low with regard to both factors. The east African savannas have relatively high populations of both cattle and humans, whereas the remaining African savannas show wide variation within a range of up to 20 people km^{-2} and 120 animals km^{-2}. The pastoral pressure would appear to be very high in east Africa, but this cannot be assessed in ranching terms, because livestock may be moved large distances to match seasonal feed availability (e.g., Little, 2003).

An assessment of percentage HANPP (Imhoff et al., 2004b) for the global savanna ecoregions reinforces the patterns observed in cattle density and human rural population density (Figure 1.9). Among the large ecoregions, percentage HANPP is highest for the Western Sahelian, Guinean, and Victoria Basin savannas, but it is greater than 20% throughout much of the East and West African savanna ecoregions. Percentage of HANPP is around 10% or less in the Southern African savanna ecoregions, and it is below 5% in

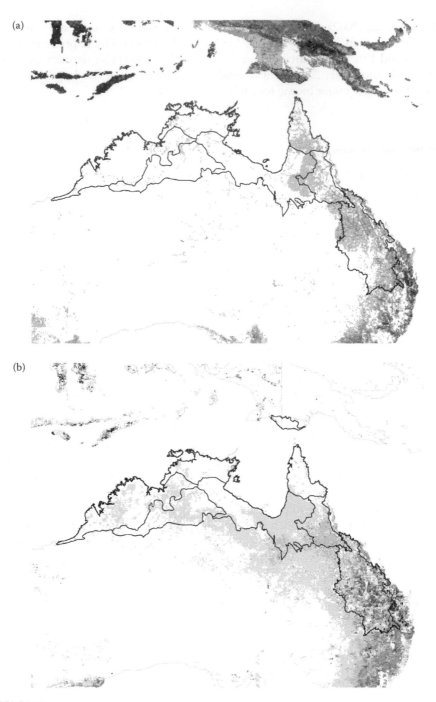

FIGURE 1.7
Density of human rural population (a, c, and e; persons/km²) and cattle density (b, d, and f; animals/km²) for Australian, South American, and African savanna ecoregions.

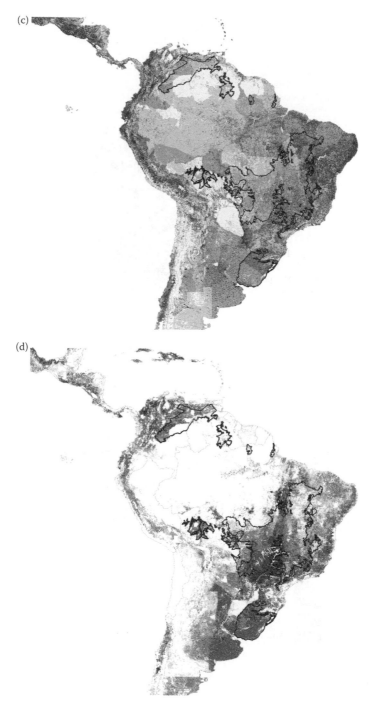

FIGURE 1.7 Continued.

(e)

(f)

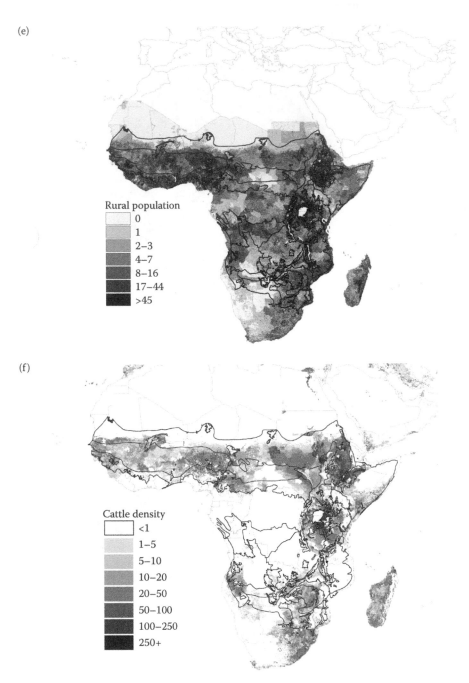

FIGURE 1.7 Continued.

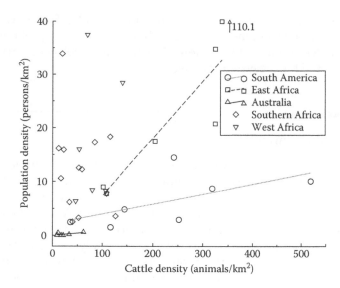

FIGURE 1.8

The potential anthropogenic pressure response space for global savanna ecoregions characterized by rural population density and cattle density.

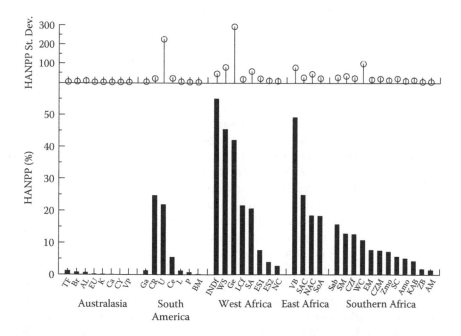

FIGURE 1.9

Percentage appropriation of net primary productivity by humans for savanna ecoregions and spatial standard deviations. The latter indicates the evenness or unevenness of the HANPP within an ecoregion.

South America and Australasia in all but the cerrado. The more arid and climatically variable regions utilize a much higher percentage of net primary production (NPP) than other regions, but this does not necessary indicate higher potential for ecosystem damage.

Temporal Dynamics

Historically, the regional savannas have received very different levels of attention from the remote sensing community. West African savannas have played a pivotal role in the development of time series analysis of remotely sensed data through early developmental work on the Advanced Very High Resolution Radiometer Normalized Difference Vegetation Index (AVHRR NDVI) (e.g., Tucker et al., 1986) and subsequent analysis (Tucker et al., 1991) and controversy about desertification (Olsson, 1993; Anyamba and Tucker, 2005). Analysis of time series of vegetation indices continues to be important in key savanna ecoregions (e.g., Ferreira and Huete, 2004; Ferreira et al., 2003; Anaya et al., 2009; Sjöstrom et al., 2009). However, aside from global scale biome assessments (e.g., Potter et al., 2008), savanna ecoregions have been examined individually rather than collectively. Therefore, the final element in this analysis is to characterize and compare the average temporal profiles of the MODIS EVI for the global savanna ecoregions over the period 2002–2007. For this analysis, only full retrieval pixels (i.e., no quality flags due to cloud or other factors) were used. This is carried out for the savanna land cover classes in each savanna ecoregion—the savanna profile is representative of the original natural vegetation.

The time series profiles starkly contrast the dynamics of the five different savanna regions and the individual savanna ecoregions (Figure 1.10). For Australia, the seasonal peaks in green vegetation are sharp, and overall EVI and peak values are reduced for the inland savananas versus Arnhem Land and Cape York, which have the strongest monsoon (Figure 1.10). By contrast, there are large differences in the seasonal profiles of EVI for the South American savannas (Figure 1.10), which reflect a wide variation in latitude and the spatial characteristics of the ecoregion in relation to the latitudinal rainfall patterns. The seasonal pattern moves from a broad peak centered around the June to September period for the Llanos and a narrower peak around June for the Guianian savanna (both north of the equator); to a broad peak around the December to March period for the cerrado, Beni-Moxos, and Pantanal; and a sharp, narrow peak around February for the Uruguayan savanna, thus reflecting more temperate rainfall patterns. The rate of land use change and degree of fragmentation in the cerrado (now as much as 50% converted; see Bustamante et al., Chapter 19) means that the profile for this ecoregion is subject to more contamination from agricultural land use, because it represents an average over 6 years.

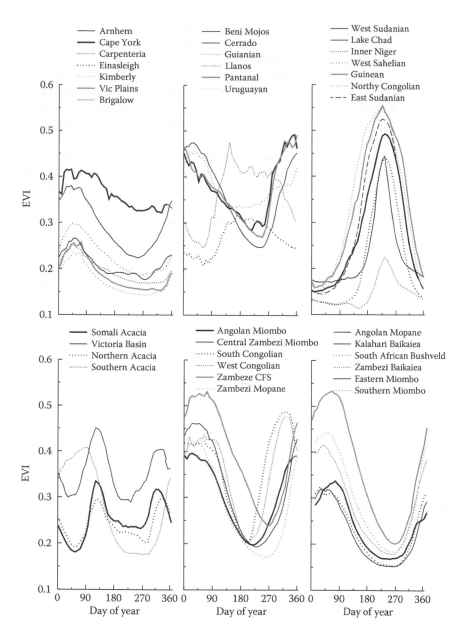

FIGURE 1.10
Time series profiles of EVI by day of year for the savanna land cover class within savanna ecoregions averaged over the period 2002–2007. Data are grouped by major savanna regions: Australia, South America, West Africa, East Africa, and Southern Africa woodlands and drylands.

The West and Southern African savanna ecoregions show very strong seasonality in their EVI profiles (Figure 1.10). With three exceptions, the seasonal signals are strongly synchronous and vary only in magnitude. The Southern and Central Congolian ecoregions have an earlier seasonal peak in EVI, whereas the Zambezi coastal flooded savanna has an extended declining phase. For West African ecoregions, the width and amplitude of the peak declines with distance from the coast in a manner similar to the Australian savanna ecoregions.

The east African savannas are characterized by a bimodal EVI profile reflecting two rainfall seasons. There is only a faint echo of this bimodality in the Southern Acacia ecoregion, which more closely resembles the Southern African ecoregions. The Southern African ecoregions can be divided into (1) the forest mosaics and miombo woodlands with significant woody cover and (2) the dryer mopane woodlands, baikaiea, and bushveld, which show a reduced seasonal strength and amplitude of the EVI peak. The baikiaea and bushveld are on the margins between savanna and shrubland, but as Figure 1.3 illustrates, the savannas as defined by the ecoregions used here are characterized by mosaics of forest, woodland, grassland, and shrubs.

A comparison of EVI profile patterns across the major regions highlights the different seasonal patterns of photosynthetic activity between Australia, South America, and Africa, mediated by the interaction between climate (and rainfall in particular), latitudinal relationship to the oceans and monsoonal influences, type and density of woody vegetation, and dynamics of the understorey grassland. The profile amplitudes give a further indication of the relatively low human and livestock carrying capacity of the Australian savannas.

Conclusions

The global savanna biome has significant definitional issues and is substantially made up of mosaics of forest, grassland, woodland, and shrubland along with classical open treed grasslands. Savannas may be considered at two levels of aggregation:

1. As an association of trees, shrubs, and grasses at patch to landscape scale with diversity influencing local-scale ecosystem function
2. As a collection of woodland, open woodlands, sparse woodlands, and grasslands in mixtures at a range of densities and scales of patch disaggregation across broad regions and subcontinental areas more broadly classified by the presence of large-scale-climate–fire–herbivory dynamics

In this chapter, we have provided a spatially explicit characterization of global tropical and subtropical savannas, as defined by the WWF Global Ecoregions (Olson et al., 2001), and a comparison of seasonal profiles of photosynthetic capacity, as represented by the EVI. In doing this, we have excluded Asian dry forests and temperate savannas in Australia and North America and included some areas predominantly covered by shrubland or grassland. The diversity of climate, vegetative cover, human and livestock footprints, and seasonal dynamics is clearly illustrated. As such, the similarities and differences between spatiotemporal patterns and distributions of the key biophysical and anthropogenic properties of the savanna biome can be readily and simply observed. The subsequent chapters in this book will provide analysis and synthesis that explain and delve deeper into much of this diversity.

Acknowledgments

We thank Mark Friedl for access to the MODIS Land Cover data. We also thank Dr. Gerold Kier for making the species richness data of global ecoregions publically available. Map data on rural populations were downloaded from http://www.fao.org/geonetwork/srv/en/main.home and from the SEDAC server of CIESIN. The senior author wishes to thank William Wint of Oxford University for providing full access to the Global Gridded Livestock Database.

References

Anaya, J. A., E. Chuvieco, and A. Palacios-Orueta, 2009. Aboveground biomass assessment in Colombia: A remote sensing approach. *Forest Ecology and Management* 257, 1237–1246.

Anyamba, A. and C. J. Tucker, 2005. Analysis of Sahelian vegetation dynamics using NOAA-AVHRR NDVI data from 1981–2003. *Journal of Arid Environments* 63, 596–614.

Asner, G. P., S. R. Levick, T. Kennedy-Bowdoin, D. E. Knapp, R. Emerson, J. Jacobson, M. S. Colgan, and R. E. Martin, 2009. Large-scale impacts of herbivores on structural diversity of African savannas. *Proceedings of the National Academy of Sciences* www.pnas.ord/cgi/doi/10.1073/pnas.0810637106

Bartholme, E. and A. S. Belward, 2005. GLC2000: A new approach to global land cover mapping from Earth observation data. *International Journal of Remote Sensing* 26, 1959–1977.

Bassett, T. J. and D. Crummey (eds), 2003. *African Savannas. Global Narratives and Local Knowledge of Environmental Change*. James Currey, Oxford, 270pp.

Beerling, D. J. and C. P. Osborne, 2006. The origin of the savanna biome. *Global Change Biology* 12, 2123–2013.

Bicheron, P., P. Defourney, C. Brockmann, L. Schouten, C. Vancutsem, M. Huc, S. Bontemps, et al. 2008. GLOBCOVER. *Products Description and Validation Report.* Medias France/POSTEL, Toulouse, France, 47pp.

Bond, W. J., F. I. Woodward, and G. F. Midgely, 2005. The global distribution of ecosystems in a world without fire. *New Phytologist* 165, 525–538.

Boone, R. B., J. M. Lackett, K. A. Galvin, D. S. Ojima, and C. J. Tucker III, 2007. Links and broken chains: Evidence of human-caused changes in land cover of remotely sensed images. *Environmental Science and Policy* 10, 135–149.

Bucini, G. and N. P. Hanan, 2007. A continental-scale analysis of tree cover in African savannas. *Global Ecology and Biogeography* 16, 593–605.

Carmona-Moreno, C., A. Belward, Ph. Caperan, et al. 2005. Characterizing interannual variations in global fire calendar using data from Earth observing satellites. *Global Change Biology* 11, 1537–1555.

Center for International Earth Science Information Network (CIESIN), Columbia University; United Nations Food and Agriculture Programme (FAO); and Centro Internacional de Agricultura Tropical (CIAT). 2005. Gridded Population of the World: Future Estimates (GPWFE). Socioeconomic Data and Applications Center (SEDAC), Columbia University, Palisades, NY. Available at http://sedac.ciesin.columbia.edu/gpw (downloaded 11/03/08).

Dansereau, P., 1957. *Biogeography: An Ecological Perspective.* Ronald Press, New York.

Defourny, P., C. Vancutsem, P. Bicheron, et al. 2006. GLOBCOVER: A 300 m global land cover product for 2005 using ENVISAT MERIS time series. *ISPRS Commission VII Mid-term Symposium "Remote Sensing: From Pixels to Processes,"* Enschede, the Netherlands, May 8–11, 2006.

Du Toit, J. T. and D. H. M. Cumming, 1999. Functional significance of ungulate diversity in African savannas and the ecological implications of the spread of pastoralism. *Biodiversity and Conservation* 8, 1643–1661.

Dyer, R., P. Jacklyn, I. Partridge, J. Russell-Smith, and R. Williams, 2001. *Savanna Burning. Understanding and Using Fire in Northern Australia.* Tropical Savanna CRC, Darwin, 136pp.

Eswaran, H., E. Van den berg, and P. Peich, 1993. Organic carbon in soils of the world. *Soil Science Society of America Journal* 57, 192–194.

Fairhead, J. and M. Leach, 1996. *Misreading the African Landscape: Society and Ecology in a Forest–Savanna Mosaic.* Cambridge University Press, Cambridge, 374pp.

Ferreira, L. G. and A. R. Huete, 2004. Assessing the seasonal dynamics of the Brazilian Cerrado vegetation through the use of spectral vegetation indices. *International Journal of Remote Sensing* 25, 1837–1860.

Ferreira, L. G., H. Yoshioka, A. Huete, and E. E. Sano, 2003. Seasonal landscape and spectral vegetation index dynamics in the Brazilian Cerrado: An analysis within the large-scale biosphere-atmosphere experiment in Amazonia (LBA). *Remote Sensing of Environment* 87, 534–550.

Friedl, M. A., D. K. McIver, J. C. F. Hodges, et al. 2002. Global land cover mapping from MODIS: Algorithms and early results. *Remote Sensing of Environment* 83, 287–302.

Friedl, M. A., D. Sulla-Menashe, B. Tan, et al. 2010. MODIS collection 5 global land cover: algorithm refinements and characterization of new datasets. *Remote Sensing of Environment* 114, 168–182.

Furley, P., 2004. Tropical savannas. *Progress in Physical Geography* 28, 581–598.

Giglio, L., G. R. van der Werf, J. T. Randerson, G. J. Collatz, and P. Kasibhatla, 2005. Global estimation of burned area using MODIS active fire observations. *Atmospheric Chemistry and Physics Discussions* 5, 11091–11141.

Haberl, H., K. H. Erb, F. Krausmann, et al. 2007. Quantifying and mapping the human appropriation of net primary production in earth's terrestrial ecosystems. *Proceedings of the National Academy or Sciences* 104, 12942–12947.

Hansen, M. C., R. S. DeFries, J. R. G. Townshend, and R. Sohlberg, 2000. Gloval land cover classification at 1 km spatial resolution using a classifaction tree approach. *International Journal of Remote Sensing* 21, 1331–1364.

Hansen, M. C., R. S. DeFries, J. R. G. Townshend, R. Sohlberg, C. Dimiceli, and M. Carroll, 2002. Towards an operational MODIS continuous field of percent tree cover algorithm: Examples using AVHRR and MODIS data. *Remote Sensing of Environment* 83, 303–319.

Hansen, M., R. S. DeFries, J. R .G. Townshend, M. Carroll, C. Dimiceli, and R. A. Sohlberg, 2003. Global percent tree cover at a spatial resolution of 500 meters: First results of the MODIS vegetation continuous fields algorithm. *Earth Interactions* 7, 1–15.

Harrington, G. N., A. D. Wilson, and M. D. Young (eds), 1984a. *Management of Australia's Rangeland*. CSIRO, Australia, 354pp.

Harrington, G. N., D. M. D. Mills, A. J. Pressland, and K. C. Hodgkinson, 1984b. Semi-arid woodlands. In *Management of Australia's Rangelands*, eds G. N. Harrington, A. D. Wilson, and M. D. Young. Australia, CSIRO, pp. 189–208.

Herold, M., P. Mayaux, C. E. Woodcock, A. Baccini, and C. Schmullius, 2008. Some challenges in global land cover mapping: An assessment of agreement and accuracy in existing 1 km datasets. *Remote Sensing of Environment* 112, 2538–2556.

Huete, A., K. Didan, T. Miura, E. P. Rodriguez, X. Gao, and L. G. Ferreira, 2002. Overview of the radiometric and biophysical performance of the MODIS vegetation indices. *Remote Sensing of Environment* 83, 195–213.

Imhoff, M. L., L. Bounoua, T. Ricketts, C. Loucks, R. Harriss, and W. T. Lawrence, 2004a. Global patterns in human consumption of net primary production. *Nature* 429, 870–873.

Imhoff, M. L., L. Bounoua, T. Ricketts, C. Loucks, R. Harriss, and W. T. Lawrence, 2004b. [HANPP Collection: Human Appropriation of Net Primary Productivity as a Percentage of Net Primary Productivity]. Data distributed by the Socioeconomic Data and Applications Center (SEDAC): http://sedac.ciesin.columbia.edu/es/hanpp.html [Date downloaded 030909].

Kier, G. and W. Barthlott, 2001. Measuring and mapping endemism and species richness: A new methodological approach and its application on the flora of Africa. *Biodiversity and Conservation* 10, 1514–1529.

Kier, G., J. Mutke, E. Dinerstein, et al. 2005. Global patterns of plant diversity and floristic knowledge. *Journal of Biogeography* 32, 1107–1116.

Krausmann, F., K.-H. Erb, S. Gingrich, C. Lauk, and H. Haberl, 2007. Global patterns of socioeconomic biomass flows in the year 2000: A comprehensive assessment of supply, consumption and constraints. *Ecological Economics* 65, 471–487.

Kreft, H. and W. Jetz, 2007. Global patterns and determinants of vascular plant diversity. *Proceedings of the National Academy of Sciences* 104, 5925–5930.

Lal, R., 2004. Soil carbon sequestration impacts on global climate change and food security. *Science* 304, 1623–1627.

van Langevelde, F., C. A. D. M. van der Vijver, L. Kumar, et al. 2003. Effects of fire and herbivory on the stability of savanna ecosystems. *Ecology* 84, 337–350.

Little, P. D., 2003. Rethinking interdisciplinary paradigms and the political ecology of pastoralism in East Africa. In *Global Narratives and Local Knowledge of Environmental Change: African Savannas*, (eds) T. J. Bassett and D. Crummey. James Currey, Oxford, pp. 161–177.

Lodge, G. M., R. D. B. Whalley, and G. G. Robinison, 1984. Temperate grasslands. In *Management of Australia's Rangelands*, (eds) G. N. Harrington, A. D. Wilson, and M. D. Young. CSIRO, Australia, pp. 317–329.

Loveland, T. R., B. C. Reed, J. F. Brown, et al. 2000, Development of a global land cover characteristics database and IGBP DISCover from 1 km AVHRR data. *International Journal of Remote Sensing*, 21(6–7), 1303–1365.

Maquere, V., J. P. Laclau, M. Bernoux, et al. 2008. Influence of land use (savanna, pasture, *Eucalyptus* plantations) on soil carbon and nitrogen stocks in Brazil. *European Journal of Soil Science* 59, 863–877.

McPherson, G. R., 1997. *Ecology and Management of North American Savannas*. University of Arizon Press, Tucson, 208pp.

Miles, L., A. C. Newton, R. S. DeFries, et al. 2006. A global view of the conservation status of tropical dry forests. *Journal of Biogeography* 33, 491–505.

Mistry, J., 2000. *World Savannas. Ecology and Human Use*. Pearson Education Limited, England, 344pp.

Moore, R. M. (ed), 1975a. *Australian Grasslands*. Australian National University Press, Canberra, 455pp.

Moore, R. M., 1975b. South-eastern temperate woodlands and grasslands. In *Australian Grasslands*, (ed) R. M. Moore. Australian National University Press, Canberra, pp. 169–190.

Mott, J. J. and J. C. Tothill, 1984. Tropical and subtropical woodlands. In, *Management of Australia's Rangelands*, (eds) G. N. Harrington, A. D. Wilson, and M. D. Young. Commonwealth Scientific and Industrial Research Organisation, East Melbourne, pp. 255–269.

Mouillot, F. and C. B. Field, 2005. Fire history and global carbon budget: A $1° \times 1°$ fire history reconstruction for the 20th century. *Global Change Biology* 11, 398–420.

Mutke, J. and W. Barthlott, 2005. Patterns of vascular plant diversity at continental to global scales. *Biologiske Skrifter* 55, 521–531.

Oliveira, P. S. and R. J. Marquis (eds), 2002. *The Cerrados of Brazil*. Columbia University Press, New York, 398pp.

Olson, D. M., E. Dinerstein, Wikramanayake, et al. 2001. Terrestrial ecoregions of the World: A new map of life on Earth. *Bioscience* 51, 933–938.

Olsson, L. 1993. On the causes of famine: Drought, desertification and market failure in the Sudan. *Ambio* 22, 395–403.

Palm, C., P. Sanchez, S. Ahamed, and A. Awiti, 2007. Soils: A contemporary perspective. *Annual Review of Environment and Resources* 32, 99–129.

Peel, M. C., B. L. Finlayson, and T. A. McMahon, 2007. Updated world map of the Köppen–Geiger climate classification. *Hydrology and Earth System Sciences* 11, 1633–1644.

Pennington, R. T., G. P. Lewis, and J. A. Ratter, 2006. *Neotropical Savannas and Seasonally Dry Forests. Plant Diversity, Biogeography and Conservation*. CRC Press, Boca Raton, FL, 484pp.

Potter, C., S. Boriah, M. Steinbach, V. Kumar, and S. Klooster, 2008. Terrestrial vegetation dynamics and global climate controls. *Climate Dynamics* 31, 67–78.

Ramankutty, N., and J. A. Foley, 1999. Estimating historical changes in global land cover: Croplands from 1700 to 1992. *Global Biogeochemical Cycles* 13, 997–1027.

Reich, P., 2005. Global Soil Regions Map. Revised September 2005. US Department of Agriculture. Natural resources Conservation Service. Soil Survey Division. World Soil resources. http://soils.usda.gov/use/worldsoils. Accessed May 7, 2009.

Riano, D. J. A. Moreno Ruiz, D. Isidoro, and S. L. Ustin, 2007. Global spatial patterns and temporal trends of burned area between 1981 and 2000 using NOAA–NASA Pathfinder. *Global Change Biology* 13, 40–50.

Robinson, T. P., G. Franceschini, and W. Wint, 2007. The Food and Agriculture organization's Gridded Livestock of the World. *Veterinaria Italiana* 43, 745–751.

Rossiter, R. C. and P. G. Ozanne, 1975. South-western temperate forests woodlands and heaths. In *Australian Grasslands*, ed. R. M. Moore, Canberra, Australia, Australian National University Press, pp. 169–190.

Salvatore, M., F. Pozzi, E. Ataman, B. Huddleston, and M. Bloise, 2005. *Mapping Global Urban and Rural Population Distributions*. Food and Agriculture Organization of the United Nations, Rome.

Sarmiento, G. (Solbrig, O. T., Translator), 1984. *The Ecology of Neotropical Savannas*. Harvard University Press, Cambridge, MA, 235pp.

Schaaf, C. B., F. Gao, A. H. Strahler, et al. 2002. First operational BRDF, Albedo and Nadir reflectance products from MODIS, *Remote Sensing of Environment* 83, 135–148.

Shaw, N. H. and M. J. T. Norman, 1975. Tropical and sub-tropical woodlands and grasslands. In *Australian Grasslands*, ed. R. M. Moore, Canberra, Australia, Australian National University Press, pp. 169–190.

Sinclair, A. R. E., C. Packer, S. A. R. Mduma, and J. M. Fryxell (eds), 2008. *Serengeti III: Human Impacts on Ecosystem Dynamics*. University of Chicago Press, Chicago IL, 512pp.

Sjöstrom, M., J. Ardo, L. Eklundh, et al. 2009. Evaluation of satellite based indices for gross primary production estimates in a sparse savanna in the Sudan. *Biogeoscience* 6, 129–138.

Tothill, J. C. and C. Gillies, 1992. *The Pasture Lands of Northern Australia. Their Condition, Productivity and Sustainability*. Tropical Grassland Society of Australia, Occasional Publication No. 5, Brisbane, 106pp.

Tucker, C. J., C. O. Justice, and S. D. Prince, 1986. Monitoring the grasslands of the Sahel 1984–1985. *International Journal of Remote Sensing* 7, 1571–1581.

Tucker, C. J., W. W. Newcomb, S. O. Los, and S. D. Prince, 1991. Mean and inter-annual variation of growing-season normalized difference vegetation index for the Sahel 1981–1989. *International Journal of Remote Sensing* 12, 1133–1135.

2

Tree–Grass Interactions in Savannas: Paradigms, Contradictions, and Conceptual Models

Niall P. Hanan and Caroline E. R. Lehmann

CONTENTS

Introduction

The savannas (mixed woody–herbaceous systems in drought-seasonal regions) are one of the most fascinating and enigmatic of the world's biomes. The relatively recent appearance and rapid global expansion of the savannas occurred only 6 to 8 million years ago (6–8 M BP; Beerling and Osborne, 2006), after the earlier evolution of C_4 grasses (15–35 M BP; Sage, 2004), to now cover more than 20% of the Earth's land surface (Hill et al., Chapter 1, this volume; Scholes and Archer, 1997). The C_4 expansion was associated with changing paleo-climate, increasing evidence of fire in the Earth system, and the evolution and numerical expansion of large herbivore guilds specialized in grazing (Pagani et al., 1999; Keeley and Rundel, 2005; Beerling and Osborne, 2006). In many semiarid regions, the increase in C_4 dominance displaced woody angiosperms that were drought tolerant but sensitive to the frequent fires which the C_4 grasses brought with them. Climate, atmospheric CO_2 concentrations, fire, and herbivory have all been associated with, and may partially explain, the historical savanna expansion; and these factors remain critical in determining their modern-day geographic extent (Beerling and Osborne, 2006; Osborne, 2008).

The new savanna biome also supported the biological and social evolution of an additional species, one that has long been the primary ecosystem engineer of the savannas, *Homo sapiens*. During the millennia since the last ice age, humans in Africa and elsewhere have promoted savanna expansion and persistence through deliberate use of fire (Bond, 2008). Indeed, huge areas of the world classified on vegetation maps as some form of semiarid woodland or dry forests (e.g., the dipterocarp forests of south and southeast Asia, the miombo woodlands of Africa, and the dry forests of Central America) are, in reality, "managed as savannas." That is, regardless of the original composition and structure of the vegetation in these regions, human management practices now deliberately reduce woody cover, increase grass production for domestic herbivores, and increase the prevalence of fires. If tree–grass mixtures, grazing, and fire are the defining characteristics of savannas, then these areas are truly savannas!

Anthropogenic climate change, agricultural conversion, and the changing prevalence of fire and herbivore populations in future decades and centuries may lead to the decline of savannas in some areas or the transformation of tropical forests to savannas in other areas (Higgins et al., 2007; Barlow and Peres, 2008; Bond, 2008; Knapp et al., 2008). For example, savannas currently experience higher rates of land use transformation than the tropical forests (Grace et al., 2006).

In this chapter, we review the sometimes-controversial and often-contradictory ecological perspectives on how and why trees (note that throughout this chapter, our use of "tree" is intended to include "shrub" and "tree" growth forms) and grasses coexist in savanna regions, referring the reader to existing literature on this theme. We then discuss a simple conceptual framework that incorporates elements of the main savanna perspectives and allows for qualitative predictions relating how climate, human management, and natural disturbances impact savanna vegetation dynamics.

Tree–Grass Coexistence and Determinants of Savanna Structure

Several alternative conceptual and numerical models for tree–grass coexistence and the determinants of woody structure in savannas have been advanced during many years of research. In recent years, a number of excellent reviews on these questions have been published (e.g., Scholes and Archer, 1997; Jeltsch et al., 2000; House et al., 2003; Sankaran et al., 2004; Bond, 2008). These sources can provide the reader with more in-depth discussion of savanna conceptual models than what is provided here. What has become clear, however, is that savanna ecology has undergone something of a paradigm shift: Widespread acceptance of the classical "Walter Hypothesis," where tree–grass coexistence depends on deep soil water to which only trees have access (niche-separation; Walker and Noy-Meir, 1982), has been largely replaced by an emphasis on the "Bottleneck Hypotheses," where disturbance by fire and herbivory or intermittent droughts increase tree mortality, prevent seedling establishment, and act as bottlenecks that restrict and reduce woody density and cover (Higgins et al., 2000; Wiegand et al., 2006; Miller and Chesson, 2009).

Both the root niche partitioning hypothesis and the bottleneck hypotheses have some empirical support in the literature, but neither is without contradictory observations (Sankaran et al., 2004). Particularly intriguing from a theoretical perspective is that the alternative models for tree–grass coexistence in savannas are based on diametrically opposing assumptions with regard to the competitive relationships (dominance) of trees or shrubs *vis-à-vis* grasses. The Walter Hypothesis assumes that grasses have a competitive advantage over trees such that, for trees to persist, they must have exclusive access to water in deep soil layers which the grasses do not access. By contrast, the bottleneck models for persistence of tree–grass systems assume that trees will out-compete, and effectively exclude, grasses unless disturbances cause tree mortality (e.g., fire, elephants) or limited, but temporally variable, rainfall makes tree establishment a rare event ("punctuated establishment"; Wiegand et al., 2004, 2006).

Different perceptions regarding the importance of competition, niche separation, and punctuated establishment, relative to disturbances like fire and herbivory, in controlling savanna structure and promoting coexistence, reflect the fact that most studies are site specific and small scale (Lehmann et al., 2009). Few studies have examined these processes over gradients of climate and soil (Frost et al., 1986; Williams et al., 1996; Fensham et al., 2005; Sankaran et al., 2005, 2008; Higgins et al., 2007; San José and Montes, 2007; Lehmann et al., 2008). Fewer still have fully explored the patterns emerging across savannas at the global scale (Lehmann et al., 2009).

Savanna Paradigms and Contradictions

Although no savanna ecologist would contest that disturbances are important in controlling savanna structure and that disturbance and resource bottleneck ideas have dominated recent literature on savanna ecology, we cannot yet assume that savannas would be closed systems in the absence of disturbance. On the contrary, recent evidence (Sankaran et al., 2005) suggests that resource availability is still a critical element in controlling maximum woody cover. Figure 2.1 shows a continental scale analysis of vegetation structure in African savannas compiled to explore how the relative importance of water, nutrients, fire, and herbivory varies across broad environmental

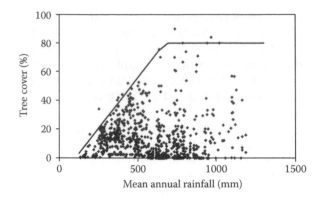

FIGURE 2.1
Change in tree cover of African savannas as a function of mean annual rainfall (MAP). These data from 854 sites suggest an upper bound to tree cover in sites with 150–650 mm MAP. (Reproduced from Sankaran, M. et al. 2005. *Nature* 438, 846–849. With permission.)

gradients (data from 854 sites spanning the continent). The analysis shows that, while tree cover ranges from 0 to >90% across the whole continent, within a narrow rainfall range (100–650 mm mean annual precipitation), *maximum* tree cover is constrained by mean annual precipitation. More particularly, this upper bound on tree cover increases linearly with rainfall but shows no relationship to fire frequency, levels of herbivory, soil textures, or soil nutrients (Sankaran et al., 2005). Although the combined effects of (we presume) fire, herbivory, and other disturbances result in a range of realized tree cover well below the bound (Bucini and Hanan, 2007; Sankaran et al., 2008), the data suggest that water availability constrains maximum tree cover in these arid and semiarid savannas.

The observed relationship between mean rainfall and maximum tree cover is entirely consistent with the predictions of the Walter Hypothesis, wherein more rainfall (in particular, more intense rainfall events that initiate saturated flow) provides more opportunity for water to percolate through the shallow soil layers into the deep soil where (the Walter hypothesis assumes) only tree roots have access. In the Walter Hypothesis, and the mathematical models based on it (e.g., Walker et al., 1981; Walker and Noy-Meir, 1982; van Langevelde et al., 2003), more rainfall each year generally means more percolation to deep roots, supporting increased woody biomass, density, and cover.

However, the root niche separation hypothesis remains inconsistent with the common observation that savanna soils are, in many locations, too shallow for root zone separation between trees and grasses, and root observations where tree and grass roots share soil volume with no clear spatial differentiation or exclusive niche (Belsky, 1990, 1994; Seghieri, 1995; Mordelet et al., 1997; Smit and Rethman, 2000; Hipondoka et al., 2003; though we also note examples where a deep tree root zone has been observed: Soriano and

Sala, 1983; Knoop and Walker, 1985; Sala et al., 1989; Pelaez et al., 1994; Weltzin and McPherson, 1997). Perhaps more worrisome from a conceptual perspective is that the Walter Hypothesis ignores the seedling and sapling stage of trees, before deep roots have been established: Competition is assumed to occur between grasses and adult trees with deep roots. For tree seedlings to establish and survive, however, they must do so in competition with grasses in the shared root zone before deep roots are established. Whether or not deep roots later promote adult growth and survival, seedlings and saplings must be competitively strong, or neutral, with regard to the grasses surrounding them to survive. Though rarely, if ever, discussed in such terms, the Walter Hypothesis, therefore, implies a shift from strong to weak competitor between seedlings and adult life stages. Such a shift in competitive ranking seems counterintuitive: Adult trees would seem more likely to be strongly competitive with grasses than seedlings (Walter implicitly assumes the reverse), because adults have multiple years during which to develop extensive root systems.

There is no question that disturbances, particularly by fires and herbivores, are important in reducing actual tree cover below the observed maximum in most locations (see, e.g., the majority of points in Figure 2.1). We must, therefore, accept, to some degree, the bottleneck concept with implicit assumption that trees are strong competitors, contrary to assumptions underlying the Walter Hypothesis. The somewhat-neglected balanced competition mechanism (Scholes and Archer, 1997; where tree–tree competition limits maximum tree density or cover) is interesting in this regard in that it might explain the rainfall-related trend in maximum cover, while being consistent with the bottleneck models in assuming that the adult trees are strong competitors to grasses. However, although we know that tree root canopies can extend laterally much further than the aerial canopy (e.g., Seghieri, 1995; Schenk and Jackson, 2002), it does not seem likely that tree–tree competition is responsible for limiting tree cover at the low rainfall, low tree cover sites (e.g., below, say, 20% maximum tree cover in Figure 2.1). At these low rainfall levels, the punctuated establishment mechanism (Wiegand et al., 2006) and the role of favored sites (e.g., depressions or runoff barriers where water can collect and favor woody plant growth and survival; Jeltsch et al., 1998) might, separately or in concert, limit maximum tree cover as a dynamic equilibrium between adult longevity and the availability of favored locations and times for juvenile establishment.

In most of this discussion, we have concentrated on tropical savannas. However, an additional interesting issue with regard to the Walter hypothesis and the role of root niche differentiation in both tree–grass coexistence and limitations to maximum woody cover is the distinction between tropical and temperate mixed woody–herbaceous systems. Rainfall in many temperate savannas, including those in Mediterranean climate and much of the South West United States and Great Basin, is concentrated in winter when plant activity (and thus transpiration) is temperature limited and soil evaporation is low. In such situations, there is an increased likelihood of deep percolation of water and, thus, an increasing potential benefit of deep roots to woody plants. By contrast, in tropical savannas, the rainy season is concurrent with plant activity and high temperatures, such that water resources are often quickly utilized in transpiration and evaporation, reducing the frequency of saturation of the upper soil layers and subsequent deep percolation. Thus, we might hypothesize that the competitive advantage conferred by deep soil resources may be more important in temperate tree–grass systems than in tropical savannas.

A final interesting component in the tree–grass interaction story is the phenomenon of shrub encroachment in response to grazing. The shrub encroachment phenomenon has been offered as evidence for the Walter Hypothesis, as reduced grass cover and grass transpiration may make additional water available to the trees in both shallow and deep soil (e.g., Walker and Noy-Meir, 1982). However, by indirect mechanisms, shrub encroachment related to grazing is also consistent with the demographic bottleneck models, as reduced grass biomass reduces fire frequency and intensity, thus reducing seedling mortality (Higgins et al., 2000). Thus, again, we are not able to reject either class of model despite their contradictory underlying assumptions regarding competitive ranking of trees and grasses.

Savannas, Climate, and Disturbance: A Conceptual Framework

It seems clear that no model, on its own, can explain the observed patterns of tree cover in the tropical savannas and that hybrid models are needed to explain both the resource-driven change in potential cover and the disturbance-driven changes in realized cover (Sankaran et al., 2004; Bucini and Hanan, 2007). However, such combinations of conceptual models must first reconcile, in some way, the contradictory assumptions with regard to the competitive ranking of savanna trees (seedlings, saplings, and adults) and grasses. Merging of niche separation models and the bottleneck models requires some resolution of this problem, as the former assumes grasses are superior competitors, whereas the latter assumes that trees are superior competitors. On the other hand, merging of the balanced competition idea with bottleneck models involves no such complication, but it may not apply in lower rainfall and maximum woody-cover situations where other mechanisms (frequency of favored establishment sites and times) may be more important.

Figure 2.2 summarizes our knowledge of the factors affecting savanna vegetation dynamics, including the importance of water resources (approximated by mean annual precipitation for a given soil type); woody demographics and tree–grass competition; and the influence of fire, grazers, and browsers on

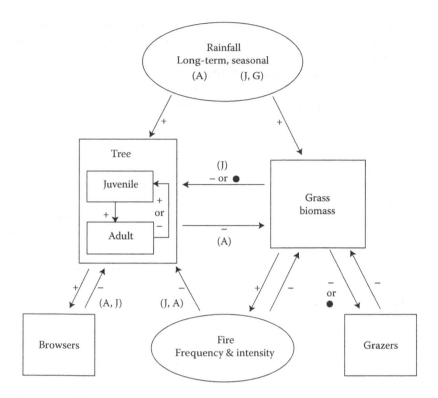

FIGURE 2.2
Interrelationships among climate, vegetation, and disturbances in tropical savannas. Only postulated main effects are shown. Arrows show expected interactions among system components, with symbols showing • neutral, – negative, and + positive effects. Thus, grazers reduce grass biomass (G), but increasing grass biomass can be neutral or deter grazers (neutral (•) or negative (–) when increasing biomass decreases forage quality or grazers favor predator avoidance over access to forage). Similarly, grass biomass may compete for soil moisture with juvenile trees (J), but adult trees (A) reduce soil moisture for grass growth. Fire has negative effects on juvenile trees, but it may also reduce long-term adult survival, and browsers (and wood harvest) impact growth and/or survival of both adult and juvenile trees.

trees and grasses. We assume, as observed in empirical data (Figure 2.1), that some combination of resource-related factors limits maximum woody cover below canopy closure when rainfall is less than approximately 650 mm/year (e.g., intra-specific competition between trees in more dense situations; spatial and temporal variability in establishment probability in drier sites). Grass production, seedling emergence, and survival are strongly affected by inter-annual variability in rainfall. Woody ("tree") and herbaceous ("grass") species interact primarily through competition for water, with grass competing for water with (and reducing survival of) juvenile trees and adult trees competing for water with (reducing production of) grasses. Grazers can reduce grass biomass and, indirectly, reduce competition between juvenile trees and grasses and fire (frequency or intensity). Fires kill juvenile trees, and they can also reduce longevity of older trees. Browsers (and wood harvest by humans) directly affect adult and juvenile trees through reductions in biomass, inhibition of reproduction, and direct mortality.

We used Figure 2.2 and estimates of disturbance intensity to estimate the sign, and relative magnitude, of fire and herbivory impacts (alone and in combination) on savanna structure for a West African example (Table 2.1). Although the *effects* shown in Table 2.1 are scaled arbitrarily to indicate qualitative (not quantitative) tendencies, we are confident of their sign. We assume, for now, that direct disturbance effects on trees (i.e., browse and fire) are equivalent but that indirect effects (grass biomass effects on fire and resource competition) are relatively weaker than the direct effects.

In the example in Table 2.1, we increased browse intensity and reduced grazer intensity in drier systems mimicking shifts in herd species composition toward smaller-body sheep or goats that occur in pastoral West Africa. We also assumed that fire intensity, and thus mortality of both adult and juvenile trees, would increase as grass biomass ("fuel load") increases with rainfall, and decreases when grazers reduce fuel load, as generally observed in savannas.

Using the combination of effects shown in Table 2.1, we graphically depict the postulated impacts of disturbance on woody populations in Figure 2.3a. Disturbance regimes *fH* and *Fh* represent our understanding of direct single-factor impacts (i.e., woody mortality or growth inhibition caused by fire or browsers) and indirect single-factor impacts (i.e., reductions in grass competition after fire or grazing). Not unexpectedly, the direct effects are always negative (fire and browse suppressing woody populations), but the indirect effects are positive (favoring woody populations) (see Table 2.1). Multifactor impacts (fire + herbivory; *FH*) are nonadditive (i.e., not the simple sum of *fH* and *Fh*), because an additional indirect effect of grazers comes into play in the presence of fire (reduced fuel load with grazers). In the "subsistence pastoral" example of Figure 2.3a, the management conditions (fire and grazing practices) favoring and suppressing woody cover differ markedly across the rainfall gradient. In drier regions, we anticipate that fire and large herbivore exclusion will most favor woody cover, whereas the combination of herbivory and fire will most suppress it. By contrast, in wetter savannas fire suppression with herbivory will favor woody cover, whereas fire without herbivory will suppress woody cover.

Figure 2.3b and c shows two alternate scenarios for the impacts of large herbivores and, by extension, human management on woody structure in savannas. In Figure 2.3b, we show an example typical of commercial livestock systems focused on cattle production as found, for example, in Australian and South American savannas and in more intensively managed rangelands of Africa (e.g., in parts of South Africa and Botswana). In cattle-intensive systems, direct herbivory impacts on tree mortality are low (since cattle tend to avoid browse), and herbivory impacts on woody structure

TABLE 2.1

Effects of Fire and Herbivory on Woody Vegetation across a Tropical Savanna Rainfall Gradient (e.g., West African Sahel and Sudanian Savannas), Where Fire is Positively Correlated and Overall Grazing Intensity is Negatively Correlated, with Mean Annual Precipitation (MAP)

Rain	Grass Competition	F (Fire)	H (lg herbivore)	G (Grazer)	B (Browser)	Effect on Trees
fh						
300	me	.	.	.	.	•
750	me	.	.	.	.	•
1200	me	.	.	.	.	•
Fh						
300	me (+0.5)	lo −1	.	.	.	−0.5
750	me (+0.5)	me −2	.	.	.	−1.5
1200	me (+0.5)	hi −3	.	.	.	−2.5
fH						
300	me (+0.5)	.	hi	me	hi −2	−1.5
750	me (+0.5)	.	me	me	me −1	−0.5
1200	me (+0.5)	.	lo	lo	lo •	+0.5
FH						
300	me (+1)	lo −1 (+0.5)	hi	me	hi −2	−1.5
750	me (+1)	me −2 (+0.5)	me	me	me −1	−1.5
1200	me (+1)	hi −3 (+0.5)	Lo	lo	lo •	−1.5

Note: In this example, subsistence practices and herd species mixtures mean that a higher proportion of total large herbivores are mixed feeders (e.g., sheep and goats that both graze and browse) in drier savannas, and a higher proportion are grazers (e.g., cattle) in mesic systems, thus changing the impact of large herbivores in the woody community. Three example rain levels (mean annual precipitation, mm) are provided for each of the four treatments. Relative estimates of ambient disturbance intensity are provided for each rainfall level (where hi = high, me = medium, and lo = low). Disturbance treatments (upper case letters; Fh, fH, and FH) show anticipated impacts (assessed on an arbitrary scale that increases with disturbance intensity), where − = reduction, • = neutral, + = increase in woody cover, basal area, and so on. Indirect effects are shown in parentheses (e.g., decreased grass competition (+0.5) when fire or grazers reduce grass biomass in Fh and fH; decreased fire [fuel load] when grazers are present in FH). Indirect effects are assumed weak (50%) relative to direct fire or browser effects and constant with rainfall. Direct fire and browser impacts are scored equally. *fh*: no fires or large herbivores; *Fh*: ambient fire, no large herbivores; *fH*: no fire, ambient herbivory; and *FH*: ambient fire, ambient herbivory.

occur primarily through indirect effects on grass biomass, with resulting reductions in competition with trees (seedlings and/or adults), reduced fuel load, and mortality during fires. Since indirect effects tend to be positive (Figure 2.2), the impact of cattle grazing tends to be positive (solid lines in

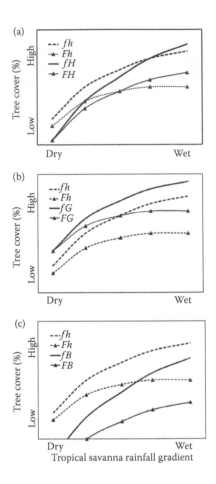

FIGURE 2.3
A conceptual model of the relative impacts of fire and different types of large mammalian herbivory on woody vegetation structure across the tropical savanna rainfall gradient (approximately 300–1200 mm MAP). (a) Herbivory (*H*) by mixed herds of domestic grazers (e.g., cattle) and mixed grazers/browsers (e.g., sheep and goats), with herd makeup approximating typical patterns of grazing in subsistence pastoral systems of Africa, with increased importance of small ruminants in drier savannas; (b) herbivory (*G*) by grazing specialists (e.g., cattle) as might occur in commercial ranching systems in savannas worldwide; and (c) herbivory (*B*) by browse specialists (e.g., elephants). Woody vegetation is depicted as woody cover, but it could equally be basal or any other metric area. Lower case (*f, h*) indicates no disturbance by fire, large herbivores, respectively; *F, H, G,* and *B* indicate disturbance by fire, mixed feeders, grazers, and browsers, respectively. *Note:* dotted (no herbivory) lines are the same in examples (a–c).

Figure 2.3b). Fire suppression increases recruitment and survival in wetter systems but has less impact in drier systems, where grass fuel-loads are lower and fires less frequent and intense. Grazed systems without fire will have highest woody cover, whereas ungrazed systems with fire are likely to have lowest woody cover, across the full rainfall gradient.

Figure 2.3c shows the contrasting, and rather more hypothetical, example of savannas across a rainfall gradient where large herbivores are primarily browsers. Although elephants are mixed feeders, they can have particularly severe impacts on woody populations (Stuart-Hill, 1992; Dublin, 1995; Baxter and Getz, 2005), and thus Figure 2.3c can be said to represent systems where elephants are common. In this case, the presence of large herbivores (browsers) generally suppresses the woody population, but in the wetter systems fires may be equally or more influential in woody suppression. Highest woody cover in this (browser) scenario occurs with the combination of fire and herbivore suppression, whereas burned and browsed savannas are likely to have lowest woody cover, across the full rainfall gradient.

Conclusions

The physical, physiological, demographic, and anthropogenic relationships and interactions shown in Figure 2.2 represent some of the most important processes contributing to tropical savanna vegetation dynamics and woody community structure. Soil type and biogeochemistry are not represented and would to some extent modify the relationships in Figure 2.2, primarily through changes in water availability, distribution in the soil (infiltration and retention), and changing nutrient availability, thereby leading to shifting productivity and competitive relationships of grasses relative to trees (Parton et al., Chapter 23). At continental scales across Africa, such soil-related differences appear to be relatively small (Bucini and Hanan, 2007; Sankaran et al., 2008), perhaps because changing soil types are often associated with shifting species composition and functional convergence (Hanan et al., 2010, this publication). At global scales, on the other hand, differences in lithology, age and weathering of soils, and divergent phytogenesis seem to result in rather different patterns of woody community structure in the savannas of Africa, Australia, and South America (Lehmann et al., 2009).

Fire as a management tool in the savannas can be expected to be consistent in its suppression of the woody community, primarily through mortality of smaller (juvenile) size classes (Hanan et al., 2008; Figure 2.3). The magnitude of this suppression, however, depends on grass fuel production and, thus, on climate (generally represented reasonably well by mean annual precipitation) and fuel losses to herbivores and degradation pathways. Anthropogenic manipulation of large herbivore populations in savannas used for livestock production (and indeed in savannas conserved for wild herbivore populations) can, on the other hand, have strongly divergent impacts on woody community structure, depending on herd density and the relative composition of grazers versus browsers and mixed feeders (Figures 2.2 and 2.3). Further, the interaction between large herbivores and fires is likely to vary in nonlinear, but predictable, ways depending on the relative importance of direct effects (browse-related tree mortality) and indirect effects (graze-related fire suppression and competitive release of woody plants) and how these vary across the rainfall gradient (Figures 2.2 and 2.3).

The conceptual model presented here (Figure 2.2) provides a framework for the qualitative prediction of climate, fire and herbivore impacts, and their interactions across the typical savanna rainfall gradients (Figure 2.3). More quantitative analyses using analytical or simulation models must still, however, grapple with the fundamental contradiction among the major classes of the savanna model (those based on root niche partitioning versus those based on demographic bottlenecks) relating to assumed competitive strength of grasses relative to juvenile and adult trees and shrubs. In many cases, these assumptions are implicit in the structure of savanna models, with little or no explicit recognition that this, therefore, precludes objective assessment of the relative validity of proposed tree–grass coexistence mechanisms and how they vary across climate and other environmental gradients.

Our implicit assumptions embodied in Figure 2.2 are that juvenile trees are weakly competitive with grasses; and juveniles are subject to mortality relating to resource availability, fire, and browser impacts. Adult trees, on the other hand, are strongly competitive with grasses by virtue of their extensive and perennial root systems (with or, in many cases, without access to an exclusive deep-root niche). Maximum woody cover (the bounding line in Figure 2.1) is likely then a function of bottlenecks to juvenile recruitment relating to mean and variability of water resources and competitive interactions: Rainfall variability and favored establishment sites may provide the bottlenecks in drier savannas, changing to juvenile suppression by grasses and adult trees in the wetter savannas. Actual woody cover (the cloud of points in Figure 2.1) then arises through the temporally stochastic and spatially variable influences of fire, herbivory, and human management (including wood harvest) that modify the resource-based maximum.

Acknowledgments

The ideas presented in this chapter have developed over several years with support from the U.S. National Science Foundation (Biocomplexity in the Environment Program and Division of Environmental Biology) and by activities under the Savanna Working Group (WG 49) of the "Research Network for Vegetation Function" supported by the Australian Research Council and Landcare Research, New Zealand.

References

Barlow, J. and C. A. Peres. 2008. Fire-mediated dieback and compositional cascade in an Amazonian forest. *Philosophical Transactions of the Royal Society of London Series B* 363, 1787–1794.

Baxter, P. W. J. and W. M. Getz. 2005. A model-framed evaluation of elephant effects on tree and fire dynamics in African savannas. *Ecological Applications* 15, 1331–1341.

Beerling, D. J. and C. P. Osborne. 2006. The origin of the savanna biome. *Global Change Biology* 12, 2023–2031.

Belsky, A. J. 1990. Tree/grass ratios in East African savannas: A comparison of existing models. *Journal of Biogeography* 17, 483–489.

Belsky, A. J. 1994. Influence of trees on savanna productivity: Tests of shade, nutrients and treegrass competition. *Ecology* 75, 922–932.

Bond, W. J. 2008. What limits trees in C4 grasslands and savannas? *Annual Review of Ecology, Evolution and Systematics* 39, 641–659.

Bucini, G. and N. P. Hanan. 2007. A continental scale analysis of tree cover in African savannas. *Global Ecology and Biogeography* 16, 593–605.

Dublin, H. 1995. Vegetation dynamics in the Serengeti-Mara Ecosystem: The role of elephants, fire and other factors. In *Serengeti II: Dynamics, Management, and Conservation of an Ecosystem,* (eds) A. R. E. Sinclair and P. Arcese. University of Chicago Press, Chicago, pp. 71–90.

Fensham, R. J., R. J. Fairfax, and S. R. Archer. 2005. Rainfall, land use and woody vegetation cover change in semi-arid Australian savanna. *Journal of Ecology* 93, 596–606.

Frost, P. G., E. Medina, J. C. Menaut, et al. 1986. *Response of Savannas to Stress and Disturbance*. Paris, IUBS.

Grace, J., J. San Jose, P. Meir, H. S. Miranda, and R. A. Montes. 2006. Productivity and carbon fluxes of tropical savannas. *Journal of Biogeography* 33, 387–400.

Hanan, N. P., W. B. Sea, G. Dangelmayr, and N. Govender. 2008. Do fires in savannas consume woody biomass? A comment on approaches to modeling savanna dynamics. *American Naturalist* 171, 851–856.

Higgins, S. I., W. J. Bond, E. C. February, et al. 2007. Effects of four decades of fire manipulation on woody vegetation structure in savanna. *Ecology* 88, 1119–1125.

Higgins, S. I., W. J. Bond, and S. W. Trollope. 2000. Fire, resprouting and variability: A recipe for grass–tree coexistence in savanna. *Journal of Ecology* 88, 213–229.

Hipondoka, M. H. T., J. N. Aranibar, C. Chirara, M. Lihavha, and S. A. Macko. 2003. Vertical distribution of grass and tree roots in arid ecosystems of Southern Africa: Niche differentiation or competition? *Journal of Arid Environments* 54, 319–325.

House, J. I., S. Archer, D. D. Breshears, and R. J. Scholes. 2003. Conundrums in mixed woody–herbaceous plant systems. *Journal of Biogeography* 30, 1763–1777.

Jeltsch, F., S. J. Milton, W. R. J. Dean, N. Van Rooyen, and K. A. Moloney. 1998. Modelling the impact of small-scale heterogeneities on tree–grass coexistence in semi-arid savannas. *Journal of Ecology* 86, 780–793.

Jeltsch, F., G. E. Weber, and V. Grimm. 2000. Ecological buffering mechanisms in savannas: A unifying theory of long-term tree–grass coexistence. *Plant Ecology* 161, 161–171.

Keeley, J. E. and P. W. Rundel. 2005. Fire and the Miocene expansion of C4 grasslands. *Ecology Letters* 8, 683–690.

Knapp, A. K., J. M. Briggs, S. L. Collins, et al. 2008. Shrub encroachment in North American grasslands: Shifts in growth form dominance rapidly alters control of ecosystem carbon inputs. *Global Change Biology* 14, 615–623.

Knoop, W. T. and B. H. Walker. 1985. Interactions of woody and herbaceous vegetation in a Southern African savanna. *Journal of Ecology* 73, 235–253.

Lehmann, C. E. R., L. D. Prior, R. J. Williams, and D. M. J. S. Bowman. 2008. Spatio-temporal trends in tree cover of a tropical mesic savanna are driven by landscape disturbance. *Journal of Applied Ecology* 45, 1304–1311.

Lehmann, C. E. R., J. Ratnam, and L. B. Hutley. 2009. Which of these continents is not like the other? Comparisons of tropical savanna systems: Key questions and challenges. *New Phytologist* 181, 508–511.

Miller, A. D. and P. Chesson. 2009. Coexistence in disturbance-prone communities: How a resistance–resilience trade-off generates coexistence via the storage effect. *American Naturalist* 173, E30–E43.

Mordelet, P., J. C. Menaut, and A. Mariotti. 1997. Tree and grass rooting patterns in an African humid savanna. *Journal of Vegetation Science* 8, 65–70.

Osborne, C. P. 2008. Atmosphere, ecology and evolution: What drove the Miocene expansion of C4 grasslands? *Journal of Ecology* 96, 35–45.

Pagani, M., K. H. Freeman, and M. A. Arthur. 1999. Late Miocene atmospheric CO_2 concentrations and the expansion of C4 grasses. *Science* 285, 876–879.

Pelaez, D. V., R. A. Distel, R. M. Boo, O. R. Elia, and M. D. Mayor. 1994. Water relations between shrubs and grasses in semiarid Argentina. *Journal of Arid Environments* 27, 71–78.

Sage, R. F. 2004. The evolution of C4 photosynthesis. *New Phytologist* 161, 341–370.

Sala, O. E., R. A. Golluscio, W. K. Lauenroth, and A. Soriano. 1989. Resource partitioning between shrubs and grasses in the Patagonian Steppe. *Oecologia* 81, 501–505.

San José, J. J. and R. A. Montes. 2007. Resource apportionment and net primary production across the Orinoco savanna–woodland continuum, Venezuela. *Acta Oecologica* 32, 243–253.

Sankaran, M., N. P. Hanan, R. J. Scholes, et al. 2005. Determinants of woody cover in African savannas. *Nature* 438, 846–849.

Sankaran, M., J. Ratnam, and N. P. Hanan. 2004. Tree–grass coexistence in savannas revisited—insights from an examination of assumptions and mechanisms invoked in existing models. *Ecology Letters* 7, 480–490.

Sankaran, M., J. Ratnam, and N. P. Hanan. 2008. Woody cover in African savannas: The role of resources, fire and herbivory. *Global Ecology and Biogeography* 17, 236–245.

Schenk, H. J. and R. B. Jackson. 2002. Rooting depths, lateral root spreads and below-ground/above-ground allometries of plants in water-limited ecosystems. *Journal of Ecology* 90, 480–494.

Scholes, R. J. and S. R. Archer. 1997. Tree–grass interactions in savannas. *Annual Review of Ecology and Systematics* 28, 517–544.

Seghieri, J. 1995. The rooting patterns of woody and herbaceous plants in a savanna; Are they complementary or in competition? *African Journal of Ecology* 33, 358–365.

Smit, G. N. and N. F. G. Rethman. 2000. The influence of tree thinning on the soil water in a semi-arid savanna of southern Africa. *Journal of Arid Environments* 44, 41–59.

Soriano, A. and O. Sala. 1983. Ecological strategies in a Patagonian arid steppe. *Vegetation* 56, 9–15.

Stuart-Hill, G. C. 1992. Effects of elephants and goats on the Kaffrarian succulent thicket of the eastern Cape, South Africa. *Journal of Applied Ecology* 29, 699–710.

van Langevelde, F., C. A. D. M. van de Vijver, L. Kumar, et al. 2003. Effects of fire and herbivory on the stability of savanna ecosystems. *Ecology* 84, 337–350.

Walker, B. H., D. Ludwig, C. S. Holling, and R. M. Peterman. 1981. Stability of semi-arid savanna grazing systems. *Journal of Ecology* 69, 473–498.

Walker, B. H. and I. Noy-Meir. 1982. Aspects of the stability and resilience of savanna ecosystems. In *Ecology of Tropical Savannas*, eds B. J. Huntley and B. H. Walker, Springer-Verlag, Berlin, pp. 556–590.

Weltzin, J. F. and G. R. McPherson. 1997. Spatial and temporal soil moisture resource partitioning by trees and grasses in a temperate savanna, Arizona, USA. *Oecologia* 112, 156–164.

Wiegand, K., F. Jeltsch, and D. Ward. 2004. Minimum recruitment frequency in plants with episodic recruitment. *Oecologia* 141, 363–372.

Wiegand, K., D. Saitz, and D. Ward. 2006. A patch-dynamics approach to savanna dynamics and woody plant encroachment—Insights from an arid savanna. *Perspectives in Plant Ecology Evolution and Systematics* 7, 229–242.

Williams, R. J., G. A. Duff, D. M. J. S. Bowman, and G. D. Cook. 1996. Variation in the composition and structure of tropical savannas as a function of rainfall and soil texture along a large-scale climatic gradient in the Northern Territory, Australia. *Journal of Biogeography* 23, 747–756.

Section II

Carbon, Water and Trace Gas Fluxes in Global Savannas

3

Disturbance and Climatic Drivers of Carbon Dynamics of a North Australian Tropical Savanna

Lindsay B. Hutley and Jason Beringer

CONTENTS

Introduction

Australian tropical savannas occur at annual rainfalls greater than 600 mm and extend from the Gulf of Carpentaria in north Queensland, across the northern half of the Northern Territory (NT), to the Kimberly region of northwest Western Australia. This is a savanna biome of almost 2 million km² and one of the world's most ecologically intact savanna regions (Mackey et al., 2007). Australian savanna ecosystems experience one of the most seasonal of the world's savanna climates (Prior and Bowman, 2005) with 90–95% of rainfall between November and April (Figure 3.1). This seasonality is driven by the annual movement of the Intertropical Convergence Zone, which drives the Australian summer monsoon. Soils of the region are ancient, highly weathered, and leached and are dominated by low-nutrient Kandosol (equivalent to Palexeralf using U.S. Taxonomy) and affiliated types. Soil and climate characteristics are generally unsuitable for broad-acre cropping, and grazing is the dominant land use with some horticulture. In the Northern Territory and Kimberly region, indigenous lands and conservation

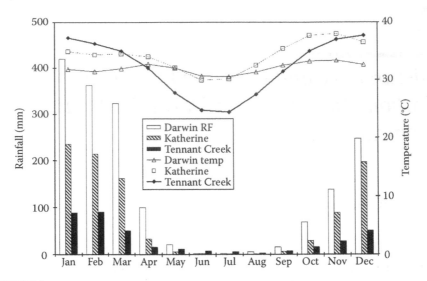

FIGURE 3.1
Mean monthly rainfall (left side y-axis) and maximum temperature (right side y-axis) for stations within the savanna biome of the Northern Territory. Locations include Darwin (12.5°S), Katherine (14.7°S), and Tennant Creek (19.6°S), illustrating the highly seasonal rainfall distribution and decline with distance inland. Temperature is aseasonal near the northern coast, becoming more seasonal with distance inland. (Data from the Australian Bureau of Meteorology, www.bom.gov.au)

reserves occupy 40% of the savanna, contributing to the intactness of these landscapes.

These savannas are woody dominant and can be described, based on tree cover, as open forest through to woodland savanna *sensu* (Specht, 1981). Tree cover tends to decline with rainfall, site quality (soil nutrient status, soil depth, and drainage), and fire severity (Lehmann et al., 2008). In the NT, this results in a steady decline in tree cover with distance from the coast as the influence of the monsoon weakens (Cook and Heerdegen, 2001; Figure 3.2). These savannas typically consist of three strata: an overstorey dominated by evergreen *Eucalyptus* and *Corymbia* tree species; a mid-storey of pan tropical semi- to fully deciduous tree and shrub species; and an understorey dominated by C4 annual and perennial grasses. Although biomass tends to be modest when compared with temperate woodlands and forests of southern Australia, due to their extensive area, these savannas store about one third of Australia's terrestrial carbon (Williams et al., 2004). Given the size of this carbon pool and the global significance of north Australia's savanna in terms of intactness and biodiversity, it is important to gain an understanding of the patterns and processes driving carbon, water, energy, and nutrient cycling in these systems. Long-term flux data are, thus, required to describe seasonal and interannual patterns of carbon, water, and energy exchange with the atmosphere to examine variation in gross primary productivity (GPP),

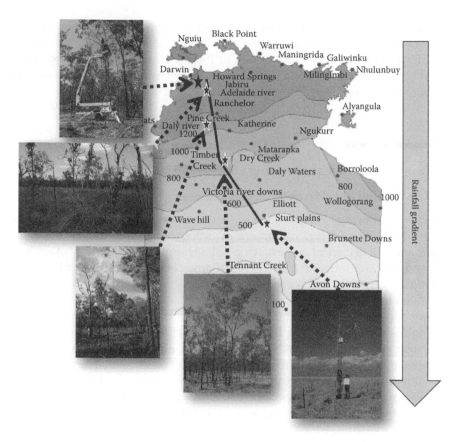

FIGURE 3.2
(**See color insert following page 320.**) Locations and photographs of savanna flux tower sites, Northern Territory, Australia. This region features extensive open-forest and woodland savanna ecosystems with scattered seasonal swamps and monsoon vine forest patches. The network of sites provides data describing mass and energy exchange across this biogeographical gradient.

ecosystem respiration (R_e), net primary productivity (NPP), net ecosystem productivity (NEP), and, ultimately, the net biome productivity (NBP), which includes losses due to the disturbance process and represents the long-term carbon balance. Such data are essential for meaningful parameterization of dynamic vegetation models and coupled, mesoscale climate models that will contribute to improved predictions of climate change impacts at regional scales. Of particular significance in this region is the role of fire in ecosystem processes, given that up to 50% of the savanna biome in north Australia can be burnt in a fire season (Russell-Smith et al., 2003a).

This chapter will review results from the long-term Howard Springs flux site near Darwin, Northern Territory, to examine seasonal and interannual patterns of carbon flux and impacts from fire. Other studies addressing productivity across north Australia have been compiled to put results from the Howard Springs site in a regional context. Past disturbances that were likely to have influenced current patterns of carbon flux are then examined. Long-term data are also used to look at the relative importance of climatic drivers on productivity using artificial neural network models, with climate change scenarios presented for this vegetation type.

Carbon Dynamics at the Howard Springs Flux Site

Site Description and Method

The Howard Springs eddy covariance site was established in the late 1990s within the Howard River catchment in order to examine fluxes of carbon, water, and energy from a mesic tropical savanna as a function of season, interannual variation, and fire severity (Eamus et al., 2001). The eddy covariance (EC) method is a micrometeorological approach to directly measure the integrated mass and energy exchange between a uniform surface (e.g., plant canopy, soil surface, and water body) and the atmosphere (Baldocchi, 2008). During the daytime, the CO_2 flux represents the net exchange of carbon (net ecosystem exchange, NEE or NEP) and comprises uptake via canopy photosynthesis (GPP) and losses from ecosystem respiration (R_e). The R_e includes both autotrophic (roots, stems, and leaves) and heterotrophic respiration (termites, soil microorganisms); nocturnal CO_2 flux is entirely due to R_e; and nocturnal fluxes are used to develop models of R_e as a function of temperature and/or soil moisture, which can be extrapolated to daytime periods. The GPP is calculated as the sum of NEP and estimated R_e (Law et al., 2002). Fluxes are calculated continuously at 30 or 60 min intervals, providing high-resolution temporal sampling that is not possible using inventory approaches. The method requires flat, homogenous vegetation with errors arsing from systematic instrument errors, uncertainty associated with nocturnal CO_2 fluxes due to reduced turbulence. At an annual time step, errors in productivity terms derived from eddy covariance are typically less than 20% and less than 10% at ideal sites (flat, homogenous terrain) with little missing data (Baldocchi, 2008).

Campaign measurements were conducted within the Howard River catchment using this method between 1996 and 1999 (Cook et al., 1998; Eamus et al., 2001). These observations described seasonal patterns of latent heat (LE) and net CO_2 flux (F_c) but were not continuous and did not incorporate fire events in a long-term flux record. In 2001, a permanent site was established ($12°29.712'S$, $131°09.003'E$) with near-continuous measurements of sensible heat (H), LE, and F_c. Climatology, vegetation, and soils have been described in detail by Hutley et al. (2000) and Beringer et al. (2007) and are briefly summarized here. The site is approximately 25 km from the northern coast of the NT with an annual rainfall of approximately 1750 mm that is strongly summer dominant (Figure 3.1). The overstorey is dominated by two evergreen species, *Eucalyptus tetrodonta*

(F. Muell.) and *E. miniata* (Cunn. Ex Schauer), which form a canopy of about 50% cover and account for approximately 70% of the total tree basal area of 9.7 m^2 ha^{-1} (±0.11 SE) at the site (Figure 3.2). The understorey is dominated by annual and perennial grasses (*Sorghum* and *Heteropogon* spp.) and the nitrogen-fixing cycad, *Cycas armstrongii* (Miq.) (Figure 3.2). Soils are extensively weathered and laterized Kandosols (after Isbell, 1996), which are weakly acidic and low in nutrient status. This flux site is representative of frequently burnt, high rainfall, coastal savanna that occupies an area of some 100,000 km^2 of the Northern Territory. Related structural and floristic assemblages also occur across north-western Western Australia and the Gulf of Carpentaria region of northern Queensland (Fox et al., 2001), and this is an important vegetation type across this region. Fire is an annual event within the Howard River catchment; although fire has been managed to a degree within the flux footprint (~300 ha) and a range of fire severities has resulted, from suppression to moderate-to-high-intensity fires (maximum fire-line intensity of 3610 kW m^{-1}) and up to 80% canopy scorch (Beringer et al., 2007).

Seasonal and Interannual Carbon Dynamics

On the basis of above- and below-ground carbon stock estimates and fluxes of GPP, NEP, and R_e derived from the eddy covariance measurements, a carbon balance for the Howard Springs site can be constructed (Figure 3.3). Fluxes can also be examined over seasonal and annual timeframes, with data from 2001 to 2006 given in Figure 3.4. Also included within this time series are fire events that occurred each year with a range of fire severities. A strong seasonal trend in NEP is evident, driven largely by rainfall patterns and leaf phenology. During the wet season months of February and March, NEP was negative (denoting a net uptake relative to the atmosphere) and averaged −3 g C m^{-2} d^{-1} (negative values denote a sink), coincident with maximal LAI. Wet season fluxes (December to April) were responsible for 72% of the mean annual NEP of −3.6t C ha^{-1} $year^{-1}$. Fire events transform the ecosystem from a weak, dry season carbon sink of approximately −1 g m^{-2} d^{-1} into a net carbon source to the atmosphere post fire with the duration and magnitude dependent on fire severity (Figure 3.4).

Impacts of Fire on NEP

Given a fire frequency of 2 out of every 3 years or greater for these savannas (Russell-Smith et al., 2003b), the impact of fire on these ecosystems is significant. Fire impacts range from damage to plant tissues to impacts on recruitment and stand structure (Murphy et al., 2009a, b) through to mesoscale impacts on the Australian monsoon (Lynch et al., 2007). Increases in the frequency of high-severity fire will increase tree mortality and result in reduced productivity and long-term declines in stored carbon (Cook et al., 2005; Liedloff and Cook, 2007). These impacts assume even greater significance,

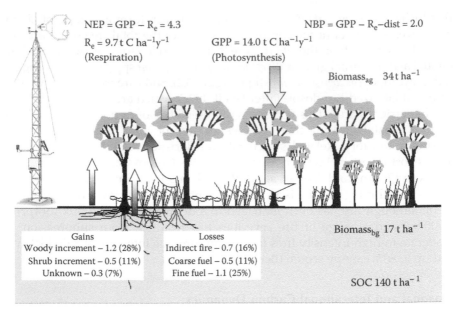

FIGURE 3.3
Carbon pools and fluxes at the Howard Springs Fluxnet site. Net carbon gains from woody increment and losses via fire impacts (fuel consumption and canopy reconstruction) are given in t C ha⁻¹ year⁻¹ and as a % of mean NEP for the site. (Derived from Beringer, J. et al. 2007. *Global Change Biology* 13, 990–1004. © 2007. With permission from Springer; Chen X., D. Eamus, and L. B. Hutley, 2003. *Oecologia* 137, 405–416. © 2003. With permission from Wiley-Blackwell. With permission.)

given predictions of increases of up to 30% in seasonal cumulative fire danger in parts of northern Australia under predicted climate change (Williams et al., 2001). Below we illustrate short-term (monthly), canopy-scale impacts of fire by examining NEP dynamics before and after fire events. In terms of carbon balance, there is a direct loss of carbon from a site via combustion of both fine and coarse fuels, and estimation of this loss is possible via pre- and postfire fuel load measurements (Russell-Smith et al., 2009) (Figure 3.3). There is also an indirect impact of fire resulting from the scorching of the canopy and subsequent leaf drop, with the impact being a function of fire severity. The resulting postfire period of reduced canopy functioning is effectively a period of lost productivity when compared with an unburned canopy. At this site, canopy reconstruction was typically complete within 60 days of a fire event. The NEP of the site remained a source during this recovery period due to enhanced maintenance and construction respiration associated with canopy reconstruction after burning and scorch (Cernusak et al., 2006), the degree of which is determined by fire severity (Beringer et al., 2003). This regenerative carbon cost is difficult to assess at an ecosystem scale; however, a modeling approach was developed by Beringer et al. (2007) using long-term flux data to examine canopy dynamics during fire-free and

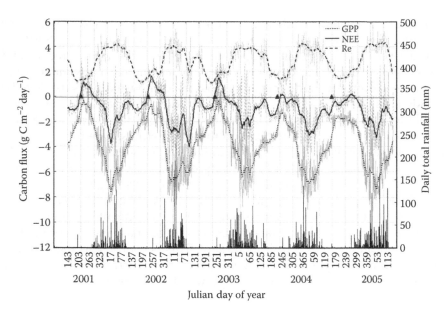

FIGURE 3.4

Daily rainfall, NEP, R_e, and GPP from the Howard Springs flux tower between 2001 and 2006. Heavy dark lines are 30-day moving average with daily average values shown in gray. Fire events are shown by the solid triangles. (Taken from Beringer, J. et al. 2007. *Global Change Biology* 13, 990–1004. With permission.)

fire-affected periods. Over a 5-year period, the integrated carbon cost associated with canopy reconstruction and lost productivity postfire was 0.7 t C ha⁻¹ year⁻¹ (Figure 3.3). This is ~16% of the annual NEP and can be as high as 25% for high-severity burns. Cernusak et al. (2006) measured leaf-scale photosynthesis of postfire, regenerating leaves at the Howard Springs site, and found respiration in excess of photosynthesis for immature foliage, which was consistent with tower-based observations of an ongoing source of CO_2 postfire.

Understanding the magnitude of this loss is especially significant in these savannas, given the high-fire frequency and the fact that the evergreen *Eucalyptus* and *Corymbia* dominated canopy is physiologically active throughout the dry, fire season. These trees have a remarkable ability to regenerate foliage after fire, an effective adaptation to tolerating frequent fire but a trait that imposes a significant construction cost and consumption of carbon reserves. Changes to fire regime can, thus, have a significant impact on savanna structure and function if stands are exposed to repeated, high-severity burning over time, which results in a gradual loss in regenerative capacity (Murphy et al., 2009b).

Mean NEP over 5 years at the Howard Springs flux site including the indirect impacts of fire through postfire reconstruction costs was –3.6 t C ha⁻¹ year⁻¹ (±0.17 SD) (Beringer et al., 2007). Chen et al. (2003) constructed a carbon balance using inventory and chamber methods across a range of analogous sites and reported an NEP of −3.8 t C ha^{-1} year^{-1}. This estimate of NEP has been derived from data that include fire events for each year of record and indirect effects of fires, as detailed earlier. For a complete assessment of fire impacts on productivity, direct carbon loss via combusted fuels needs to be included, as tower-based measures using eddy covariance methods are unable to accurately account for this loss during a fire event. All losses, including lost productivity and carbon lost to the atmosphere from combustion of biomass, must be subtracted from NEP to provide an estimate of NBP. Including the indirect costs productivity of 0.7 t C ha^{-1} year^{-1} and loss via combustion of 1.5 t C ha^{-1} year^{-1} over the 5-year period gives an NBP of −2.0 t C ha^{-1} year^{-1} (Figure 3.3), indicating that carbon accumulation occurred at this site over the period of measurement despite frequent burning.

Pathways for this sink are likely to be woody biomass increment via tree growth and understorey shrub increment. This is contrary to the assumption for greenhouse gas accounting purposes, that Australian savannas are essentially carbon neutral, with a stable woody component and ecosystem scale productivity due largely to annual grass growth, which is then consumed via subsequent dry season fires or is decomposed (Department of Climate Change, 2007). Two tree increment studies have been conducted at the Howard Springs flux site, with Chen et al. (2003) reporting an increment (i.e., carbon sink) of 1.6 t C ha^{-1} year^{-1} (1999–2000, which included monitored trees subjected to a cool fire, <2000 kW m^{-1}) and Cernusak et al. (2006), who reported a value of 1.2 t C ha^{-1} year^{-1} (2004–2005). Understorey shrubs have thickened noticeably on the site as well and account for approximately 1 t C ha^{-1} year^{-1}. These two pathways (woody growth) roughly account for the observed NBP at this site, although this sequestration may decline over time; as other factors will limit productivity, such as competition for light (especially in the wet season at maximum canopy fullness), available N and P, and available moisture. Other sink pathways that have not yet been quantified could include storage in soils, export through dissolved organic carbon, and fluxes of volatile organic carbon compounds (VOCs).

Spatial Variability in Savanna Productivity

The Howard Springs site has provided data describing carbon stocks and fluxes for high rainfall, frequently burnt savannas of the NT. There have been numerous other studies examining aspects of productivity, tree growth and/or tree cover change, and fire in the savannas of north Australia that also provide insights into carbon fluxes across a broader biogeographic range (Table 3.1). Most, but not all, studies report an increase in tree cover and carbon accumulation. Burrows et al. (2002) observed a net sequestration of 0.53 t C ha^{-1} year^{-1} via stem increment of existing tree stocks in grazed,

TABLE 3.1

Literature Estimates of Australian Savanna Productivity

Savanna Type	Site Features	Productivity (t C ha^{-1} year^{-1})	Method	Source
Mesic open-forest savanna, Gunn Point, NT	20 km from coastline, fire 2 in 3 y, vegetation impacted by Cyclone Tracey, 1974	NEP – 2.8	Eddy covariance, campaigns 1996–1998	Eamus et al. (2001)
Woodland savanna, central Queensland	Subtropical to dry tropical rangeland, grazed, low fire frequency	B_i – 0.53	Long-term (14 years) tree growth increment	Burrows et al. (2002)
Mesic open-forest savanna, Gunn Point and Darwin region, NT	Within 50 km coastline, fire 2 in 3 year, vegetation impacted by Cyclone Tracey, 1974	GPP – 20.2 R_e – 17.0 NPP – 11.0 NEP – 3.8	Inventory, chamber methods, above- and below-ground C pools	Chen et al. (2003)
Open-forest and woodland savanna, Kapalga, Kakadu National Park, NT	Fire treatments: biennial, early dry season fire	B_i – 0.5	Long-term (14 years) tree growth increment, stand structure	Cook et al. (2005)
Open-forest and woodland savanna, Munmarlary, Kakadu National Park, NT	Long-term fire treatments: exclusion, biennial early dry season, annual early dry season, and annual late dry season hot fires	B_i (no fire) – 1.4 B_i (hot fire) – 1.0	Long-term (23 years) tree growth increment	Russell-Smith et al. (2003)
Open-forest and woodland savanna, Western Arnhem Land, NT	WALFA fire abatement area	NBP – 1.0→1.5	Literature sources and WALFA fire emissions data	Williams et al. (2004)
Open-forest and woodland savanna, north Queensland		%cover ~ ~30%	Historical rainfall and vegetation records, remote sensing	Fensham and Holman (1999)

continued

TABLE 3.1 (continued)

Literature Estimates of Australian Savanna Productivity

Savanna Type	Site Features	Productivity (t C ha^{-1} year^{-1})	Method	Source
Mesic open-forest savanna, Howard Springs Flux site, NT	35 km from coastline, annual fire regime (1.0–3.5 MW m^{-1}), some impact from Cyclone Tracey, 1974	GPP – 13.6 R$_e$ – 10.0 NEP – 3.6 NBP – 2.0	Eddy covariance, continuous measurement 2001—ongoing, tree growth, component respiration studies	Beringer et al. (2007)
Open-forest savanna and seasonal floodplain, Kakadu National Park, NT	Structural changes to savanna and floodplain tree cover inside and outside a 116 km^2 Buffalo enclosure in Kakadu NP	%cover ~ +20%	Long-term (40 y) tree cover change using aerial photography	Bowman et al. (2008)
Open-forest and woodland savanna, Nitmiluk, Litchfield and Kakadu National Parks, NT	Long-term structural and floristic assessments across three National Parks, plots sampled from a range of savanna types to sandstone escarpment heaths. Ambient fire regime	B_i (savanna) – 0.08 B_i (heath) – 0.12	Long-term (10 y) tree growth increment	Murphy et al. (2009a, b)
Open-forest and woodland savanna, Kakadu National Park, NT	Ambient fire regime for KNP, fire 2 in 3 years	B_i – 0.1	Estimated from allometry and long-term (40 y) tree cover change using aerial photography	Lehmann et al. (2008)

Note: Studies include those detailing a full carbon balance through to those examining spatial and temporal patterns of woody dynamics only. Negative values represent a net uptake of carbon. B_i is tree biomass increment.

semiarid savanna woodlands of central Queensland, attributed to modified fire regimes and grazing pressure enhancing woody regrowth. A similar phenomenon was documented by Lewis (2002), who compiled paired site photographs of tree cover change over 100 years in the Victoria River District in the NT, a region of significant pastoral activity. By contrast, Fensham et al. (1999) found that tree dynamics of north Queensland savanna were more influenced by drought, with extended dry periods reducing tree cover by almost 30% due to enhanced tree mortality. Fire-grazing interactions were not the dominant driver of this change in woody carbon pools (Fensham et al., 2005), and climate-change-induced shifts in rainfall amount and/or distribution may well be a significant driver of long-term savanna structure, especially in low rainfall savanna.

A number of tree cover studies have been conducted in Kakadu National Park (KNP), NT, a catchment scale reserve dominated by open forest and woodland savanna vegetation (Table 3.1). These studies suggest a mosaic of tree cover change, either thickening (e.g., Cook et al., 2005; Russell-Smith et al., 2003b), limited or no change (Murphy et al., 2009a, b; Lehmann et al., 2008), or loss (Murphy et al., 2009a). Donohue et al. (2009) examined trends in rainfall, perennial (evergreen trees) and recurrent (grass, herbs, and deciduous species) cover, over a 26-year period (1981–2006) using AVHRR imagery. Changes in the fraction of photosynthetically active radiation absorbed by vegetation (fPAR) was computed and at a continental scale, woody cover vegetation increased by 21%. In the savanna biome, there was a complex of increases and declines in tree cover, similar to that observed in ground-based studies (Table 3.1), with significant areas of decline evident in central Queensland and increases along the north coast of the NT and KNP.

Disturbance Regimes and Carbon Turnover

These studies suggest that for large areas of NT savannas, at least, there is evidence of landscape scale tree thickening, as most studies report an increase in woody biomass. Tree dynamics will be determined by site quality and the long- and short-term disturbance regime (fire regime, grazing pressure, and cyclone return interval for coastal savannas). Fire is a major factor common to almost all of the studies detailed in Table 3.1, with sink strength reduced by 50% or more if fire was frequent and/or severe. It is also apparent that carbon flow and turnover in these savannas is rapid, especially in high rainfall regions (>1000 mm). Beringer et al. (2007) demonstrated that these savannas are not carbon neutral and carbon accumulation is possible, despite the occurrence of frequent fire. However, accumulated carbon can be rapidly lost to the atmosphere as demonstrated by Cook et al. (2005) and their analysis of extreme fire events during the Kapalga experiment in KNP.

Cyclones are also an important component of the savanna disturbance regime and carbon cycle, especially for coastal, mesic savanna. Cook and Goyens (2008) investigated the impact of Cyclone Monica, which crossed the NT coast in April 2006, on above-ground biomass. This cyclone affected approximately 7000 km² of savanna across the NT, resulting in a high proportion of tree mortality. Given the location of the Howard Springs flux site, an examination of past disturbance from cyclones is pertinent. Simple analysis of the size class distribution of woody vegetation at the Howard Springs site can provide evidence of past disturbance events (Figure 3.5). A wide distribution of tree sizes is present at this site, with trees of significant size and height (up to 30 m). Of note are two significant recruitment events that are apparent in the distribution (Figure 3.5a). If a mean growth rate of 2 mm year^{-1} is assumed across size classes (Murphy et al., 2009b; Prior et al., 2004), a tree age distribution can be constructed based on tree diameter at breast height (DBH). Both recruitment events closely match the time since the two most significant cyclones to impact this Darwin region over the last 100 years, namely Cyclone Tracey (35 years ago) and The Great Hurricane (111 years ago, Figure 3.5b). It is likely that such disturbance and its subsequent recovery is a contributing factor to the observed carbon sink at the Howard Springs site.

High rates of net carbon accumulation coupled with disturbance from fire (frequent, low to moderate impact) and cyclone occurrence (infrequent, high impact) result in a high turnover rate of carbon, estimated by Chen et al. (2003) to be 5 years for biomass and approximately 20 years for the ecosystem as a whole. Understanding these dynamics is important for an informed assessment of the potential of savanna for short- or long- term carbon sequestration. Although carbon can rapidly accumulate, it can also be rapidly lost unless managed for fire, in particular (Grace et al., 2006). In addition, the major determinants of savanna carbon dynamics, namely climate drivers, fire, and cyclone regime (frequency and severity), are all likely to change over the next 100 years (Hennessey et al., 2004); and understanding these processes is essential for future savanna management.

Savanna Productivity under Climate Change

It has been estimated that the terrestrial biosphere has been a net sink of approximately 2.6 Gt C year^{-1} and is currently offsetting CO_2 emissions from fossil fuels combustion (6.4 Gt C year^{-1}) for the 1990s (IPCC, 2007). The northern Australian savannas are likely to be a mosaic of sources and sinks (Table 3.1; Donahue et al., 2009), with an upper bound for sink strength in the vicinity of 3–4 t C ha^{-1} year^{-1} (Beringer et al., 2007). There is uncertainty in the spatial and temporal nature of this sink and, in particular, there is uncertainty as to how savanna carbon balance may be affected by future climate

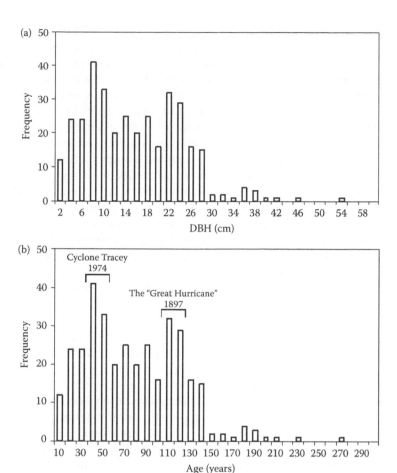

FIGURE 3.5

(a) Size class distribution of all *Eucalyptus* and *Corymbia* trees at the Howard Springs flux site >2 cm DBH. Tree sizes converted to age, (b) by assuming a mean stem increment of 2 mm year⁻¹ across all size classes. Two recruitment events are evident, exactly corresponding to Cyclone Tracey (34 years ago) and the Great Hurricane (111 years ago), which have impacted the site and subsequent carbon dynamics. (After Prior, L. D. and D. M. J. S. Bowman, 2005. *Australian Journal of Botany* 53, 379–39 and Murphy, B. P., J. Russell-Smith, and L. D. Prior, 2009b. *Global Change Biology* 10.111/j.1365-2486.2009.01933.x.)

change. Projected climate changes for a region such as the Howard Springs flux site by 2070 is an increase in temperature of +4.9°C and changes in precipitation amount of ±20% (Hennessy et al., 2004). Changes in wet season timing have also been projected with onset and retreat times expected to change by as much as one month by 2070.

Potential impact of this climate change on the carbon and water budgets of savannas can be examined using an Artificial Neural Network (ANN) approach. The ANN enables sophisticated statistical modeling of extremely complex functions such as land-atmosphere exchanges of mass and energy as measured by an eddy covariance system. The model is trained using specified input and output data, with input variables weighted based on those that have the greatest influence on output variables. A specific neural networking model is generated for each set of input and output variables, and it is, thus, possible to change the input variables to simulate climate change by input data to investigate effects of such changes on output variables (Basheer and Hajmeer, 2000). An ANN model was trained using data from the Howard Springs flux tower (2001–2006), with input variables of air temperature, soil temperature, vapor pressure deficit, soil moisture, wind speed, and radiation to predict outputs of GPP, R_e, NEP, and ET. Climatic inputs were changed as given in Table 3.2 to simulate climate change scenarios, with model runs undertaken for single and then multiple changes to parameter values. The ANN predicted the likely response of carbon and water fluxes as a function of these changes to input variables.

Single parameter changes (temperature, soil moisture, or wet season onset times) resulted in a range of ecosystem responses (Table 3.2). Increases in available moisture (+20%) resulted in a large increase in NEP (+25.2%), underlining the water limited nature of this ecosystem. The NEP also increased (+18.4%) with temperature (+4.9°C) due to the small change in GPP and a significant reduction in R_e under high temperatures, largely due to deceased soil respiration. Both chamber and flux tower measurements of respiration show an optimal temperature above which respiration becomes limited at this site. This contrasts with many temperate systems, where R_e increases exponentially with temperature (Schulze and Freibauer, 2005); and most process-based climate models (e.g., Cox et al., 2000) assume an exponential relationship between soil respiration and temperature. Simulations examining temperature responses typically result in elevated respiration, which for arid and seasonally dry systems may not always be the case, as there is a significant soil water limitation (Chen et al., 2002). When temperature and moisture were simultaneously increased, the predicted response was not additive and NEP increased by only +9.3%. For the +4.9°C and –20% soil moisture scenario, a dramatic decline in NEP resulted (–39%) largely driven by a reduced GPP (–25%), a result that would not have been predicted by the use of single response scenarios alone. The response to changes in wet season onset and retreat was also investigated, and productivity during these periods was most sensitive to changes in vapor pressure deficit (VPD). Of significance is the warmer, drier scenario, which was likely to result in increased fire occurrence and drive further declines in productivity, as predicted here. The use of an ANN is a different approach from the use of a process-based model in that ANN does not assume prior relationships between variables. Therefore, these results are extremely powerful in quantifying possible sensitivities and identifying areas that need further research, such as the nonexponential response of ecosystem respiration to

TABLE 3.2

Changes in Energy Balance (H and ET) and Component Carbon Fluxes (GPP, R_e, and NEP) for Each of the Climate Change Scenarios for 2030 and 2070

T (°C)	Soil Moisture	Wet Season	H (%)	H (MJ m⁻² d⁻¹)	ET (%)	ET (MJ m⁻² d⁻¹)	GPP (%)	GPP (t C ha⁻¹ year⁻¹)	R_e (%)	R_e (t C ha⁻¹ year⁻¹)	NEP (%)	NEP (t C ha⁻¹ year⁻¹)
Current			0	4.3	0	7.9	0	14	0	9.7	0	4.3
+1.2	—	—	3.2	0.1	3	0.2	−0.3	0	−0.1	0	−0.5	0
+4.9	—	—	15.4	0.7	17.2	1.4	1.8	0.3	−6.4	−0.6	18.4	0.8
—	+10%	—	−3.9	−0.2	4	0.3	5.1	0.7	1	0.1	14	0.6
—	−10%	—	3.3	0.1	−5.1	−0.4	−5.2	−0.7	−2.3	−0.2	−11.2	−0.5
—	+20%	—	−9.0	−0.4	7.2	0.6	9.4	1.3	1.5	0.1	25.2	1.1
—	−20%	—	6.7	0.3	−10.5	−0.8	−9.5	−1.3	−5.5	−0.5	−17.6	−0.8
—		Onset 1 month early	−5.6	−0.2	4.5	0.4	4.2	0.6	2.9	0.3	6.8	0.3
—		Retreat 1 month late	−4.2	−0.2	2.7	0.2	2.2	0.3	2.5	0.2	1.7	0.1
+4.9	+20%	—	2.7	0.1	19.4	1.5	1.1	0.2	−2.3	−0.2	9.3	0.4
+4.9	−20%	—	30.2	1.3	−7.2	−0.6	−23.0	−3.2	−14.7	−1.4	−39.2	−1.7

Note: For each variable, the % change from the baseline observational period (2001–2006) is given. Mean values for each variable in the "Current" row are taken from Beringer et al. 2007. *Global Change Biology* 13, 990–1004.

temperature. This is critical for determining future carbon and water budget responses under projected climate change.

Conclusions

This chapter has examined fluxes from a particular site, the Howard Springs flux site, and measurements suggest that this savanna type is capable of carbon uptake, even when subjected to fires of varying severity. Comparison of this site with other savanna studies in north Australia indicates a range of responses across the savanna biome, with most studies reporting accumulation of carbon in woody vegetation pools. These changes have been attributed to recovery from disturbance, interactions between fire regime and grazing, variation in rainfall, and elevated CO_2. Climate change may result in a productivity decline toward the end of the current century, depending on rainfall patterns.

Eddy covariance studies provide detailed data describing relationships between environmental drivers, vegetation structure, and fluxes. These data can be combined with long-term ecological studies of vegetation change (e.g., Cook et al., 2005; Murphy et al., 2009a, b) or demographic models (Cook and Liedloff, Chapter 17 of this book) for a more complete understanding of the short-term drivers of long-term change. Flux data are also essential to calibrate dynamic vegetation models (e.g., Barrett, Chapter 19 and Carter et al., Chapter 22 of this book) that form key components of land-atmosphere models, as described in Section VI of this book. Flux tower networks (Figure 3.2) provide spatiotemporal data that will help constrain carbon budget models and provide meaningful predictions of savanna fluxes as a function climate and fire regimes at fine to regional spatial scales. Currently, our broader understanding and modeling of these systems is limited by observations of patterns and processes. Models are required to predict the impacts of climate change and impacts of altered disturbance regimes to provide a basis for greenhouse gas accounting and assessment of the carbon trading potential of savanna environments.

References

Baldocchi, D. D., 2008. Breathing of the terrestrial biosphere: Lessons learned from a global network of carbon dioxide flux measurement systems. *Australian Journal of Botany* 56, 1–26.

Basheer, I. A. and M. Hajmeer, 2000. Artificial neural networks: Fundamentals, computing, design and application. *Journal of Microbiological Methods* 43, 3–31.

Beringer, J., L. B. Hutley, N. J. Tapper, A. Coutts, A. Kerley, and A. P. O'Grady, 2003. Fire impacts on surface heat, moisture and carbon fluxes from a tropical savanna in north Australia. *International Journal of Wildland Fire* 12, 333–340.

Beringer, J., L. B. Hutley, N. J. Tapper, and L. A. Cernusak, 2007. Savanna fires and their impact on net ecosystem productivity in north Australia. *Global Change Biology* 13, 990–1004.

Bowman, D. M. J. S., J. E. Riley, G. S. Boggs, C. E. R. Lehmann, and L. D. Prior, 2008. Do feral buffalo *Bubalus bubalis* explain the increase of woody cover in savannas of Kakadu National Park, Australia? *Journal of Biogeography* 35, 1976–1988.

Burrows, W. H., B. K. Henry, P. V. Back, et al. 2002. Growth and carbon stock change in eucalypt woodlands in northeast Australia: Ecological and greenhouse sink implications *Global Change Biology* 8, 769–784.

Cernusak, L. A, L. B. Hutley, J. Beringer, and N. J. Tapper, 2006. Stem and leaf gas exchange and their responses to fire in a north-Australian tropical savanna. *Plant, Cell and Environment* 29, 632–646.

Chen, X., D. Eamus, and L. B. Hutley, 2002. Seasonal patterns of soil carbon dioxide efflux from a wet–dry tropical savanna of northern Australia. *Australian Journal of Botany* 50, 43–51.

Chen X., D. Eamus, and L. B. Hutley, 2003. Carbon balance of a tropical savanna in northern Australia. *Oecologia* 137, 405–416.

Cook, G. D. and R. G. Heerdegen, 2001. Spatial variation in the duration of the rainy season in monsoonal Australia. *International Journal of Climatology* 21, 723–732.

Cook, G. D., A. C. Liedloff, R. W. Eager, et al. 2005. The estimation of carbon budgets of frequently burnt tree stands in savannas of northern Australia, using allometric analysis and isotopic discrimination. *Australian Journal of Botany* 53, 621–630.

Cook, G. D. and C. M. A. C. Goyens, 2008. The impact of wind on trees in Australian tropical savannas: Lessons from Cyclone Monica. *Australian Ecology* 33, 462–470.

Cook, P. G., T. H. Hatton, D. Pidsley, A. L. Herczeg, A. Held, and A. O'Grady, 1998. Water balance of a tropical woodland ecosystem, northern Australia: A combination of micro-meteorological, soil physical and groundwater chemical approaches. *Journal of Hydrology* 210, 161–177.

Cox, P. M., R. A. Bettes, C. D. Jones, S. A. Spall, and I. J. Totterdell, 2000. Acceleration of global warming due to carbon cycle feedbacks in a coupled climate model. *Nature* 408, 184–187

Department of Climate Change, 2007. *Australian Methodology for the Estimation of Greenhouse Gas Emissions and Sinks 2006—Agriculture.* Commonwealth of Australia, Canberra.

Donohue, R. J., T. R. McVicar, and M. L. Roderick, 2009. Climate-related trends in Australian vegetation cover as inferred from satellite observations, 1981–2006. *Global Change Biology* 15, 1025–1039.

Eamus, D., L. B. Hutley, and A. P. O'Grady, 2001. Daily and seasonal patterns of carbon and water fluxes above a north Australian savanna. *Tree Physiology* 21, 977–988.

Fensham, R. J., R. J. Fairfax, and S. R. Archer, 2005. Rainfall, land use and woody vegetation cover change in semi-arid Australian savanna. *Journal of Ecology* 93, 596–606.

Fensham, R. J. and J. E. Holman, 1999. Temporal and spatial patterns in drought-related tree dieback in Australian savanna. *Journal of Applied Ecology* 36, 1035–1050.

Fox, I. D., V. J. Neldner, and G. W. Wilson, 2001. *The Vegetation of the Australian Tropical Savannas*. Environmental Protection Agency, Brisbane.

Hennessy, K., C. Page, J. Bathols, K. McInnes, B. Pittock, R. Suppiah, and K. Walsh, 2004. *Climate Change in the Northern Territory*, Climate Impact Group, CSIRO Atmospheric Research, Canberra.

IPCC, 2007. *Climate Change 2007: The Physical Science Basis. Contribution of Working Group I to the Fourth Assessment Report of the Intergovernmental Panel on Climate Change*, (eds) S. D. Solomon, M. Qin, Z. Manning, M. Chen, K.B. Marquis, M. Avery, Tignor, and H. L. Miller. Cambridge University Press, Cambridge.

Isbell, R. F., 1996. *Australian Soil Classification*. CSIRO Publishing, Melbourne.

Law, B. E., E. Falge, and L. Gu, et al. 2002. Environmental controls over carbon dioxide and water vapor exchange of terrestrial vegetation. *Agricultural and Forest Meteorology* 113, 97–120.

Lehmann, C. R., L. D. Prior, R. J. Williams, and D. M. J. S. Bowman, 2008. Spatiotemporal trends in tree cover of a tropical mesic savanna are driven by landscape disturbance. *Journal of Applied Ecology* 45, 1304–1311.

Lewis, D., 2002. Slower than the eye can see. Environmental change in northern Australia's cattle lands: A case study from the Victoria River District, Tropical Savannas Cooperative Research Centre, Charles Darwin University, Northern Territory. Darwin.

Liedloff, A. C. and G. D. Cook, 2007. Modelling the effects of rainfall variability and fire on tree populations in an Australian tropical savanna with the Flames simulation model. *Ecological Modelling* 201, 269–282.

Lynch, A. H., D. Abramson, K. Görgen, J. Beringer, and P. Uotila, 2007. Influence of savanna fire on Australian monsoon season precipitation and circulation as simulated using a distributed computing environment. *Geophysical Research Letters* 34, L20801. DOI: 10.1029/2007GL030879.

Mackey, B. G, J. C. Z. R. Woinarski, H. Nix, and B. Trail. 2007. *The Nature of Northern Australia: Its Natural Values, Ecology, and Future Prospects*. ANU Electronic Press, Canberra.

Murphy, B. P., J. Russell-Smith, F. A. Watt, and G. D. Cook, 2009a. Fire management and woody biomass carbon stocks in mesic savannas. In *Culture, Ecology and Economy of Fire Management in North Australian Savannas*, (ed.), J. Russell-Smith Charles Darwin University Press, Darwin, Australia.

Murphy, B. P., J. Russell-Smith, and L. D. Prior, 2009b. Frequent fires reduce tree growth in northern Australian savannas: Implications for tree demography and carbon sequestration. *Global Change Biology* 10.1111/j.1365–2486.2009.01933.x

Prior, L. D., D. Eamus, and D. M. J. S. Bowman, 2004. Tree growth rates in north Australian savanna habitats: Seasonal patterns and correlations with leaf attributes. *Australian Journal of Botany* 52, 303–313.

Prior, L. D. and D. M. J. S. Bowman, 2005. Why do evergreen trees dominate the Australian seasonal tropics? *Australian Journal of Botany* 53, 379–399.

Russell-Smith, J., C. Yates, A. Edwards, et al. 2003a. Contemporary fire regimes of northern Australia, 1997–2001: Change since Aboriginal occupancy, challenges for sustainable management. *International Journal of Wildland Fire* 12, 283–297.

Russell-Smith, J., P. J. Whitehead, G. D. Cook, and J. L. Hoare, 2003b. Response of *Eucalyptus*-dominated savanna to frequent fires: Lessons from Munmarlary, 1973–1996. *Ecological Monographs* 73, 349–375.

Russell-Smith, J., B. P. Murphy, C. P. Meyer, et al. 2009. Improving estimates of savanna burning emissions for greenhouse accounting in northern Australia: Limitations, challenges, applications. *International Journal of Wildland Fire* 18, 1–18.

Schulze, E. D. and A. Freibauer, 2005. Carbon unlocked from soils. *Nature* 437, 205–206.

Specht, R. L., 1981. Foliage projective cover and standing biomass. In *Vegetation Classification in Australia*, (eds) A. N. Gillison and D. J. Anderson, 10–21. CSIRO Publishing, Canberra.

Williams, A. A. J., D. J. Karoly, and N. J. Tapper, 2001. The sensitivity of Australian fire danger to climate change. *Climatic Change* 49, 171–191.

Williams, R. J., L. B. Hutley, G. D. Cook, J. Russell-Smith, A. Edward, and X. Chen, 2004. Assessing the carbon sequestration potential of mesic savannas in the Northern Territory, Australia: Approaches, uncertainties and potential impacts of fire. *Functional Plant Biology* 31, 415–422.

4

Functional Convergence in Ecosystem Carbon Exchange in Adjacent Savanna Vegetation Types of the Kruger National Park, South Africa

Niall P. Hanan, Nicolas Boulain, Christopher A. Williams, Robert J. Scholes, and Sally Archibald

CONTENTS

Introduction

Approximately one-eighth of the global land surface is covered by savannas and open tropical woodlands. Of that total, about 60% is in Africa, where savannas are the dominant land cover south of the Sahara (Scholes and Hall, 1996). Savannas are characterized by the coexistence of a tree or shrub canopy of varying density with a lower canopy of annual or perennial herbs, often composed primarily of grasses, with variable contribution by forbs (Sankaran et al., 2004). In Africa, the savanna regions are of great economic and ecological importance to the human populations of the region, being the home and primary subsistence resource of the majority of the population. Future changes in land use and management practices (agricultural and grazing systems, fire frequency, and wood harvest) may result in the partial or complete modification of extensive areas of the remaining seasonal savannas, with direct impact on ecosystem functioning, water, and energy balance. Changes in weather patterns or increased variability associated with global climate change may also impact the productivity and sustainability of the savanna areas. Climate change could also modify tree–grass interactions sufficient to induce changes in community structure and the associated carbon, water, and energy relations of the vegetation (Bond, 2008).

Differences in canopy structure, physiology, and phenology related to species composition, climate and disturbance history, and local land use mean that land surface–atmosphere exchanges of mass and energy are affected not only by large scale weather but also by more subtle and spatially complex differences in surface biophysics (Pielke et al., 1998). Understanding savanna ecosystem processes, and predicting likely responses of savannas in the future, requires better understanding of how the savanna land surface interacts with the atmosphere and how the interactions are controlled by various aspects of vegetation structure, physiology, and phenology. An improved understanding of the processes controlling savanna functioning would also facilitate better representation of savanna ecosystems in models that simulate biophysical and biogeochemical processes at a range of spatiotemporal scales.

Here we analyze eddy flux measurements over adjacent fine-leaf and broad-leaf savannas in southern Africa to examine how species composition differences, occurring in response to differences in soil texture, influence ecosystem-scale patterns of carbon uptake and release via photosynthesis and respiration. We observe convergence in vegetation structure, phenology, and function (i.e., ecosystem scale photosynthesis and respiration) in these two savannas that occurs despite sometimes large differences in soil water availability, soil physics, and chemistry. We explore how this functional convergence may be mediated by soil water potential (SWP), which follows similar temporal trends and seems to provide comparable controls and limitations on carbon dynamics in the two ecosystems.

The Skukuza Flux Tower

The Skukuza eddy flux site is located 13 km west of Skukuza Camp, Kruger National Park, South Africa (25°1.18′S, 31°29.81′E, at an altitude of 420 m; Figure 4.1a, b). There is a minor gravel tourist road 1.5 km to the west and a tarmac road with light tourist traffic 5 km to the north. To the east and south, the nearest roads are ~8 km away. The topography is gently undulating (maximum 100 m altitude range over distances >2 km) in all directions for >10 km, with local slopes less than 4%.

The climate is semiarid subtropical, with hot, rainy summers and warm dry winters. Long-term average rainfall at Skukuza village (~20 km north

(a)

Kruger National park

Skukuza KNP flux tower

Skukuza

Kilometers
0   25   50

(b)

FIGURE 4.1
(See color insert following page 320.) (a) Location of the Skukuza flux tower within the Kruger National Park (KNP), South Africa, and KNP within South Africa and the African continent and (b) photograph of the Skukuza flux tower showing associated vegetation.

east of the tower site) is 550 mm year^{-1} (Scholes et al., 2001), but actual rainfall during the 6 years of data reported here has included growing seasons with cumulative rainfall both well above and well below the average.

The vegetation at the site consists of a catenal ecotone between *Combretum apiculatum* dominated savanna on the ridgetop and *Acacia nigrescens* dominated savanna from the midslope position to the valley bottom. The *Combretum* savanna is about 6 m tall, with ~30% of tree canopy cover. *C. apiculatum* makes up about 70% of the tree leaf area, with *Sclerocarya birrea* contributing a further 20%. The grass layer is dominated by perennial, tuft-forming grasses such as *Pogonarthria squarosa*. The *Acacia* savanna is about 10 m tall, also with ~30% of tree canopy cover. *A. nigrescens* contributes 50% to the leaf area, whereas *S. birrea* contributes 25%. The grasses here are more palatable, with *Panicum maximum* being prominent (Scholes et al., 2001).

This mosaic of *Acacia* and *Combretum* savanna types covers an area of about 50,000 ha within the Kruger National Park (Gertenbach, 1983). In a functional classification of African savannas (Scholes and Walker, 1993), the *Combretum* and *Acacia* woodlands represent the semiarid savannas (<~700 mm MAP) on infertile and fertile soils, respectively. The *Combretum* savanna is structurally and functionally representative of the (often congeneric) broad-leafed semiarid savannas of East and West Africa, whereas the *Acacia* savanna is representative of the fine-leafed savannas. Thus, the two woodland savanna types near Skukuza are representative of two of the most important savanna types of Southern Africa and, indeed, of Africa as a whole.

The soil of the ridgetop *Combretum* savanna is a Haplustalf, less than 1 m deep, with a topsoil sand content of 83% and a clay content of 8%, increasing to 10% in the subsoil. The catenal transition is abrupt, occurring over a distance of about 5 m and is typically marked by a band of Quartzipsamments. The soils of the lower slopes of the catena are alfisols verging on vertisols, ~60-cm deep, with 25% clay (64% sand) in the topsoil and 31% clay in the subsoil. The area is grazed by a wide range of mammalian herbivores (from 10 kg antelope to 1500 kg elephants) with a combined stocking rate of around 5000 kg km^{-2}. All natural processes occur essentially unhindered, including fire and tree uprooting by wind and elephants.

Eddy Covariance and Environmental Measurements

A 22-m scaffold tower located at the boundary between the *Acacia* and *Combretum* savannas was instrumented for eddy covariance measurements in April 2000. The tower was deliberately positioned on the transition between the two savanna types in a strategy that traded the opportunity for more continuous measurements in a single vegetation type for the ability to sample two distinct ecosystems using a single tower and instrument package.

For this strategy to be successful, it is important that dominant wind directions provide adequate sample frequencies from the two ecosystems. Thus, the tower location was selected with fetch directions over *Acacia* and *Combretum* savannas corresponding to the prevailing south-east and north-west wind directions for the region. Wind direction data collected on-site confirm that both savanna types are well sampled during all months and times of day (Table 4.1), with a slightly higher proportion of measurements over the *Combretum* fetch than the *Acacia* fetch, mostly reflecting the larger acceptance angle for the *Combretum* fetch.

The eddy covariance technique provides a method by which the vertical transport of a scalar (CO_2, momentum, heat, and water vapor) can be measured by correlating the fluctuations in scalar concentration with fluctuations in vertical wind speed (Kanemasu et al., 1979; Baldocchi et al., 1988; Verma, 1990). The eddy covariance instrumentation included a three-dimensional sonic anemometer (Wind Master Pro, Gill Instruments) to measure the components of the horizontal and vertical winds and sonic temperature and a fast-response closed-path infra-red gas analyzer (IRGA; LiCor 6262) to measure CO_2 and H_2O mixing ratios. The sonic anemometer and IRGA intake were mounted at 16.5 m above the ground on a boom mounted 2 m east of the tower, so that instruments and hardware did not interfere with the prevailing wind directions from the sampled *Acacia* and *Combretum* fetch directions. The gas analyzer and logging systems were mounted in a weatherproof enclosure at 16 m on the tower, requiring a 4.5-mm internal diameter Teflon sample tube to bring air samples to the IRGA with flow rates of ~6 lpm, ensuring turbulent flow. *In situ* IRGA CO_2 and H_2O calibration was carried out at intervals using WMO standard CO_2 calibration gases (calibrated to 0.1 ppm by NOAA-CMDL, Boulder) and a dewpoint generator (LiCor 610). Raw measurements were logged at 10 Hz on a laptop computer using a high-capacity removable disk and processed off-line at a later date.

TABLE 4.1

Distribution of Wind Directions (% of 30 min Averages) Measured on the Skukuza Flux Tower During Calendar Year 2000, Showing the Fraction of Time Periods When the Eddy Flux Measurements Sample the *Combretum* Savanna ($225° < WDIR < 75°$; Extending 1.0–1.5 km in These Directions) and the *Acacia* Savanna ($105° < WDIR < 195°$; Extending >2 km)

	Combretum	Acacia	Boundary
Rainy season	41.6	37.5	20.9
Dry season	55.5	27.5	17.0
Daylight	53.5	28.8	17.8
Dark	43.2	36.9	19.9

Note: Distributions are shown for the rainy season months (October to April) and dry season months (May to September; averaging all hours), and for daylight (6 am–6 pm) and nighttime (6 pm–6 am; averaging all months).

The high frequency (10 Hz) scalar and vertical wind-speed data were processed in the laboratory to determine covariances representing mean fluxes of CO_2, water, and heat. Processing steps were consistent with recommendations of Lee et al. (2004), including spike filtering, planar rotation of 3-D wind speeds to normalize effects of terrain or instrument tilt (Wilczak et al., 2001), and lag corrections to account for time delay of air samples between tube intake and IRGA. Frequency response corrections were also assessed to account for flux underestimation related to instrument separation, tube attenuation, and gas analyzer response, using empirical cospectral adjustment to match the H cospectrum (Eugster and Senn, 1995; Su et al., 2004). Fluxes were calculated as half-hour averages, using the convention that fluxes into the atmosphere from the land surface are reported as positive, with a final filtering to remove measurements obtained during calm and stable conditions (omitting data if friction velocity, $u^* < 0.1$ m s^{-1}) when eddy covariance measurements become unreliable.

Standard environmental monitoring instruments were installed on the main tower, including measurements of wind speed and direction (Climatronics F460), rainfall (Texas instruments TE525), air temperature and humidity (Vaisala HMP45C), barometric pressure (Vaisala Barocap), incident and reflected or emitted total solar (Kipp and Zonen CM14), and long-wave radiation (Kipp and Zonen CG2). Most meteorological instruments were mounted at 22 m to avoid interference from the physical structure of the tower. Air temperature and humidity instruments were, however, mounted at 16.5 m to measure at the reference height selected for the eddy covariance measurements. Measurements were stored as 30-min averages on a data logger located in the instrument enclosure. Power for the eddy covariance and ancillary systems was provided by six 75-W solar panels with five deep cycle 100 Ah batteries.

Flux and meteorological measurements on the main tower were complemented by measurements within the canopy airspace and soils of the *Acacia* and *Combretum* fetch directions. Soil temperature and humidity profiles were continuously measured and stored as 30-min averages using soil temperature probes (Campbell Sci., CS108) and time domain reflectometer probes (TDR; Campbell Sci., CS615). Two profiles sampling soil temperature and volumetric soil moisture (SM) in locations under a tree canopy and between tree canopies were installed in each of the two fetch directions, with 4 or 5 instruments distributed in each profile from near soil surface to the weathered rock interface at ~60–70 cm depth. In addition to these SM profiles under and between trees in the *Acacia* and *Combretum* savannas, we deployed additional SM instruments (Delta-T ML2x) alongside SWP instruments (Delta-T EQ2) to develop *in situ* soil water release curves (i.e., the relationship between SM and SWP, which differs strongly depending on soil texture; Figure 4.2a; van Genuchten and Nielson, 1985; Mualem, 1992). At these sites, a texture-based parameter estimation technique (Wosten et al., 1998) for the soil water release curve was unsuccessful (Figure 4.2a), emphasizing the importance of

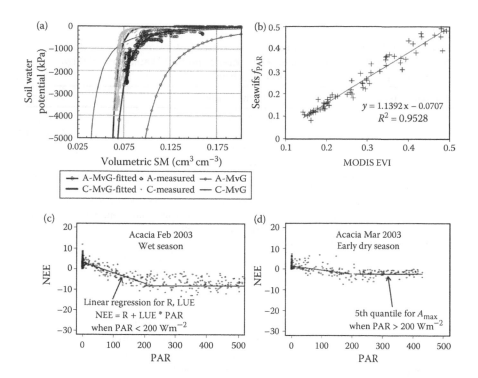

FIGURE 4.2
Calibration relationships used in this analysis and illustration of methods used to estimate monthly CO_2 flux statistics from 30-min flux measurements. (a) Soil water release curves for the *Combretum* (C) and *Acacia* (A) savannas at the Skukuza flux site. *In situ* measurements are shown using symbols with nonlinear regression relationships using the Mualem–van Genuchten equation see Table 4.2 for fitted parameter values. A priori Mualem–van Genuchten lines are also shown for each soil type, using parameters estimated based on measured soil texture (sand, silt, clay, and organic matter content) and relationships provided by Wosten et al. (1998); (b) relationship between monthly mean (April 2000 to December 2005) SeaWifs fractional PAR interception (f_{PAR}) in the 7×7 km block centered on the Skukuza flux tower and MODIS EVI measurements in the same area. This relationship was used to convert high resolution (250 m) MODIS EVI into f_{PAR} estimates in the separate *Acacia* and *Combretum* fetch directions (Table 4.1); (c) and (d) wet and early dry season examples of monthly flux characterization methodology showing linear regressions used for R, LUE, and 5th percentile method for A_{max}.

in situ derivation of the soil water release curve. All soil and canopy airspace instruments were controlled and logged using independent data loggers, instrument enclosures, and power systems installed in the *Acacia* and *Combretum* fetch directions.

For this study, the seasonal and interannual fractional PAR interception (f_{PAR}) time series in the *Acacia* and *Combretum* savannas surrounding the Skukuza flux tower were defined using MODIS EVI data at 250-m pixel resolution for the measurement period (April 2000 to December 2005). The

TABLE 4.2

Parameters of Mualem-van Genuchten Soil Water Retention Curves Fitted to
In Situ Measured Volumetric Water (Θ) Content and Soil Water Potential (ψ)
at the Skukuza *Combretum* and *Acacia* Savanna Sites

Fitted Parameter	Combretum	Acacia
Θ_r	−5.142384E–003	4.353160E–002
Θ_s	1.696341E–001	3.369754E–001
N	1.077373	1.288210
A	7.041393	7.853948E–002

Source: Data from van Genuchten, M. T. 1980. *Soil Science Society of America Journal* 44,
         892–898. Mualem, Y. 1992. *International Workshop on Indirect Methods for Estimating the
         Hydraulic Properties of Unsaturated Soils*, eds. M. T. van Genuchten, F. J. Leij, and L. J.
         Lund, pp. 15–36. Riverside, University of California.
Note: Mualem–van Genuchten equation

$$\Psi = -\left\{\left[\frac{(\Theta - \Theta_r)n}{(\Theta_s - \Theta_r)}\right]^{(1/1-n)} - 1\right\}^{(1/n)} \Big/ \alpha$$

MODIS f_{PAR} data were not yet available at the higher spatial resolutions
necessary to separate the two fetch directions. To overcome this limitation,
we developed a coarser resolution (1 km) relationship (Figure 4.2b) between
f_{PAR} measured by the SeaWifs instrument and MODIS EVI in a 7 × 7 km block
centered on the flux tower as provided by the Oak Ridge Distributed Active
Data Archive Center (ORNL, 2010). We used this calibration relationship to
estimate f_{PAR} from 250 m MODIS EVI in the same 7 × 7 km (28 × 28 pixels)
and then extracted pixels assigned to the *Combretum* (51 pixels) and *Acacia*
(78 pixels) based on direction and distance from the tower. An average f_{PAR}
was thus obtained for the two savannas types and for each month of the
April 2000 to December 2005 study period.

Summary of Earlier Studies at the Skukuza Flux Tower

Water limitation prevails at the Skukuza savanna, influencing all manner of
ecological processes. The site experiences pronounced seasonal variation in
water availability associated with the November to April rains, coincident
with high insolation and temperatures, and contrasting the drought conditions during the remainder of the year (Scholes et al., 2001; Williams et al.,
2009). Leaf area generally tracks rainfall with about a one-month lag.
However, in addition to these seasonal dynamics, rainfall pulses drive the
dynamics of respiration, evapotranspiration, and photosynthesis at the
storm-interstorm timescale (Williams et al., 2009). Respiration reaches a
maximum 1–3 days after rainfall events, whereas maximum photosynthesis

occurs up to 9 days after soil wetting. This delay is thought to be due to the upregulation of the photosynthetic machinery required after a long intra-seasonal drought. Both the extent of the respiration and photosynthesis pulse and the rate at which these processes respond appear to be a complex response to the history of the system: the extent and the period of time that has elapsed since the last wetting event (Williams et al., 2009).

It is challenging to represent the ecophysiological mechanisms that underlie such pulse responses. Nonetheless, Kutsch et al. (2008) provided an in-depth analysis of seasonal and water status controls on canopy conductance as it responds to vapor pressure deficit at the Skukuza site. They reported evidence of stomatal acclimation to short-term water deficits. Being at the canopy scale, it is unclear whether this pattern is a function of vegetation dynamics and dieback that modifies the landscape scale stomatal density and vegetation composition (i.e., grasses and herbs senesce but tree leaf area remains active) or whether it is, in fact, a bona fide physiological regulation. The temporally conservative water use efficiency (Kutsch et al., 2008) seems to argue for the latter.

Limited understanding of such complexities motivates the use of ANN not only for gap-filling of flux tower data but also to advance process understanding. Archibald et al. (2009) used ANN to gap-fill 5 years of Skukuza flux data (not separating by fetch) and found that annual NEE was highly variable: NEE was negative (carbon uptake) in 2 of the 5 years (2003–04 = –138 g carbon m^{-2}, 2004–05 = –83 g m^{-2}), strongly positive (carbon source) in 2 years (2001–02 = 155 g m^{-2}, 2002–03 150 g m^{-2}), and near neutral in 1 year (2000–01 = 42 g m^{-2}). Fractional PAR interception (f_{PAR}) explained more of this interannual variation than mean annual rainfall or number of growing season days. Although f_{PAR} is largely driven by SM—especially for the grasses—some of the dominant tree species show seasonal patterns that are disconnected from the SM profile (Archibald and Scholes, 2007). *Sclerocarya birrea* and *Acacia nigrescens* put out leaves before the first rain when the SM is at a minimum. *Combretum apiculatum*, which accounts for 60% of the tree cover on sandy soils, maintains a green canopy well into the dry season when SM is declining. This is perhaps why carbon fluxes seem to track leaf display more closely than SM. All of the above studies combined the *Acacia* and *Combretum* measurements despite the contrasting soil-vegetation conditions. The separate behavior of the two systems is worthy of further exploration, as elaborated here.

Carbon Flux Characterization at Monthly Time Scales

Given our strategy to sample two distinct vegetation types with a single tower and eddy flux system, we expect to obtain discontinuous measurements of fluxes from the *Acacia* and *Combretum* savannas at the Skukuza tower site.

However, as shown in Table 4.1, the distribution of wind directions during wet and dry season and between day and night provide adequate samples from the two systems for separate analysis of the flux behavior in the two systems.

With regard to CO_2 uptake and release characterization, we defined three summary parameters, calculated at monthly time-step from the 30-min eddy flux measurements sampled over the *Acacia* and *Combretum* fetch directions and relating to the photosynthetic uptake and respiratory release processes. These summary parameters are (Figure 4.2c, d):

1. Canopy scale initial light use efficiency (LUE, $\mu mol\ W^{-1}$) defined as the slope of the relationship between net ecosystem exchange (NEE, $\mu mol\ m^{-2}\ s^{-1}$) and incident photosynthetically active radiation (PAR, $W\ m^{-2}$) at low radiation levels (PAR < 200 $W\ m^{-2}$) (note: more negative LUE implies increased photosynthetic canopy activity, because negative flux implies canopy uptake)

2. Maximum canopy scale net CO_2 assimilation (A_{max}, $\mu mol\ m^{-2}\ s^{-1}$) defined as the 5th percentile net ecosystem exchange during midday high radiation conditions (PAR > 200 $W\ m^{-2}$) (more negative A_{max} implies increased canopy uptake)

3. Canopy scale respiration (R, $\mu mol\ m^{-2}\ s^{-1}$), including heterotrophic and autotrophic CO_2 losses, defined as the intercept (PAR = 0 $W\ m^{-2}$) of the light response relationship fitted for the LUE parameter (above) and representing nighttime CO_2 emissions from the canopy (autotrophic and heterotrophic respiration)

To estimate the above parameter values for the *Acacia* and *Combretum* fetch directions, we used all available quality controlled measurements for each calendar month (69 months between April 2000 and December 2005), partitioned them between the two systems using wind direction measurements, and then further filtered them using the high and low incident PAR criteria.

We analyzed the sensitivity of the monthly scale ecosystem-level flux parameters to f_{PAR}, SM, and SWP using least-squares linear regression, with multiple regression and interactions. Model selection, using the Akaike Information Criterion (AIC, Akaike, 1973), was used to test the value of different independent variables, determine appropriate model complexity, and assess whether the *Acacia* and *Combretum* sites should be considered separately or combined.

Results

Time series of measured canopy and soil variables in the *Acacia* and *Combretum* savannas are shown in Figure 4.3a–c. As expected from field observations of leaf area index and phenology in the vicinity of the flux

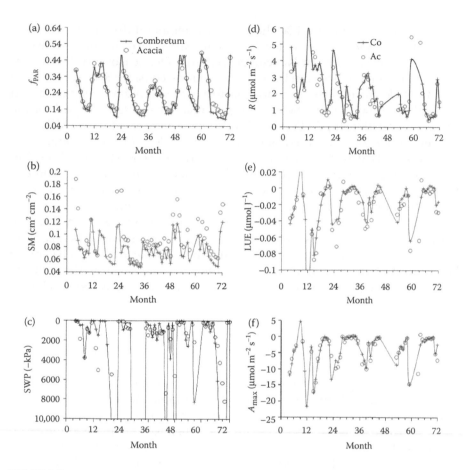

FIGURE 4.3
Time series of measured canopy and soil parameters (a–c) and fitted flux parameters (d and e) for the Skukuza tower site *Acacia* and *Combretum* savannas (month 1 = January 2000). (a) MODIS f_{PAR} averages for each fetch direction calculated from 250 m data averaged over 54 cells for the *Combretum* fetch and 72 cells for the *Acacia* fetch; (b) monthly average volumetric SM content (4 probes –2 under a tree canopy, 2 between tree canopies) in 3–16 cm depth range in each fetch; (c) soil water potential (SWP) calculated from the SM measurements using *in situ* soil water release curves (SWP values <–10 MPa occurring in late dry season are not shown, as these reflect high sensitivity of SWP to SM beyond the calibration set); (d) estimated canopy scale respiration (R); (e) canopy scale light use efficiency (LUE); and (f) maximum canopy scale CO_2 uptake (A_{max}), where R, LUE, and A_{max} are defined in the text.

tower, the time series of canopy structure (i.e., f_{PAR}; Figure 4.3a) are very similar between the adjacent savannas, though with some tendency for slightly higher values in the *Acacia* system, especially in the dry season (e.g., June to September). Volumetric SM time series (Figure 4.3b), on the other hand, clearly show the impact of soil texture, with generally higher SM in the finer textured *Acacia* system than in the more sandy *Combretum* savanna. When converted to SWP (Figure 4.3c), however, the temporal differences between *Acacia* and *Combretum* are much less clear, particularly in the wet season (i.e., SWP near zero kPa in November to April). Dry season SWP estimates diverge rather more between the two sites, but we consider SWP < –5 MPa to be unreliable due to the exponential nature of the calibration relationship in dry soils and the fact that field water potential measurements used for the calibration relationship in Figure 4.2 did not measure less than –4 MPa.

Time series of the fitted monthly time-scale CO_2 flux parameters (Figure 4.3d–f) also show remarkable similarity in ecosystem-level respiration (R), light use efficiency (LUE), and daytime maximum photosynthetic uptake (A_{max}). There is some indication that R decreases more rapidly at the end of the rainy season in the *Acacia* system, but, in general, any differences are small with no clear pattern emerging. This convergence in function between the two savannas is not surprising, given very similar phenological, leaf area, and f_{PAR} patterns; but it confirms the importance of rainfall (similar between sites), more so than soils (different between sites), as the primary limitation on vegetation structure, leaf area dynamics, and carbon exchange in these drought seasonal, semiarid systems.

Figures 4.4 and 4.5 show the relationships between monthly R, LUE, and A_{max} and measures of canopy structure (f_{PAR}), volumetric SM, and SWP. There is likely to be considerable covariance between f_{PAR}, soil water content measures, and other biological and physical variables, because all vary seasonally. Thus, simple correlations with CO_2 flux parameters do not necessarily imply causality. However, in Figure 4.4, we show the linear relationships with CO_2 exchange parameters to illustrate the similarity in sensitivity to canopy structure and divergence with SM. When SM is converted into SWP (Figure 4.5), we see that the divergence between sites is reduced, but the relationship is more scattered and curvilinear in drier soils even when plotted with SWP on a log scale.

Table 4.3 shows the best (i.e., most statistically parsimonious) models for each CO_2 flux parameter assessed using AIC. Coefficient of determination (r^2) and root mean square error (RMS) are also shown. We fitted models independently for each CO_2 flux parameter and each savanna type (*Acacia, Combretum*) and tested models of varying complexity (simple regressions on f_{PAR}, SM, or SWP; then multiple regression combinations; and models with and without parameter interactions). We then combined data from the two sites and fitted the same models for the two sites combined. We used the AIC to determine which model best described the behavior of each of the three fluxes.

In the case of R, we found that the two savanna types were consistently different, and more complex models provided significantly better overall fit. The AIC indicated that the separate regressions shown in Figure 4.4 for R versus f_{PAR} were indeed significantly different and were improved by addition of separate SM and interaction (i.e., $f_{PAR}*SM$) effects. Thus, *Acacia* and *Combretum* sites are reported separately in Table 4.3 for R; and coefficients are reported for f_{PAR}, SM, and the interaction between them. By contrast, for LUE,

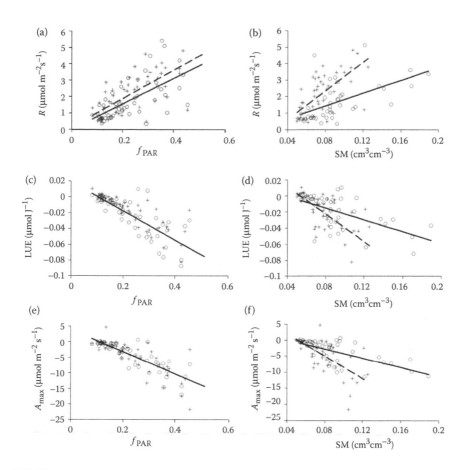

FIGURE 4.4

Relationships between fitted CO_2 flux parameters, canopy structure, and SM. (a and b) Ecosystem respiration (R), (c and d) light use efficiency (LUE), (e and f) maximum net ecosystem exchange (A_{max}) plotted against canopy structure (f_{PAR}, left column) and volumetric soil water content (SM, right column). Data show convergence between Acacia and Combretum savannas in the relationships with f_{PAR} but divergent relationships with regard to volumetric SM (SM), due to differences in soil texture. Linear regressions (Table 4.3) are shown for Acacia f_{PAR} and SM relationships (solid line) and Combretum (dashed line) when they are statistically distinct between vegetation types or for the combined sites when not distinct (i.e., c and e).

we found that f_{PAR} alone seems to be an adequate descriptor, with no significant benefit from adding either SM or SWP, and the two savannas behaved similarly. For A_{max}, we found that both f_{PAR} and SM were significant factors when considering the sites separately but that SWP allowed the two sites to be considered together.

Figure 4.6 shows graphically the final ("best") relationships, as detailed in Table 4.3, describing the environmental controls and correlates on R, LUE, and A_{max} at the Skukuza site. The ecosystem respiration variable (R)

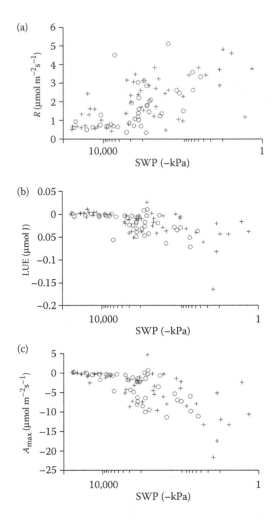

FIGURE 4.5
Relationships between fitted CO_2 flux parameters and $\log_{10}$ soil water potential (SWP). (a) Ecosystem respiration (R), (b) light use efficiency (LUE), (c) maximum net ecosystem exchange (A_{max}). Unlike relationships for SM (Figure 4.4), SWP shows convergence again between *Acacia* and *Combretum* relationships. SWP was inferred from SM measurements using an *in situ* calibration relationship (Figure 4.2). Linear regressions are omitted for SWP, which remains curvilinear even with SWP on a log scale.

has separate relationships for the *Acacia* and *Combretum* savannas, each of which includes sensitivity to f_{PAR}, SM, and the impact of their interaction. The LUE relationship is common for both savanna types, with strong increases in photosynthetic activity (i.e., more negative LUE) at higher f_{PAR} but no significant sensitivity to SM variables. The A_{max} relationship is also common to both savannas types, with less negative SWP (i.e., nearer zero,

TABLE 4.3

Model Selection Results Describing Monthly Time-Scale CO_2 Flux Parameters Estimated from 30-min Eddy Flux Measurements for the *Acacia*, *Combretum*, and Combined Savanna Types at the Skukuza, South Africa, Flux Tower

Site and Parameter	Intercept	$\sigma\text{-}f_{PAR}$	$\sigma\text{-SM}$	$\sigma\text{-SWP}$	$\Sigma\text{-}(f_{PAR} * SM)$	r^2	N	RMSE
Combretum–R	−2.988	15.678	53.24	n/a	−127.18	0.56	49	0.876
Acacia–R	−3.046	15.044	38.70	n/a	−100.57	0.50	41	0.798
Combined–LUE	0.024	−0.196	n/a	n/a	n/a	0.56	90	0.017
Combined–A_{max}	−1.908	−26.227	n/a	0.535	n/a	0.64	90	2.697

Note: Results for the best models, as identified using AIC, are presented here. Functional responses of ecosystem respiration (R), light use efficiency (LUE), and maximum daylight net ecosystem exchange (A_{max}) were tested with regard to potential independent variables, including fractional PAR interception (f_{PAR}, –), volumetric soil water content (SM, cm^3 cm^{-3}), and log_{10} soil water potential (SWP, –kPA), and interactions between f_{PAR} and either SM or log_{10} (SWP). Columns in the table report the intercept and slope (σ) terms for linear models, the coefficient of determination (r^2), number of observations contributing to the regressions (N), and the root mean square error (RMSE) of the fitted model.

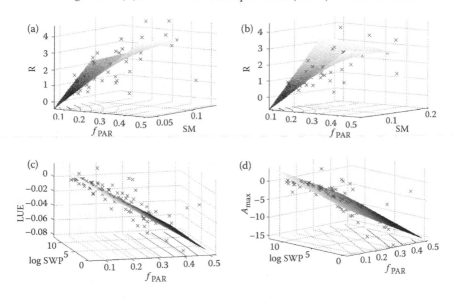

FIGURE 4.6

Fitted response surfaces for monthly CO_2 flux parameters relating to fractional PAR interception (f_{PAR}), volumetric soil water content (SM; cm^3 cm^{-3}), soil water potential (SWP; $log_{10}[-kPa]$) in the *Acacia* and *Combretum* savannas at the Skukuza eddy flux site (see Table 4.3 for model details). (a and b) Ecosystem respiration (R; μmol m^{-2} s^{-1}) in the *Combretum* and *Acacia* savannas, respectively; (c) light use efficiency (LUE; μmol J^{-1}) in the combined sites; and (d) maximum daytime net CO_2 exchange (A_{max}; μmol m^{-2} s^{-1}) in the combined sites. The original observations used to fit these models are shown in each plot (note that some data points are obscured by the fitted response surfaces). Contours on the lower surface show isolines of the dependent variable in two-dimensional independent variable space. Lines parallel to an axis indicate no sensitivity to that variable (e.g., LUE is not sensitive to SWP).

for wetter soils) leading to increasing (i.e., more negative) maximum net ecosystem exchange.

Discussion

The apparent convergence in ecosystem-level CO_2 flux behavior between the *Acacia* and *Combretum* savannas at the Skukuza eddy flux site in South Africa occurs despite quite large differences in soil texture and chemistry (Scholes et al., 2001). Our initial supposition was that the soil differences would mean that the vegetation of the *Acacia* savanna would generally operate under more negative and, therefore, more physiologically stressful SWP. However, in reality, volumetric water content (SM) is generally higher in the finer texture *Acacia* soil, such that the two vegetation communities maintain similar SWP. Given higher-available N in the *Acacia* soils and, generally, higher N content of tree and, to a lesser extent, grass leaves of the *Acacia* system, the convergence in ecosystem-level CO_2 flux behavior implies a higher N-use efficiency in the *Combretum* system. If SWP provides more direct physiological controls on plant photosynthesis than nitrogen in these systems, and both autotrophic (plant) and heterotrophic (primarily soil) respiration, this may explain the emergence of similar temporal trends in CO_2 flux behaviors (Figure 4.3). Given finite, and similar, input of water to these adjacent plant communities via rainfall, it is interesting to ask whether such functional convergence, depending primarily on rainfall rather than on nutrients or other limiting factors, should be considered a common feature in semiarid plant communities.

Despite superficial similarities in the temporal trends in CO_2 flux parameters (R, LUE, and A_{max}; Figure 4.3), differences emerge in the models that best describe them. In particular, LUE presented the simplest relationship: *Acacia* and *Combretum* sites could be combined and modeled using only f_{PAR}. The best model describing A_{max} required f_{PAR} with addition of a SM term: When the two savanna types were modeled separately, SM was the favored SM variable, but using SWP allowed the two sites to be combined, and the combined model was statistically most favored. For ecosystem respiration (R), however, we found that a more complex model incorporating f_{PAR}, SM, and an interaction term was necessary, with separate relationships for each site providing the best models.

This sequence of relatively simple to more complex functional models to describe the fitted CO_2 flux parameters likely reflects the combination of processes involved in each. The LUE, estimated here as the sensitivity of net ecosystem exchange to increasing PAR at low PAR, is a uniquely photosynthetic process, with little or no influence from autotrophic or heterotrophic respiration. Further, since LUE is estimated for low light conditions (generally early morning or late afternoon), the influence of soil water stress is reduced. The A_{max}, on the other hand, is computed as a function of midday net ecosystem exchange and, therefore, includes photosynthetic uptake and respiratory release terms, albeit dominated by the larger midday photosynthesis component. However, A_{max}, via both changing photosynthetic capacity and R, is much more likely to be impacted by changing soil water availability and water stress. Finally, R is composed of both autotrophic and heterotrophic terms where vegetation structure plays a role and where soil properties other than SM may be particularly important in controlling soil respiration (e.g., nutrient biochemistry relating to soil type, organic matter as a substrate for respiration). In this case, SWP does not allow the two sites to be combined, and the best model included separates SM and f_{PAR}-SM interaction terms, perhaps reflecting the complex of heterotrophic and autotrophic factors driving ecosystem respiration.

These results have implications for carbon cycle modeling, indicating that spatiotemporal changes in photosynthesis parameters are relatively easily modeled as functions of f_{PAR} and that SWP may integrate differences arising due to soil type. However, ecosystem respiration appears to have more complex relationships with f_{PAR} and SM, and in this case, SWP does not unify the two sites that are statistically different.

The functional convergence of the Skukuza *Acacia* and *Combretum* savannas confirms the importance of water as a primary limitation on these drought seasonal systems, despite very different soil physical and chemical situations. The factors that dictate the rather radical change in species composition between the *Acacia* and *Combretum* systems, however, remain unclear. We might conclude that, given relatively small differences in SWP, factors such as soil chemistry or nutrient status may be involved. Future work should continue to examine water and nutrient budgets for the *Acacia* and *Combretum* savannas and relationships to woody plant establishment and growth: are seedlings, saplings, and adults of one species capable of growing in the neighboring soil? What physical or physiological process so strongly favors one species over another in the two soil types where, we conclude, SWP is generally surprisingly similar?

Acknowledgments

The instrumentation of the Skukuza flux tower was made possible by an initial grant from NASA's Terrestrial Ecology Program; subsequent support from the U.S. National Science Foundation, NASA, and NOAA to Hanan; and support from the South African Council for Scientific and Industrial Research (CSIR) to Scholes. Invaluable field assistance was provided by Walter Khubeka and Ian McHugh, and we are grateful for the generous scientific and technical support of Kruger National Park Scientific Services.

References

Akaike, H. 1973. Information theory as an extension of the maximum likelihood principle. *Second International Symposium on Information Theory*, Budapest. Akademiai Kiado.

Archibald, S. A., A. Kirton, van der Merwe, et al. 2009. Drivers of interannual variability in net ecosystem exchange in a semi-arid savanna ecosystem, South Africa. *Biogeosciences* 6, 251–266.

Archibald, S. and R. J. Scholes. 2007. Leaf green-up in a semi-arid African savanna—separating tree and grass responses to environmental cues. *Journal of Vegetation Science* 18, 583–594.

Baldocchi, D. D., B. B. Hicks, and T. P. Meyers. 1988. Measuring biosphere-atmosphere exchanges of biologically related gases with micrometeorological techniques. *Ecology* 69, 1331–1340.

Bond, W. J. 2008. What limits trees in C4 grasslands and savannas? *Annual Review of Ecology, Evolution and Systematics* 39, 641–659.

Eugster, W. and W. Senn. 1995. A cospectral correction model for measurement of turbulent NO_2 flux. *Boundary Layer Meteorology* 74, 321–340.

Gertenbach, W. P. D. 1983. Landscapes of the Kruger National Park. *Koedoe* 23, 35–44.

Kanemasu, E. T., M. L. Wesely, B. B. Hicks, and J. L. Heilman. 1979. Techniques for calculating energy and mass fluxes. *Modification of the Aerial Environment of Crops*, eds. B. J. Barfield and J. F. Gerber, pp. 156–183. American Society of Agricultural Engineers, St. Joseph, MI.

Kutsch, W. L., N. P. Hanan, R. J. Scholes, et al. 2008. Response of carbon fluxes to water relations in a savanna ecosystem in South Africa. *Biogeosciences* 5, 1797–1808.

Lee, X. H., Q. Yu, X. M. Sun, J. D. Liu, Q. W. Min, Y. F. Liu, and X. Z. Zhang. 2004. Micrometeorological fluxes under the influence of regional and local advection: a revisit. *Agricultural and Forest Meteorology* 122, 111–124.

Mualem, Y. 1992. Modeling the hydraulic conductivity of unsaturated porous media. In *International Workshop on Indirect Methods for Estimating the Hydraulic Properties of Unsaturated Soils*, eds. M. T. van Genuchten, F. J. Leij and L. J. Lund, pp. 15–36. Riverside, University of California.

ORNL. 2010. Oak Ridge National Laboratory Distributed Active Archive Center (ORNL DAAC), MODIS subsetted land products, Collection 5. Available on-line [http://daac.ornl.gov/MODIS/modis.html] from ORNL DAAC, Oak Ridge Tennessee, USA, accessed February 2010.

Pielke, R. A., R. Avissar, M. Raupach, et al. 1998. Interactions between the atmosphere and terrestrial ecosystems: Influence on weather and climate. *Global Change Biology* 4, 461–475.

Sankaran, M., J. Ratnam and N. P. Hanan. 2004. Tree–grass coexistence in savannas revisited—insights from an examination of assumptions and mechanisms invoked in existing models. *Ecology Letters* 7, 480–490.

Scholes, R. J., N. Gureja, M. Giannecchinni, et al. 2001. The environment and vegetation of the flux measurement site near Skukuza, Kruger National Park. *Koedoe* 44, 73–83.

Scholes, R. J. and D. O. Hall. 1996. The carbon budget of tropical savannas, woodlands and grasslands. In *Global Change: Effects on Coniferous Forests and Grasslands*, ed. A. I. Breymeyer, D. O. Hall, J. M. Melillo and G. I. Agren. John Wiley & Sons, Chichester.

Scholes, R. J. and B. H. Walker. 1993. *An African Savanna: Synthesis of the Nylsvley Study*. Cambridge University Press, Cambridge.

Su, H. B., H. P. Schmid, C. S. B. Grimmond, C. S. Vogel, and A. J. Oliphant. 2004. Spectral characteristics and correction of long-term eddy-covariance measurements over two mixed hardwood forests in non-flat terrain. *Boundary Layer Meteorology* 110, 213–253.

van Genuchten, M. T. 1980. A closed-form equation for predicting the hydraulic conductivity of unsaturated soils. *Soil Science Society of America Journal* 44, 892–898.

van Genuchten, M. T. and D. R. Nielson. 1985. On describing and predicting the hydraulic properties of unsaturated soils. *Annales Geophysicae* 3, 615–628.

Verma, S. B. 1990. Micrometeorological methods for measuring surface fluxes of mass and energy. *Remote Sensing Reviews* 5, 99–115.

Wilczak, J. M., S. P. Oncley, and S. A. Stage. 2001. Sonic anemometer tilt correction algorithms. *Boundary Layer Meteorology* 99, 127–150.

Williams, C. A., N. P. Hanan, R. J. Scholes, and W. Kutsch. 2009. Complexity in water and carbon dioxide flux following rain pulses in an African savanna. *Oecologia* 161, 469–480.

Wosten, J. H. M., A. Lilly, A. Nemes, and C. Le Bas. 1998. Using existing soil data to derive hydraulic parameters for simulation models in environmental studies and in land use planning. Winand Staring Centre, Wageningen, p. 106.

5

Flux Dynamics in the Cerrado and Cerrado–Forest Transition of Brazil

George Louis Vourlitis and Humberto Ribeiro da Rocha

CONTENTS

Introduction

The savanna vegetation of Brazil is referred to as *cerrado*, which is the Portuguese word for "dense," "thick," or "closed." Cerrado covers approximately 1.5–2 million km^2, or 20–25% of the total land cover of Brazil, and is the second largest vegetation type after the Amazonian forest (Furley and Ratter, 1988; Fearnside, 2000). Contiguous cerrado covers over 10 states, from Piauí and Maranhão to the north; Minas Gerais, São Paulo, and Mato Grosso do Sul to the south; and Rondônia to the west (Figure 5.1), corresponding to a 20° range in latitude and 0–1800 m range in elevation (Ratter et al., 1997). Biodiversity, including known plants, animals, and fungi, is estimated to be 160,000 species (Oliveira and Marquis, 2002).

Cerrado Vegetation

Several cerrado classification schemes have been presented; however, most authors recognize distinctive physiognomies of cerrado based on the height,

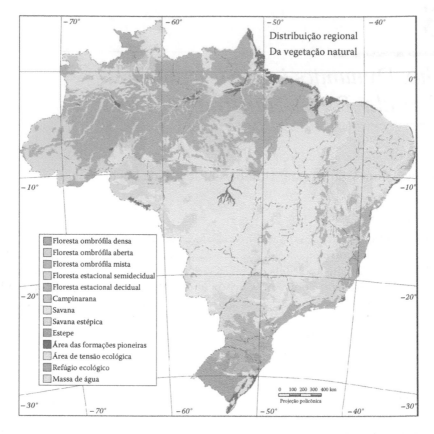

FIGURE 5.1
(See color insert following page 320.) The distribution of vegetation of Brazil. Savanna (cerrado) is shown in tan, and Area de Tensao Ecologica (transitional forest) is shown in gray-blue. (Accessed from the Instituto Brasileiro de Geografia e Estatística [http://www.ibge.gov.br/ servidor_arquivos_geo/], verified on March 30, 2009.)

cover, and/or density of trees (Figure 5.2). Cerradão ("big cerrado") is composed of trees with an average height of 10 m and a density of 2360 plants/ ha (Table 5.1). Basal area is approximately 35 m^2/ha, and average leaf area index (LAI) varies between 3.5 m^2/m^2 in the wet season and 2.7 m^2/m^2 in the dry season. Cerradão makes up 11% of cerrado area and is generally found on upland soils of intermediate and low fertility (Eiten, 1972; Furley and Ratter, 1988). Cerrado composed of both trees and shrubs is referred to as cerrado *sensu stricto* (Figure 5.2), which covers 70% of cerrado land area (Table 5.1). Cerrado *sensu stricto* (*s.s.* also known as restrito) may have a density and basal area similar to cerradão, but LAI is 2–3 times lower, reflecting the shorter, less vertically stratified canopy (Table 5.1). Cerrado *s. s.* is also restricted to well-drained upland soils that tend to be infertile with a low pH and high aluminum saturation (Eiten, 1972; Furley and Ratter, 1988).

FIGURE 5.2
Cerrado physiognomy. (From Miranda, A. C. et al. 1996. In *Amazonian Climate and Deforestation*, (ed.) J. H. C. Gash, C. A. Nobre, J. M. Roberts, and R. L. Victoria, J. M. Wiley and Sons, New York, pp. 353–364. With permission from J. M. Wiley and Sons.)

Cerrado vegetation with a mixture of trees, shrubs, and grasses is termed *campo cerrado* (Figure 5.2), corresponding to 18% of the cerrado land area combined (Table 5.1). Woody cover in campo cerrado is usually <30–40% (Eiten, 1972), and the height and density of trees and shrubs is on average 4 m and 1566 plants/ha, respectively, reflecting the higher importance of grasses (Table 5.1). As the height, density, and basal area of woody plants declines, grasses become relatively more abundant, and there is a concomitant decline in LAI. Campo cerrado composed of scattered shrubs and small trees is recognized as *campo sujo* (dirty field), whereas in *campo limpo* (clean field), woody vegetation is essentially absent (Goodland, 1971; Eiten, 1972). Campo cerrado is often found on soils that are less well drained, more acidic, and higher in saturated aluminum than cerrado *s. s.* and cerradão (Lopes and Cox, 1977; Furley and Ratter, 1988).

Influence of Climate and Soils

The factors that affect the distribution of cerrado are subject to debate; however, seasonal variation in precipitation, soil fertility and drainage, and fire are considered the most important (Goodland, 1971; Eiten, 1972; Furley and Ratter, 1988). Total annual precipitation in cerrado is lower and/or more seasonal than Amazonian and transitional forest (Figure 5.3). Cerrado and transitional forest have a 4–5 month dry season between May and September, whereas rainforest experiences substantially smaller (1–2 month) dry season duration, and cerrado receives 300–500 mm less rainfall than transitional forest and nearly 1000 mm less rainfall than Amazonian rainforest. Within the cerrado domain climate, variation may be large, with average annual temperature varying between 18 and 28°C and total annual precipitation

TABLE 5.1

Characteristics of Cerrado Ecosystems

Characteristic	Cerradão	Sensu Stricto	Campo	Campo Sujo	Campo Limpo	Reference
Area (10^6 ha)	17.1	108.2	10.3	7.9	9.4	6
Density (#/ha)	2360 (1546–3215)	2911 (911–8198)	1566 (203–2928)	764 (678–849)	ND	1,4–5,7,8–9
Height (m)	10 (6–18)	6 (3–10)	4 (3–6)	3 (1–5)	ND	1,3–5
Basal area (m^2/ha)	35.1 (16.1–68.1)	43.2 (9.7–121.2)	19.5 (1.7–49.2)	3.6 (3.0–4.3)	ND	1,4–5,7,9
Leaf area index (m^2/m^2)	3.5/2.7	1.5/0.7	0.5/0.4	0.3/0.2	ND	3,12,14,16
Aboveground C (Mg/ha)	45.9	9.7 (5.0–15.9)	3.9 (3.0–4.3)	2.8 (1.7–4.7)	2.1 (0.8–3.6)	2,4,6,11,13
Belowground C (Mg/ha)	23.0	23.4 (20.6–25.2)	33.2	15.8 (15.1–17.4)	9.4 (7.6–12.3)	4,6,13
Litter C (Mg/ha)	3.1 (2.5–4.0)	2.0 (1.3–2.6)	0.7	0.5 (0.1–1.0)	0.1 (0.05–0.15)	2,4,8,13,15
Soil C-50 cm depth (Mg/ha)	45 (26–61)	37 (17–67)	ND	ND	ND	17
Soil C-100 cm depth (Mg/ha)	ND	315 (257–373)	ND	211	246	13

Note: Data are the mean and the range (values in parentheses) from values compiled from the literature from the references indicated below. Data for density and basal area include trees and shrubs whereas data for height is for trees only. Data for leaf area index are mean wet season/dry season values. ND = no data obtained.

References include: (1) Goodland (1971); (2) Kauffman et al. (1994); (3) Miranda et al. (1996); (4) de Castro and Kauffman (1998); (5) Haase (1999); (6) Fearnside (2000); (7) Batalha et al. (2001); (8) Roscoe et al. (2000); (9) Ledru (2002); (10) Ribeiro and Tabarelli (2002); (11) Barbosa and Fearnside (2005); (12) Hoffmann et al. (2005); (13) Grace et al. (2006); (14) Bucci et al. (2008); (15) Sanches et al. (2008a); (16) Sanches et al. (2008b); and (17) Maia et al. (2009).

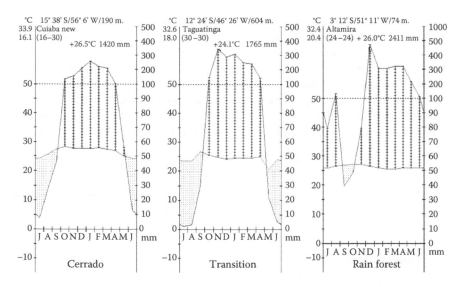

FIGURE 5.3
Climate diagrams for cerrado (Cuiaba, Mato Grosso), the rain forest–cerrado transition (Taguatinga, Tocantins), and rainforest (Altamira, Pará). Climate records are for 1974–1994 (temperature) and 1965–1994 (precipitation) for Cuiaba, 1961–1990 for Taguatinga, and 1967–1990 for Altamira. (Data from www.globalbioclimatics.org [verified January 29, 2009].)

varying between 800 and 2000 mm (Oliveira-Filho and Ratter, 2002). Soil properties vary considerably between forest and cerrado vegetation types, and most cerrado soils are dystrophic with low pH and organic matter and high aluminum saturation (Lopes and Cox, 1977). Fire frequency is thought to be a key control in the distribution of cerrado and type of cerrado. Forest trees are sensitive to fire (Hoffmann et al., 2003), whereas many cerrado trees and shrubs appear to be adapted to recurring fire (Hoffmann, 2002; Miranda et al., 2002). Frequent fire favors the development and persistence of grass-dominated cerrado such as *campo limpo* or *campo sujo*, and many authors suggest that campo forms of cerrado are simply early successional stages, whereas cerrado *s. s.* and cerradão represent late-successional and climax vegetation, respectively (Pivello and Coutinho, 1996). However, since fire properties are substantially affected by fuel loads, vegetation composition, microclimate, season, and land use and human activity, the effects of fire on the distribution of cerrado vegetation type is complex (Miranda et al., 2002).

Estimation of Carbon Pools and Carbon and Water Fluxes

Data compiled from the literature was used to estimate the C pools and C and water fluxes for cerrado; however, in many cases data were extracted from published tables and figures, which may add uncertainty in the pools and fluxes reported here. In addition, estimates reported for "cerradão" were, in part, derived from field studies conducted in the forest-cerrado transition in northern Mato Grosso, which is a heterogeneous mixture of cerrado *s. s.*, floristically unique cerradão, and *mata seca* ("dry forest"), which is common on more fertile soils in this region (Ackerly et al., 1989). Other data for "cerradão" were derived from seasonally flooded forests located in the forest-cerrado transition of Tocantins state and the Pantanal of Mato Grosso; and since cerradão is thought to be restricted to upland soils, these forests are probably better thought of as transitional forests or *mata ciliar* ("gallery forest"). However, cerradão is broadly defined in the literature (Goodland, 1971; Eiten, 1972; Furley and Ratter, 1988; Ackerly et al., 1989); thus, the term *cerradão* used here refers to both transitional forests and woodlands, including *mata seca*, *mata ciliar*, and floristically unique forms of cerradão.

Carbon Pools

Aboveground C pool size ranges from 45.9 Mg/ha in cerradão to 2.1 Mg/ha in *campo limpo* (Table 5.1), a pattern that reflects a decline in tree abundance along this physiognomic gradient (Goodland, 1971; de Castro and Kauffman, 1998). A similar pattern was observed for surface litter C storage, which ranged from 3.1 Mg/ha in cerradão to 0.01 Mg/ha in *campo limpo* (Table 5.1). Carbon storage in roots in the surface 2 m soil layer is less dependent on aboveground physiognomy, reflecting differences in rooting depth and/or biomass allocation among cerrado vegetation types (de Castro and Kauffman, 1998; Grace et al., 2006). Root biomass C is highest in campo cerrado (33.2 Mg/ha) and cerrado *s. s.* (23.4 Mg/ha) and lowest in *campo sujo* (15.8 Mg/ha) and *campo limpo* (9.4 Mg/ha).

Soil C storage varies markedly across the physiognomic gradient, and cerrado dominated by trees and shrubs has higher soil C than campo landforms. Estimates of C in the upper 50 cm soil layer range from 45 Mg/ha in cerradão to 37 Mg/ha in cerrado *s. s.*, but if estimates include the upper 1 m soil layer, soil C values increase considerably to 315 Mg/ha for cerrado *s. s.* and to 211 and 246 Mg/ha for *campo sujo* and *campo limpo*, respectively (Table 5.1). Even if only the upper 50 cm soil layer is considered, soil C storage is either equal to (cerradão) or substantially larger (cerrado *s. s.*) than aboveground C storage in these systems.

Leaf Gas Exchange

There is a significant reduction in maximum photosynthesis and stomatal conductance during the dry season for a variety of cerrado tree species (Table 5.2).

TABLE 5.2

Estimates of Maximum Net Photosynthesis (P_{max}) and Stomatal Conductance (g_s) of Evergreen (E), Semi-deciduous (SD), and Deciduous (D) Cerrado and Transitional Forest (cerradão) Tree Species Compiled from the Literature

Species	Habit	P_{max} ($\mu mol\ m^{-2}\ s^{-1}$)		g_s ($mol\ m^{-2}\ s^{-1}$)		Reference
		Wet	Dry	Wet	Dry	
Connarus suberosus	E	8.6	2.2	0.30	0.09	Prado et al. (2004)
Didymopanax vinosum	E	15.4	15.8	1.10	0.17	Prado et al. (2004)
Duguetia furfuracea	E	8.7	5.8	1.10	0.17	Prado et al. (2004)
Miconia albicans	E	11.9	6.7	0.63	0.09	Prado et al. (2004)
		24.9	14.8	0.11	0.02	Monteiro & Prado (2006)
Miconia ferruginata	E	13.0	10.4	0.19	0.14	Franco et al. (2005)
Miconia ligustroides	E	10.1	8.8	0.53	0.10	Prado et al. (2004)
Ouratea hexasperma	E	9.2	6.2	0.13	0.10	Franco et al. (2005)
*Quiina pteridophylla**	E	9.3	5.5	0.14	0.06	Miranda et al. (2005)
Pitocarpha rotundifolia	E	8.7	8.7	0.86	0.26	Prado et al. (2004)
Roupala Montana	E	11.6	2.4	0.98	0.06	Prado et al. (2004)
	E	10.0	7.4	0.16	0.10	Franco et al. (2005)
Schefflera macrocarpa	E	16.4	13.4	0.23	0.16	Franco et al. (2005)
Sclerolobium paniculatum	E	13.5	11.9	0.19	0.16	Franco et al. (2005)
Styrax camporum	E	8.6	7.5	0.35	0.12	Prado et al. (2004)
*Tovomita schomburgkii**	E	7.0	6.3	0.10	0.06	Miranda et al. (2005)
		7.9	4.3	0.16	0.05	Sendall (2006)
Vochysia elliptica	E	8.7	9.1	0.16	0.14	Franco et al. (2005)
*Brosimum lactescens**	SD	11.0	8.4	0.20	0.18	Miranda et al. (2005)
		7.5	4.3	0.18	0.05	Sendall (2006)
Casearia sylvestris	SD	8.4	6.7	0.44	0.06	Prado et al. (2004)
Copaifera langsdorffii	SD	6.7	4.5	0.49	0.08	Prado et al. (2004)
*Dinizia excelsa**	SD	12.4	6.1	0.22	0.09	Miranda et al. (2005)
Eriotheca gracilipes	SD	8.9	6.8	0.53	0.16	Prado et al. (2004)
Erythroxylum campestre	SD	9.4	4.6	0.40	0.09	Prado et al. (2004)

continued

TABLE 5.2   (continued)

Estimates of Maximum Net Photosynthesis (P_{max}) and Stomatal Conductance (g_s) of Evergreen (E), Semi-deciduous (SD), and Deciduous (D) Cerrado and Transitional Forest (cerradão) Tree Species Compiled from the Literature

Species	Habit	P_{max} (μmol m^{-2} s^{-1}) Wet	Dry	g_s (mol m^{-2} s^{-1}) Wet	Dry	Reference
Erythroxylum suberosum	SD	9.2	5.2	0.38	0.10	Prado et al. (2004)
Gochnatia floribunda	SD	11.9	8.2	1.10	0.07	Prado et al. (2004)
Hancornia speciosa	SD	13.2	8.0	0.17	0.13	Lobo et al. (2008)
Memora axillaris	SD	7.8	3.4	0.28	0.06	Prado et al. (2004)
Tibouchina stenocarpa	SD	11.7	7.9	1.2	0.37	Prado et al. (2004)
Annona coriacea	D	9.4	4.5	0.35	0.10	Prado et al. (2004)
Bahuinia rufa	D	10.2	4.1	0.55	0.08	Prado et al. (2004)
Bowdichia virgiloides	D	6.3	8.1			Kanegae et al. (2000)
*Bowdichia virgiloides**	D	5.2	4.1			Kanegae et al. (2000)
Campomanesia xanthocarpa	D	6.7	4.5	0.49	0.08	Prado et al. (2004)
Kielmeyera coriacea	D	10.5				Prado & Moraes (1997)
			11.5			Franco et al. (2005)
		10.5				Prado & Moraes (1997)
		5.5	8.4			Nardoto et al. (1998)
Solanum lycocarpum	D	13.6	9.8	0.63	0.26	Prado et al. (2004)
Stryphnodendron adstringens	D	14.1	7.2	0.58	0.11	Prado et al. (2004)
Stryphnodendron obovatum	D	10.4	4.5	0.23	0.04	Prado et al. (2004)
Evergreen (mean ± sd)		11.3 ± 4.3	8.2 ± 3.9	0.41 ± 0.36	0.11 ± 0.06	
Semi-deciduous (mean ± sd)		9.8 ± 2.1	6.2 ± 1.7	0.47 ± 0.34	0.12 ± 0.09	
Deciduous (mean ± sd)		9.3 ± 3.1	6.7 ± 2.7	0.47 ± 0.15	0.11 ± 0.08	
Average (mean ± sd)		10.3 ± 3.5	7.2 ± 3.2	0.44 ± 0.32	0.12 ± 0.07	

Note: Species names with an asterisk (*) are from cerradão.

When water stress occurs, plants may close their stomata due to an increase in evaporative demand and a decline in leaf water potential, which decreases internal leaf CO_2 concentration and leaf photosynthesis (Meinzer et al., 1993). This pattern appears to be consistent with many cerrado and transitional forest tree species, regardless of growth form (Franco, 1998; Prado et al., 2004; Franco et al., 2005; Miranda et al., 2005; Monteiro and Prado, 2006). However, although photosynthetic rates, specific leaf area, and stomatal conductance are strongly affected by water availability, these traits are also negatively correlated with leaf age (Miyaji et al., 1997; Lobo et al., 2008), which increases into the dry season when leaf shedding occurs (Sanches et al., 2008a). Thus, leaf age and the development of drought stress are not independent, and it is unclear how much of the decline in leaf photosynthesis and stomatal conductance is due to the direct effects of drought or an increase in leaf age.

Seasonal variations in nutrient availability, specifically N and P, are important in controlling leaf gas exchange (Franco et al., 2005). In cerrado, nutrients are released from decomposing organic matter at the beginning of the wet season and rapidly assimilated by plants (Bustamante et al., 2006). Leaf N and P concentration reaches a maximum at the beginning of the wet season for deciduous trees and toward the peak of the wet season for evergreen trees, which coincides with the seasonal period when photosynthetic rates, specific leaf area, and stomatal conductance are highest (Franco, 1998; Prado et al., 2004; Franco et al., 2005; Monteiro and Prado, 2006).

Carbon and Water Fluxes

Seasonal variations in net CO_2 exchange (NEE) measured from eddy covariance are substantially larger for cerrado than cerradão (Figure 5.4; Table 5.3). Cerrado is a consistent net sink for atmospheric CO_2 during the wet season and a small net source of CO_2 to the atmosphere during the dry season (Figure 5.4a). *Campo sujo* appears to be a smaller net CO_2 sink than cerrado *s. s.* during the wet season, but it is unclear whether these differences are driven by vegetation type, interannual variation in climate, and/or a small sample size. Average (±95% CI) wet season rates of NEE for cerrado are -0.51 ± 0.49 gC m^{-2} d^{-1} during the wet season and 0.32 ± 0.52 gC m^{-2} d^{-1} during the dry season (negative values depict net ecosystem C uptake; Table 5.2). Over the annual cycle, cerrado is a slight net sink for CO_2 although the mean value is not statistically different from zero (Table 5.3).

The seasonal amplitude of NEE is substantially smaller for cerradão (Figure 5.4b). Cerradão tends to be a small net C sink during the wet season months of November to February but a small net source of C during the dry season and seasonal transition periods. The C loss during the transitional periods

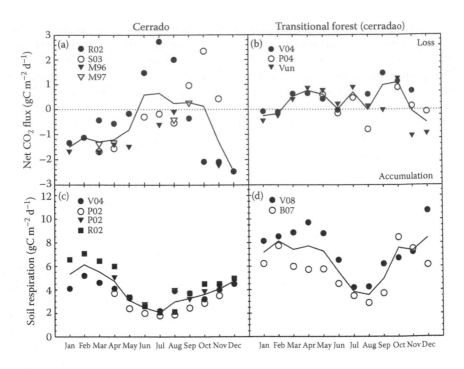

FIGURE 5.4
Net ecosystem CO_2 exchange for (a) cerrado, (b) transitional forest, (c) soil respiration for cerrado, and (d) transitional forest. Symbols depict values obtained from the literature, whereas the solid line depicts average calculated from the values displayed in the figures. (Data from (R02) da Rocha, H. R. et al. 2002. *Biota Neotropica* 2, 1–11; (S03) Santos, A. J. B. et al. 2003. *Functional Ecology* 17, 711–719; (M96) Miranda, A. C. et al. 1996. In *Amazonian Climate and Deforestation*, (ed.) J. H. C. Gash, C. A. Nobre, J. M. Roberts, and R. L. Victoria, J. M. Wiley and Sons, New York, pp. 353–364; (M97) Miranda, A. C. et al. 1997. *Plant, Cell and Environment* 20, 315–328; (V04) Vourlitis, G. L. et al. 2004. *Ecological Applications* 14, S89–S113; (P04) Priante Filho, N. et al. 2004. *Global Change Biology* 10, 863–876; (Vun) Vourlitis, G. L. et al., unpublished data; (P02) Pinto, A. S. et al. 2002. *Journal of Geophysical Research* 107 (D20) 8089; (V08) Valentini, C. M. A. et al. 2008. *Journal of Geophysical Research-Biogeosciences* 113, G00B10; and (B07) Borges, O. 2007. M.Sc. Thesis, Universidade Federal de Mato Grosso, Instituto de Ciências Exatas e da Terra, Programa de Pós-graduação em Física e Meio Ambiente, Cuiabá, Mato Grosso, Brasil, pp. 77.)

appears to be a relatively consistent phenomenon and may be due to seasonal variations in phenology and allocation, such as leaf aging that may occur during the wet–dry season transition (Borchert, 1994), or to transient increases in decomposition that occur after the first rainfall event in October (Sanches et al., 2008b). Cerradão is a small net source of CO_2, but the annual NEE is not significantly different from zero (Table 5.3). Unfortunately, it is unclear whether the seasonal pattern and/or magnitude of NEE is typical for cerradão, because there are no other published NEE data from other cerradão sites; however, the annual NEE of other seasonal tropical forests with slightly

TABLE 5.3

Average (±95% confidence interval) Daily Net Ecosystem CO_2 Exchange, Latent Heat Flux, and the Bowen Ratio for Cerrado and Transitional Forest (cerradão) during Wet (October–April) and Dry (May–September) Season and Annual Time Periods

	Wet	Dry	Annual	Reference
Net Ecosystem CO_2 Exchange (gC m^{-2} d^{-1})				
Cerrado	−0.51 ± 0.49	0.23 ± 0.52	−0.19 ± 0.42	1–3,5
Transitional forest	0.16 ± 0.30	0.32 ± 0.32	0.19 ± 0.24	6–7,12
Latent Heat Exchange (MJ m^{-2} d^{-1})				
Cerrado	8.83 ± 0.67	3.68 ± 0.66	6.95 ± 0.92	5,8,10
Transitional forest	8.54 ± 0.50	7.26 ± 0.43	8.03 ± 0.34	4,6,9–12

Note: Confidence intervals were calculated by bootstrapping values over 1000 iterations. Gaps in annual time series were filled by linear interpolation.

References include: (1) Miranda et al. (1996), (2) Miranda et al. (1997), (3) da Rocha et al. (2002), (4) Vourlitis et al. (2002), (5) Santos et al. (2003), (6) Priante et al. (2002), (7) Vourlitis et al. (2004), (8) Oliveira et al. (2005), (9) Biudes (2008), (10) da Rocha et al. (2009), (11) Vourlitis et al. (2008), and (12) Vourlitis et al., unpublished data.

higher annual rainfall is similar in magnitude and sign to that reported here (Goulden et al., 2004; Miller et al., 2004).

Seasonal variations in soil respiration are coincident with seasonal variations in rainfall, and transitional forest has higher rates of soil respiration than cerrado (Figure 5.4c and d). Soil respiration is reportedly 4–6 gC m^{-2} d^{-1} for cerrado and 8 gC m^{-2} d^{-1} for cerradão during the wet season, whereas during the dry season, soil respiration is only 2 and 4 gC m^{-2} d^{-1} for cerrado and cerradão, respectively. Seasonal variations in soil water availability can cause rapid fluctuations in microbial growth and activity, physiological processes such as ion uptake, enzyme activity and concentration, tree growth, and root mortality, and trigger a variety of phenological processes that alter resource allocation patterns in tropical trees (Borchert, 1994; McGroddy and Silver, 2000; Valentini et al., 2008). All of these responses will affect rates of heterotrophic and autotrophic respiration and, ultimately, soil CO_2 efflux.

Seasonal variations are also apparent in the rate of latent heat exchange (L_e) and the Bowen ratio (β), which is the ratio of sensible heat exchange (H) and L_e (H/L_e). For cerrado, rates of L_e are approximately two times higher during the wet season than the dry season (Figure 5.5a), resulting in a nearly fourfold difference in the β between the wet and dry seasons (Figure 5.5c). Seasonal variations in L_e are also apparent in cerradão (Figure 5.5b); however, wet-season L_e is only about 20% higher than dry season L_e (Table 5.2), and β is consistently <1 regardless of season (Figure 5.5d). Rates of L_e are statistically similar for cerrado and cerradão during the wet season; however, cerradão has a significantly higher L_e during the dry season (Table 5.3).

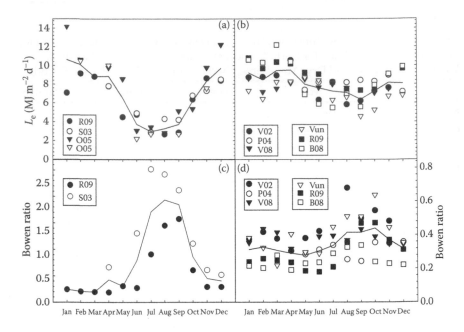

FIGURE 5.5
Latent heat exchange (L_e) for (a) cerrado, (b) transitional forest, (c) the Bowen ratio calculated as the ratio of sensible heat exchange divided by L_e for cerrado, and (d) transitional forest. Symbols depict values obtained from the literature, whereas the solid line depicts average calculated from the values displayed in the figures. (Data from (R09) da Rocha, H. R. et al. 2009. *Journal of Geophysical Research-Biogeosciences* 114, G00B12; (S03) Santos, A. J. B. et al. 2003. *Functional Ecology* 17, 711–719; (O05) Oliveira, R. S. et al. 2005. *Functional Ecology* 19, 574–581; (V02) Vourlitis, G. L. et al. 2002. *Water Resources Research* 38(6), 30; (P04) Priante Filho, N. et al. 2004. *Global Change Biology* 10, 863–876; (V08) Vourlitis, G. L. et al. 2008. *Water Resources Research* 44, W03412 (Vun) Vourlitis, G. L. et al., unpublished data; and (B08) Biudes, M. S. 2008. Balanço de energia em área de vegetação monodominante de Cambará e pastagem no norte do Pantanal. Tese (doutorado)—Universidade Federal de Mato Grosso, Faculdade de Agronomia e Medicina Veterinária, Pósgraduação em Agricultura Tropical, pp. 167.] Note the scale change from figures (c) and (d).

Differences in the seasonal patterns of NEE and L_e for cerrado and cerradão can be explained by differences in the importance of woody vegetation, rooting depth, and/or LAI. Many cerrado trees have root systems that can extract water to soil depths >4 m; however, as the density of trees declines and the importance of grass and shrubs increases, temporal trends in LAI, photosynthesis, and transpiration are more closely tied to surface soil water dynamics (Nepstad et al., 1994; Franco, 2002; Priante et al., 2004; Oliveira et al., 2005). A greater reliance on surface water causes a rapid decline in leaf gas exchange, NEE, L_e, and LAI as drought ensues, and a rapid increase in these processes was rainfall increases in the wet season (Santos et al., 2003; Priante et al., 2004; Oliveira et al., 2005). Such tight coupling between rainfall, surface soil water availability, and ecosystem C and H_2O cycling is less pronounced in transitional forest. Indeed, relatively high dry-season rates of sap flow and L_e for transitional forest have been found to be entirely supported by ground water that is >5 m below the soil surface (Vourlitis et al., 2008). A greater reliance of trees on deep water may decouple LAI and canopy CO_2 exchange to rainfall and surface water dynamics. For example, temporal trends in LAI may lag behind rainfall or soil water content by several weeks (Nepstad et al., 1994; Sanches et al., 2008a), which, in turn, can cause concomitant lags in canopy photosynthesis, ecosystem respiration, and sap flow (Vourlitis et al., 2004, 2008).

Comparison to Other Tropical Ecosystems

The severity of seasonal drought stress is an important control on the structure (i.e., tree/grass cover, LAI) and mass and energy exchange of cerrado and transitional forest, a pattern that appears to scale across many ecosystems of the Brazilian tropics. For example, Vourlitis et al. (2005) reported that the annual amplitude of NEE (calculated as the difference in the mean monthly maximum and minimum NEE) across the major Brazilian tropical ecosystems was linearly correlated with annual rainfall and dry-season duration. Extending this analysis with additional data, the annual amplitude of NEE declines significantly ($r^2 = 0.65$; $p < 0.001$) as annual rainfall increases and increases significantly ($r^2 = 0.52$; $p < 0.005$) as dry-season duration increases (Figure 5.6a and b). Cerrado sites with the lowest annual rainfall and longest dry-season duration exhibit the largest seasonal variation in NEE, whereas humid evergreen forest sites with the highest annual rainfall and the shortest dry-season duration exhibit the smallest annual amplitude in NEE (Figure 5.6a and b). Most of the annual variation in NEE appears to be driven by the variation in the annual maximum NEE, which declines significantly as annual rainfall increases ($r^2 = 0.63$; $p < 0.001$) and increases significantly ($r^2 = 0.80$; $p < 0.001$) with dry-season duration (Figure 5.6c and d). Thus, sites with lower annual rainfall or longer dry-season duration are relatively large net sources during times of drought. In contrast, rates of the annual minimum NEE are apparently independent of annual rainfall (Figure 5.6e) and less highly correlated with dry-season duration ($r^2 = 0.39$; $p < 0.01$; Figure 5.6e), indicating that cerrado, transitional forest, and humid evergreen forest have similar rates of minimum NEE (i.e., maximum net CO_2 uptake) when seasonal drought is alleviated. These results suggest that drought arising from climate change may substantially reduce the potential for tropical forests to sequester atmospheric CO_2 and may provide a glimpse of how an increase in seasonality from climate change might alter the future NEE of the Amazon Basin.

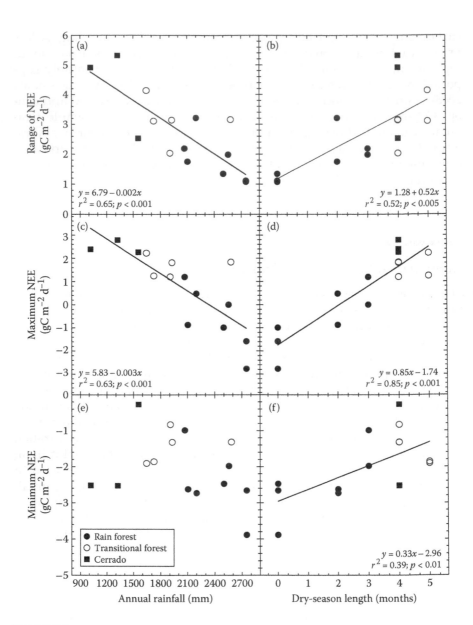

FIGURE 5.6

The annual range of net ecosystem CO_2 exchange (NEE) calculated as the difference between the maximum and minimum monthly value (a and b), the annual maximum value of NEE (c and d), and the annual minimum value of NEE (e and f) as a function of total annual rainfall (a, c, and e) and dry season duration (b, d, and f) for cerrado (closed-squares), transitional forest (open-circles), and humid evergreen forest (closed-circles). (Modified from Vourlitis, G. L. et al. 2005. *Earth Interactions* 9: 9–027, online at http://EarthInteractions.org)

Effects of Disturbance on Mass and Energy Exchange

Cerrado is heavily impacted by human activities such as fire, cattle ranching, agriculture, and production of charcoal, steel, and sugar cane for ethanol fuel and sugar (Pivello and Coutinho, 1996; Ratter et al., 1997; Fearnside, 2000; Miranda et al., 2002; Brannstrom et al., 2008). Fire caused by humans or lightning has been the most common agent of disturbance in cerrado for thousands of years (Miranda et al., 2002). Most fires in cerrado are surface fires that occur during the dry season and consume nearly all of the aboveground biomass in grass-dominated cerrado (*campo limpo* and *campo sujo*) and 50–70% of aboveground biomass in cerrado s. s. and cerradão (Kauffman et al., 1994; Pivello and Coutinho, 1996; de Castro and Kauffman, 1998; Miranda et al., 2002); however, soil C stocks appear to be more stable (Roscoe et al., 2000). Many cerrado trees have thick, highly suberized bark that offers some resistance from fire; thus, cerrado trees are typically not killed by fire (Hoffmann, 2002; Miranda et al., 2002). However, human activities can significantly modify the natural disturbance regime, affecting cerrado structure and function (Pivello and Coutinho, 1996; Miranda et al., 2002). In particular, land-use practices that increase fire frequency can lead to the development and persistence of the more grass-dominated forms of campo cerrado (Pivello and Coutinho, 1996). Unfortunately, there are limited data to assess rates of C sequestration during fire recovery. In *campo sujo*, rapid growth of grasses during the initial post-fire recovery phase can neutralize combustion losses of C after approximately 12 months (Santos et al., 2003). It is important to determine whether this pattern holds in other types of cerrado vegetation types.

Conversion of cerrado to agriculture has mixed effects on ecosystem C storage. On the one hand, conversion of cerrado s. s. to agriculture can lead to substantial net C loss from biomass replacement and a decline in soil C stock (Fearnside, 2000). However, the nature of land use and soil type can have a large effect on the potential for soil C loss. For example, tilling can reduce soil C storage relative to virgin cerrado, whereas no-till management can lead to an increase in soil C storage (Resck et al., 2000). Cerrado and forest soils that are originally low in SOM often exhibit an increase in SOM after land conversion, whereas soils that are originally higher in SOM often exhibit a reduction in SOM with pasture conversion (Neill and Davidson, 2000). Degraded pastures with reduced productivity have significantly lower soil C storage; whereas pastures that experience moderate grazing pressure and/or receive inputs of lime, fertilizers, irrigation, or leguminous plants can exhibit the same or higher soil C stocks than virgin cerrado (Resck et al., 2000; Maia et al., 2009). Given that approximately 50% of all planted pastures are considered degraded (Klink and Moreira, 2002), it is likely that there is an overall soil C loss from cerrado conversion to pasture.

Conclusions

Cerrado varies considerably in physiognomy, from grass-dominated campo to tree-dominated cerrado *s. s.* and cerradão. These variations are due, in part, to variations in climate, soil physical and chemical properties, and disturbance regimes. In turn, variations in physiognomy affect aboveground and soil C storage; and, in general, aboveground plant and soil C pool size declines as the importance of trees and shrubs diminishes. However, root C storage is less variable across the cerrado structural gradient, presumably due to variations in rooting depth and belowground C allocation.

Large variation in climate and rainfall distribution lead to concomitant variation in mass and energy exchange. Annual variations in mass and energy exchange are large for cerrado, regardless of vegetation type, but less so for cerradão. During the wet season, cerrado is a net sink of atmospheric CO_2, and transitional forest is approximately in balance, whereas both are comparable sources of water vapor to the atmosphere. During the dry season, both cerrado and cerradão were in balance, and the L_e of transitional forest is significantly higher than cerrado. These seasonal differences in mass and energy exchange may reflect differences in rooting depth and the reliance on deep water reserves between cerrado and cerradão. Seasonal differences in the pattern of mass and energy exchange notwithstanding, data available to date indicate that the NEE of cerrado and cerradão is approximately in balance over the annual cycle.

Cerrado is heavily impacted by a variety of human activities that have the capacity to alter ecosystem C storage, cycling, and biodiversity. Ecosystem recovery from fire appears to be rapid, resulting in a transient loss of C as long as there is no change in fire frequency or other processes (vegetation substitution) that affect stand recovery. However, land-cover change can result in a net decline in ecosystem C storage, depending on land-use practices and the nature of land-cover change. Losses of aboveground C storage are typical when woody vegetation is substituted for herbaceous vegetation, whereas losses of soil C are more complex and are influenced by soil type and management strategy. Given the rapid rate of cerrado conversion, more work is needed to quantify how land-cover change and land-use practices alter the mass and energy exchange of these diverse ecosystems.

References

Ackerly, D. D., W. W. Thomas, C. A. C. Ferreira, and J. R. Pirani. 1989. The forest–cerrado transition zone in southern Amazonia: Results of the 1985 Projecto Flora Amazônica Expedition to Mato Grosso. *Brittonia* 41, 113–128.

Barbosa, R. I. and P. M. Fearnside. 2005. Above-ground biomass and the fate of carbon after burning in the savannas of Roraima, Brazilian Amazonia. *Forest Ecology and Management* 216, 295–316.

Batalha, M. A., W. Mantovani, and H. N. de Mesquita Junior. 2001. Vegetation physiognomies in south-eastern Brazil. *Brazilian Journal of Biology* 61, 475–483.

Biudes, M. S. 2008. Balanço de energia em área de vegetação monodominante de Cambará e pastagem no norte do Pantanal. Tese (doutorado)—Universidade Federal de Mato Grosso, Faculdade de Agronomia e Medicina Veterinária, Pós-graduação em Agricultura Tropical. pp. 167.

Borchert, R. 1994. Soil and stem water storage determine phenology and distribution of tropical dry forest trees. *Ecology* 75, 1437–1449.

Borges, O. 2007. Efluxo de CO_2 do solo em floresta de transição Amazônia cerrado e am área de pastagem. MSc Thesis, Universidade Federal de Mato Grosso, Instituto de Ciências Exatas e da Terra, Programa de Pós-graduação em Física e Meio Ambiente, Cuiabá, Mato Grosso, Brasil, pp. 77.

Brannstrom, C., W. Jepson, A. M. Filippi, D. Redo, Z. Xu, and S. Ganesh. 2008. Land change in the Brazilian Savanna (Cerrado), 1986–2002: Comparative analysis and implications for land-use policy. *Land Use Policy* 25, 579–595.

Bucci, S. J., F. G. Scholz, G. Goldstein, W. A. Hoffmann, F. C. Meinzer, A. C. Franco, T. Giambelluca, F. Miralles-Wilhelm. 2008. Controls on stand transpiration and soil water utilization along a tree density gradient in a Neotropical savanna. *Agricultural and Forest Meteorology* 148, 839–849.

Bustamante, M. M. C., E. Medina, G. P. Asner, G. B. Nardoto, and D. C. Garcia-Montiel. 2006. Nitrogen cycling in tropical and temperate savannas. *Biogeochemistry* 79, 209–237.

da Rocha, H. R., H. C. Freitas, R. Rosolem, et al. 2002. Measurements of CO_2 exchange over a woodland savanna (Cerrado Sensu stricto) in southeast Brazil. *Biota Neotropica* 2, 1–11.

da Rocha, H. R., A. O. Manzi, O. M. Cabral, et al. 2009. Patterns of water and heat flux across a biome gradient from tropical forest to savanna in Brazil *Journal of Geophysical Research-Biogeosciences* 114, G00B12, doi:10.1029/2007JG000640.

de Castro, E. A. and J. B. Kauffman. 1998. Ecosystem structure in the Brazilian cerrado: A vegetation gradient of aboveground biomass, root mass and consumption by fire. *Journal of Tropical Ecology* 14, 263–283.

Eiten, G. 1972. The cerrado vegetation of Brazil. *The Botanical Review* 38, 201–341.

Fearnside, P. M. 2000. Global warming and tropical land use change: Greenhouse gas emissions from biomass burning, decomposition, and soils in forest conversion, shifting cultivation, and secondary vegetation. *Climatic Change* 46, 115–158.

Franco, A. C. 1998. Seasonal patterns of gas exchange, water relations and growth of *Roupala montana*, an evergreen savanna species. *Plant Ecology* 136, 69–76.

Franco, A. C. 2002. Ecophysiology of woody plants. In *The Cerrados of Brazil*, ed. P. S. Oliveira, and R. J. Marquis, Columbia University Press, New York, pp. 178–200.

Franco, A. C., M. Bustamante, L. S. Caldas, et al. 2005. Leaf functional traits of Neotropical savanna trees in relation to seasonal water deficit. *Trees* 19, 326–335.

Furley, P. A. and J. A. Ratter. 1988. Soil resources and plant communities of the central Brazilian cerrado and their development. *Journal of Biogeography* 15, 97–108.

Goodland, R. 1971. A physiognomic analysis of the Cerrado vegetation of central Brasil. *The Journal of Ecology* 59, 411–419.

Goulden, M. L., S. D. Miller, H. R. da Rocha, et al. 2004. Diel and seasonal patterns of tropical forest CO_2 exchange. *Ecological Applications* 14, S42–S54.

Grace, J., J. S. José, P. Meir, H. S. Miranda, and R. A. Montes. 2006. Productivity and carbon fluxes of tropical savannas. *Journal of Biogeography* 33, 387–400.

Haase, R. 1999. Litterfall and nutrient return in seasonally flooded and non-flooded forest of the Pantanal, Mato Grosso, Brazil. *Forest Ecology and Management* 117, 129–147.

Hoffmann, W. A. 2002. Direct and indirect effects of fire on radial growth of cerrado savanna trees. *Journal of Tropical Ecology* 18, 137–142.

Hoffmann, W. A., E. R. da Silva Jr, G. C. Machado, et al. 2005. Seasonal leaf dynamics across a tree density gradient in a Brazilian savanna. *Oecologia* 145, 307–316.

Hoffmann, W. A., B. Orthen, P. Kielse, and V. do Nascimento. 2003. Comparative fire ecology of tropical savanna and forest trees. *Functional Ecology* 17, 720–726.

Kauffman, J. B., D. L. Cummings, and D. E. Ward. 1994. Relationships of fire, biomass and nutrient dynamics along a vegetation gradient in the Brazilian Cerrado. *The Journal of Ecology* 82, 519–531.

Klink, C. A. and A. G. Moreira. 2002. Past and current human occupation and land use. In *The Cerrados of Brazil*, (ed.) P. S. Oliveira and R. J. Marquis, Columbia University Press, New York, NY, pp. 69–90.

Ledru, M. P. 2002. Late quatrenary history and evolution of the cerrados as revealed by palynological records. In *The Cerrados of Brazil*, (ed.) P. S. Oliveira and R. J. Marquis, Columbia University Press, New York, NY, pp. 33–50.

Lobo, F., A. de, J. H. Campelo Jr., C. E. R. Ortiz, I. C. Lucena, and G. L. Vourlitis. 2008. Gas exchange and leaf and fruiting phenology of Mangabeira in response to irrigation. *Brazilian Journal of Plant Physiology* 20, 1–10.

Lopes, A. S. and F. R. Cox. 1977. Cerrado vegetation in Brazil: An edaphic gradient. *Agronomy Journal* 69, 828–831.

Maia, S. M. F., S. M. Ogle, C. E. P. Cerri, and C. C. Cerri. 2009. Effect of grassland management on soil carbon sequestration in Rondônia and Mato Grosso states, Brazil. *Geoderma* 149, 84–91.

McGroddy, M. and W. L. Silver. 2000. Variations in belowground carbon storage and soil CO_2 flux rates along a wet tropical climate gradient. *Biotropica* 32, 614–624.

Meinzer, F. C., G. Goldstein, N. M. Holbrook, P. Jackson, and J. Cavalier. 1993. Stomatal and environmental control of transpiration in a lowland tropical forest tree. *Plant, Call and Environment* 16, 429–436.

Miller, S. D., M. L. Goulden, M. C. Menton, et al. 2004. Biometric and micrometeorological measurements of tropical forest carbon balance. *Ecological Applications* 14, S114–S126.

Miranda, A. C., H. S. Miranda, J. Lloyd, et al. 1996. Carbon dioxide fluxes over a cerrado *sensu stricto* in central Brazil. In *Amazonian Climate and Deforestation*, (ed.) J. H. C. Gash, C. A. Nobre, J. M. Roberts, and R. L. Victoria, J. M. Wiley and Sons, New York, pp. 353–364.

Miranda, A. C., H. S. Miranda, J. Lloyd, et al. 1997. Fluxes of carbon, water, and energy over Brazilian cerrado: An analysis using eddy covariance and stable isotopes. *Plant, Cell and Environment* 20, 315–328.

Miranda, H. S., M. M. C. Bustamante, and H. S. Miranda. 2002. The fire factor. In *The Cerrados of Brazil*, (ed.) P. S. Oliveira, and R. J. Marquis, Columbia University Press, New York, NY, pp. 51–68.

Miranda, E. J., G. L. Vourlitis, N. P. Filho, et al. 2005. Seasonal variation in the leaf gas exchange of tropical forest trees in the rain forest-savanna transition of the southern Amazon Basin. *Journal of Tropical Ecology* 21, 451–460.

Miyaji, K., W. S. da Silva, and P. T. Alvim. 1997. Longevity of leaves of a tropical tree, *Theobroma cacao*, grown under shading, in relation to position within the canopy and time of emergence. *New Phytologist* 135, 445–454.

Monteiro, J. A. F. and C. H. B. A. Prado. 2006. Apparent carboxylation efficiency and relative stomatal and mesophyll limitations of photosynthesis in an evergreen cerrado species during water stress. *Photosynthetica* 44, 39–45.

Nardoto, G. B., M. P. Souza, and A. C. Franco. 1998. Estabelecimento e padrões sazonais de produtividade de *Kielmeyera coriacea* (Spr) Mart. nos cerrados do Planalto Central: Efeitos do estresse hídrico e sombreamento. *Revista Brasileira de Botânica* 21, 1–12.

Neill, C. and E. A. Davidson. 2000. Soil carbon accumulation or loss following deforestation for pasture in the Brazilian Amazon. In *Advances in Soil Science: Global Climate Change and Tropical Ecosystems*, (ed.) R. Lal, J. M. Kimble, and B. A. Stewart, CRC Press, Boca Raton, FL, pp. 197–212.

Nepstad, D. C., C. R. de Carvalho, E. A. Davidson, et al. 1994. The role of deep roots in the hydrological and carbon cycles of Amazonian forests and pastures. *Nature* 372: 666–669.

Oliveira, P. S. and R. J. Marquis. 2002. *The Cerrados of Brazil*. Columbia University Press, New York, NY, p. 398.

Oliveira-Filho, A. T. and J. A. Ratter. 2002. Vegetation physiognomies and woody flora of the cerrado biome. In *The Cerrados of Brazil*, (ed.) P. S. Oliveira and R. J. Marquis, Columbia University Press, New York, NY, pp. 91–120.

Oliveira, R. S., L. Bezerra, E. A. Davidson, et al. 2005. Deep root function in soil water dynamics in cerrado savannas of central Brazil. *Functional Ecology* 19, 574–581.

Pinto, A. S., M. M. C. Bustamante, K. Kisselle, et al. 2002. Soil emissions of N_2O, NO, and CO_2 in Brazilian Savannas: Effects of vegetation type, seasonality, and prescribed fires. *Journal of Geophysical Research* 107 (D20) 8089, doi:10.1029/2001JD000342.

Pivello, V. R. and L. M. Coutinho. 1996. A qualitative successional model to assist in the management of Brazilian cerrados. *Forest Ecology and Management* 87: 127–138.

Prado, C. H. B. A. and J. A. P. V. Moraes. 1997. Photosynthetic capaciity and specific leaf mass in twoenty woody species of cerrado vegetaton under field conditions. *Photosynthetica* 33, 103–112.

Prado, C. H. B. A., Z. Wenhui, M. H. C. Rojas, and G. M. Souza. 2004. Seasonal leaf gas exchange and water potential in a woody cerrado species community. *Brazilian Journal of Plant Physiology* 16, 7–16.

Priante Filho, N., G. L. Vourlitis, M. M. S. Hayashi, et al. 2004. Comparison of the mass and energy exchange of a pasture and a mature transitional tropical forest of the southern Amazon Basin during a seasonal transition. *Global Change Biology* 10, 863–876.

Ratter, J. A., J. F. Ribeiro, and S. Bridgewater. 1997. The Brazilian Cerrado Vegetation and Threats to its Biodiversity. *Annals of Botany* 80, 223–230.

Resck, D. V. S., C. A. Vasconcellos, L. Vilela, and M. C. M. Macedo. 2000. Impact of conversion of Brazilian cerrados to cropland and pastureland on soil carbon pool and dynamics. In *Advances in Soil Science: Global Climate Change and Tropical Ecosystems*, (ed.) R. Lal, J. M. Kimble, and B. A. Stewart, CRC Press, Boca Raton, FL, pp. 169–196.

Ribeiro, L. F. and M. Tabarelli. 2002. A structural gradient in cerrado vegetation of Brazil: Changes in woody plant density, species richness, life history and plant composition. *Journal of Tropical Ecology* 18, 775–794.

Roscoe, R., P. Buurman, E. J. Velthorst, and J. A. A. Pereira. 2000. Effects of fire on soil organic matter in a 'cerrado sensu-stricto' from Southeast Brazil as revealed by changes in $\delta^{13}C$. *Geoderma* 95, 141–160.

Sanches, L., N. L. Reis, J. S. Nogueira, M. S. Biudes, and G. L. Vourlitis. 2008a. Índice área foliar em floresta de transição amazonia cerrado em diferentes métodos de estimative. *Revista de Ciência e Natura* 30, 57–69.

Sanches, L., C. M. A. Valentini, O. B. P. Júnior, et al. 2008b. Seasonal and interannual litter dynamics of a tropical semideciduous forest of the southern Amazon Basin, Brazil, *Journal of Geophysical Research-Biogeosciences* 113, G04007, doi:10.1029/2007JG000593.

Santos, A. J. B., G. T. D. A. Silva, H. S. Miranda, A. C. Miranda, and J. Lloyd. 2003. Effects of fire on surface carbon, energy and water vapour fluxes over campo sujo savanna in central Brazil. *Functional Ecology* 17, 711–719.

Sendall, K. M. 2006. Seasonal variation in rates of photosynthesis of a transitional tropical forest of Mato Grosso, Brazil. Thesis, Masters Program in Biology, Biological Sciences Department, California State University, San Marcos, CA, p. 62.

Valentini, C. M. A., L. Sanches, S. R. Paula, et al. 2008. Soil respiration and above-ground litter dynamics of a tropical transitional forest in northwest Mato Grosso, Brazil. *Journal of Geophysical Research-Biogeosciences* 113, G00B10, doi:10.1029/2007JG000619.

Vourlitis, G. L., N. Priante-Filho, M. M. S. Hayashi, J. de Sousa Nogueira, F. T. Caseiro, and J. H. Campelo, Jr. 2002. Seasonal variations in the evapotranspiration of a transitional tropical forest of Mato Grosso, Brazil. *Water Resources Research* 38(6), 30.

Vourlitis, G. L., N. Priante-Filho, M. M. S. Hayashi, et al. 2004. Effects of meteorological variations on the net CO_2 exchange of a Brazilian transitional tropical forest. *Ecological Applications* 14, S89–S113.

Vourlitis, G. L., J. de S. Nogueira, N. Priante-Filho, et al. 2005. The sensitivity of diel CO_2 and H_2O vapor exchange of a tropical transitional forest to seasonal variation in meteorology and water availability. *Earth Interactions* 9, 9–027, online at http://EarthInteractions.org

Vourlitis, G. L., J. de S. Nogueira, F. de A. Lobo, et al. 2008. Energy balance and canopy conductance of a tropical semi-deciduous forest of the southern Amazon Basin. *Water Resources Research* 44, W03412, doi: 10.1029/2006WR005526.

6

Woody Plant Rooting Depth and Ecosystem Function of Savannas: A Case Study from the Edwards Plateau Karst, Texas

Marcy E. Litvak, Susan Schwinning, and James L. Heilman

CONTENTS

Introduction

Savannas are characterized by the co-dominance of grasses and deep rooted shrubs or trees (e.g., Walter, 1971) in seasonally dry environments with a limited and fluctuating water supply. The two dominant plant functional types have contrasting strategies of drought avoidance, which determine the ecosystem-wide responses of savannas to precipitation variability. Grasses typically avoid drought by focusing their physiological activity in periods when soil water availability is high and become dormant as soon as the shallow soil, the sole source of their water supply, dries out. Shrubs and trees typically avoid drought by switching to deeper water sources, below the root zone of the grasses. Although this pattern has been observed in savannas, there are also variations on the theme, depending on two key factors: rooting depth and water availability in the lower root zone of the trees.

Rooting depth is a trait that can vary greatly among tree species; for example, the genera of mesquite, tamarix, and eucalypt (Stromberg et al., 2007; Scott et al., 2006; Mitchell et al., 2009) have some of the most deeply rooted species known. These species are often called *phreatophytes*, implying access to groundwater, though in many cases the phreatophytic life style is "facultative" (Smith et al., 1997), and trees do well even without groundwater (Potts et al., 2008; Brunel, 2009). On the other end of the spectrum, many gymnosperms, including juniper, are described as having an extensive shallow root system rather than a deep root system (e.g., Hall, 1952; Dealy, 1990).

Rooting depth is also affected by soil condition and geology. For example, roots in sandy soils tend to grow deeper, due, in part, to the greater infiltration depth of water in sandy soil and also because of less mechanical resistance to root growth (Collins and Bras, 2007). Then there are systems, including many savanna systems, in which the soil is shallow and underlain by fractured bedrock. Based on a global analysis of rooting depths, Schenk (2008) concluded that the rooting depths of trees in these systems can far exceed those of trees in deep soils (7.9 m vs. 2.2 m in geometric mean rooting depth). Some observations appear to support this generalization. For example, Jackson et al. (1999), working in the eastern Edwards Plateau, found oak roots (*Quercus fusiformis*) in caves 20 m below ground; and in one frequently cited anecdote, a mesquite root (*Prosopis velutina*) was found at 60 m depth in an Arizona mine. However, there are also examples to the contrary. For example, roots of eastern white cedar on cliff faces extended only 30 cm into the bedrock (Matthes-Sears, 1995); and in the limestone karst of the Yucatan Peninsula, Mexico, roots of a shrub community penetrated only 3 m into a fractured limestone matrix (Querejeta et al., 2006, 2007), even though a permanent water table was available at 9–20 m. Clearly, bedrock is not a uniform medium for root proliferation, in some cases providing enhanced opportunities for deep root development and in others prohibiting the formation of deep roots.

Rooting depth, whether due to genetic differences or imposed by the environment, affects the function of savannas and woodlands in important ways. Where roots are deep, they are likely to tap into more permanent water sources (e.g., groundwater or perched water tables), as seen in the study by Jackson et al. (1999). This should buffer gross primary production to precipitation variability and support the carbon sequestration potential of wooded ecosystems, albeit potentially at the cost of reduced groundwater recharge, stream flow, and a receding water table (Jackson et al., 2005). Where deep root proliferation is blocked or where no water subsidy is available at depth, we would expect ecosystem fluxes to be more tightly coupled to precipitation, and persistent droughts would cause a severe decrease in primary production. Accordingly, we would also not expect trees in such systems to inordinately impact regional water balance.

Here we present tower-based measurements of net ecosystem exchange from three study sites on the eastern Edwards Plateau, a karst region in central Texas (Figure 6.1). Our goal is to examine whether groundwater significantly subsidizes the water consumption of trees in the region, a reasonable hypothesis given that the hydrology of karst areas is complex and frequently

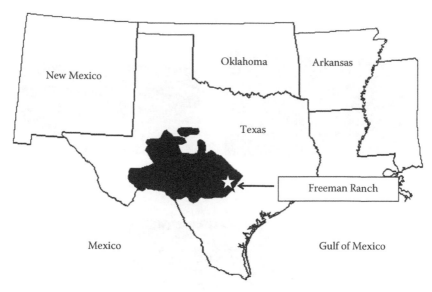

FIGURE 6.1
Location of the Edwards Plateau (shaded) and the Freeman Ranch.

involves perched water tables, springs, and underground streams. The study site was not far (approximately 50 km) from the cave sites where Jackson et al. (1999) and McElrone et al. (2004, 2007) conducted their examination of roots in caves. Both sites have similar overstory composition, dominated by Ashe juniper (*Juniperus ashei*) and Plateau live oak (*Quercus fusiformis*). One important difference is, however, that our study site has no known caves.

We examine a 3-year continuous record (2005–2007) of eddy covariance measurements of carbon, water, and energy fluxes across three study sites with varying degrees of woody cover (grassland, savanna, and woodland) (Figure 6.2) to focus on estimating the role that woody species play in regulating carbon, water, and energy exchange and to assess whether and how the presence of woody plants changes sensitivity to precipitation. The 3 years span a wide range of moisture conditions, with annual precipitation in 2 consecutive years of the study being 13 and 19% below the 100 year mean of 913 mm, whereas precipitation in the third year was 66% above the mean and it was one of the wettest years on record (see Figure 6.3).

Site Descriptions

The three sites, located within 2.5 km of each other, include a grassland, a live oak-Ashe juniper forest, and a juniper or honey mesquite (*Prosopis glandulosa*) savanna on the Freeman Ranch, a 1701-ha research area on the eastern edge of the Edwards Plateau just north of San Marcos, Texas, owned

FIGURE 6.2
(See color insert following page 320.) (a) View from the top of the woodland flux tower. (b) View from the top of the savanna flux tower. (c) Grassland tower pasture.

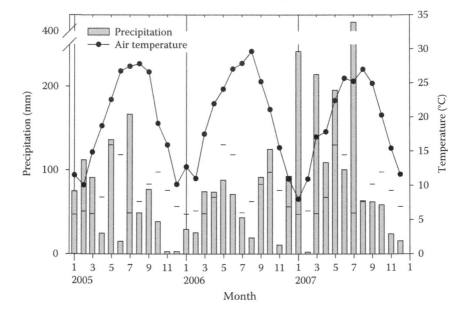

FIGURE 6.3
Monthly precipitation totals and mean daily temperatures for 2005–2007. Dashes represent the historical mean precipitation for each month.

by Texas State University, San Marcos. The entire ranch is in the Edwards Aquifer recharge zone. Extensive portions of the Plateau are dominated by live oak (*Quercus virginiana*)-Ashe juniper (*Juniperus asheii*) savannas interspersed with perennial, mixed C_3/C_4 grasslands. In the last 150 years, both Ashe juniper and honey mesquite have expanded rapidly into grasslands and savannas in this region mainly due to suppression of wildfires and introduction of livestock (Archer et al., 1995; Van Auken, 2009). Soils on the Plateau are shallow and rocky on slopes but deeper in broad valleys and flats. The limestone bedrock is fractured, and imbedded in the limestone are clay lenses and layers of marl and caliche, where water can be stored. *J. asheii* is primarily associated with the rocky and shallow soils of the limestone karst on the eastern edge of the Plateau. Mesquite is excluded or reduced to a shrub-like habit where soils are ≤1.5 m thick. In general, mesquite is not very common on the eastern Edwards Plateau, and it is more abundant in the deep prairie soils to the east and south of this region.

The grassland site is an intermittently grazed, perennial C_3/C_4 pasture with widely interspersed, small (shrublike) honey mesquite trees inside the opening of an oak-juniper savanna (Table 6.1, Figure 6.2). The soil at the site is a Comfort stony clay (clayey-skeletal, mixed, superactive, thermic Lithic Argiustoll) with an approximately 20-cm deep A horizon overlying an approximately 20-cm thick Bt1 horizon containing chert fragments. A Bt2

TABLE 6.1

Location, Vegetation, and Soil Characteristics of Each Tower Site

Site	Location	Elevation (m)	Dominant Species	Soils	LAI $(m^2 m^{-2})$
Grassland	29°55.80'N, 98°0.60'W	230	*Bothriochloa ischaemum* (C4) *Stipa leucotricha* (C3)	Comfort stony clay (~40 cm)	0.5
Savanna	29°56.97'N, 97°59.77'W	272	*Juniperus ashei* *Prosopis glandulosa* *Bothriochloa ischaemum* (C4) *Stipa leucotricha* (C3)	Rumple-Comfort dominated by Rumple gravelly clay loam (1.5–2.0 m)	1.1
Woodland	29°56.40'N, 97°59.40'W	232	*Juniperus ashei* *Quercus virginiana*	Comfort stony clay (~10–20 cm)	2.0

Source: Adapted from ORNL DAAC (Oak Ridge National Laboratory Distributed Active Archive Center). 2009. MODIS subsetted land products. Collection 5. Available on-line (http://www.daac.ornl.gov/MODIS/modis.html) from ORNL DAAC, Oak Ridge, Tennessee. Accessed May 1, 2009.

Note: Leaf area index (LAI) for each site is the average of the maximum value for years 2005–2007. Leaf area index at the grassland site was estimated from replicate clipped samples. The herbaceous component of LAI at the savanna site was estimated from replicate clipped samples. Juniper LAI was based on allometric equations of Hicks and Dugas (1998). Mesquite LAI was estimated from destructive sampling following Ansley et al. (1998). Leaf area index at the woodland site is from the subset MODIS Land Product from Oak Ridge National Laboratory Distributed Active Archive Center.

horizon starting at a depth of approximately 50 cm contains chert slabs with cracks and gaps between slabs filled with soil. Grasses are dominated by King Ranch bluestem [*Bothriochloa ischaemum* (L. Keng.), an introduced C_4 species that has become invasive throughout the Plateau, and Texas wintergrass (*Stipa leucotricha* Trin. & Rupr.)], a C_3 species. The vegetation also includes a high percentage of forbs, which is typical of Plateau grasslands.

The savanna site consists of clusters of Ashe juniper and honey mesquite interspersed among intermittently grazed grassland that is similar to that at the grassland site (Table 6.1, Figure 6.2). This site is typical of Plateau grasslands with deeper soils undergoing encroachment (Eggemeyer and Schwinning, 2009). In 2006, woody cover was estimated as 48% based on lidar measurements. Both *J. ashei* and *P. glandulosa* have been increasing at this site since 1970, and there are numerous juvenile trees of both species. The soil is a Rumple-Comfort association dominated by Rumple gravelly clay loam (clayey-skeletal, mixed, active, thermic Typic Argiustoll), and it is deeper than at the other two sites, about 1.5–2.0 m.

The woodland site is a continuous closed canopy of live oak and Ashe juniper that is typical of ecosystems occupying intermittent drainages and upland slopes on the Plateau (Table 6.1, Figure 6.2). Mean diameter at breast height (DBH) and basal area (BA) of live oak are 0.17 m and 23.5 m^2 ha^{-1}, respectively. Most of the Ashe juniper in the woodland is multi-trunked with an estimated density of 2850 trunks ha^{-1}. Mean DBH and BA for juniper are 0.08 and 18.4 m^2 ha^{-1}, respectively. Above ground biomass was estimated to be 5.6 kg m^{-2} for juniper and 5.8 kg m^{-2} for oak, based on the allometric equation of Jenkins et al. (2003). As in the grassland, the soil at the woodland site is Comfort stony clay but with only approximately 0.1 m soil cover over fractured rock. Rock outcrops occupy approximately 50% of the surface, and most of the soil is covered by oak and juniper leaf litter. The herbaceous understory consists chiefly of isolated clumps of cedar sedge (*Carex planostachys*). Historical aerial photos suggest that although *J. asheii* has been present at this site for at least the past 70 years, it has increased in density and cover over this time.

Methods

Identical tower-based eddy covariance systems were used to measure fluxes of CO_2, sensible heat (H), and latent heat (λE) at all three sites. Wind velocity and air temperature fluctuations were measured acoustically using CSAT-3 sonic anemometers (Campbell Scientific Inc., Logan, UT), whereas CO_2 and water vapor density fluctuations were measured using LI-7500 open-path infrared gas analyzers (LI-COR Inc., Lincoln, NE). Anemometers and gas analyzers were placed at 3 m at the grassland, 6 m at the savanna, and 15 m at the woodland; and outputs were sampled at 10 Hz. Gas analyzers were calibrated periodically using a gas of known CO_2 concentration and a dew-point generator (LI-610, LI-COR).

The surface energy balance is described by the equation

$$R_n - G - S = H + \lambda E,$$

where R_n is net radiation, G is soil heat flux, S is heat storage in aboveground biomass and in the canopy air space, H is sensible heat flux, and λE is latent heat flux. Collectively, the three terms to the left of the equal sign represent the available energy, the net amount of energy on hand for partitioning between the turbulent fluxes of H and λE. Net radiation was measured with model Q7.1 net radiometers (REBS, Seattle, WA) placed at the same elevations as the anemometers and gas analyzers. Soil heat flux was determined by soil heat flux plates and calorimetry (Liebethal et al., 2005). Sensible and latent heat fluxes were obtained from eddy covariance. In the woodland, heat storage in aboveground biomass and in the canopy air space was calculated using temperature and humidity profiles (Heilman et al., 2009). Storage fluxes in savanna and grassland biomass were ignored.

All fluxes were calculated as 30-min averages. Eddy covariance calculations included spike removal, natural wind coordinate rotation (Lee et al., 2004), adjustments for variations in air density due to water vapor (Webb et al., 1980; Ham and Heilman, 2003), and corrections for frequency response (Massman, 2000). We forced energy balance closure with the Bowen ratio conservation approach discussed by Twine et al. (2000) to facilitate comparison of evapotranspiration (ET) between the three land covers. The Bowen ratio conservation method (Twine et al., 2000) is a widely used method of forcing closure, and it has been extensively used in woodland ecosystems (Oliphant et al., 2004; Scott et al., 2004; Barr et al., 2006; Steinwand et al., 2006; Kosugi et al., 2007). The principal assumption of this method is that eddy covariance correctly measures the Bowen ratio, allowing it to be used to partition the missing energy between sensible and latent heat fluxes. Wolf et al. (2008) showed that eddy covariance estimates of the Bowen ratio were accurate, based on comparisons with traditional, gradient-based estimates, when all relevant corrections (density, frequency response, etc.) were applied to the eddy covariance data. Data rejected during a filtering process for each site include during precipitation events, when wind direction was from behind the tower, if friction velocity (u^*) was below specific thresholds determined for each site (Reichstein et al., 2005) or for high deviation in stationarity tests. Three profiles of soil water content (Decagon ECHO probes) and soil temperature (Type T thermocouples) were measured in each site (at 0–5 cm, 5–10 cm, 10–20 cm depth). Additional measurements at each site include precipitation (Texas Electronics tipping bucket TE-525), air temperature and relative humidity profiles (Vaisala HMP 45C), and photosynthetically active radiation (Licor SI-190). Gaps in meteorological data and turbulent fluxes were filled, and net ecosystem exchange of carbon was partitioned into the components gross ecosystem production and ecosystem respiration using the online tools of Reichstein (http://www.bgc-jena.mpg.de/bgc-mdi/html/eddyproc/index.html).

Results and Discussion

Energy Balance

Available energy $(R_n - G - S)$ was highest in the woodland site, due to greater absorption of shortwave radiation, reduced emission of longwave radiation by the forest canopy, and lower soil heat flux. Partitioning of available energy between H and λE was controlled by water availability and turbulence. Bowen ratios in the savanna were generally lower than those at the other sites, perhaps due to greater water storage or retention in the root zone consistent with deeper soils at this site relative to the other two sites. Sensible heat flux is controlled by aerodynamic conductance, which increases with turbulence, whereas λE is controlled mainly by canopy (stomatal) conductance, especially when water becomes limiting (Rost and Mayer, 2006). In absolute terms, the woodland generated 20% more H than the grassland or savanna in 2005, 22% more in 2006, and 29–47% more in 2007 (Table 6.2).

Carbon and Water Balance

In terms of carbon balance, the presence of woody species increased the ability of these savannas to sequester carbon. Over the entire 3-year record, cumulative carbon gain in the grassland was 308 g C m^{-2}, compared with 893

TABLE 6.2

Annual Totals of Available Energy $(R_n - G - S)$, Latent Heat Flux (λE), and Sensible Heat Flux (H) for the Grassland, Savanna, and Woodland in 2005–2007

Year	Site	$R_n - G - S$ (GJ m^{-2} year^{-1})	λE (GJ m^{-2} year^{-1})	H (GJ m^{-2} year^{-1})	$\lambda E / (R_n - G - S)$	$H/(R_n - G - S)$
2005	Grassland	3.58	1.79	1.79	0.50	0.50
	Savanna	3.73	2.01	1.72	0.54	0.46
	Woodland	4.03	1.98	2.15	0.49	0.51
2006	Grassland	3.51	1.44	2.07	0.41	0.59
	Savanna	3.66	1.57	2.09	0.43	0.57
	Woodland	4.04	1.50	2.54	0.37	0.63
2007	Grassland	3.20	2.05	1.15	0.64	0.36
	Savanna	3.56	2.25	1.31	0.63	0.37
	Woodland	3.82	2.13	1.69	0.56	0.44

Source: Adapted from Twine, T. E. et al. 2000. *Agricultural and Forest Meteorology* 103, 279–300.
Note: λE and H are the fractions of available energy. Energy balance closure was forced here using the procedure of Twine, et al. (2000).

and 978 g C m^{-2} in the savanna and woodland, respectively. The sites with woody species are also slightly more resilient to drought and less responsive to variability in precipitation. In the grassland, NEP was very sensitive to differences in annual precipitation (Table 6.3), as expected of an ecosystem dominated by shallow-rooted herbaceous species. The overall carbon sink in the grassland (NEP) decreased by 22% from 2005 to 2006 as precipitation inputs decreased by 50 mm; then, it quadrupled in 2007 as precipitation doubled. Most of the increase in carbon gain can be explained by an increase in total carbon uptake through photosynthesis (GPP) (Table 6.3). In the woodland site, NEP was more similar across years and actually slightly less in the wet year compared with the two dry years, as respiration (R) in the woodland increased disproportionately more than total carbon uptake (GPP) in 2007 (Table 6.3). The savanna sink was similar to the woodland sink, except in 2006, the second year of the drought, when NEP was about halfway between that of the grassland and woodland sites. One interpretation is that in 2005, the savanna was able to utilize stored water carried over from the previous year, thus inflating carbon gain with regard to the grassland. In 2006, this reserve was likely depleted, and the intermediate value of NEP between that of the forest and the grassland mirrored the intermediate functional type composition of the savanna. The savanna sink increased by 80% from 2006 to 2007, in response to the large increase in precipitation in 2007.

There is very little evidence to support a permanent deep source of water for the woody species in our savanna and woodland. In 2006 and 2007, annual ET did not exceed precipitation inputs in any of the sites (Table 6.3). Although

TABLE 6.3

Annual Totals of Net Ecosystem Production (NEP), Gross Primary Production (GPP), Ecosystem Respiration (R), and Evapotranspiration (ET) for the Grassland, Savanna, and Woodland in 2005, 2006, and 2007

Year	Site	NEP (g C m^{-2} y^{-1})	GPP (g C m^{-2} y^{-1})	R (g C m^{-2} y^{-1})	R/GPP	ET (mm y^{-1})	Precip (mm y^{-1})
2005	Grassland	−64	719	655	0.92	732	790
	Savanna	−389	1285	896	0.70	820	
	Woodland	−337	1529	1192	0.78	806	
2006	Grassland	−50	569	520	0.91	587	741
	Savanna	−183	978	795	0.81	638	
	Woodland	−353	1282	929	0.72	610	
2007	Grassland	−194	1291	1097	0.85	816	1536
	Savanna	−326	1643	1316	0.81	861	
	Woodland	−289	1576	1287	0.82	849	

Note: We used the standard sign convention for NEE with NEE > 0, indicating a net loss of CO$_2$ to the atmosphere (source) and NEE < 0 indicating net CO$_2$ uptake by the ecosystem (sink). R and GEP are always positive. Also shown is the fraction of total carbon lost as respiration per total carbon gained through photosynthesis (R/GEP).

annual ET is slightly higher than precipitation inputs in the woodland and savanna in 2005 (Table 6.3), these values of ET were likely inflated relative to precipitation inputs, as the trees utilized stored water carried over from the previous year. In 2004, total precipitation was approximately 30% higher than the historical average, particularly toward the end of the year (where 560 mm fell in October and November alone). Further, woodland ET exceeded grassland ET by only 10%, 4%, and 4% in 2005, 2006, and 2007, respectively (Table 6.3), far less than the upward of 60%, which is typically observed when trees have access to groundwater (e.g., Engel et al., 2005; Scott et al., 2006; Paco et al., 2009). Differences in ecosystem ET were, in part, related to differences in available energy, which was greatest in the woodland (Table 6.2), and second, due to likely differences in the amount of water retained in the root zone after rainfall. Proportionally more available energy was partitioned into ET at the savanna site, the ecosystem with the deepest soil (~2 m of relatively clay-rich soils compared with thin soil over fractured bedrock at the woodland site, Table 6.1), and, thus, highest water storage capacity in the root zone.

Nevertheless, both the savanna and woodland had access, at least intermittently, to water sources in either the soil (deeper soils at the savanna site) or bedrock (at the woodland site in the rooting zone of the trees) that was not available to the grasses at the grassland site. During the very dry months of February and August in 2006 after months of below-average precipitation, the woodland and, to a lesser extent, the savanna were taking up carbon, whereas the grassland was largely dormant (Figure 6.4). In 2007, when precipitation inputs were large, net carbon uptake patterns were nearly identical among the three sites (Figure 6.4). Overall, the patterns we see in carbon and water balance over the 3-year period suggest that woody species increase the ability of Edwards Plateau landscapes to sequester carbon, but not necessarily at the expense of increased water use.

Sensitivity to Precipitation Pulses and Drying Cycles

In all three land covers, ecosystem fluxes of carbon and water were highly sensitive to precipitation pulses and subsequent drying cycles throughout the year. Focusing specifically on a 100-day period in 2005, both NEE of CO_2 and ET or ETo declined rapidly at all sites during drying cycles (Figure 6.5), regardless of the amount of woody cover. The response shown in Figure 6.5 is typical of what occurred during most drying cycles, with the Bowen ratio ($\beta = H/\lambda E$). approaching 5 at the grassland and woodland. However, during severe dry periods, β of both the savanna and woodland exceeded that of the grassland, reaching values greater than 20, due to higher turbulence at those sites. Characteristic of all three sites were large reductions in λE when the soil dried between rainfall events, accompanied by large increases in H. This response suggests that all three land covers relied primarily on water from recent rainfall events and a limited amount of stored water rather than on a permanent water source at depth.

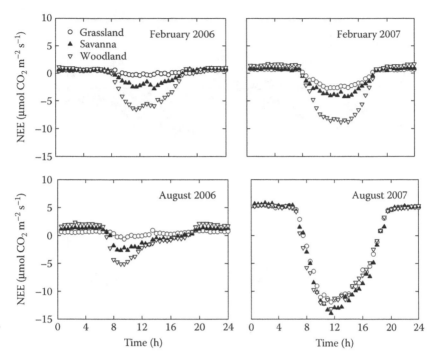

FIGURE 6.4

Mean monthly diurnal NEE for grassland, savanna, and woodland for February and August 2006 and 2007. Numbers represent means ±1 SE.

The response of both NEE and ET/ET_0 to precipitation pulses in the savanna and woodland sites typically outpaced the response in grassland sites (Figure 6.5). This suggests that the trees at our sites, unlike the grasses, remain physiologically active during drought and were able to use precipitation immediately. In addition, NEE and ET/ET_0 of the wooded systems remained higher for the rest of the summer, suggesting that the series of very large rainfall events in September not only recharged the shallow soil and the grass root zone but also recharged the deeper soil or bedrock in the root zone of the trees. Rapid subsurface flow is characteristic of karst systems and distinguishes karst from systems with deep soils, where the bottom of the tree root zone receives water only through drainage from saturated soil layers above. This rapid recharge of the entire tree root zone is likely to contribute to the observed sensitivity of the karst ecosystems to precipitation, even in tree-dominated landscapes.

Plant Water Status

The three dominant tree species at our study sites contrast in functional type characteristics, thus they probably contributed differentially to overall

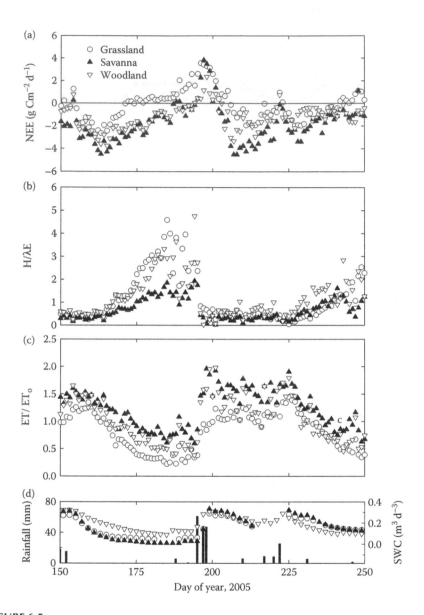

FIGURE 6.5
Daily totals of (a) net ecosystem exchange of CO_2, (b) Bowen ratios ($H/\lambda E$), (c) evapotranspiration (ET) normalized by reference ET (ET_o), and (d) rainfall and soil water content for all three sites during a 100-day period in 2005. This pattern suggests that all ecosystems relied primarily on water from recent rainfall events and a limited amount of stored water rather than on a permanent water source at depth.

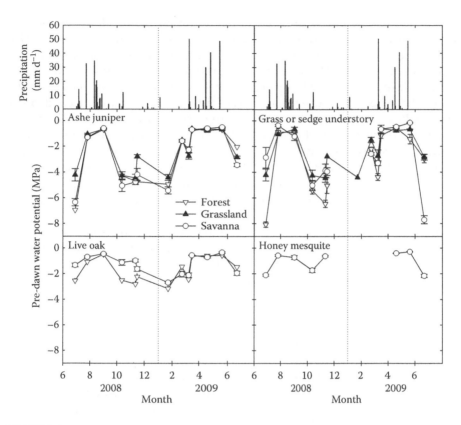

FIGURE 6.6
Water potentials taken at pre-dawn at the three eddy flux study sites from June 2008 to June 2009. Live oaks and juniper trees in the grassland site were surrounding the opening, but they were outside the fetch area of the tower. Missing points are due to unavailable samples during dormancy periods (grasses and mesquite).

ecosystem fluxes. Pre-dawn water potentials measured from June 2008 to June 2009 highlight some of these differences (Figure 6.6). This period was recognized as the worst drought since the 1950s in South Central Texas (specified as D4 exceptional drought condition by the National Weather Service) and provides further insight into the water sources available to these species and their drought tolerance.

Juniper exhibited the greatest fluctuations in pre-dawn water potentials, closely following the patterns seen in grasses and cedar sedge. This indicates that juniper did not likely have greater access to water than the understory species and is extraordinarily drought tolerant. By contrast, oaks and mesquites did exhibit pre-dawn water potentials below −4 MPa. Although it is possible that the two species maintained higher water potentials because they were significantly deeper rooted than juniper and had access to less depleted water sources than the −6 MPa water source of juniper trees, this is unlikely. First, McElrone et al. (2004) have shown that (unlike juniper) live oaks lose approximately 90% of their hydraulic conductivity at −2 MPa. Velvet mesquite (*Prosopis velutina*), a close relative of honey mesquite, in one study showed a 40–50% loss in stem hydraulic conductivity at −2 MPa and a 60–80% loss at −4 MPa (Hultine et al. 2004). Thus, a water source at −4 MPa would have been of no use to oaks and of little use to mesquite. Second, other studies have shown that leaf gas exchange rates in oak are rapidly down-regulated during drought (Owens and Schreiber, 1992), more rapidly than in juniper, which is able to maintain a low but steady rate of gas exchange. This would suggest that GPP of the woodland site is increasingly carried by juniper as conditions get drier and could explain why the woodland is more water use efficient and NEP is slightly more buffered to precipitation variability than the other two sites.

At the savanna site, juniper and mesquite rooting depths were very similar. An excavation study conducted in 2006 showed that roots of mature individuals of juniper and mesquite extended to the bottom of the soil layer, at about 2 m depth although juniper root distribution was more skewed toward the surface when compared with that of mesquite (Schwinning, unpublished). In 2006, the pre-dawn water potentials of the two species at this site varied in almost perfect synchrony throughout the growing season (Eggemeyer and Schwinning, 2009), and with the exception of saplings <1 m tall, which stayed above −2.5 MPa in both species. After an extended dry period, large mesquite trees took up relatively more water from the deeper soil layers compared with large juniper trees, based on a comparison of stem water isotope ratios. This is consistent with the observation that mesquite had greater root allocation to deeper soil layers, but this did not prevent gas exchange rates in mesquite from decreasing to the severely low rates observed in juniper in the driest month of August. Similar to the woodland site, the contribution of juniper to savanna GPP increased with the progression of drought conditions.

Overall, tree species coexisting in the same site experienced very similar constraints of root development, even in species as dissimilar in rooting habit as Ashe juniper and honey mesquite. However, these species differed in their physiological response to variable water availability, with oaks and mesquite being more active during wet periods and juniper being more active in dry periods.

Conclusion

Although an earlier study observed roots from the codominant live oak and Ashe juniper trees of the eastern Edwards Plateau to be deep enough to reach into perched water tables or underground streams (Jackson et al., 1999; McElrone et al., 2004), the trees at our site clearly had no such access. During extended drought periods, these species frequently reached their limits of drought tolerance and drastically reduced gas exchange. This occurred although at least two of the dominant tree species at the site, live oak and mesquite, are understood to have the capacity to form deep roots (Jackson et al., 1999). Edaphic constraints, in the form of largely unbroken limestone layers, prohibited the development of deep tap roots. This suggests a strong control of karst geology over the water sources available to plants. Contrary to the earlier study (Jackson et al., 1999; McElrone et al., 2004), which showed local geology facilitating access to a deep permanent water source, at our site, local geology constrained deep root development and limited access to water.

Questions of water access by trees are extremely sensitive. The general public often sees the water demands of municipalities threatened by tree encroachment (see Tenneson, 2008). The fix on the Edwards Plateau is seen in clear-cutting, which is encouraged by government subsidies without regard to local geology and the likelihood that trees have access to groundwater at a given site. Several studies to date have shown that the relationships between woody cover and stream flow or groundwater levels are complex and case dependent (Wilcox et al., 2005; Wilcox and Thurow, 2006; Wilcox et al., 2008). Huxman et al. (2005) wrote an insightful review of ecohydrological consequences of woody encroachment and developed general rules to evaluate circumstances in which woody encroachment is likely to impact water yield and those in which it is not. Large portions of the Edwards Plateau, even those described as being in the "recharge zone" of the Edwards Aquifer, as our study site, do not appear to provide much opportunity for water theft (Tenneson, 2008); and the large amount of public funds allocated to support brush control may not have the desired effect on groundwater recharge, but rather, at least temporarily, reduce rates of carbon sequestration.

Another indication that water sources for trees on the Edwards Plateau are quite limited is the conspicuous switch in savanna composition between the Blackland Prairie, where mesquite is common and juniper uncommon, and the Edwards Plateau, where juniper and mesquite abundances are reversed (Eggemeyer and Schwinning, 2009). The karst landscape appears to select shallow-rooted, drought-tolerant tree species and is against deep-rooted phreatophytes, contrary to what would be expected if access to groundwater was truly made easier by the karst geology. Querejeta et al. (2006, 2007) obtained similar results in the karst of the Yucatan Peninsula, showing that shrubs and trees were predominantly rooted in the top 1 m of the bedrock.

References

Ansley, R. J., B. A. Trevno, and P. W. Jocoby. 1998. Interspecific competition in honey mesquite: Leaf and whole plant response. *Journal of Range Management* 51, 345–352.

Archer, S. R., D. S. Schimel, and E. H. Holland. 1995. Mechanisms of shrubland expansion: Land use, climate or CO_2. *Climate Change* 29, 91–99.

Barr, A. G., K. Morgenstern, T. A. Black, J. N. McCaughey, and Z. Nezic. 2006. Surface energy balance closure by the eddy-covariance method above three boreal forest stands and implications for the measurement of the CO_2 flux. *Agricultural and Forest Meteorology* 140, 322–337.

Brunel, J. P. 2009. Sources of water used by natural mesquite vegetation in a semi-arid region of northern Mexico. *Hydrological Sciences Journal-Journal Des Sciences Hydrologiques* 54(2): 375–381.

Collins, D. B. G., and R. L. Bras. 2007. Plant rooting strategies in water-limited ecosystems. *Water Resources Research* 43, ARTN W06407, doi:10.1029/2006WR005541.

Dealy, J. E., 1990. Western juniper. In *Silvics of North America, U. S. Department of Agriculture Handbook 654, Vol. 1, Conifers*, eds. R. M. Burns and B. H. Honkala (Technical Coordinators). U.S. Government Printing Office, Washington, DC; 113.

Eggemeyer, K. D. and S. Schwinning. 2009. Biogeography of woody encroachment: Why is mesquite excluded from shallow soils? *Ecohydrology* 2(1), 81–87.

Engel, V., E. G. Jobbagy, M. Stieglitz, M. Williams, and R. B. Jackson. 2005. Hydrological consequences of eucalyptus afforestation in the argentine pampas. *Water Resources Research* 41, W10409.

Hall, M. T. 1952. Variation and hybridization in Juniperus. *Annals of the Missouri Botanical Garden* 39, 1–64.

Ham, J. M. and J. L. Heilman. 2003. Experimental test of density and energy-balance corrections on CO_2 flux as measured using open-path eddy covariance. *Agronomy Journal* 95, 1393–1403.

Heilman, J. L., K. J. McInnes, J. F. Kjelgaard, M. K. Owens, and S. Schwinning. 2009. Energy balance and water use in a subtropical karst woodland on the Edwards Plateau, Texas. *Journal of Hydrology* 373, 426–435.

Hicks, R. A. and W. A. Dugas. 1998. Estimaing ashe juniper leaf area from tree and stem characteristics. *Journal of Range Management* 51, 633–637.

Hultine, K. R., R. L. Scott, W. L. Cable, and D. G. Williams. 2004. Hydraulic redistribution by a dominant, warm desert phreatophyte: seasonal patterns and response to precipitation pulses. *Functional Ecology* 18, 530–538.

Huxman, T. E., B. P. Wilcox, D. D. Breshears, et al. 2005. Ecohydrological implications of woody plant encroachment. *Ecology* 86, 308–319.

Jackson, R. B., E. G. Jobbagy, R. Avissar, et al. 2005. Trading water for carbon with biological sequestration. *Science* 310 (5756), 1944–1947.

Jackson, R. B., L. A. Moore, W. A. Hoffman, W. T. Pockman, and C. R. Linder. 1999. Ecosystem rooting depth determined with caves and DNA. *Proceedings of the National Academy of Sciences of the United States of America* 96(20), 11387–11392.

Jenkins, J. C., D. C. Chojacky, L. S. Heath, and R. A. Birdsey. 2003. National-scale biomass estimators for United States tree species, *Forest Science* 49, 12–35.

Kosugi, Y., S. Takanashi, H. Tanaka, et al. 2007. Evapotranspiration over a Japanese cypress forest. I. Eddy covariance fluxes and surface conductance characteristics for 3 years. *Journal of Hydrology* 337, 269–283.

Lee, X., J. Finnigan, and U. K. T. Paw. 2004. Coordinate systems and flux bias error. In, *Handbook of Micrometeorology A Guide for Surface Flux Measurement and Analysis*, ed. X. Lee, W. Massman, and B. Law, Kluwer Academic Publishers, pp. 33–66.

Liebethal, C., B. Huwe, and T. Foken. 2005. Sensitivity analysis for two ground heat flux calculation approaches. *Agricultural and Forest Meteorology* 132, 253–262.

Massman, W. J. 2000. A simple method for estimating frequency response corrections for eddy covariance systems. *Agricultural and Forest Meteorology* 104, 185–198.

Matthes-Sears, U. and D. W. Larson. 1995. Rooting characteristics of trees in rock—a study of *Thuja-occidentalis* on cliff faces. *International Journal of Plant Sciences* 156(5), 679–686.

McElrone, A. J., W. T. Pockman, J. Martinez-Vilalta, and R. B. Jackson. 2004. Variation in xylem structure and function in stems and roots of trees to 20 m depth. *New Phytologist* 163(3), 507–517.

McElrone, A. J., J. Bichler, W. T. Pockman, C. R. Linder, R. N. Addington, and R. B. Jackson. 2007. Aquaporin-mediated changes in hydraulic conductivity of deep tree roots accessed via caves. *Plant Cell and Environment* 30(11), 1411–1421.

Mitchell P. J., E. J. Veneklaas, H. Lambers, and S. S. O. Burgess. 2009. Partitioning of evapotranspiration in a semi-arid eucalypt woodland in south-western Australia. *Agricultural Meteorology* 149, 25–37.

Oliphant, A. J., C. S. B. Grimmond, H. N. Zutter, et al. 2004. Heat storage and energy balance fluxes for a temperate deciduous woodland. *Agricultural and Forest Meteorology* 126, 185–201.

ORNL DAAC (Oak Ridge National Laboratory Distributed Active Archive Center). 2009. MODIS subsetted land products. Collection 5. Available on-line (http://www.daac.ornl.gov/MODIS/modis.html) from ORNL DAAC, Oak Ridge, Tennessee. Accessed 2005–2007.

Owens, M. K. and M. C. Schreiber. 1992. Seasonal gas-exchange characteristics of two evergreen trees in a semi-arid environment. *Photosynthetica* 26(3), 389–398.

Paco, T. A., T. S. David, M. O. Henriques, et al. 2009. Evapotranspiration from a Mediterranean evergreen oak savannah: The role of trees and pasture. *Journal of Hydrology* 369, 98–106.

Potts, D. L., R. L. Scott, J. M. Cable, T. E. Huxman, and D. G. Williams. 2008. Sensitivity of mesquite shrubland CO_2 exchange to precipitation in contrasting landscape settings. *Ecology* 89(10), 2900–2910.

Querejeta, J. I., L. M. Egerton-Warburton, and M. F. Allen. 2007. Hydraulic lift may buffer rhizosphere hyphae against the negative effects of severe soil drying in a California Oak savanna. *Soil Biology & Biochemistry* 39(2), 409–417.

Querejeta, J. I., H. Estrada-Medina, M. F. Allen, J. Jiménez, and R. Ruenas. 2006. Utilization of bedrock water by *Brosimum alicastrum* trees growing on shallow soil atop limestone in a dry tropical climate. *Plant and Soil* 287 (1–2), 187–197.

Reichstein, M., E. Falge, D. Baldocchi, et al. 2005. On the separation of net ecosystem exchange into assimilation and ecosystem respiration: Review and improved algorithm. *Global Change Biology* 11, 1424–1439.

Rost, J. and H. Mayer. 2006. Comparative analysis of albedo and surface energy balance of a grassland site and an adjacent Scots pine forest. *Climate Research* 30, 227–237.

Schenk, H. J. 2008. Soil depth, plant rooting strategies and species' niches. *New Phytologist* 178(2), 223–225.

Schwinning, S. 2008. The water relations of two evergreen tree species in a karst savanna. *Oecologia* 158(3), 373–383.

Scott, R. L., E. A. Edwards, W. J. Shuttleworth, T. E. Huxman, C. Watts, and D. C. Goodrich. 2004. Interannual and seasonal variation in fluxes of water and carbon dioxide from a riparian woodland ecosystem. *Agricultural and Forest Meteorology* 122, 65–84.

Scott, R. L., T. E. Huxman, D. G. Williams, and D. C. Goodrich. 2006. Ecohydrological impacts of woody-plant encroachment: Seasonal patterns of water and carbon dioxide exchange within a semiarid riparian environment. *Global Change Biology* 12, 311–324.

Smith, S. D., R. K. Monson, and J. E. Anderson. 1997. *Physiological Ecology of North American Desert Plants*. Springer-Verlag, Berlin.

Steinwand, A. L., R. F. Harrington, and D. Or. 2006. Water balance for Great Basin phreatophytes derived from eddy covariance, soil water, and water table measurements. *Journal of Hydrology* 329, 595–605.

Stromberg, J. C., S. J. Lite, R. Marler, et al. 2007. Altered stream-flow regimes and invasive plant species: The Tamarix case. *Global Ecology and Biogeography* 16, 381–393.

Tenneson, M. 2008. When juniper and woody plants invade, water may retreat. *Science* 322, 1630–1631.

Twine, T. E., W. P. Kustas, J. M. Norman, et al. 2000. Correcting eddy-covariance flux underestimates over a grassland. *Agricultural and Forest Meteorology* 103, 279–300.

Van Auken, O. W. 2009. Causes and consequences of woody plant encroachment into western North American grasslands. *Journal of Environmental Management* 90, 2931–2942.

Walter, H. 1971. Natural savannahs as a transition to the arid zone. In: *Ecology of Tropical and Subtropical Vegetation*. Oliver and Boyd, Edinburgh, pp. 238–265.

Webb, E. K., G. I. Pearman, and R. Leuning. 1980. Correction of flux measurements for density effects due to heat and water vapour transfer. *Quarterly Journal of the Royal Meteorological Society* 106, 85–100.

Wilcox, B. P., M. K. Owens, R. W. Knight, and R. K. Lyons. 2005. Do woody plants affect streamflow on semiarid karst rangelands? *Ecological Applications* 15, 127–136.

Wilcox, B. P. and T. L. Thurow. 2006. Emerging issues in rangeland ecohydrology: Vegetation change and the water cycle. *Rangeland Ecology & Management* 59, 220–224.

Wilcox, B. P., P. I. Taucer, C. L. Munster, M. K. Owens, B. P. Mohanty, J. R. Sorenson, and R. Bazan. 2008. Subsurface stormflow is important in semiarid karst shrublands. *Geophysical Research Letters* 35 L10403, doi:10.1029/2008GL033696.

Wolf, A., N. Saliendra, K. Akshalov, D. A. Johnson, and E. Laca. 2008. Effects of different eddy covariance correction schemes on energy balance closure and comparisons with the modified Bowen ratio system. *Agricultural and Forest Meteorology* 148, 942–952.

7

The Dynamics of Energy, Water, and Carbon Fluxes in a Blue Oak (Quercus douglasii) Savanna in California

Dennis Baldocchi, Qi Chen, Xingyuan Chen, Siyan Ma, Gretchen Miller, Youngryel Ryu, Jingfeng Xiao, Rebecca Wenk, and John Battles

CONTENTS

> All his leaves
> Fall'n at length
> Look, he stands
> Trunk and bough
> Naked strength.

THE OAK, ALFRED LORD TENNYSON

Introduction

Oak trees and their savanna woodlands have played many important roles in the history, development, and ecology of California and the American West (Pavlik et al., 1991; Tyler et al., 2006). Starting with the Spanish Mission period till today, cattle have grazed the oak savanna, producing beef for our dinner, leather for our shoes, and tallow for our soap. In the mid-nineteenth century, 49'ers mined the oak savanna for gold and used its wood for cooking, heating, and building mine shafts. Today, the oak savanna provides many ecological services and benefits to the region. Oak savannas vegetate the watersheds of the many rivers stemming from the Sierra Nevada and Coastal mountain ranges. In doing so, they protect the soils of this hilly terrain and provide habitat for wildlife and acorns that have sustained the Native American population and wildlife for millennia. From a hydrologic perspective, runoff from these watersheds provides water for a large fraction of California's multibillion dollar agricultural economy and its population of more than 35 million inhabitants.

Despite their intrinsic value, California's oak savannas are suffering in today's world. Regeneration failure (Tyler et al., 2008), exotic diseases (e.g., sudden oak death syndrome) (Rizzo and Garbelotto, 2003), and invasive species (Seabloom et al., 2006) are among the key problems that this biome faces. Land use change is another issue threatening oak savannas. Many oak savannas are being converted to vineyards or being carved into ranchettes, thereby fragmenting the landscape (Huntsinger et al., 1997; Heaton and Merenlender, 2000). In the future, California's oak savannas may experience a shift or reduction in their native range with global warming (Kueppers et al., 2005; Loarie et al., 2008).

One way to assess the health of an ecosystem is to study the dynamics of its breathing (Baldocchi, 2008). For instance, investigators can quantify how an ecosystem's metabolism responds to environmental perturbations by measuring carbon, water vapor, and energy fluxes between an ecosystem and the atmosphere, quasi continuously and over multiple years. Then they can use these flux measurements to validate and parameterize ecosystem carbon and water cycle models, which, in turn, can be used to predict and diagnose ecosystem behavior and integrate fluxes in time and space (Baldocchi and Wilson, 2001; Moorcroft, 2006).

In this chapter, we present information on energy, water, and carbon fluxes of a Californian blue oak (*Quercus douglasii*) savanna. To acquire this information, we use a multifaceted approach that reflects a contemporary convergence in the fields of micrometeorology, ecosystem ecology, and eco-hydrology. Energy, water vapor, and carbon dioxide exchange data are derived from seven years of nearly continuous eddy covariance measurements over a blue oak (*Quercus douglasii*) savanna near Ione, California (latitude: 38.43 N; longitude: 120.96 W). The mass and energy flux data shown here are evaluated on daily, annual, and multi-year time scales. We interpret these fluxes with a suite of ancillary measurements, pertaining to climate, canopy structure, and physiological function at the tree and leaf scales and the underlying soil processes. Finally, we upscale the carbon dioxide flux information across the blue oak biome with the aid of remote sensing. The following overarching question is asked and answered in this chapter: How do factors associated with leaf, tree, canopy function and structure, climate, and soil conspire to enable these woodlands to sustain themselves in the extreme and variable Mediterranean-type climate of California?

The information produced in this chapter adds a new biophysical component to former ecological analyses of California oak savanna (Griffin, 1988; Barbour and Minnich, 2000). Flux measurements from savannas in the Mediterranean-type climate of California also provide information that is relevant for understanding the function and structure of European oak savannas, known as dehesa and montado (Joffre et al., 2007; Pereira et al., 2007), because they share many similarities. Finally, savanna woodlands are situated at one end of the North American climate-niche gradient, so these measurements provide critical information for upscaling carbon fluxes across the continent using a combination of flux data, remote sensing, and climate data (Xiao et al., 2008).

Biogeography

The Californian oak savanna consists of evergreen and deciduous species. This biome is situated on the coastal and Sierra Nevada mountain foothills that circumscribe the outlying border of the Great Central Valley. Its spatial extent occupies over 45,000 km² of land area, which represents about 11% of the state of California (Tyler et al., 2006).

The climate space of *Quercus douglasii* includes regions where the mean annual air temperature ranges between 14 and 16°C, and the mean annual precipitation ranges between 400 and 800 mm per year (Griffin, 1988; Major, 1988). Colder temperatures and wetter conditions favor conifers (Stephenson, 1998), and warmer temperatures and drier conditions favor annual grasslands and native perennial grasses (Heady, 1988; Major, 1988).

Although the mean climate conditions provide constraints on the biogeography of blue oak, the oak savanna must endure inter-annual and intra-annual extremes in precipitation. Annual precipitation, based on a 50-year record at Camp Pardee (latitude: 38.250°N; longitude, 120.867°W), ranges between 200 and 1200 mm per year, and it experiences a standard deviation of 196 mm per year; typically, surplus rainfall occurs during late winter and early spring of *El Niño* years and deficit rainfall occurs during *La Niña* years. Moreover, the rainfall regime across the California oak savanna region is highly seasonal (Major, 1988; Ma et al., 2007). In the late autumn and early winter (November through February), seasonal rainfall commences. This prompts the grasses to germinate and grow, whereas the deciduous oaks are leafless and dormant. In spring (March through May), the grasses and trees function simultaneously; the grasses experience exponential growth, then reproduce, and oak trees leaf out and achieve full canopy. Between June and November, the region is rainless, causing the grasses to die, whereas the trees assimilate carbon and transpire water.

Consequently, there is a tight anticorrelation between soil moisture and soil temperature and a positive correlation between soil temperature and intercepted solar radiation. Soil temperature ranges between 5 and 35°C, whereas soil moisture ranges between 40% and 5%, and mean midday visible sunlight ranges between 300 and 1500 μmol m^{-2} s^{-1}. The highest soil temperatures occur when the soil is driest and the sunshine is maximum.

Structure and Function

Canopy Architecture

From structural and functional perspectives, Mediterranean tree-grass savanna ecosystems are distinct from many other ecosystems. These systems are (1) heterogeneous in space; (2) they consist of a mix of contrasting plant growth forms (herbaceous and woody, deciduous and evergreen, and annual and perennial); (3) they possess physiological and architectural features to endure and tolerate extreme soil water deficits, which enable them to survive harsh environmental conditions; (4) they often reside on undulating topography; and (5) they are grazed regularly and burn periodically.

We have quantified the structural features of the oak woodland with a variety of direct and remote sensing methods, including airborne LIDAR, IKONOS, and AVIRIS images and ground-based measurements of light transmission (Chen et al., 2006, 2008a). The openness and heterogeneity of the canopy is best visualized with a two-dimensional plot of tree location and crown diameter (Figure 7.1). On average, the trees are about 9.4 ± 4.3 m tall; the oak and pine trees cover about 63% of the landscape and constitute an LAI of 0.7. Additional tree demographic and structural information is listed in Table 7.1.

Leaf and Tree Physiology

To understand how leaves and plants modulate carbon and water fluxes, we made frequent measurements of leaf photosynthesis, stomatal conductance, pre-dawn water potential, and tree transpiration. Leaf gas exchange measurements, based on an environmentally controlled cuvette system, were made over the course of a growing season to deduce information on photosynthetic capacity and stomatal conductance model parameters (Xu and Baldocchi, 2003). These oak leaves achieve extremely high values of maximum carboxylation velocity, V_{cmax} (110–120 μmol m^{-2} s^{-1}) during the early period of their growing season, when moisture is ample (Xu and Baldocchi, 2003); V_{cmax} is the key model parameter of the Farquhar–von Caemmerer–Berry photosynthesis model (Farquhar et al., 1980). Then V_{cmax} and leaf nitrogen drop dramatically, and in concert, as soil moisture is depleted.

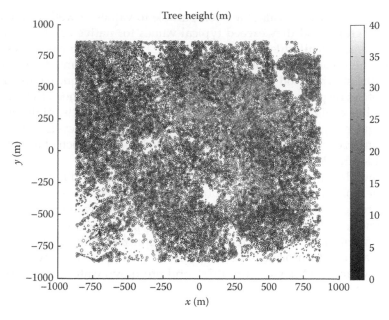

FIGURE 7.1
(See color insert following page 320.) A map of tree location, height, and crown size of a blue oak savanna, near Ione, CA. The horizontal dimensions of the figure are 1000 m × 1000 m, and the flux measurement tower is at the center. Canopy structure was quantified using an airborne LIDAR during the summer of 2009.

TABLE 7.1

Properties of Individual Trees and Oak Woodland Stand for a Seasonally Grazed, Blue Oak Woodland, Near Ione, CA

Metric	Mean	Std
Crown area (m²)	39.23	41.2
Crown radius (m)	3.18	1.54
Tree height (m)	0.41	4.33
Trunk height (m)	1.75	1.35
Crown height (m)	7.66	4.56
Basal area (m²)	0.0744	0.084
Stem volume (m³)	0.734	1.23
Stem biomass	(kg)	440.43	739.6
Leaf area (m²)	38.32	64.36
LAI	0.706	0.408
Stems per hectare	144	
Fraction oaks (%)	50.1	
Fraction pine (%)	13.6	
Fraction ground (%)	36.3	

The maximum V_{cmax} values are comparable to values of well-watered and fertilized crops, and they exceed typical values for native oak trees in temperate biomes (Wullschleger, 1993; Wilson et al., 2001). Blue oak leaves must develop high-capacity, light harvesting, and carbon assimilation systems during the short period when soil water is available to acquire enough carbon to offset respiratory costs for the rest of the year. They need the mechanics to achieve high rates of photosynthesis. In other words, high levels of V_{cmax} require high levels of Rubisco, which requires high levels of nitrogen, which, in turn, requires the production of thick leaves (Reich et al., 1997; Xu and Baldocchi, 2003).

Summer drought in California is particularly severe and is more extreme than the conditions that vegetation may face in other Mediterranean-climate regions. For example, we measured pre-dawn water potentials values as low as −6.0 MPa (Xu and Baldocchi, 2003), far below the conventional wilting point of −1.5 MPa (Sperry, 2000). In contrast, oaks growing on the coastal range of California experience less negative pre-dawn water potentials, down to −4.0 MPa (Griffin, 1973), and pre-dawn water potentials measured on Mediterranean oaks tend to be in the order of −3.0 MPa (Joffre et al., 2007).

There are many issues regarding the effects of moderate and severe drought on photosynthesis (Cornic, 2000). For example, does stomata adjust to soil water deficits although photosynthetic capacity remains unchanged? Or is there a down-regulation in photosynthetic capacity with soil moisture deficits? Some researchers argue that modest soil water deficits induce stomatal closure, which, in turn, reduces the internal CO_2 concentration and subsequently down-regulates photosynthesis (Cornic, 2000); this physiologically induced reduction in photosynthesis does not imply a change in photosynthetic capacity. An opposing line of evidence, derived from studies on Mediterranean oaks in Europe and California, suggests that oaks exposed to chronic soil water deficits down-regulate their photosynthetic capacity, decrease their mesophyll conductance, and dissipate excess photon flux energy via the xanthophyll cycle (Chaves et al., 2002; Flexas et al., 2007). Our data show that seasonal changes in V_{cmax} are negatively correlated with pre-dawn water potential, suggesting a down-regulation of photosynthetic capacity with drought. This drought-mediated reduction in V_{cmax} comes in concert with progressive stomatal closure. However, the drought-induced reduction in photosynthesis is also associated with a progressive reduction in the ratio between internal and atmospheric CO_2 (Xu and Baldocchi, 2003), which is consistent with the conclusions of Cornic (2000). Further, we find that stomatal conductance is a constant fraction of the product between leaf photosynthesis, relative humidity, and the reciprocal of CO_2 concentration (Collatz et al., 1991) for both wet and dry soil moisture conditions.

Although water is a precious commodity to blue oaks during the summer, there are some inefficiencies and leaks in the system. Sapflow occurs at night and the magnitude of nocturnal transpiration increases with vapor pressure deficit, indicating that stomatal closure is not complete at night (Fisher et al., 2007).

Canopy Scale: Energy and Water Fluxes

Solar radiation is the main energy source to the ecosystem and the microclimate. On an annual basis, the site receives about 6.59 ± 0.14 GJ m^{-2} year^{-1} of global solar radiation (R_g) (Table 7.2). Year-to-year variation in R_g ranged between 6.46 and 6.70 MJ m^{-2} year^{-1}. This value is consistent with historical measurements at two climate stations in sunny northern California (Major, 1988). This amount of solar energy has the potential to produce about 3600 gC m^{-2} year^{-1} of photosynthesis, if the vegetation had a year-round growing season, ample soil moisture, and a photosynthetic efficiency of 2% (Ruimy et al., 1996).

Net radiation is defined as the sum of the gains and losses of incoming shortwave and longwave radiation and is a function of albedo and emissivity. On an annual basis, 3.17 ± 0.098 GJ m^{-2} of net radiation is available to the woodland (Table 7.2). In comparison, the neighboring annual grassland net radiation budget ranged between 2.1 and 2.3 GJ m^{-2} during 2001 and 2002 (Baldocchi et al., 2004; Ryu et al., 2008). About 1 GJ m^{-2} more energy is available to the oak savanna, compared with the neighboring grassland, as the savanna is optically darker and the cooler transpiring leaves cause it to radiate less longwave energy.

Theoretically, net radiation is partitioned into latent- and sensible-heat exchange and conductive heat exchange into the soil. On an annual basis, sensible heat flux of the savanna woodland equals 1.87 ± 0.22 GJ m^{-2}, almost double that of latent heat exchange (0.96 ± 0.13 GJ m^{-2}). Soil heat flux is close to zero on an annual basis.

Across the sunny Mediterranean climate of California, potential evaporation typically exceeds annual precipitation (Hidalgo et al., 2005). At our field

TABLE 7.2

Annual Sums of Energy Flux Densities Over an Oak Savanna in California

Variable	Location	Year						
		2001–02	2002–03	2003–04	2004–05	2005–06	2006–07	Avg
R_g (GJ m^{-2})	Above canopy	6.70	6.48	6.69	6.46	6.47	6.76	6.59
R_n (GJ m^{-2})	Above canopy	3.25	3.22	3.17	3.08	3.04	3.29	3.17
λE (GJ m^{-2})	Above canopy	0.93	1.02	0.79	1.15	1.02	0.83	0.96
H (GJ m^{-2})	Above canopy	2.06	1.87	2.11	1.62	1.66	1.91	1.87
G (GJ m^{-2})	Below canopy	0.03	0.04	0.06	0.02	0.03	0.03	0.04

Note: R_g is solar radiation, R_n is net radiation, λE is latent heat flux, H is sensible heat flux, and G is soil heat flux.

TABLE 7.3

Annual Sums of Water Balance Components

	Year						
Variable	2001–02	2002–03	2003–04	2004–05	2005–06	2006–07	Avg
Ppt (mm year^{-1})	493	509	465	628	884	411	565
E_p (mm year^{-1})	1412	1421	1427	1394	1411	1510	1429
E_t (mm year^{-1})	381	416	320	469	416	339	390
E_o (mm year^{-1})	243	254	197	276	222	189	230
E_u (mm year^{-1})	137	162	123	193	194	149	160
E_u/E_t (%)	36	39	39	41	47	44	41

Note: Ppt is annual precipitation, E_p is potential evaporation, E_t is total evaporation, E_o is overstory evaporation, and E_u is understory evaporation.

site, annual potential evaporation was 1429 mm (Table 7.3). This sum vastly exceeds mean annual actual evaporation (390 mm per year) and rainfall (565 mm per year) and is about 400 mm greater than potential evaporation from an annual grassland (Ryu et al., 2008). On the other hand, these eddy covariance measurements of actual evaporation compare well with a 17-year water balance study performed in an oak woodland watershed, northeast of Sacramento, California; actual evaporation was estimated to equal 368 ± 69 mm per year (Lewis et al., 2000).

Some eco-hydrological models compute seasonal trends in actual evaporation as a function of equilibrium evaporation (E_{eq}); $E_{Equation}$ is a function of available energy and these potential evaporation rates are down-regulated in a linear fashion with regard to decreasing volumetric soil moisture (Rodriguez-Iturbe, 2000; Chen et al., 2008b). We quantified the functional relationship between evaporation and volumetric soil moisture applying the Markov Chain Monte Carlo method to 3 years of evaporation data (Chen et al., 2008b). Under well-watered conditions, the ratio between actual and equilibrium evaporation (E_a/E_{eq}) is about 0.65. This $E_a/E_{Equation}$ value is much lower than the typical values for wet, irrigated, and extensive vegetation, 1.26, also known as the Priestley–Taylor coefficient (Priestley and Taylor, 1972). The ratio, E_a/E_{eq}, declines when the root-weighted volumetric water content drops below 0.12 cm^3 cm^{-3} and the ratio approaches zero as volumetric soil moisture approaches 0.06 cm^3 cm^{-3}.

A growing number of Mediterranean savanna studies are indicating the importance of deep water sources to sustain oak trees during the long, hot dry summer (Baldocchi et al., 2004; Joffre et al., 2007). To address the hypothesis that oaks tap deeper sources of water, we draw on other data— the change in the soil moisture after the rains ceased and the grass died. Between day 150 and 309, our soil moisture budget detected a loss of 48 mm of water from the upper 0.60 m layer of the soil profile. In contrast, our eddy covariance measurements of evaporation indicate that 114 mm of water evaporated from the landscape during this period and of this total, 20 mm of water was lost from the dry grass layer in the understory. Based on this water budget, we concluded that 66 mm of water, or 57% of the total, came from other sources. We originally surmised that a significant fraction of moisture probably came from below the fractured shale layer, supporting the measurements of Lewis and Burgy (1964). More recently, we established three wells at the site and have monitored their changes in a water table. These new measurements support the hypothesis that oak trees are tapping deep sources of water during the summer. Diurnal fluctuations in the depth to the groundwater table are observed during the oak active season and disappear when senescence of the oaks is complete. The depth to the water table increases during the daylight hours, when the trees are transpiring, and decreases during the nighttime, indicating recharge of the aquifer. We found that groundwater uptake (ET_g) of data for seven days in July 2007 was between 0.25 and 0.45 mm d^{-1} on using a method developed by Vincke and Thiry (2008). In contrast, the eddy covariance system measured total daily ET at rates between 0.50 and 0.84 mm d^{-1}, indicating that approximately 55% of transpired water came from deep sources. Similar patterns have been observed in other dryland or riparian ecosystems (Loheide et al., 2005).

Canopy Scale: Carbon Fluxes

The Blue oak woodlands in California are small, net carbon sinks on an annual basis (-92 ± 48 gC m^{-2} year^{-1}); for perspective, the mean net carbon exchange over six years, derived from more than 500 site years of measurements across the FLUXNET network, is -183 gC m^{-2} year^{-1} (Baldocchi, 2008).

This relatively small net carbon flux is the residual between two relatively large and offsetting carbon fluxes. On average, gross primary productivity creates a carbon sink of 1031 gC m^{-2} year^{-1}, and ecosystem respiration produces a carbon source of 939 gC m^{-2} year^{-1} (Ma et al., 2007) (Table 7.4). Consequently, inter-annual variability in net carbon exchange is explained by a tight relationship with spring-time cumulative rainfall, which strongly modulates cumulative photosynthesis and respiration.

Net (*NEE*) and gross (*GPP*) carbon update and ecosystem respiration (R_{eco}) experience pronounced seasonal variation (Figure 7.2). Peak carbon fluxes occur during the spring, shortly after leaf expansion when soil moisture is ample; *GPP*, R_{eco}, and *NEE* peak at rates near 6, 4, and -3 gC m^{-2} d^{-1}, respectively. Over a year, this ecosystem operates, technically, as an evergreen ecosystem. The grass is green and photosynthesizing during the dormant winter period; and the trees are photosynthesizing during the dry summer period when the grass is dead.

TABLE 7.4

Annual Net Ecosystem Exchange of CO_2 (*NEE*, gC m^{-2} year^{-1}), Gross Primary Productivity (*GPP*), and Ecosystem Respiration (R_{eco}) and Uncertainties (in parentheses)

Hydrological Year	NEE (gC m^{-2} year^{-1})		GPP (gC m^{-2} year^{-1})		R_{eco} (gC m^{-2} year^{-1})	
2001–2002	−144	(±50)	888	(±75)	744	(±48)
2002–2003	−116	(±50)	1091	(±65)	975	(±49)
2003–2004	−52	(±62)	899	(±56)	847	(±41)
2004–2005	−143	(±47)	1360	(±66)	1217	(±56)
2005–2006	−35	(±44)	1113	(±53)	1078	(±52)
2006–2007	−78	(±56)	904	(±65)	773	(±44)
2007–2008	−79	(±48)	797	(±46)	718	(±42)
Mean ± STD	−92	±43	1007	±193	907	±189

Source: Data from Ma, S. et al. 2007. *Agricultural and Forest Meteorology* 147, 151–171.

Note: Hydrological year starts on October 23 (day 296) and ends on October 22 (day 295) of the next year.

Episodic rain events during the summer period can cause huge pulses in soil respiration (Xu and Baldocchi, 2004; Baldocchi et al., 2006). Two mechanisms are possible for producing enhanced respiration rates after summer rainfall events. One is a physical displacement of soil, air, and CO_2 by the

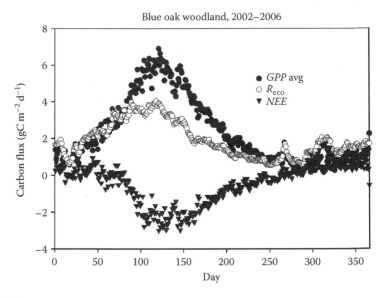

FIGURE 7.2

Seasonal variation in gross primary productivity, ecosystem respiration, and net ecosystem exchange. These data are presented for daily integrals averaged over the period between 2002 and 2006.

downward moving front of water in the soil. However, this effect is short lived and the volume of air in the soil profile is relatively small. The other effect is attributed to a rapid activation of heterotrophic respiration (Xu et al., 2004; Baldocchi et al., 2006). The size of these respiration pulses diminish with successive rain events as the size of the labile carbon pool diminishes.

We recently produced an independent carbon budget of the stand by measuring litterfall, fine root production, and stem and root growth (Figure 7.3). The stand maintains 37 Mg C ha^{-1} in above ground biomass and 12 Mg C ha^{-1} in below ground biomass. Net primary productivity of the oak woodland is defined as the sum of the differences in carbon gains and losses. Above ground net primary productivity of the oak woodland equals 235 gC m^{-2} year^{-1}. This sum is comprised of bole increment growth (44 gC m^{-2} year^{-1}), litterfall (37 gC m^{-2} year^{-1}), fine root production (59 gC m^{-2} year^{-1}), grass (84 gC m^{-2} year^{-1}), and root net primary production (59 gC m^{-2} year^{-1}). Overall, this biometric carbon budget compares well with an independent estimate of *NPP* (283 gC m^{-2} year^{-1}), derived from our eddy covariance measurements (Ma et al., 2007).

Using a combination of flux data from the entire AmeriFlux network, environmental drivers from MODIS sensor on the TERRA satellite, and regression tree analysis (Xiao et al., 2008), we produced maps of *GPP* (Figure 7.4a) and *NEE* (Figure 7.4b) for the blue oak woodland domain of California. The area-averaged fluxes of *GPP* and *NEE* were 932 and −150 gC m^{-2} year^{-1}, respectively. When summed up on an area-basis, the net and gross carbon fluxes equaled −8.6 and 53.8 TgC year^{-1}, respectively.

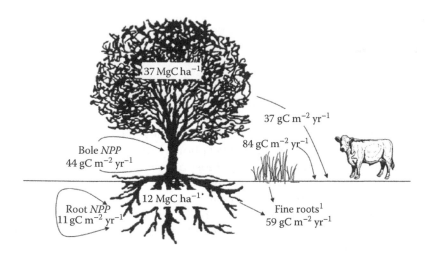

FIGURE 7.3
Summary of carbon budget for oak woodland near Ione, CA, for the 2007–2008 hydrological growing season. Boxes represent pools (MgC ha^{-1}); arrows represent fluxes (gCm^{-2} year^{-1}).

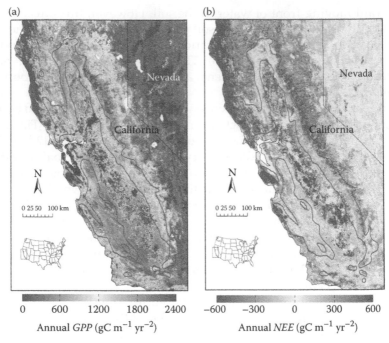

(a)
(b)

0   600   1200   1800   2400   −600   −300   0   300   600
Annual *GPP* (gC m^{-1} yr^{-2})   Annual *NEE* (gC m^{-1} yr^{-2})

FIGURE 7.4
(See color insert following page 320.) Spatial map of net ecosystem carbon exchange for the
blue oak biome: (a) is for annual gross primary productivity (*GPP*) and (b) is for net annual
ecosystem carbon exchange (*NEE*). (Adapted from Xiao, J. et al. 2008. *Agricultural and Forest
Meteorology* 148, 1827–1847; and derived using data are from the MODIS sensor and the
AmeriFlux network.)

Discussion

To survive and thrive, the oak trees in the California savanna must coordinate how much water they use to gain enough carbon to offset respiratory
needs and evaporative demand. This task is achieved by a series of multiple
constraints (Figure 7.5). First, stand-level photosynthesis must exceed ecosystem respiration. However, the act of photosynthesis comes with an additional
cost, water loss via transpiration and evaporation. From a biophysical stand-
point, the ecosystem cannot evaporate more water than is available (through
precipitation minus runoff and infiltration). To balance water supply and
use, individual plants and assemblages of plants can exploit a variety of
plausible mechanisms. On short time scales (hours to season), physiological
adjustments, such as down-regulation by stomata closure (Cornic, 2000) or
by changes in hydraulic conductance (Sperry, 2000) can be invoked to reduce
transpiration. However, this drought-induced stomatal closure inhibits

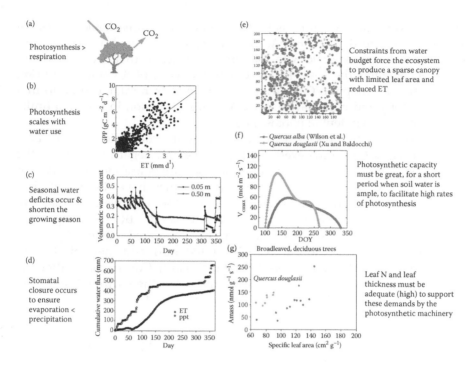

FIGURE 7.5
(See color insert following page 320.) An eco-hydrological explanation for the structure and function of a blue oak savanna and its sustenance. For the oak savannas to survive in highly seasonal wet/dry, cool/hot climate, they must comply with several linked constraints. (a) First, annual photosynthesis must exceed ecosystem respiration; (b) the oaks experience difficulty in achieving high rates of photosynthesis during the growing season, because the lack of summer rains causes volumetric soil moisture to drop below 5% and induces physiological stress; (c) further, the act of photosynthesis comes with a cost—evaporation; (d) consequently, annual evaporation must be constrained such that it does not exceed annual precipitation, or the trees invest in deep roots that tap ground water; (e) the landscape can reduce its total evaporation by establishing a partial canopy; (f) it can assimilate enough carbon if it adjusts and maximizes its photosynthetic capacity during the short spring wet period; and (g) high rates of photosynthesis are not free and come with producing thick leaves, rich in nitrogen and Rubisco.

photosynthesis. Thus, the plant must develop leaves with a high photosynthetic capacity in order to assimilate enough carbon during the short hydrological growing season. This task is accomplished by producing leaves with high nitrogen content (Xu and Baldocchi, 2003; Ma et al., 2007). On longer timescales (years to millennia), morphological and ecological adaptations or modifications can lead to reduced transpiration. For instance, a plant can (1) develop smaller leaves that convect heat more efficiently (Taylor, 1975; Baldocchi and Xu, 2007); (2) produce more reflective leaves that reduce its solar heat load (Gates, 1980); (3) adopt drought-deciduous behavior so it can drop leaves when soil moisture deficits are severe (Jolly and Running,

2004); or (4) produce deep roots that tap otherwise unreachable sources of water (Lewis and Burgy, 1964). At the landscape scale, the LAI that the canopy establishes must scale with the amount of water that is available and nutrient content of the leaves (Baldocchi and Meyers, 1998; Eamus and Prior, 2001). An oak canopy in California with 500 mm of rainfall cannot form a closed canopy and must have a low LAI. This limitation in leaf area limits light capture and potential photosynthesis.

Based on the data presented so far, we draw the conclusion that the blue oak savanna of California exerts both positive and negative effects on the climate system. On the positive side, oak savannas are modest sinks for carbon, taking up about 100 gC m^{-2} year^{-1}, so they have the potential to mitigate anthropogenic carbon emissions. On the negative side, they are optically darker than surrounding open grasslands. By absorbing more energy and by being aerodynamically rougher, compared with nearby grasslands, they evaporate more water (about 100 mm), and they inject more sensible heat into the atmosphere (0.5 GJ m^{-2} year^{-1}). Consequently, their air temperature is about 0.84 C degrees warmer than the surrounding grasslands (Baldocchi et al., 2004). In California, water yield is a critical ecosystem service, so the negative effects of extra water use by the oak woodlands may outweigh their positive effect on the carbon budget. However, we should not discount the roles of oak woodlands for grazing, wildlife habitat, biodiversity, recreation, and aesthetic value.

Acknowledgments

This research was supported in part by the Office of Science (BER), U.S. Department of Energy, Grant No. DE-FG02-03ER63638, and through the Western Regional Center of the National Institute for Global Environmental Change under Cooperative Agreement No. DE-FC02-03ER63613. Other sources of support included the Kearney Soil Science Foundation, the National Science Foundation, and the Californian Agricultural Experiment Station. We also express thanks to Ted Hehn, our long-term lab technician, and Russell Tonzi for access to his ranch.

References

Baldocchi, D. 2008. TURNER REVIEW No. 15. "Breathing" of the terrestrial biosphere: Lessons learned from a global network of carbon dioxide flux measurement systems. *Australian Journal of Botany* 56, 1–26.

Baldocchi, D. D. and T. Meyers. 1998. On using eco-physiological, micrometeorological and biogeochemical theory to evaluate carbon dioxide, water vapor and trace gas fluxes over vegetation: A perspective. *Agricultural and Forest Meteorology* 90, 1–25.

Baldocchi, D. D., J. Tang, and L. Xu. 2006. How switches and lags in biophysical regulators affect spatio-temporal variation of soil respiration in an oak–grass Savanna. *Journal Geophysical Research, Biogeosciences* 111, G02008, doi:10.1029/2005JG000063.

Baldocchi, D. D. and L. Xu. 2007. What limits evaporation from Mediterranean oak woodlands—The supply of moisture in the soil, physiological control by plants or the demand by the atmosphere? *Advances in Water Resources* 30, 2113–2122.

Baldocchi, D. D., L. Xu, and N. Kiang. 2004. How plant functional-type, weather, seasonal drought, and soil physical properties alter water and energy fluxes of an oak-grass savanna and an annual grassland. *Agricultural and Forest Meteorology* 123, 13–39.

Baldocchi, D. D. and K. B. Wilson. 2001. Modeling CO_2 and water vapor exchange of a temperate broadleaved forest across hourly to decadal time scales. *Ecological Modeling* 142, 155–184.

Barbour, M. G. and B. Minnich. 2000. California upland forests and woodlands. In *North American Terrestrial Vegetation*, 2nd Edition, eds. M. Barbour and W. Billings. Cambridge University Press, Cambridge, UK.

Chaves, M. M., J. S. Pereira, J. Maroco, et al. 2002. How plants cope with water stress in the field. Photosynthesis and growth. *Annals of Botany* 89, 907–916.

Chen, Q., D. Baldocchi, P. Gong, and M. Kelly. 2006. Isolating individual trees in a savanna woodland using small footprint lidar data. *Photogrammetric Engineering and Remote Sensing* 72, 923–932.

Chen, Q., D. Baldocchi, P. Gong, and T. Dawson. 2008. Modeling radiation and photosynthesis of a heterogeneous savanna woodland landscape with a hierarchy of model complexities. *Agricultural and Forest Meteorology* 148, 1005–1020.

Chen, X. Y., Y. Rubin, S. Y. Ma, and D. Baldocchi. 2008. Observations and stochastic modeling of soil moisture control on evapotranspiration in a Californian oak savanna. *Water Resources Research* 44, W08409, doi:10.1029/2007WR006646.

Collatz, G. James, J. Timothy Ball, C. Grivet, and Joseph A. Berry. 1991. Physiological and environmental regulation of stomatal conductance, photosynthesis and transpiration: a model that includes a laminar boundary layer. *Agricultural and Forest Meteorology* 54, 107–136.

Cornic, G. 2000. Drought stress inhibits photosynthesis by decreasing stomatal aperture—not by affecting ATP synthesis. *Trends in Plant Science* 5, 187–188.

Eamus, D. and L. Prior 2001. Ecophysiology of trees of seasonally dry tropics: Comparisons among phenologies. *Advances in Ecological Research* 32, 113–197.

Farquhar, G. D., S. V. Caemmerer, and J. A. Berry. 1980. A biochemical-model of photosynthetic CO_2 assimilation in leaves of C-3 species. *Planta* 149, 78–90.

Fisher, J. B., D. D. Baldocchi, L. Misson, T. E. Dawson, and A. H. Goldstein. 2007. What the towers don't see at night: Nocturnal sap flow in trees and shrubs at two AmeriFlux sites in California. *Tree Physiology* 27, 597–610.

Flexas, J., M. Ribas-Carbo, A. Diaz-Espejo, J. Galmes, and H. Medrano. 2007. Mesophyll conductance to CO_2: Current knowledge and future prospects. *Plant, Cell & Environment* 31, 602–621.

Gates, D. M. 1980. *Biophysical Ecology*. Dover, Mineola, NY.

Griffin, J. R. 1973. Xylem sap tension in three woodland oaks of central California. *Ecology* 54, 152–159.

Griffin, J. R. 1988. Oak woodland. In *Terrestrial Vegetation of California*, eds. M. G. Barbour and J. Major. California Native Plant Society, John Wiley, New York.

Heady, H. F. 1988. Valley grassland. In *Terrestrial Vegetation of California*, eds. M. G. Barbour and J. Major. California Native Plant Society.

Heaton, E. and A. M. Merenlender. 2000. Modeling vineyard expansion, potential habitat fragmentation. *California Agriculture* 54, 12–20.

Hidalgo, H. G, D. R. Cayan, and M. D. Dettinger. 2005. Sources of variability of evapotranspiration in California. *Journal of Hydrometeorology* 6, 3–19.

Huntsinger, L., L. Buttolph, and P. Hopkinson. 1997. Ownership and management changes on California hardwood rangelands: 1985 to 1992. *Journal of Range Management* 50, 423–430.

Joffre, R., S. Rambal, and C. Damesin. 2007. Functional Attributes in Mediterranean-type Ecosystems. In *Functional Plant Ecology*, eds. F. I. Pugnaire, pp. 286–312, CRC Press/Taylor and Francis, Boca Raton, FL, 726pp.

Jolly, W. M. and S. W. Running. 2004. Effects of precipitation and soil water potential on drought deciduous phenology in the Kalahari. *Global Change Biology* 10, 303–308.

Kueppers, L. M., M. A. Snyder, L. C. Sloan, E. S. Zavaleta, and B. Fulfrost. 2005. Modeled regional climate change and California endemic oak ranges. *Proceeedings of the National Academy of Sciences* 102, 16281–16286.

Lewis, D. C. and R. H. Burgy 1964. Relationship between oak tree roots + groundwater in fractured rock as determined by tritium tracing. *Journal of Geophysical Research* 69, 2579.

Lewis, D., M. J. Singer, R. A. Dahlgren, and K. W. Tate. 2000. Hydrology in a California oak woodland watershed: A 17-year study. *Journal of Hydrology* 240, 106–117.

Loarie, S. R., B. E. Carter, K. Hayhoe, et al. 2008. Climate change and the future of Califorina's endemic flora. *PLOS one* 3, e2502.

Loheide, S. P., J. J. Butler, and S. M. Gorelick 2005. Estimation of groundwater consumption by phreatophytes using diurnal water table fluctuations: A saturated-unsaturated flow assessment. *Water Resources Research* 41, W07030, doi:10.1029/2005WR003942.

Ma, S., D. D. Baldocchi, L. Xu, and T. Hehn 2007. Inter-annual variability in carbon dioxide exchange of an oak/grass savanna and open grassland in California. *Agricultural and Forest Meteorology* 147, 151–171.

Major, J. 1988. California climate in relation to vegetation. In *Terrestrial Vegetation of California*, eds. M. Barbour and J. Major. Native Plant Society of California, Sacramento, CA.

Moorcroft, P. R. 2006. How close are we to a predictive science of the biosphere? *Trends in Ecology & Evolution Twenty years of TREE—Part 2* 21, 400–407.

Pavlik, B. M., P. C. Muick, S. G. Johnson, and M. Popper. 1991. *Oaks of California*. Cachuma Press, Los Olivos, CA.

Pereira, J. S., J. Mateus, L. Aires, G. Pita, C. Pio, J. David, V. Andrade, J Banza, T. S. David, T. A. Paco, and A. Rodrigues. 2007. Net ecosystem carbon exchange in three contrasting Mediterranean ecosystems—the effect of drought. *Biogeosciences* 4, 791–802.

Priestley, C. H. B. and R. J. Taylor. 1972. On the assessment of surface heat flux and evaporation using large-scale parameters. *Monthly Weather Review* 100, 81–92.

Reich, B. Peter, Michael B. Walters, and David S. Ellsworth. 1997. From tropics to tundra: Global convergence in plant functioning. *Proceedings of the National Academy of Sciences* 94, 13730–13734.

Rizzo, D. M. and M. Garbelotto. 2003. Sudden oak death: endangering California and Oregon forest ecosystems. *Frontiers in Ecology and the Environment* 1, 197–204.

Rodriguez-Iturbe, I. 2000. Ecohydrology: A hydrologic perspective of climate–soil–vegetation dynamics. *Water Resources Research* 36, 3–9.

Ruimy, A., G. Dedieu, and B. Saugier. 1996. TURC: A diagnostic model of continental gross primary productivity and net primary productivity. *Global Biogeochemical Cycles* 10, 269–285.

Ryu, Y., D. D. Baldocchi, S. Ma, and T. Hehn. 2008. Interannual variability of evapotranspiration and energy exchange over an annual grassland in California. *Journal of Geophysical Research* 113, D09104, doi:10.1029/2007JD009263.

Seabloom, E. W., J. W. Williams, D. Slayback, D. M. Stoms, J. H. Viers, and A. P. Dobson. 2006. Human impacts, plant invasion, and imperiled, plant species in California. *Ecological Applications* 16, 1338–1350.

Sperry, J. S. 2000. Hydraulic constraints on plant gas exchange. *Agricultural and Forest Meteorology* 104, 13–23.

Stephenson, N. L. 1998. Actual evapotranspiration and deficit: Biologically meaningful correlates of vegetation distribution across spatial scales. *Journal of Biogeography* 25, 855–870.

Taylor, S. E. 1975. Optimal leaf form. In *Perspectives of Biophysical Ecology*, eds. D. M. Gates and R. B. Schmerl. Springer-Verlag, New York.

Tyler, C. M., F. W. Davis, and B. E. Mahall. 2008. The relative importance of factors affecting age-specific seedling survival of two co-occurring oak species in southern California. *Forest Ecology and Management* 255, 3063–3074.

Tyler, C. M., B. Kuhn, and F. W. Davis. 2006. Demography and recruitment limitations of three oak species in California. *Quarterly Review of Biology* 81, 127–152.

Vincke, C. and Y. Thiry. 2008. Water table is a relevant source for water uptake by a Scots pine (*Pinus sylvestris* L.) stand: Evidences from continuous evapotranspiration and water table monitoring. *Agricultural and Forest Meteorology* 148, 1419–1432.

Wilson, K. B., D. D. Baldocchi, and P. J. Hanson. 2001. Leaf age affects the seasonal pattern of photosynthetic capacity and net ecosystem exchange of carbon in a deciduous forest. *Plant, Cell and Environment* 24, 571–583.

Wullschleger, S. D. 1993. Biochemical limitations to carbon assimilation in C3 plants–a retrospective analysis of the A/Ci curves from 109 species. *Journal of Experimental Botany* 44, 907–920.

Xiao, J. Z. Qianlai, D. D. Baldocchi, et al. 2008. Estimation of net ecosystem carbon exchange for the conterminous United States by combining MODIS and AmeriFlux data. *Agricultural and Forest Meteorology* 148, 1827–1847.

Xu, L. and D. D. Baldocchi. 2003. Seasonal trend of photosynthetic parameters and stomatal conductance of blue oak (*Quercus douglasii*) under prolonged summer drought and high temperature. *Tree Physiology* 23, 865–877.

Xu, L. and D. D. Baldocchi. 2004. Seasonal variation in carbon dioxide exchange over a Mediterranean annual grassland in California. *Agricultural and Forest Meteorology* 123, 79–96.

Xu, L., D. D. Baldocchi, and J. Tang. 2004. How soil moisture, rain pulses and growth alter the response of ecosystem respiration to temperature. *Global Biogeochemical Cycles* 18, DOI 10.1029/2004GB002281.

Section III

Remote Sensing of Biophysical and Biochemical Characteristics in Savannas

8

Quantifying Carbon in Savannas: The Role of Active Sensors in Measurements of Tree Structure and Biomass

Richard M. Lucas, Alex C. Lee, John Armston, Joao M. B. Carreiras,
Karin M. Viergever, Peter Bunting, Daniel Clewley, Mahta Moghaddam,
Paul Siqueira, and Iain Woodhouse

CONTENTS

Introduction

The Distribution of Carbon in Savannas

Savannas occupy approximately 20% (depending on definition) of the global land surface and occur largely in the tropics and subtropics. Most occupy seasonal environments where average monthly temperatures exceed 18°C and annual precipitation is relative low (<600 mm) and seasonal. Savannas are structurally diverse ecosystems in terms of the varying proportions, spatial distributions and architectures of tree, shrub, and grass cover. Where savannas are wooded, tree density varies primarily with mean annual rainfall; and these ecosystems merge with closed canopy forest and semidesert in moister and drier regions, respectively.

The distribution, state, and dynamics of savannas are influenced primarily by precipitation, fire (frequency and intensity), and herbivore grazing. Pressures from human activities are also modifying the state and functioning of savannas, and associated losses of vegetation (and hence carbon) may be complete, partial (e.g., through selective thinning, fuelwood extraction), or attributable to indirect intervention (e.g., grazing). Regeneration after clearance and fires, woody thickening, and rainforest encroachment are common processes leading to uptake of carbon. Since these events and processes may affect vast areas, the impacts on regional to global carbon cycles may be substantial and potentially exacerbated by climate change. Knowledge of the amount, distribution, and turnover of carbon in savannas is, therefore, becoming increasingly important.

Remote Sensing of Savannas

To quantify carbon in woody vegetation (for which biomass is the common surrogate), both passive (primarily optical) and active remote sensing data have been exploited. Optical sensors are sensitive only to the above ground and two-dimensional extent and structure of vegetation and have, therefore, been used mainly for classifying land cover (e.g., Stuart et al., 2006). Quantitative retrieval of biophysical attributes (e.g., canopy cover) has proved challenging, largely due to variability in tree crown size and structure, location relative to neighbors, and background reflectance contributions. Woody biomass has necessarily been estimated using relationships established with canopy measures (e.g., Suganuma et al., 2006).

Active sensors differ in that they generate, transmit and receive their own electromagnetic (primarily spectral and microwave) energy, which can penetrate vegetation to varying degrees and provide information on the amount and three-dimensional distribution of structures contained. Key sensors are Synthetic Aperture Radar (SAR) and Light Detection and Ranging (LiDAR); and, in this chapter, their role in quantifying the carbon state and dynamics of savannas at scales ranging from within plants to entire regions is presented. The considerable benefits of integrating data from both active and passive (optical) sensors are also highlighted. Ground-based sensors (e.g., Terrestrial Laser Scanners or TLS) are reviewed, as these are increasingly providing avenues for retrieving detailed information on vegetation structure and biomass.

Common Descriptors of Savannas

Direct quantitative retrieval of carbon in vegetation is challenging, as the interaction of LiDAR pulses or SAR microwaves is complex and often specific with some, but not all, structural elements. For this reason, retrieval is often achieved due to secondary relationships between returned power or LiDAR waveform moments and vegetation structural elements (e.g., height). Users of remote sensing data also often expect to retrieve measures equivalent to those obtained routinely through ground survey that can then be translated to an estimate of carbon. However, additional or alternative field-based measures are needed for the development of retrieval algorithms, as active sensors provide a different (and sometimes more detailed) perspective on vegetation structure. For this reason, this first section reviews commonly used field-based descriptors of savannas and then explains why these might not always be appropriate for interpreting remote sensing data.

Vegetation Structure

At the individual tree level, quantitative field-based measures relating to structure include stem diameter (e.g., at 30 or 130 cm), height, and cover (surface/projected areas, volumes and/or widths of crowns). More qualitative assessments include growth form and condition (e.g., fire damage or severity). Many measures are tied to individual trees but are often not geographically referenced in three-dimensional space. At the stand level, common descriptors are simply scaled from individual tree measurements and include stem number density or size (basal area or diameter/height distributions) and/or canopy cover. Combinations of attributes, such as height and canopy cover, are often more informative descriptors that have been exploited to standardize classifications of savannas (e.g., Specht and Specht, 1999).

Although many measures are comparable within and between regions and are universally recognized and understood, they often do not adequately capture the multi-dimensional variability in structure typical of savannas. For this reason, measures may not relate to what is sensed by either passive or active instruments. For example, measures of the size, orientation, density, and moisture content of the stems, branches and/or foliage, or the clumping of leaves in crowns and crowns in stands for overstorey and understorey vegetation are often not quantified by standard field measures; yet, these attributes influence the interaction of spectral energy and microwaves with savanna canopies. The spatial distribution and size of field plots may also be insufficient to encompass variation associated with a number of sensor parameters (e.g., variation in incidence angle of the SAR system). The mismatch between what is measured through field-based survey and observed by ground, airborne, or spaceborne sensors, therefore, often hinders reliable retrieval of structural attributes using active and also passive remote sensing data.

Vegetation Biomass (Carbon)

Biomass, which relates to carbon, refers to the amount of dry matter (e.g., g) per unit volume of material whereas biomass density is a unit area measurement (e.g., Mg ha^{-1}). Total or component biomass is typically estimated by applying allometric equations that relate field-based measurements of individual tree size to biomass (as determined through destructive harvesting). A common limitation is that these equations are not available for many species and not always applicable (although often applied) outside of the environmental envelope and size range for which they were originally intended. For this reason, errors in biomass estimation, particularly when progressively scaled from individual trees to regions, are introduced and propagated.

In savannas, vegetation biomass is distributed unevenly in three-dimensional space, with variable allocation to the aboveground (leaves, branches, and/or trunks) and below ground components (roots). The biomass can be contained within a small number of large individual trees or distributed (albeit unevenly) across stands with varying stem number density and proportions of open space (both vertically and horizontally). This is in contrast to many closed forests for which the majority of remote sensing–based retrieval algorithms have been developed, where biomass is more evenly distributed within the volume space. This variability in both the distribution of biomass and open spaces within the volume of the stand is difficult to quantify on the ground and also complicates retrieval from remote sensing data. The proportion of dead relative to live biomass in vegetation is also greater compared with many closed forests and varies both spatially and temporally, and in response to environmental events and processes such as fire and drought (Fensham et al., 2003).

Active Remote Sensing

LiDAR

LiDAR is a relatively new technology designed originally for topographic mapping but has since proved to have application in retrieving vegetation structure and biomass. From the 1990s, LiDAR provided discrete single or dual return profiling and scanning capability, and these systems are now readily accessible through commercial operators. Since 2004, full waveform LiDAR sensors, which record the entire waveform of the backscatter pulse instead of just the hardware-determined range of individual echoes, have been available on experimental ground-based (e.g., the Echidna Validation Instrument or EVI; Lovell et al., 2003; Strahler et al., 2008) and airborne (e.g., LVIS/SLICER; Blair et al., 1999) platforms. Such systems are also operating in space (e.g., the Geoscience Laser Altimeter System [GLAS] carried on the Ice, Cloud, and land Elevation Satellite [ICESat]; Schutz et al., 2005). Full wave-form airborne LiDAR can operate using footprint sizes ranging from <10 cm to >70 m and permit detailed information on vegetation structure to be derived at multiple spatial scales. In contrast to discrete return systems, indi-vidual echoes from diffuse targets with low reflectivity can often be recov-ered. This potentially increases the density of the point cloud in savannas where trees are typically of low stem number density and support sparse, clumped, and often inclined foliage. Additional parameters can also be retrieved including the amplitude, width, height and number of echoes (Figure 8.1) and physical quantities (e.g., the backscatter cross-section).

Technical specifications for a range of discrete and full waveform LiDAR are given in Baltsavias (1999), Lim et al. (2003), and Mallet and Bretar (2009),

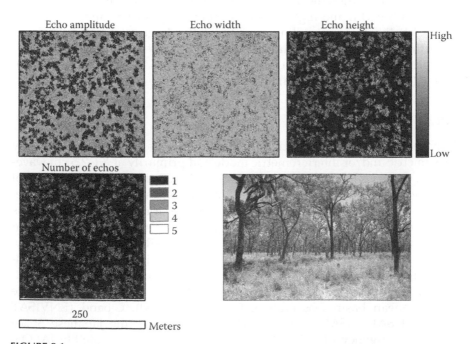

FIGURE 8.1
Small footprint full-waveform LiDAR (Riegl LMS-Q560) data for wooded savannas near Charter Towers, Queensland, decomposed into echo amplitudes, widths, height, and number. These derived data are not provided by discrete return systems.

with these systems differing mainly in terms of footprint size, post spacing, pulse energy, and pulse repetition frequency (PRF). More robust methods for extracting quantitative measures of vegetation structure and biomass from LiDAR data have also developed as systems have evolved (e.g., Koetz et al., 2006).

Many of the algorithms and indices (e.g., for height distribution and cover retrieval) seem generic. However, variable accuracies are obtained between algorithms when applied outside of the regions or sensor or acquisition configurations for which they were developed. Such variability has so far been attributable largely to differences in foliage clumping and leaf angle distribution, crown shape, the location and spatial arrangement of trees within footprints of varying dimensions, local terrain, and surface reflectivity affecting the vertical and horizontal distribution of returns. For this reason, there is increasing focus on the use of radiative transfer models to better understand interaction of LiDAR pulses with canopy architecture (Ni-Meister et al., 2001; North et al., 2008). By taking a more physically based approach, the requirement for extensive field data sets coincident with LiDAR surveys may be minimized, particularly as sensor and acquisition configurations change. Inversion of these models has also shown promise for parameter retrieval (e.g., Koetz et al., 2006).

For regional studies, the use of LiDAR data has been limited largely by the small swath width and high costs of airborne acquisitions and, in the case of ICESat, the lower sampling density of the large footprint waveform. Nevertheless, options for scaling retrieved attributes to the wider landscape using combinations of data sets (e.g., LiDAR and SAR) have been demonstrated (e.g., Lucas et al., 2006a).

SAR

Airborne SAR has been operating for several decades in single, dual, or fully polarimetric and/or interferometric mode and primarily at X (~2.5 cm wavelength), C (~6 cm), L (~25 cm), and/or P-band (~68 cm). After experimental missions with SeaSAT and the Shuttle Imaging Radar (SIR-A and B), spaceborne SAR data became more routinely available for regional observations from the 1990s, commencing with single channel C-band (the European Resources Satellite [ERS-1] and the Canadian RADARSAT-1) and L-band (the Japanese Earth Resources Satellite [JERS-1] SAR). In the late 2000s, the technology had advanced to satellite sensors with polarimetric capability operating at X-band (TerraSAR-X and the Constellation of small Satellites for the Mediterranean basin Observation; COSMO-SkyMed), C-band (ENVISAT Advanced SAR [ASAR], RADARSAT-2), and L-band (the Japanese Space Exploration Agency's [JAXA] Advanced Land Observing Satellite [ALOS] Phased Array L-band SAR [PALSAR]). Data from these satellite systems also provided opportunities for repeat-pass interferometry. The specifications of these sensors are summarized in Zhou et al. (2009).

As with LiDAR, most SAR-based studies have focused on closed-canopy forests, but algorithms developed for classification or biophysical retrieval are only partly applicable to savannas where the more open canopy and greater variability in the location, size, and number density of woody components (e.g., stems) leads to a greater diversity of microwave interactions between vegetation components and/or ground surfaces.

Several studies have used modifications of SAR simulation models to decompose the reflected power, or the scattering cross section per unit area (Sigma nought; σ^0), into contributory scattering mechanisms, thereby gaining a better understanding of the potential of SAR for retrieving structural attributes. Models are typically parameterized using ground measurements of attributes that directly influence σ^0 (e.g., size, number density, moisture content, and orientation of plant components; ground roughness and moisture content; Moghaddam and Saatchi, 1999; Woodhouse, 2006a). Common outputs include the relative contribution to σ^0 of the various scattering mechanisms within and between vegetation and the ground surface (e.g., single bounce, double bounce, and volume scattering). Sensitivity analyses can also be conducted to better understand the relative contribution of specific variables (e.g., stem size and density) to the σ^0. For savannas, and using the model of Durden et al. (1989), Lucas et al. (2004) highlighted that σ^0 recorded at C-band HV and L-band HH and HV was primarily the result of microwave interactions with the leaves and small (<1 cm diameter) branches, trunks, and larger branches, respectively.

Ground-Based Sensors

Ground-based (terrestrial) laser scanners are a recent innovation that has revolutionized the characterization of woody vegetation. Examples include the Echidna Validation Instrument (EVI; Strahler et al., 2008) and scanners provided by commercial companies including Leica, Riegl, and Trimble. These sensors have the capacity to provide structural information at the tree and stand level (Figure 8.2a) that is difficult to capture from field measurements, including degrees of foliage clumping, crown and canopy gap fraction and openness, LAI, stand foliage profile, leaf or branch angle distributions, wood volume, and biomass. A permanent record of forest structure is provided that can be used as a baseline against which to assess change (e.g., in permanent plots), and data can be reanalyzed using revised or new algorithms. These data can be compared with airborne LiDAR data (Figure 8.2b and c), thereby allowing better interpretation and calibration of these. However, for this purpose, accurate spatial georeferencing of point clouds is essential. Better parameterization of SAR simulation models (e.g., in terms of quantifying branch angle/densities) may also be achieved. Even so, careful quantification of errors is necessary, as these are propagated when derived measures are subsequently related to remote sensing data.

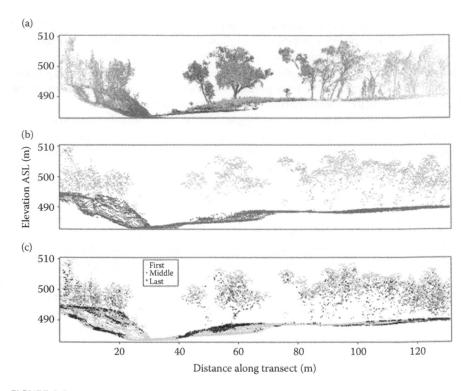

FIGURE 8.2
(a) A first return point cloud acquired by a Leica ScanStation-2 Terrestrial Laser Scanner (TLS); and (b) a first return point cloud derived from Riegl LMS-Q560 airborne LiDAR waveform data and registered to the TLS point cloud. Darker points correspond to higher return intensity. All returns detected in the airborne LiDAR waveform data are shown in (c). The elevation is above sea level (ASL).

Retrieving Structural Attributes from Active Sensors

Vegetation Height

From LiDAR data, vegetation height is generally measured as the elevation difference between first (signal start) and last (centroid of the ground) returns. Retrieval accuracies are typically <1 m, depending upon the footprint size, post spacing, crown shape, and whether individual trees or stands are being considered. Where smaller (0.2–3.0 m) footprint systems are used, height estimates are less reliable, particularly where a dense overstorey or understorey reduces the likelihood of pulses returning from the ground. Further, the shape and dimensions of the crowns can compromise retrieval, with underestimation commonly associated with trees of excurrent (pine-like) growth forms and lower post spacings. Better height retrieval at the plot level is generally obtained using LiDAR with larger (10–70 m) footprint sizes due to the greater likelihood of receiving returns from the ground (Mallet and Bretar, 2009). However, where very large footprints are used (e.g., with ICESat GLAS), variations in topographic relief and canopy openness can compromise height retrieval (Harding and Carabajal, 2005).

The Interferometric SAR (InSAR) also allows retrieval of tree and stand height. This method combines two SAR measurements from antennas displaced in space and/or time. The phase difference between the two returned signals for corresponding points in an image pair is then measured from which an altitude is calculated. This provides a three-dimensional model of the Earth's surface that is commonly used as a Digital Elevation Model (DEM). Many InSAR viewing geometries exist, but across-track single-pass InSAR from high frequency SAR is typically used for vegetation height retrieval. All wavelengths of radiation penetrate, to some extent, into the vegetation canopy, although attenuation occurs as a function of structural variability. In addition, the phase centre, which gives rise to the topography, is generally within several meters of canopy tops at high (X- and C-band) frequencies for dense, closed canopies. The retrieved 3D surface model, therefore, approximates the near top of the vegetation canopy and provides a Digital Surface Model (DSM). A DSM is essentially a DEM with vegetation bias, but this bias can be used to estimate vegetation height by subtracting the known ground surface under the forest (e.g., derived from a ground survey or LiDAR-derived DEM) from the DSM. Alternatively, a DEM can be created with lower frequency (e.g., L- or P-band) InSAR when a survey-derived DEM is unavailable. Lower frequency microwaves typically have phase centers near the ground, although variations occur as a function of forest and scattering behavior.

In contrast to LiDAR measurements, which measure range at the point of interception of the vegetation canopy, the vertical location of the InSAR-retrieved height is determined indirectly through analysis of the relative scattering contributions from the stems, branches, leaves, and the ground surface in combination with radar characteristics such as frequency (and wavelength), polarization, and incidence angle. The use of InSAR in savannas is, however, complicated by sensor parameters (e.g., spatial resolution) and also the relative openness of the canopy, as the high values of σ^0 from the ground surface effectively lower the scattering phase centre. For this reason, the height retrieved at X-band may be lower compared with that at C-band (Viergever et al., 2008; Figure 8.3).

Polarimetric interferometry (PolInSAR) uses the polarimetric information from single frequency InSAR to distinguish between canopy and ground returns. Since this technique relies on SAR penetration to the ground surface to obtain contributions from the canopy and the ground, fully polarimetric L-band is considered to be most suitable among today's available data sets. The canopy consists mainly of randomly orientated elements (leaves, branches) and acts as a random scattering volume with no preference for a specific

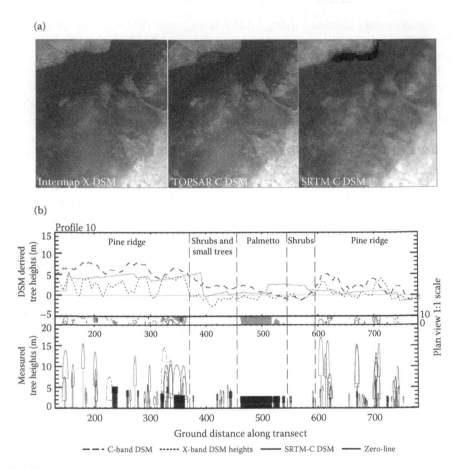

FIGURE 8.3
(a) High-frequency single-pass digital surface models (DSMs) showing the height of sparse savanna woodlands and closed-canopy gallery forest within a Belizean savanna, generated using X-band (Intermap, airborne) and C-band (AIRSAR, airborne and Shuttle Radar Topography Mission [SRTM], spaceborne) data. (b) Savanna woodland stand height profiles compared with field measurements. (Adapted from Viergever, K. M. et al. 2009. *Proceedings of IGARSS 2009*, Cape Town, 12–17 July © 2009. IEEE. With permission.)

polarization response. In contrast, ground surface and trunk-ground interactions have a distinctive polarimetric response (Woodhouse, 2006b). This differentiation between crown and ground scattering enables forest height (DSM) and ground height (DEM) retrieval, which is typically undertaken through inversion of a forest backscattering model [e.g., the Random Volume over Ground (RVoG) model; Treuhaft and Siqueira, 2000, Papathanassiou and Cloude, 2001]. The RVoG models a forest as a statistically homogeneous medium with randomly orientated and distributed scattering elements in the canopy layer. Inputs include the measured coherence matrix (containing all polarizations), which is used for determining the penetration depth and correlation magnitude as a function of the polarization. Hence, by varying polarization, the height of the true ground surface (highest correlation) and the canopy volume (lowest correlation contributor) can be determined.

Stem Location, Diameter, and Number Density

Using LiDAR data, common approaches to density estimation have relied on tree crown delineation and counting procedures. Although applicable in open forests where trees are relatively isolated, such approaches have limitations where canopies are closed and/or multi-layered. For identifying stem locations in savannas using discrete return LiDAR, Lee and Lucas (2007) developed a Height-Scaled Crown Openness Index (HSCOI), which considers the relative penetration of the LiDAR pulse through canopies. Low HSCOI values indicated poor penetration associated with a high density of branches (emanating from a stem) or the presence of the stem itself. A particular advantage of the approach was that multiple stems forming combined crowns could be identified (Figure 8.4), and crowns could also be delineated. For a range of forest types and growth stages, accuracies in stem location averaged approximately 70% (range 49–89%). Across wider areas, a close correspondence has been reported between lower frequency L and P-band HH data and the number density and/or basal area by Armston et al. (2008) and Viergever (2008), respectively.

Crown and Canopy Descriptors

A number of studies have used either discrete return or full waveform LiDAR to quantify crown cover, including by summing the planimetric area of delineated crowns (Lee, 2008) or using empirical relationships (Lee and Lucas, 2007). Measures relating to canopy cover (e.g., Foliage Projected Cover) have further considered the fraction of the signal reflected by the target and compensated for differences in canopy and ground reflectivity. Corrections

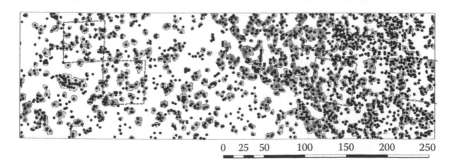

0   25   50   100   150   200   250

FIGURE 8.4
Tree crown and stem location map derived from the HSCOI index derived from discrete return LiDAR data. (Adapted from Lee, A. C. and R. M. Lucas 2007. *Remote Sensing of Environment,* 111, 493–518.)

that account for the sensor not resolving gaps smaller than the footprint size of the LiDAR pulse have also been applied (Armston et al., 2009). Spatial indices of cover as a function of the distribution of returns through the vertical profile (e.g., Hopkinson, 2007) have potential for characterizing savannas.

Combined Retrieval of Structural Attributes

From the same LiDAR data sets, information on structural attributes (e.g., height, number density, and crown cover) can be extracted either directly or indirectly to provide more comprehensive and wider ranging descriptions (Table 8.1). As an example, Tickle et al. (2006) used samples of height and canopy cover from LiDAR and species type obtained through aerial photograph interpretation to describe central Queensland savannas. Using SAR data, opportunities exist for simultaneously retrieving multiple attributes. Moghaddam and Saatchi (1999) developed a nonlinear estimation technique, where an analytically based numerical discrete component scattering model (Durden et al., 1989) is used to produce simulated values of σ^0 for a range of forest parameters by varying the model inputs. Parametric equations are then derived from the simulated data, reducing the inversion problem to one of optimization, which may be solved using a number of iterative techniques. The inversion can be undertaken successively for the canopy, stem, and ground layers. The outputs from the model include spatial data sets relating to the height, diameter, number density (of crowns and stems), and the moisture content of the vegetation and the soil. From these, biomass can be derived.

For complete characterization, more than three channels are needed, particularly for quantifying ground effects, which necessitates the use of

TABLE 8.1

Examples of Structural Measures Derived Directly and Indirectly from LiDAR Data

Directly Derived Attribute	Indirectly Derived Attribute	Derived from:
H	Diameter	H or crown dimensions, HSCOI
Crown/canopy cover	Basal area	H or crown dimensions, CGV
Crown canopy depth	Volume	CGV
Apparent canopy profile	Biomass	H, H-squared, HOME
Outer canopy ruggedness		Canopy cover, depth volume, reflectance
Gap fraction	Density	HSCOI, counts of delineated crowns
	LAI, PAI	H or crown dimensions, profile area
	FPC, PPC	Gap fraction, crown dimensions
	Forest type/species	An ensemble of derived attributes

Abbreviations: H, height; HSCOI, height-scaled crown openness index; CGV, canopy geometric volume; HOME, height of median energy return; LAI, leaf area index; PAI, plant area index; FPC, foliage projected cover; PPC, plant projected cover.

multi-frequency polarimetric data. However, one or several spaceborne SAR sensors with limited bandsets can be used if allometrics (e.g., relating tree height to canopy depth) are available, as these reduce the number of unknowns. To configure the model, *in situ* measurements of vegetation parameters (e.g., structure, moisture, and soil) are needed, and several (e.g., branch density, size, and angle distributions) can be derived from ground-based lasers.

Retrieval of Biomass

Above-Ground Biomass

Using LiDAR data of sufficient post spacing (e.g., >1 return per m^2), and at the individual tree level, biomass can be estimated using allometric equations, as diameter (which generally relates to height) and height itself are commonly used as independent variables (Mallet and Bretar, 2009). However, since equations are typically species or genus specific, the association of each tree identified with a species type (e.g., as classified using hyperspectral data) leads to better estimation (Lucas et al., 2008a). Even so, such approaches are increasingly limited when the canopy becomes closed and/or multi-layered. For estimating biomass at the stand level, most studies have focused on establishing relationships with LiDAR measures, including height, height squared, canopy cover and depth volume, or the Height of Median Energy Return (HOME; Table 8.1). As an example, Lee (2008) and Lucas et al. (2006a) established a relationship between measures of the distribution of pulse strikes within the vertical profile and field-estimated biomass (Figure 8.5).

A limitation of using airborne LiDAR to estimate biomass is that the extent of coverage is limited. Data from the ICESat GLAS, which samples globally, have shown promise in estimating the biomass of closed forests (Lefsky et al., 2005), although retrieval is complicated by the large footprint size of the sensor, especially in areas of high relief (Harding and Carabajal, 2005; Lefsky et al., 2005). For this reason, many have focused more on the use of SAR for regional mapping. The most common approach to biomass retrieval from SAR has been to establish empirical relationships between field-based estimates and single channel data. Most studies of closed canopy forests have shown that σ^0 increases with biomass up to a saturation point approximating 20 Mg ha^{-1}, 60 Mg ha^{-1}, and 100 Mg ha^{-1} for C, L, and P-band, respectively, although levels vary within and between forest types. Similar increases in σ^0 are observed in the more open savannas, but there is evidence that only a proportion of the woody biomass is observed, particularly at lower frequencies (Lucas et al., 2006b).

Scatter in the relationship between σ^0 and biomass is typically observed. This scatter is greater for savannas of complex structure and is attributable to variability in the size, dielectric and geometric properties of the woody

(a)

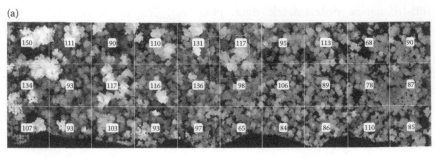

(b)

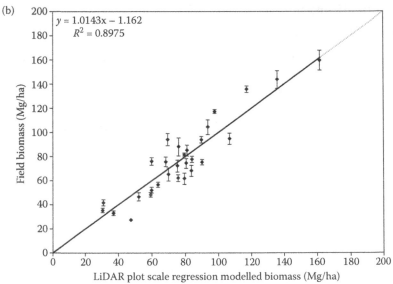

FIGURE 8.5

(a) Stand level biomass (Mg/ha) estimated from LiDAR data and (b) actual versus predicted biomass. (Adapted from Lucas, R. M. et al. 2006a. *Remote Sensing of Environment* 100, 388–406.)

components, and sensor parameters (e.g., viewing geometry). To provide confidence limits for estimation, Le Toan (2008) proposed a Bayesian approach that utilized *a priori* knowledge of forest biomass to quantify uncertainties such that:

$$P(B \,|\, \gamma^0) \propto P(\gamma^0 \,|\, B) \cdot P(B) \qquad (8.1)$$

where $P(B|\gamma^0)$ is the probability of biomass given a value of backscatter (gamma nought; γ^0), $P(\gamma^0|B)$ is the probability of γ^0 given a value of biomass, and $P(B)$ is the probability distribution function for biomass. In the case of woody savannas, *a priori* information such as age class maps or biomass distributions derived from LiDAR data (Lucas et al., 2006a) can be used to derive $P(B)$.

Component Biomass

The retrieval of component biomass from LiDAR has been addressed indirectly through estimation of structural attributes including stem height (for excurrent growth forms) and foliage volume (Wulder et al., 2007). Another approach has been to apply allometric equations to the maximum heights of trees located with the HSCOI and classified to species using co-registered hyperspectral data (Lucas et al., 2008a).

On the basis that microwaves of different frequency and polarization interact with different components of the forest volume, Saatchi et al. (2007) developed a semi-empirical quadratic equation relating crown and stem biomass to polarimetric L- and/or P-band SAR data. The general form of the regression equation was

$$\log(W) = a_0 + a_1\sigma_{HV}^0 + a_2(\sigma_{HV}^0)^2 + b_1\sigma_{HH}^0 + b_2(\sigma_{HH}^0)^2 + c_1\sigma_{VV}^0$$
$$+ c_2(\sigma_{VV}^0)^2, \tag{8.2}$$

where W is the total above ground biomass or the biomass of the respective components (i.e., stem, branch, and foliage) and σ^0_{HH}, σ^0_{HV}, and σ^0_{VV} represent the radar backscatter intensity measurements at HH, HV, and VV polarizations, respectively (dB scale). By including terms for HH, HV, and VV polarizations, the range of scattering mechanisms represented by polarimetric measurements was considered. The influence of surface topography was recognized. Although the approach was originally developed for boreal forests, the coefficients can be modified for savannas using appropriate ground calibration.

Quantifying Changes in Carbon

Active sensors provide unique opportunities for quantifying changes in carbon, primarily through assessment of state (e.g., stages of regeneration or senescence) or time-series comparison of data or retrieved biophysical attributes (e.g., biomass). For example, changes in the height or crown area of individual trees or the amount of stand canopy volume may be quantified by comparing LiDAR data acquired several years apart (Wulder et al., 2007; Lucas et al., 2008). Object-based comparisons using, for example, delineated tree crowns or stands are generally more interpretable, although may be time consuming to implement. Detecting changes in structure and biomass using SAR data has proved more difficult, because σ^0 also varies over time with environmental factors (e.g., changes in precipitation, soil moisture) or sensor parameters (e.g., geolocation, calibration, and incidence angle).

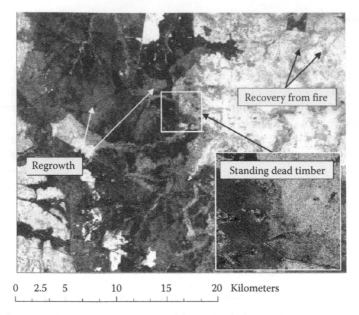

FIGURE 8.6
(See color insert following page 320.) Composite of Landsat-derived FPC and ALOS L-band HH and HV data in RGB. Areas of dead standing timber (green; with high HH σ^0, low FPC, and HV σ^0; note that confusion exists with rough surfaces, particularly when surface and soil moisture is high), regrowth after clearance (red; low HH and HV σ^0 and high FPC), and fire (orange; moderate HH and HV and high FPC) are highlighted. Inset is an AIRSAR image showing areas of standing dead timber (blue) (Adapted from Lucas, R. M., et al. 2008b. In *Patterns and Processes in Forest Landscapes: Multiple uses and sustainable management*, eds. R. Lafortezza, J. Chen, G. Sanesi and T. R. Crow, Springer, New York, pp. 213–240.)

In many cases, the use of optical remote sensing data acquired by airborne and/or spaceborne sensors over varying spatiotemporal scales can add value to the SAR data. As an example, combinations of ALOS PALSAR L-band HH, HV, and Landsat-derived Foliage Projected Cover (FPC) facilitate characterization and mapping of states of regrowth, dieback, and wildfire damage (Figure 8.6; Lucas et al., 2008b).

Future Outlook

For quantifying carbon stocks and changes in savannas, an increasing amount of data from active sensors is becoming available. As examples, the ALOS PALSAR is currently providing repeated, systematic, wall-to-wall, and global observations that facilitate regional to global mapping of biomass, forest structural types, growth stages, and detection of change. Continuation is also anticipated through the ALOS-2 L-band SAR (due for launch in 2013) with increased spatial resolution (to 3 m) and continuous interferometry. The ICESAT-2 mission (proposed from 2010 to 2013) will provide 140 m spacing of 50–70 m footprints, thereby allowing more detailed information on the heights and vertical profile of vegetation to be extracted. The proposed Deformation, Ecosystem Structure, and Dynamics of Ice (DESDynI) mission will support a combination of multiple-beam LiDAR and L-band SAR or InSAR, with the latter serving to interpolate the LiDAR-derived measures of vegetation structure into swath-wide measures. The European Space Agency's (ESA) BIOMASS P-band SAR mission is also anticipated to provide information on the global distribution of forest biomass.

Significant developments in ground-based and airborne LiDAR are expected, with a trend toward multispectral full waveform systems and system specifications providing better capability for large area mapping at finer spatial resolution. Such developments will be accompanied by advancements in hardware or software and approaches to retrieving biophysical parameters (e.g., inversion of physically-based models) and integrating data from different sources and time periods. Such advances are anticipated to lead to better interpretation and advances in the use of spaceborne SAR and optical data for environmental applications.

The importance of integrating active remote sensing data for understanding the carbon dynamics of savannas should not be understated, particularly as these provide unique perspectives on the three-dimensional structure of vegetation. Although savannas have been increasingly threatened by expansion of human populations and agriculture, they are now also being exposed to climatic change. The impacts on savannas are likely to be significant, particularly given that they occur in areas vulnerable to climatic extremes (e.g., low rainfall and drought) and induced events (e.g., extreme fires). This chapter has demonstrated that active sensors, including those proposed or in development, can make substantial contributions to assessing the state and condition of savanna ecosystems, quantifying change, and informing approaches to their long-term conservation.

Acknowledgments

The authors would like to thank the Queensland Department of Environment and Resource Management (QDERM), the Queensland Herbarium, the Australian Research Council, Definiens AG, The National Aeronautics and Space Administration (NASA), European Space Agency (ESA), and the Programme for Belize and Rio Bravo Management and Conservation Area Staff. This work has been undertaken in part within the framework of the JAXA Kyoto and Carbon Initiative. ALOS PALSAR data have been provided by JAXA EORC.

References

Armston, J., R. Denham, T. Danaher, P. Scarth, and T. Moffiet. 2009. Prediction and validation of foliage projective cover from Landsat-5 TM and Landsat-7 ETM+ imagery. *Journal of Applied Remote Sensing* 3, 033540.

Armston, J., R. Lucas, J. Carreiras, and P. Bunting. 2008. A first look at integrating ALOS PALSAR and Landsat data for assessing woody vegetation in Queensland. *Proceedings of the 14th Australasian Remote Sensing & Photogrammetry Conference*, Darwin, Australia.

Baltsavias, E. 1999. Airborne laser scanning: Existing systems and firms and other resources. *ISPRS Journal of Photogrammetry and Remote Sensing* 54, 164–198.

Blair, J., D. Rabine, and M. Hofton. 1999. The laser vegetation imaging sensor: A medium-altitude digitisation-only, airborne laser altimeter for mapping vegetation and topography. *ISPRS Journal of Photogrammetry and Remote Sensing* 54, 115–122.

Durden, S. L., J. J. van Zyl, and H. A. Zebker. 1989. Modeling and observation of the radar polarization signature of forested areas. *IEEE Transactions on Geoscience and Remote Sensing* 27, 290–301.

Fensham, R.J., R. J. Fairfax, D. W. Butler, and D. M. J. S. Bowman. 2003. Effects of fire and drought in a tropical eucalpyt savanna colonised by rainforest. *Journal of Biogeography* 30, 1405–1414.

Harding, D. J. and C. C. Carabajal. 2005. ICESat waveform measurements of within-footprint topographic relief and vegetation vertical structure. *Geophysical Research Letters* 32, L21S10.

Hopkinson, C. 2007. The influence of flying altitude and beam divergence on canopy penetration and laser pulse return distribution characteristics. *Canadian Journal of Remote Sensing* 33, 312–324.

Koetz, B., F. Morsdorf, G. Sun, J. Ranson, K. Itten, and B. Allgoewer. 2006. Inversion of a Lidar Waveform Model for Forest Biophysical Parameter Estimation. *IEEE Geoscience and Remote Sensing Letters* 3, 49–53.

Lee, A. 2008. *Utilising Airborne Scanning Laser (LIDAR) to Improve the Assessment of Australian Native Forest Structure*. PhD Thesis, Australian National University, Canberra, 339pp.

Lee, A. C. and R. M. Lucas. 2007. A LiDAR-derived canopy density model for tree stem and crown mapping in Australian forests. *Remote Sensing of Environment* 111, 493–518.

Lefsky, M. A., D. J. Harding, M. Keller, et al. 2005. Estimates of forest canopy height and aboveground biomass using ICESat. *Geophysical. Research Letters* 32, L22S02.

Le Toan, T., 2007. Technical notes to ESA on intensity inversion.

Lim, K., P. Treitz, M. Wulder, B. St-Onge, and M. Flood. 2003. LiDAR remote sensing of forest structure. *Progress in Physical Geography* 27, 88–106.

Lovell, J. L., Jupp, D. L. B., Culvenor, D. S., Coops, N. C. 2003. Using airborne and ground-based ranging LiDAR to measure canopy structure in Austrlain forests. *Canadian Journal of Remote Sensing*, 29, 607–622.

Lucas, R. M., N. Cronin, A. C. Lee, C. Witte, and M. Moghaddam. 2006a. Empirical relationships between AIRSAR backscatter and forest biomass, Queensland, Australia, *Remote Sensing of Environment* 100, 388–406.

Lucas, R. M., M. Moghaddam, and N. Cronin. 2004. Microwave scattering from mixed species woodlands, central Queensland, Australia. *IEEE Transactions on Geoscience and Remote Sensing*, 47, 2142–2159.

Lucas, R. M., A. Accad, L. Randall, and P. Bunting. 2008b. Assessing human impacts on Australian forests through integration of airborne/spaceborne remote sensing data. In *Patterns and Processes in Forest Landscapes: Multiple uses and sustainable management*, eds. R. Lafortezza, J. Chen, G. Sanesi and T. R. Crow, Springer, New York, pp. 213–240.

Lucas, R. M., N. Cronin, M. Moghaddam, et al. 2006b. Integration of Radar and Landsat-derived Foliage Projected Cover for Woody Regrowth Mapping, Queensland, Australia, *Remote Sensing of Environment* 100, 407–425.

Lucas, R. M., A. C. Lee and P. J. Bunting. 2008a. Retrieving forest biomass through integration of CASI and LiDAR data. *International Journal of Remote Sensing* 29, 1553–1577.

Mallet, C. and F. Bretar. 2009. Full-waveform topographic lidar: State of the art. *ISPRS Journal of Photogrammetry and Remote Sensing* 64, 1–16.

Moghaddam, M. and S. Saatchi. 1999. Monitoring tree moisture using an estimation algorithm applied to SAR data from boreas. *IEEE Transactions on Geoscience and Remote Sensing* 37, 901–916.

Ni-Meister, W., D. L. B. Jupp and R. Dubayah. 2001. Modeling lidar waveforms in heterogeneous and discrete canopies, *IEEE Transactions on Geoscience and Remote Sensing* 39, 1943–1958.

North, P., J. Rosette, J. Suarez and S. Los. 2008. A Monte Carlo radiative transfer model of satellite waveform lidar *SilviLaser 2008*, Sept. 17–19, 2008, Edinburgh, UK.

Papathanassiou, K. P. and S. R. Cloude. 2001. Single-baseline polarimetric SAR interferometry. *IEEE Transactions on Geoscience and Remote Sensing* 39, 2352–2363.

Saatchi, S., K. Halligan, D. Despain and R. L. Crabtree. 2007. Estimation of forest fuel load from radar remote sensing. *IEEE Transactions Geoscience and Remote Sensing* 45, 1726–1740.

Schutz, B. E., H. J. Zwally, C. A. Shuman, D. Hancock and J. P. DiMarzio. 2005. Overview of the ICESat mission. *Geophysical Research Letters* 32, L21S01.

Specht, R. L. and A. Specht. 1999. *Australian Plant Communities*. Dynamics of Structure, Growth and Biodiversity. Oxford University Press, Melbourne.

Strahler, A. H., D. L. B. Jupp, C. E. Woodcock, et al. 2008. Retrieval of forest structural parameters using a ground-based lidar instrument (Echidna). *Canadian Journal of Remote Sensing* 34, 426–440.

Stuart, N., T. Barran and C. Place. 2006. Classifying the Neotropical savannas of Belize using remote sensing and ground survey. *Journal of Biogeography* 33, 476–490.

Suganuma, H., Y. Abe, M. Taniguchi, et al. 2006. Stand biomass estimation method by canopy coverage for application to remote sensing in an arid area of Western Australia. *Forest Ecology and Management* 222, 75–87.

Treuhaft, R. and P. Siqueira. 2000. The vertical structure of vegetated land surfaces from interferometric and polarimetric radar. *Radio Science* 35, 141–177.

Viergever, K. M. 2008. *Establishing the Sensitivity of Synthetic Aperture Radar to Aboveground Biomass in Wooded Savannas*. Ph.D. Thesis, School of Geosciences, The University of Edinburgh, Edinburgh, 220pp.

Viergever, K. M., I. H. Woodhouse, A. Marino, M. Brolly, and N. Stuart. 2008. SAR Interferometry for Estimating Above-Ground Biomass of Savanna Woodlands in Belize. *IGARSS Conference Proceedings*, Boston, Massachusetts, 6–11 July 2008.

Viergever, K. M., I. H. Woodhouse, A. Marino, M. Brolly, and N. Stuart. 2009. Backscatter and interferometry for estimating above-ground biomass of sparse woodland: A case study in Belize. *Proceedings of IGARSS 2009*, Cape Town, 12–17 July 2009.

Woodhouse, I. H. 2006a. *Introduction to Microwave Remote Sensing*, Taylor & Francis, London.

Woodhouse, I. H. 2006b. Predicting backscatter-biomass and height-biomass trends using a macroecology model. *IEEE Transactions on Geoscience and Remote Sensing*, 44(4), 871–877.

Wulder, M., T. Han, J. White, T. Sweda and H. Tsuzuki. 2007. Integrating profiling LIDAR with Landsat data for regional boreal forest canopy attribute estimation and change characterization, *Remote Sensing of the Environment* 110, 123–137.

Zhou, X., N. Change, and S. Li. 2009. Applications of SAR interferometry in Earth and Environmental Science Research. *Sensors* 9, 1876–1919.

9

Remote Sensing of Tree–Grass Systems: The Eastern Australian Woodlands

Tim Danaher, Peter Scarth, John Armston, Lisa Collett, Joanna Kitchen, and Sam Gillingham

CONTENTS

Introduction

About 1.8 million square kilometers of Australia is, or was, covered by woodland vegetation communities, defined here as *ecosystems*, with widely spaced trees having an overstorey of vertically projected cover of photosynthetic foliage (FPC) of between 10% and 30%. The woodlands of Eastern Australia shown in Figure 9.1 are mostly located within the states of Queensland and New South Wales. This area includes tropical and subtropical savannas in the north and temperate and Mediterranean woodlands in the southern areas. The woodlands are located between the higher rainfall coastal areas to the east and arid areas to the west. The vegetation types within these woodland areas are quite diverse due to many factors including climate, geology, soils, presence or absence of fire, and land management practices (Lindenmayer et al., 2005). The predominant land use of these woodland areas is grazing of cattle and sheep. In all areas of northern Queensland where precipitation exceeds potential evaporation, vegetation growth is

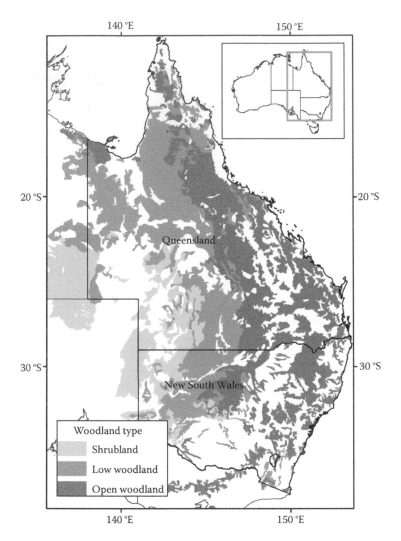

FIGURE 9.1
East Australian Woodland Communities.

water limited, and trees and grasses compete for rainfall. This has lead to widespread clearing of woody vegetation to increase pasture productivity, and significant areas of woodland have also been cleared for agriculture, particularly those on more fertile clay soils. Some ecosystems have been cleared to the extent that less than 10% of the pre-clearing area remains intact (Fensham et al., 2003). The clearing of woody vegetation has significant bio-diversity and greenhouse gas emission implications (Henry et al., 2005).

Due to the fragile nature of these vegetation systems, land management has profound long-term effects on vegetation cover, pasture species composition, and soil loss (Hill et al., 2003, 2006a). The government and community have a responsibility in balancing the production and conservation interests in these regions to avoid further degradation of these systems. There is a need for good information to inform policy decisions, assist in regulation and compliance with legislation, and assist landholders and nongovernment organizations in the management of these lands. Remote sensing, when combined with suitable ground measurements, can provide accurate information at a suitable scale and frequency.

Early work looking at vegetation cover change by (Walker et al., 1993) used maps of current and pre–European vegetation cover, in conjunction with broad national vegetation classifications and allometric equations, to estimate the number of trees that had been removed in the Murray-Darling basin. A major limitation of this approach is the lack of accurate woody cover data. This is one of the more pressing issues in operational Australian remote sensing systems and typically involves decomposing a remotely sensed vegetation proxy (e.g., NDVI) into perennial and herbaceous components, because the interaction between the seasonal and evergreen components is difficult to separate in a single date optical image (Roderick et al., 1999). Lu et al. (2003) processed AVHRR time series from 1981 to 1994 to estimate mean woody cover. More recent work using both trend analysis and decomposition of AVHRR time series has observed a "greening" of the Australian landscape, although a major persistent fPAR drop, corresponding to woody vegetation change, was observed in the Queensland woodlands (Donohue et al., 2009). These local products and a more recent MODIS-derived product (Gill et al., 2009) have demonstrated good accuracy. However, Hill et al. (2006b) in an assessment of the global MODIS LAI product in Australian ecosystems, found major overestimation of LAI in some eastern Australian open forests and woodlands.

These continental scale monitoring and mapping systems based on MODIS and AVHRR have the limitation of coarse spatial resolution. At a pixel resolution of 1km, products are difficult to interpret in the field, unable to measure moderate scale land cover changes, such as the clearing of a small paddock, and the scale is difficult to link to existing land use mapping to determine the nature of change (Wallace et al., 2006). Australia's National Carbon Accounting System (NCAS) was developed to provide accounting for greenhouse gas emissions from land use change. It uses Landsat imagery to report nationally on changes in forest extent (defined as greater than 20% canopy cover and 0.2 ha area). (Australian Greenhouse Office, 2005).

The monitoring system documented in this chapter was developed as part of a Queensland Government funded monitoring program known as the Statewide Landcover and Trees Study (SLATS). The program commenced in 1995, providing information supporting the development of vegetation management policies and accurate information informing greenhouse gas inventories. Since 2006, SLATS methods have also been used for monitoring or woody vegetation change in NSW.

Technology That Matches the Scale of Problem in Space and Time

The Queensland Government required information enabling statewide monitoring of vegetation, and at a resolution and extent that was suitable for regulation, compliance, restoration, and other management actions. The imagery resolution needed to be suitable for reliable delineation of woody vegetation changes greater than 1ha and direct calibration, validation, and interpretation in the field.

The reduction in woody vegetation cover due to clearing needs to be mapped at a minimum of 1- to 2-year time intervals. Longer periods may miss some clearing as regrowth, with significant canopy and additional pasture, and may establish within the intervening period, making it difficult to map vegetation change. Annual monitoring was considered necessary for regulation and compliance regimes. The monitoring of increasing vegetation cover, regrowth, invasion, and thickening requires use of a longer imagery time series (Röder et al., 2008a; Armston et al., 2009). In most cases, it takes several years for vegetation to increase to a level where it is possible to detect change with a high level of certainty. Climate can be a significant driver in the establishment of regrowth and is episodic (Henry et al., 2005). In summary, there was a requirement to monitor both short-term and longer-term changes occurring over decades with regular repeat acquisitions and coincident field observations, providing information to support compliance activities at particular dates.

Imagery from the Landsat series of sensors—Landsat Thematic Mapper (TM) and Landsat Enhanced Thematic mapper (ETM+)—was chosen as the major imagery for SLATS, as it provided the necessary spatial resolution (30 m) and regular repeat coverage to meet most of the requirements outlined above. In the near future, it may not be possible to obtain Landsat TM and ETM+ imagery, as these satellites are aged and operating on backup systems; and the Landsat Data Continuity Mission is not scheduled for launch until 2012 (Wulder et al., 2008). However, there are similar resolution satellites (e.g., SPOT4, SPOT5) that may help fill the gap in this important archive until a new Landsat satellite is launched.

Image Pre-Processing

Accurate detection and quantification of vegetation change over time and space requires removal of the confounding effects of geometric distortion; radiometric variability; illumination geometry; and cloud, shadow, and water contamination from imagery. These pre-processed Landsat TM and ETM data form the basis of all derived image products (see Figure 9.2). A high level

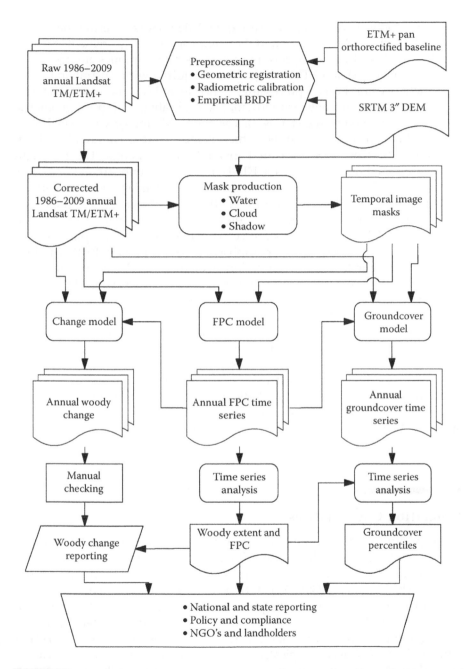

FIGURE 9.2
Flow chart of the SLATS processing system showing the largely automated processing streams
that ingest raw Landsat data to produce information products for use by stakeholders.

of automation is essential, as there are over 3500 images covering the period 1987–2008 in the SLATS archive; and there is an ongoing requirement to reprocess the whole archive as improved pre-processing techniques emerge.

Path-oriented TM and ETM+ imagery are acquired with Australian Centre of Remote Sensing Level 5 processing, which includes some systematic geometric correction and onboard radiometric calibration. Sub-pixel geometric accuracy is obtained by registering all TM and ETM+ imagery to an orthorectified ETM+ panchromatic mosaic, which is registered to Differential Global Positioning System (GPS) measured ground control points. The pre-flight radiometric calibration is used for ETM+ images; the onboard calibration is removed from TM images and an improved time-dependant vicarious calibration (de Vries et al., 2007). In the future, coefficients published in Chander et al. (2009) will be adopted, as they were determined from a longer time series of Landsat 5 TM imagery. Images are converted to top-of-atmosphere (TOA) reflectance and a three-parameter empirical model applied to further reduce variation in scene-to-scene illumination geometry and bidirectional reflectance distribution function (BRDF). The model, a modified version of Walthall's BRDF model (Walthall et al., 1985), was parameterized using the north-south and east-west overlapping regions of the ETM+ and TM images. Although generalized, the correction considerably reduces scene-to-scene variability.

Images are acquired in the dry season to minimize cloud cover and to maximize the spectral contrast between perennial vegetation and senescing annual grasses; however, almost a third of all acquired Landsat imagery contains some cloud contamination. Reflectance values affected by cloud and cloud shadow, topographic shadow, and surface water are removed using automated and semi-automated Landsat-derived masks (NRW, 2008).

Vegetation Cover Indices

The capacity to detect and monitor relative differences in vegetation cover has largely been accomplished, and it has been implemented in numerous operational programs in Australia and elsewhere using Landsat and other imagery (Wallace et al., 2006). Vegetation cover, and its change over time, is a critical factor when studying the properties and dynamics of these ecosystems, particularly when looking at issues such as land degradation, carbon cycling, climate variability (McVicar and Jupp, 1998), and global warming (Hill et al., 2005). Cover is typically defined as the percentage of green and/ or dead vegetation that is present in a particular stratum in the ecosystem or, more generally, as ground cover and overstorey cover.

Overstorey cover is a popular structural descriptor in both the remote sensing and ecological fields, and several measurements are commonly used. These are crown cover (CC), foliage projective cover (FPC), and LAI. The CC is defined as the percentage of ground area covered by the vertical projection of crowns, which in this case are assumed opaque. The FPC is defined as the vertically projected percentage cover of photosynthetic foliage of all strata or, equivalently, the fraction of the vertical view that is occluded by foliage. Overstorey FPC is defined as the vertically projected percentage cover of photosynthetic foliage from tree and shrub life forms greater than 2 m height and was the definition of woody vegetation cover adopted by SLATS. Overstorey FPC is one minus the gap probability at a zenith angle of zero, and, therefore, it has a logarithmic relationship with effective LAI (Chen et al., 1997). Since woodland plant communities are dominated by trees and shrubs with sparse foliage and irregular crown shapes, overstorey FPC is a more suitable indicator of a plant community's radiation interception and transpiration than crown cover. For the same leaf area, variation in foliage clumping and leaf orientation angle determines the FPC of individual crowns. High variation in leaf orientation angle and foliage clumping across Australian woody species has been reported as functionally reducing levels of intercepted light at high sun elevations (King, 1997). Due to these clumping effects, the relationship between FPC and CC is not linear and can be characterized by a model that accounts for the within-canopy and stand-scale clumping (Macfarlane et al., 2007).

Ground cover is a critical attribute of the landscape affecting infiltration, runoff, water, and wind erosion, and as such, it is a key indicator of land condition (Karfs et al., 2009). However, a reduction in cover does not necessarily correspond to a decline in land condition (Pickup et al., 1998), as ground cover is driven largely by climate and management. Remote sensing offers the only way to measure ground cover over large spatial extents. Extensive work by Pickup and co-authors (Pickup and Nelson, 1984; Pickup and Chewings, 1994) pioneered the monitoring of woodland ground cover by remote sensing in Australia. More recent work has focused on using remotely sensed ground cover to inform landscape leakiness models to better understand the link between cover, patchiness and condition (Ludwig et al., 2007), and the identification of changes due to grazing impacts over time (Röder et al., 2008b).

The development of algorithms to produce ground and canopy fractional vegetation cover is critical in the SLATS processing system workflow. The FPC index is used both on its own and in time series models to map woody vegetation extent, estimate woody change, and provide carbon estimates to support government policy. The ground cover index is used to assess ground cover targets in the Great Barrier Reef catchments, as input into surface water runoff and erosion models, and for use in assessing infrastructure location and pasture utilization in savanna systems.

Field Calibration and Validation

To enable the development and validation of these cover models, a distribution of calibration sites across Queensland has been assembled via an extensive ongoing field campaign since 1996. The fieldwork was originally designed to support SLATS mapping of woody vegetation extent and FPC and was expanded to also provide data to calibrate ground cover relationships since 1999. Field sites were chosen to represent an area equivalent to a block of 3 × 3 TM pixels, which are resampled to a resolution of 25 m. The sites were located in mature, undisturbed stands over a range of vegetation communities on various soil types and covering a range of structural formations from sparse, low, open woodland to tall, closed forest communities. The collection of these data is described in Armston et al. (2009).

With the most recent field methods implemented in 1999, the sites were each set up as star transects, each 100 m long, and set out at 60° intervals (Figure 9.3). The center of the star is located using a differential GPS. This is a useful configuration, as it is efficient in practice and the area it samples can be readily linked to high and moderate spatial resolution imagery, such as Landsat TM. Each of the three 100-m transects sample ground cover, FPC at mid and upper storey, and CC measurements at 1-m intervals using a vertical tube method (Specht, 1983) with intercepts classified as bare soil, rock, cryptogram, ground, tree and/or shrub green leaf, dead leaf, branch, or sky by the observer. These measurements were converted to percentage cover for each site. There are currently more than 300 sites measured with this method and more than 400 sites measured using a similar method but limited to a single 100-m transect.

In addition to the transect measurements, tree basal area was estimated using a calibrated optical wedge at all sites (Dilworth and Bell, 1971). An average from seven basal area measurements at each site was used (see Figure 9.3). There were an additional 1176 sites where the basal area was measured but

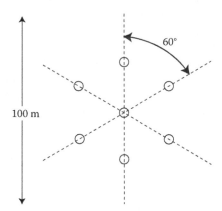

FIGURE 9.3
Schematic of the sampling design used at the field sites coincident with the LiDAR surveys. Estimates of overstorey foliage projective cover were derived from three 100 m point intercept (1 m spacing) transects oriented at 0°, 60°, and 120° from magnetic north (dotted lines). Stand basal area was calculated as the average from seven optical wedge counts (circles) located at 25 m and 75 m along each transect and the center of the nominal 1 ha field plot.

transect measurements were not recorded. The site measurements with coincident FPC and basal area measurements were used to calibrate a stand allometric equation between overstorey FPC and basal area (Armston et al., 2009). This relationship had an RSME of 7.3% and allowed the derivation of FPC allometric values for use in developing a Landsat FPC index.

FPC Model

The detection of long-term trends in woody vegetation in Queensland, Australia, from the Landsat-5 TM and Landsat-7 ETM+ sensors requires the automated prediction of overstorey FPC using over 1900 Landsat images. Statistical techniques such as multiple linear regression (MLR) are commonly used to estimate fractional vegetation cover. The MLR is similar to linear spectral unmixing when the calibration data is representative of the a priori distribution of the cover fraction. Several studies have shown that MLR can be used to predict fractional vegetation cover provided the calibration data set is proximate and covers a similar or extended range, to the validation data set (Fernandes et al., 2004).

A model for prediction of overstorey FPC based on pre-processed Landsat imagery, a climatological ancillary variable, Vapor Pressure Deficit, and field calibration data collected from over 1900 sites in Queensland, was developed by Lucas et al. (2006) using MLR. A cross-validation approach showed that the model had an adjusted R^2 of 0.80, and overstorey FPC could be predicted with an RMSE of 9.9%. Armston et al. (2009) compared several parametric models and machine learning techniques for prediction of overstorey FPC and independently validated these using FPC derived from field and airborne LiDAR data. They also concluded that all models provided a prediction accuracy of <10% RMSE; however, models also showed greater than 10% bias in plant communities with high herbaceous or understorey FPC.

Time Series Woody Extent and FPC Product

The major limitation of predictions made from any single date model of FPC is that overstorey and understorey FPC are not decoupled. The use of overstorey FPC products derived from TM or ETM+ data requires the assumption of a senescent or absent herbaceous understorey at the time of image acquisition. This has serious implications for trend analysis of woody vegetation cover, because photosynthetic herbaceous FPC will increase the variance of the sparse Landsat time series.

The woody extent and FPC product uses a time series of Landsat images to calculate the best possible prediction of FPC for each pixel within a scene, for a specified time period, for example, 1988–2008. The FPC value is assumed to be a combination of green herbaceous and woody FPC with an error component. In order to discriminate woody vegetation from nonwoody areas such as crop and pasture, an approach using simple time series statistics has been

FIGURE 9.4
(See color insert following page 320.) Example of time series approach to discriminate woody extent. (a) is the FPC derived from a 2008 Landsat TM image showing high photosynthetic activity across the image. (b) is the standard error of a linear fit to the time series with dark areas representing a low standard error. (c) This is indicative of stable perennial vegetation and is used in conjunction with the minimum of the time series to estimate the 2008 woody extent product.

developed. Pixels are classified as nonwoody (0% FPC) or woody (1–100% FPC) on the basis of thresholds applied to the time series minimum and normalized standard error of a regression line fitted to the series. Figure 9.4 shows the difference between single date FPC and the woody extent product for a mixed forest and grazing area where the greater standard error over the grassland areas has led to those areas being mapped as nonwoody. Default thresholds have been derived from a statewide data set by maximizing the Kappa statistic, a measure of omission and commission, using a global optimization technique and SLATS field observations. The final model has a Kappa statistic of 85%.

The final product is then generated using a linear robust time series regression that predicts FPC for each woody pixel based on the trend across the annual FPC imagery provided to the model. Anomalies in the time series such as fire scars have less impact on the regression when using robust techniques. The time series method enables complete map coverage by assigning predicted FPC values to pixels that would otherwise remain unclassified due to presence of cloud or shadow.

In areas where the time series FPC shows a significant increase in woody vegetation over part of the time series, the high variance due to rapidly increasing FPC will cause those areas to be mapped as nonwoody. In these cases, for example, plantation clearing followed by regrowth, the time series is split. This is done by calculating the most significant linear regression line back in time from the end of the specified time period. A *t*-test for the significance is performed for each line, and the line with the highest *t* value is taken as the split point (Figure 9.5). This procedure is applied to those areas mapped as plantation in a land use layer and to areas mapped as nonwoody by the standard algorithm in order to detect woody regrowth that has been omitted.

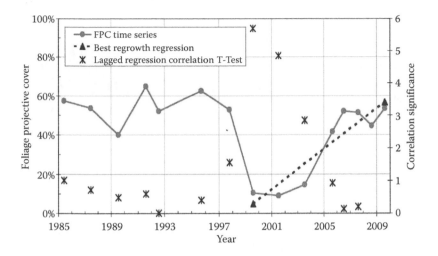

FIGURE 9.5

Example of the reset regression method used to split the FPC time series when a regrowth event is detected. Gray line is the FPC time series and asterisks show the significance of the regression from 2009 back in time to any given year. The highest *t*-value is taken to be the best fitting regression line, in this case the dashed line starting in 1999. Thus, the time series is split at this point, and the most recent section is used to calculate the woody FPC value.

Ground Cover

Three quarters of Queensland have an overstorey FPC of less than 20%, so methods of monitoring the extent of ground cover, defined as the total organic soil surface cover, are required. Monitoring of ground cover is important, because it is linked to indicators of soil loss, biodiversity, and pasture production (Booth and Tueller, 2003). Ground cover is also an indicator target adopted by regional Natural Resource Management groups and Catchment Management Authorities. Using a similar method to that used to derive the FPC index, SLATS uses a regression between the measured field sites and Landsat data to estimate fractional bare ground with an RMS prediction error of 13% (Karfs et al., 2009). Since the calibration data represents the major variability in cover types and soil backgrounds across Queensland, the resulting model can be applied across the state to provide "snapshot" estimates of ground cover.

When the full archive of Landsat data is available, the fractional estimates of ground cover provide a valuable data set for discriminating changes in land management in savanna environments. Figure 9.6 shows the paddock means of the bare ground fraction and NDVI values for two adjacent grazing trial paddocks in Queensland. The light stocking rate paddock was stocked at the long-term sustainable carrying capacity, whereas the heavily grazed paddock was stocked at twice that capacity (O'Reagain et al., 2009). The change in the amount of bare ground can be clearly seen in the later years of

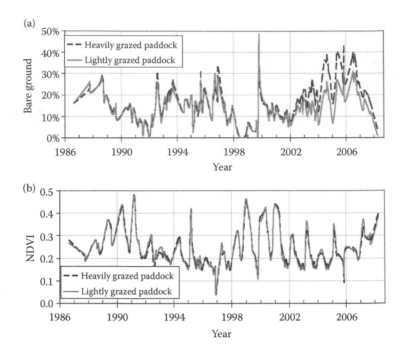

FIGURE 9.6

Landsat derived time series plots from a grazing trial in north Queensland, Australia, showing temporal trajectories of adjacent heavily and lightly grazed paddocks. Note that the grazing trial was initiated in 1997. (Adapted from O'Reagain, P. et al. 2009. *Animal Production Science* 49, 85–99.) (a) shows the paddock spatial means of the fractional bare ground product showing the spike in bare ground corresponding to a fire in 2000 and clear separation of the two paddocks, with the lightly grazed paddock containing up to 10% less bare ground in the later stages of the trial, (b) shows the paddock spatial mean NDVI values and shows very little difference between paddocks.

the trial, whereas the NDVI fails to pick up any significant difference between the paddocks. It is clear here that that ground cover is variable due to both climate and management. Since the bare ground time series for any given pixel is related to the climate history, the response of the land type to that climate and to the management history techniques is required to separate these influences. More advanced time series methods are constricted by the limited number of observations, typically only one per year over a 25-year period in the standard SLATS archive. A simple method is to rank a given land parcel relative to its adjacent neighbors on similar land types to identify grazing management related trends, as shown in Figure 9.6. Predictions of the 5th and 95th percentiles of the time series from an analysis of their statistical distribution are also operationally produced . These are used as estimates of "1 in 20 year" minimum or maximum, respectively. The time series 5th percentile can be used to assess relative animal stocking rates in drought times, whereas the time series 95th percentile can be used to identify landscape "hotspots," areas such as scalds, water points, and fence lines that have consistently lower cover than the surrounding regions.

Woody Change

The difficulties in consistently and reliably detecting woody change across a wide range of vegetation types over a large geographical area required the development of a change method that combined both the spectral changes from scene to scene and the historical variability in foliage cover. This combination of change detection methodologies was based on combining a modified form of image differencing (e.g., Mas, 1999) with a time series model using the FPC time series to detect true change (e.g., Nemani and Running, 1997). Calibration data, collected from an analysis of previous operator-interpreted and field-checked woody change data sets from previous years, were used to develop the spectral and temporal change indices.

The spectral index used a regression-based approach utilizing the extracted training data, and it was optimized to use bands and transforms that minimized the RMSE between predictions and measured change; minimized the sensitivity of the model to extreme values by calculating the mean and maximum of the partial derivatives of each band; and maximized the reduction in error variance per term added using the method of Elden (2004). All processing was completed on logarithmic transformed reflective bands to minimize the effect of illumination differences (Savvides and Vijaya Kumar, 2003). These coefficients were applied to the transformed Landsat bands and were summed to generate the regression model output.

The time series FPC data were used to build two change indices: The first is a simple difference of the corrected FPC over the change dates; and the second is a time series metric that uses a difference test based on the measured variance around a line fitted to the FPC time series. The initial step in processing the SLATS FPC product was to normalize the annual FPC estimates. Although image acquisitions are selected to correspond with the local dry season, there is still significant interaction between the woody and herbaceous components of the FPC signal. A simple histogram normalization method was applied on a scene-by-scene basis to account for these effects (Kautsky et al., 1984; Yang and Lo, 2000), resulting in all image dates having a similar FPC histogram. The corrected FPC data were then used to build the two change indices: the difference of the corrected FPC over the change dates; and the time series metric. The time series metric utilizes the standard error (s) around a line fitted to the FPC time series in a difference test. The form of the test is given by

$$t = \frac{fpc_{\text{pred}} - fpc_{\text{meas}}}{s},$$

(9.1)

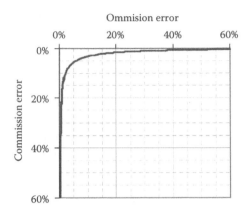

FIGURE 9.7
Receiver operating characteristics analysis of the SLATS woody change classifier. The commission error is calculated as (1—specificity) and the omission error is calculated as (1—sensitivity).

where fpc_{meas} is the FPC determined from the Landsat imagery at the change point, whereas the predicted FPC (fpc_{pred}) is the fit of the line through the FPC time series, extrapolated to the current year.

The final combined classifier was developed by training a linear combination of both the spectral and temporal indices against the calibration data. Determination of the classification threshold for the automated change mapping was based on receiver operating characteristics (ROC) analysis (e.g., Alatorre and Beguería, 2009). Figure 9.7 shows that the optimum classification model has a misclassification rate of 5.1% if omission and commission errors are balanced. The final classifier sits within the SLATS operational processing framework and is run automatically across the state. However, in order to use this classifier in an operational reporting environment, three levels of classification at the 2%, 5%, and 15% omission level are produced, which are then further interpreted, edited, and field checked by an operator. This manual interpretation stage is used to check the output of the classifier and to further improve the accuracy of the final product before it is used to produce the annual SLATS woody change figures and reports.

Discussion

The SLATS mapping and monitoring program has demonstrated how Landsat imagery can be processed to provide an integrated package of information including woody FPC; woody change; and ground cover, within the eastern Australian Woodlands (Figure 9.8). The methods described in this chapter have been applied by Queensland and New South Wales State Governments across the whole of each state, which includes most of the eastern Australian

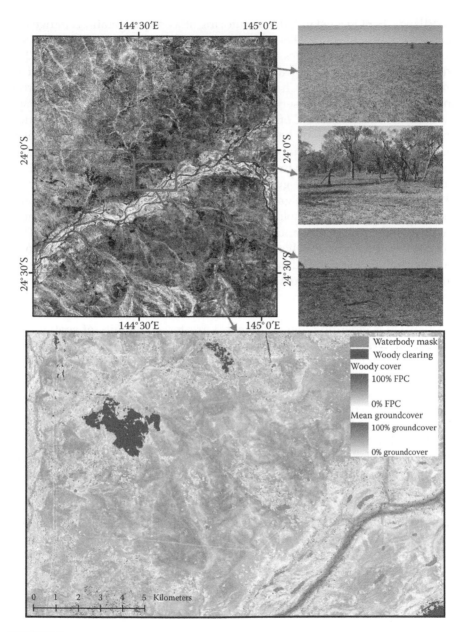

FIGURE 9.8
(**See color insert following page 320.**) Clockwise from top left. Landsat 5 TM image of a portion of the East Australian Woodlands with bands 5,4,3 displayed as RGB. Photographs show examples of open pasture, open woodland communities, and recently cleared areas. Map shows information products developed from time series analysis of Landsat imagery including waterbody location, areas of woody clearing, mean ground cover, and woody foliage projective cover. © The State of Queensland (Department of Environment and Resource Management).

woodlands. For Queensland, the monitoring of woody vegetation change has been done annually since 1999 and in periods of two to four years from 1987 to 1999. For New South Wales, annual monitoring has been done since 2006; and from 1988 to 2006, it was analyzed every two years (Department of Environment and Climate Change, 2008; Department of Natural Resources and Water, 2008). The time series FPC method has provided statewide maps showing the extent and FPC of woody vegetation. This scientific information provided by remote sensing has assisted in the development of policies relating to management of remnant or native vegetation. It has been used to update vegetation mapping utilized in regulation of vegetation clearing and provides current information for state government compliance staff to investigate possible unlawful clearing. The ground cover information is produced annually and has been used by landholders to support more sustainable management for improved productivity and reduced risk of long-term degradation. This tool can be used to objectively assess land condition across regions with comparable climate history to identify properties that are under higher grazing pressure. Assessment of existing property infrastructure (fence lines and water points) and the resultant effect on land condition can be readily undertaken using the ground cover information.

A new era is emerging where there is a greater availability of spatial data and the ongoing uncertainty of climate change. There is a burgeoning need to better understand long-term savanna vegetation dynamics and assimilate these remotely sensed products into spatial process-based models to assist stakeholders in their decision making. The significant investment in field data and image corrections by the SLATS program means that adapting the products to these changing requirements is a straightforward task.

This work was supported by significant institutional support over a long term that enabled development of data storage and processing infrastructure; image pre-processing systems; and an extensive calibration and validation program. It also relies on the ongoing availability of the Landsat imagery archive. Although SLATS provides significant information related to woody vegetation and ground cover, it is the adoption and use of this information that helps make a difference in terms of more sustainable management of the ecosystems. This was assisted by the formation of a stakeholder reference group comprising representatives from government, industry, environmental, and academic bodies that provided a forum to discuss methods and results and to disseminate this information.

References

Alatorre, L. C. and S. Beguería. 2009. Identification of eroded areas using remote sensing in a badlands landscape on marls in the central Spanish Pyrenees. *CATENA* 76(3), 182–190.

Armston, J. D., R. J. Denham, T. J. Danaher, P. F. Scarth, and T. N. Moffiet. 2009. Prediction and validation of foliage projective cover from Landsat-5 TM and Landsat-7 ETM+ imagery. *Journal of Applied Remote Sensing* 3, 033540-28.

Australian Greenhouse Office. 2005. *Greenhouse gas emissions from land use change in Australia: Results of the National Carbon Accounting System 1988–2003.* Australian Greenhouse Office, Canberra, Australia.

Booth, D. T. and P. T. Tueller. 2003. Rangeland monitoring using remote sensing. *Arid Land Research and Management* 17, 455–467.

Chander, G., B. L. Markham, and D. L. Helder. 2009. Summary of current radiometric calibration coefficients for Landsat MSS, TM, ETM+, and EO-1 ALI sensors. *Remote Sensing of Environment* 113, 893–903.

Chen, J. M., P. M. Rich, S. T. Gower, J. M. Norman, and S. Plummer. 1997. Leaf area index of boreal forests: Theory, techniques, and measurements. *Journal of Geophysical Research D* 102 (D24), 29429–29443.

de Vries, C., T. Danaher, R. Denham, P. Scarth, and S. Phinn. 2007. An operational radiometric calibration procedure for the Landsat sensors based on pseudo-invariant target sites. *Remote Sensing of Environment* 107, 414–429.

Department of Environment and Climate Change. 2008. *NSW Woody Vegetation Change 2006 To 2007 Report.* NSW Department of Environment and Climate Change, Sydney South.

Department of Natural Resources and Water. 2008. Land cover change in Queensland 2006–07: A Statewide Landcover and Trees Study (SLATS) Report. Department of Natural Resources and Water, Brisbane.

Dilworth, J. R. and J. F. Bell. 1971. *Variable Probability Sampling: Variable Plot and Three P.* OSU Book Stores.

Donohue, R. J., T. R. McVicar, and M. L. Roderick. 2009. Climate-related trends in Australian vegetation cover as inferred from satellite observations. *Global Change Biology* 15, 1025–1039.

Elden, L. 2004. Partial least-squares vs. Lanczos bidiagonalization-I: Analysis of a projection method for multiple regression. *Computational Statistics and Data Analysis* 46, 11–31.

Fensham, R. J., R. J. Fairfax, S. J. L. Choy, and P. C. Cavallaro. 2003. Modelling trends in woody vegetation structure in semi-arid Australia as determined from aerial photography. *Journal of Environmental Management* 68, 421–436.

Fernandes, R., R. Fraser, R. Latifovic, J. Cihlar, J. Beaubien, and Y. Du. 2004. Approaches to fractional land cover and continuous field mapping: A comparative assessment over the BOREAS study region. *Remote Sensing of Environment* 89, 234–251.

Gill, T. K., S. R. Phinn, J. D. Armston, and B. A. Pailthorpe. 2009. Estimating tree-cover change in Australia: Challenges of using the MODIS vegetation index product. *International Journal of Remote Sensing* 30, 1547–1565.

Henry, B., C. Mitchell, A. Cowie, O. Woldring, and J. Carter. 2005. A regional interpretation of rules and good practice for greenhouse accounting: Northern Australian savanna systems. *Australian Journal of Botany* 53, 589–605.

Hill, M. J., R. Braaten, and G. M. McKeon. 2003. A scenario calculator for effects of grazing land management on carbon stocks in Australian rangelands. *Environmental Modelling and Software* 18, 627–644.

Hill, M. J., S. H. Roxburgh, J. O. Carter, and G. M. McKeon. 2005. Vegetation state change and consequent carbon dynamics in savanna woodlands of Australia in response to grazing, drought and fire: A scenario approach using 113 years of synthetic annual fire and grassland growth. *Australian Journal of Botany* 53, 715–739.

Hill, M. J., S. J. Roxburgh, G. M. McKeon, J. O. Carter, and D. J. Barrett. 2006a. Analysis of soil carbon outcomes from interaction between climate and grazing pressure in Australian rangelands using Range-ASSESS. *Environmental Modelling and Software* 21, 779–801.

Hill, M. J., U. Senarath, A. Lee, et al. 2006b. Assessment of the MODIS LAI product for Australian ecosystems. *Remote Sensing of Environment* 101, 495–518.

Karfs, R. A., B. N. Abbott, P. F. Scarth, and J. F. Wallace. 2009. Land condition monitoring information for reef catchments: a new era. *The Rangeland Journal* 31, 69–86.

Kautsky, J., N. K. Nichols, and D. L. B. Jupp. 1984. Smoothed histogram modification for image processing. *Computer Vision, Graphics, and Image Processing* 26, 271–91.

King, D. A. 1997. The functional significance of leaf angle in Eucalyptus. *Australian Journal of Botany* 45, 619–639.

Lindenmayer, D. B., E. Beaton, R. B. Cunningham, et al. 2005. *Woodlands: A disappearing landscape.* CSIRO Publishing, Collingwood, Vic.

Lu, H., M. R. Raupach, T. R. McVicar, and D. J. Barrett. 2003. Decomposition of vegetation cover into woody and herbaceous components using AVHRR NDVI time series. *Remote Sensing of Environment* 86, 1–18.

Lucas, R. M., N. Cronin, M. Moghaddam, et al. 2006. Integration of radar and Landsat-derived foliage projected cover for woody regrowth mapping, Queensland, Australia. *Remote Sensing of Environment* 100, 388–406.

Ludwig, J. A., G. N. Bastin, V. H. Chewings, R. W. Eager, and A. C. Liedloff. 2007. Leakiness: A new index for monitoring the health of arid and semiarid landscapes using remotely sensed vegetation cover and elevation data. *Ecological Indicators* 7, 442–454.

Macfarlane, C., M. Hoffman, D. Eamus, et al. 2007. Estimation of leaf area index in eucalypt forest using digital photography. *Agricultural and Forest Meteorology* 143, 176–188.

Mas, J. F. 1999. Monitoring land-cover changes: A comparison of change detection techniques. *International Journal of Remote Sensing* 20, 139–152.

McVicar, T. R. and D. L. B. Jupp. 1998. The current and potential operational uses of remote sensing to aid decisions on drought exceptional circumstances in Australia: A review. *Agricultural-Systems* 57, 399–468.

Nemani, R. and S. Running. 1997. Land cover characterization using multitemporal red, near-IR, and thermal-IR data from NOAA/AVHRR. *Ecological Applications* 7, 79–90.

O'Reagain, P., J. Bushell, C. Holloway, and A. Reid. 2009. Managing for rainfall variability: Effect of grazing strategy on cattle production in a dry tropical savanna. *Animal Production Science* 49, 85–99.

Pickup, G., G. N. Bastin, and V. H. Chewings. 1998. Identifying trends in land degradation in non-equilibrium rangelands. *Journal of Applied Ecology* 35, 365–377.

Pickup, G. and V. H. Chewings. 1994. A grazing gradient approach to land degradation assessment in arid areas from remotely-sensed data. *International Journal of Remote Sensing* 15, 597–617.

Pickup, G. and D. J. Nelson. 1984. Use of Landsat radiance parameters to distinguish soil erosion, stability and deposition in arid central Australia. *Remote Sensing of Environment* 16, 195–209.

Röder, A., J. Hill, B. Duguy, J. A. Alloza, and R. Vallejo. 2008a. Using long time series of Landsat data to monitor fire events and post-fire dynamics and identify driving factors. A case study in the Ayora region (eastern Spain). *Remote Sensing of Environment* 112, 259–273.

Röder, A., Th Udelhoven, J. Hill, G. del Barrio, and G. Tsiourlis. 2008b. Trend analysis of Landsat-TM and -ETM+ imagery to monitor grazing impact in a rangeland ecosystem in Northern Greece. *Remote Sensing of Environment* 112, 2863–2875.

Roderick, M. L., I. R. Noble, and S. W. Cridland. 1999. Estimating woody and herbaceous vegetation cover from time series satellite observations. *Global Ecology & Biogeography* 8, 501–508.

Savvides, M. and B. V. K. Vijaya Kumar. 2003. Illumination Normalization Using Logarithm Transforms for Face Authentication. In *Lecture Notes in Computer Science (including subseries Lecture Notes in Artificial Intelligence and Lecture Notes in Bioinformatics)*.

Specht, R. L. 1983. Foliage projective covers of overstorey and understorey strata of mature vegetation in Australia. *Austral Ecology* 8, 433–439.

Walker, J., F. Bullen, and B. G. Williams. 1993. Ecohydrological changes in the Murray-Darling Basin. I. The number of trees cleared over two centuries. *Journal of Applied Ecology* 30, 265–273.

Wallace, J., G. Behn, and S. Furby. 2006. Vegetation condition assessment and monitoring from sequences of satellite imagery. *Ecological Management and Restoration* 7 (SUPPL. 1).

Walthall, C. L., J. M. Norman, J. M. Welles, G. Campbell, and B. L. Blad. 1985. Simple equation to approximate the bidirectional reflectance from vegetative canopies and bare soil surfaces. *Applied Optics* 24, 383–387.

Wulder, M. A., J. C. White, S. N. Goward, et al. 2008. Landsat continuity: Issues and opportunities for land cover monitoring. *Remote Sensing of Environment* 112, 955–969.

Yang, X. and C. P. Lo. 2000. Relative radiometric normalization performance for change detection from multi-date satellite images. *Photogrammetric Engineering and Remote Sensing* 66, 967–980.

10

Remote Sensing of Fractional Cover and Biochemistry in Savannas

Gregory P. Asner, Shaun R. Levick, and Izak P. J. Smit

CONTENTS

Introduction

Savanna ecosystems are comprised of complex three-dimensional (3-D) mixtures of woody and herbaceous vegetation, along with varying amounts of bare soil, rock, and other background features. Understanding, monitoring, and managing the spatiotemporal variation in savanna ecological processes requires a quantitative understanding of the 3-D structure and biochemical properties of the vegetation. Remote sensing provides a uniquely powerful approach to quantify vegetation properties in savannas. Given the enormous geographic extent of both temperate and tropical savannas (e.g., Huntley and Walker, 1982), combined with their heterogeneity at local scales, the need for remote sensing seems obvious. As a result, there has been continuing effort to develop and apply both airborne and satellite mapping approaches to estimate key aspects of savanna structure and biochemistry.

Characteristics used to describe savanna vegetation include horizontal fractional canopy cover, vertical profile, LAI, and biomass. Here we focus on fractional cover, because of its central role in driving and expressing savanna structure and function. We further define fractional cover in terms of basic surface material components of savannas, namely photosynthetic vegetation (PV), nonphotosynthetic vegetation (NPV), bare substrate (B), and shade (S) (Roberts et al., 1993). Fractional cover mapping can also be applied to plant growth form, plant functional types, disturbance, and species (Wessman et al., 1997, Roberts et al., 1998). We consider both perspectives in this chapter.

Relative to vegetation structure, remote sensing of plant chemical properties has received comparatively less attention in savannas, owing to the challenge of determining chemical concentrations of plant material within a highly varying medium of canopy structure (Figure 10.1). The chemicals most often studied from field and aircraft (and some spaceborne) sensors include water, leaf pigments, nitrogen, and carbon fractions. Two new, comprehensive syntheses on canopy biochemical remote sensing are provided by Ustin et al. (2009) and Kokaly et al. (2009) for pigment and nonpigment chemistry, respectively. Here we focus on issues specific to remote sensing of savanna canopy chemistry, for which there has been relatively less effort until recently.

Fractional Cover Mapping

Fractional cover mapping differs from vegetation classification or thematic mapping in that quantitative mixture models are applied to the image data to estimate the percentage lateral surface area or cover of various materials (Roberts et al., 1993). Most approaches utilize linear spectral mixture models, although a few nonlinear mixing approaches have been demonstrated in shrubland and savanna systems (e.g., Borel and Gerstl, 1994, Asner et al., 1998). Linear models simulate the reflectance signature of each image pixel (ρ_{pixel}) using an equation such as

$$\rho_{\text{pixel}} = \sum_{i=1}^{n}(C_i \cdot \rho_i) + \varepsilon \tag{10.1}$$

where C_i is the fractional cover and ρ_i is the reflectance of material (i), with the fractions (C_i) forced to sum to unity. The focus then turns to defining the spectral endmember libraries (ρ_i). Libraries can be defined in terms of surface materials (e.g., PV, NPV, B, and S), plant growth forms, plant functional types, or species. The validity and accuracy of the mixture model, thus, largely depends on the statistical separability of the spectral endmembers as well as the spectral resolution of the data (e.g., hyperspectral imagery with 100s contiguous narrow bands vs. multispectral imagery with <10 noncontiguous broad bands).

The reflectance properties of vegetation canopies are driven not simply by variation in fractional cover but also by the architecture and chemistry of the plants (Figure 10.1), which may, in turn, be influenced by local growing conditions. Factors such as LAI, leaf optical (and thus chemical) properties, and leaf orientation also influence savanna reflectance signatures measured from field, airborne, and space-based sensors (Asner et al., 2000). Combined, these factors often express the presence of plant functional types and species (Wessman et al., 1997, Roberts et al., 1998). Mixture analysis has also been

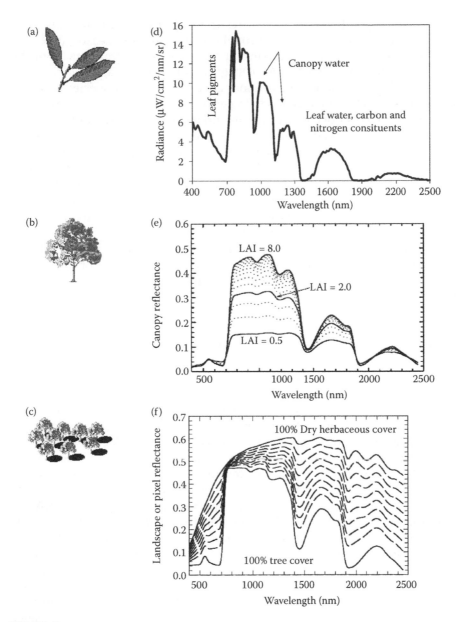

FIGURE 10.1

Scale-dependent factors controlling (a) leaf, (b) canopy, and (c) landscape variation in the spectral properties of savannas. (d) Leaf radiance spectrum shows the major scattering and absorption features in the 400–2500 nm range; (e) canopy leaf area index (LAI) affects the shape and magnitude of the canopy reflectance spectra; (f) fractional canopy cover of green tree and senescent grass is a major control over landscape-scale reflectance in savannas.

used with temporal spectral signatures that capture the different phenologies of major vegetation types throughout the growing season (trees vs.
grasses), allowing for the unmixing of time-series spectral data from the
NOAA AVHRR and the NASA MODIS (e.g., Defries et al., 2000).

Despite the work done on fractional cover mapping of species and life-
forms, most spectral mixture modeling efforts have focused on determining
the percentage cover of basic surface materials, including live and dead vegetation as well as bare soil (reviewed by Ustin et al., 2004a). Various approaches
have been tested, but the most common methods rely on Equation 10.1. With
multispectral sensors such as Landsat Thematic Mapper (TM) and Enhanced
Thematic Mapper Plus (ETM+), fractional cover mapping has provided estimates of PV, NPV, and bare soil in open woodlands and savannas (Asner
et al., 2003a; Huang et al., 2009). In practice, one often finds that the seven
multispectral channels spanning the 400–2500 nm range provided by
Landsat-type sensors can only determine two to three endmembers at best
(Theseira et al., 2002). There simply is not enough information in the 5–7
band reflectance signatures to yield more than the most basic cover types.
Moreover, although the fractional cover results derived from linear unmixing of multispectral data may be fairly precise, the accuracy is often low
(Asner and Heidebrecht, 2002). There is a usually a bias in the results caused
by variation in the brightness of the surface materials, especially in soils and
dried or dead vegetation.

Imaging spectroscopy, also known as hyperspectral imaging, differs from
multispectral imaging in that hundreds of contiguous, narrow spectral
bands are used (Kumar et al., 2001). Imaging spectroscopy, thus, provides a
means to identify more detailed spectral endmembers for fractional cover
mapping using spectral mixture analysis. The detailed reflectance endmembers provided using spectroscopic data can be manipulated in unique ways
to effectively increase sensitivity to certain plant functional types, species,
and fractional material covers in ecosystems. These techniques were largely
developed using airborne imaging spectrometers, especially NASA's
Airborne Visible and Infrared Imaging Spectrometer (AVIRIS) (Green et al.,
1998), although other airborne systems such as the Compact Airborne
Spectrographic Imager (CASI) (Babey and Anger, 1989) and HyMap (Cocks
et al., 1998) have also been very important. In Australian tropical savannas,
detailed spectral endmembers obtained from the spaceborne EO-1 Hyperion
recently enabled the large-scale assessment of PV, NPV, and bare soil fractional cover by upscaling to MODIS imagery (Guerschman et al., 2009).
Again, this study and many others (reviewed by Ustin et al., 2004b) have
relied on linear mixture modeling (Equation 10.1) as the fundamental
approach to quantifying lateral surface covers.

Fractional cover mapping of savannas and shrublands has been a particularly strong focus of terrestrial imaging spectroscopy; the high structural
variability of savannas calls for some of the most detailed spectral measurements in order to quantitatively dissect the landscape into vegetation and

soil constituents. One area of analysis has focused on species and functional types, often demonstrating an ability to determine the fractional cover of even closely related species based on their hyperspectral reflectance properties. However, the majority of studies have focused on fractional material cover, again PV, NPV, soils, and a few other surface constituents. Nonetheless, variability in spectral endmembers (Figure 10.2) can cause great uncertainty in any application of these endmembers in spectral mixture analysis (Equation 10.1). In response to this problem, Bateson et al. (2000) introduced the concept of spectral endmember "bundles," by which cover types are represented with an ensemble of spectra similar to that shown in Figure 10.2a–c. Unmixing of hyperspectral data with spectral bundles provided a means to incorporate endmember variability. Asner and Lobell (2000) incorporated the spectral bundle concept into an approach focused on the shortwave-infrared (SWIR; 2.0–2.5 µm) portion of the spectrum, which displays far lower within-bundle variability among PV, NPV, and soil spectra (Figure 10.2d–f), especially after spectral tying at 2.03 µm. This greatly reduces uncertainty in the mixture modeling results. Based on this finding, the method called Automated Monte Carlo Unmixing (AutoMCU; Asner and Heidebrecht, 2002) was developed to provide fractional cover estimates of PV, NPV, and soil along with uncertainty bounds on a pixel by pixel basis. The reported uncertainty is based on variability remaining in the SWIR spectra (Figure 10.2d–f).

Many have applied the AutoMCU approach to shrubland and savanna ecosystems. Asner and Heidebrecht (2005) used one of the few time series of AVIRIS data to show that the fundamental response of semiarid vegetation to climate variability shifts after landscape-scale changes in vegetation structure associated with desertification. In a temperate savanna of southern Texas, the United States, Martin et al. (2003) combined fractional PV and NPV cover maps derived from AVIRIS and the AutoMCU, with a field network of soil measurements, to show that encroachment of a nitrogen-fixing tree increased soil nitrogen oxide emissions in a spatially explicit manner. The stability of the AutoMCU output results from its strategic use of a portion of the reflectance spectrum in which endmembers are largely unique from one another, and in which the endmembers have relatively low levels of spectral variability. For this reason, the AutoMCU approach is being considered as a core algorithm in the planned NASA HyspIRI spaceborne imaging spectrometer mission (http://hyspiri.jpl.nasa.gov). With this mission, fractional cover mapping of surface materials will become routine throughout the world's savannas.

LiDAR for Fractional Cover Mapping

The optical remote sensing approaches discussed thus far primarily operate across two-dimensional space (x and y). From a remote sensing perspective, the third dimension of height (z) can sometimes be inferred from textural or

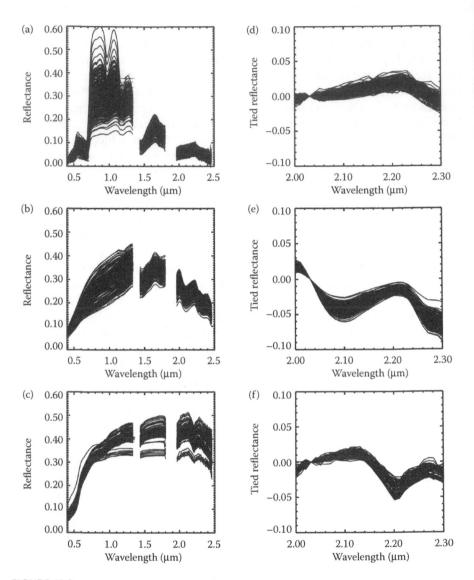

FIGURE 10.2
Variability in the spectral reflectance (0.4–2.5 mm) of (a) photosynthetic vegetation, (b) nonphotosynthetic vegetation, and (c) bare mineral soil collected across a range of arid and semiarid ecosystems in North and South America. Despite the broad variation in these spectral endmember groups, the shortwave-infrared (SWIR) region from 2.0 to 2.5 mm collapses to much less variable endmember bundles, especially after normalization by subtraction (spectral tying) at 2.03 mm (d–f subsets). (Based on previous work by Asner, G. P. and D. B. Lobell. 2000. *Remote Sensing of Environment* 74, 99–112. © 2000. With permission from Elsevier.)

angular analyses of two-dimensional optical data; (Widlowski et al., 2004) but laser altimetry, or LiDAR (light detection and ranging), provides an alternative technique to actively measure the third dimension. The LiDAR sensors can directly measure the 3-D distribution of vegetation canopies, in addition to Earth surface topography, thus providing highly accurate estimates of vegetation height, cover, and canopy structure (Lefsky et al., 2002). The LiDAR has been extensively used for vegetation characterization in forested systems; but it has also proved valuable in semiarid landscapes (e.g., Mundt et al., 2006), where cover is sparse; and distinguishing the woody fraction from that of bare ground or herbaceous layer is challenging with passive optical sensors. The LiDAR can readily distinguish the woody vegetation canopy component from that of bare ground or herbaceous material based on the return pattern of the laser pulses (Figure 10.3a). Moreover, in addition to delineating woody canopy cover, the use of LiDAR enables accurate estimations of top-of-canopy height (Figure 10.3b).

Although the spatial representation of these characteristics holds major potential for savanna ecology, exploration of the LiDAR point cloud itself can provide even deeper insight into the 3-D distribution of the woody fraction. For example, a cross-section through a LiDAR point cloud collected over savanna reveals the distribution of laser returns within the vegetation canopy itself, and not just the ground level and top-of-canopy height (Figure 10.3c). Although a sub-canopy layer is often absent in savanna settings, the point cloud distribution is still useful for determining the lower extent of the primary canopy. Exploration of the frequency of occurrence of LiDAR returns at different height intervals, for a given horizontal scale, enables the distribution of these returns to be summarized for a particular area of interest (Figure 10.3d). Such a perspective provides unique 3-D insight into the distribution of woody cover fraction and enables a new exploration of factors regulating fractional cover in savannas (Asner et al., 2009; Levick et al., 2009).

The accuracy achieved by current LiDAR systems makes them the prime choice for detailed mapping of savanna 3-D structure. However, imaging radar has also shown potential for mapping vegetation structure in a range of ecosystems (Waring et al., 1995). Current methods for retrieving vegetation structural attributes from SAR, however, have been primarily developed in forests and are poorly suited to savanna systems due to the discontinuous nature of the canopy (Viergever et al., 2008). Nonetheless, given the ability of imaging radar to penetrate cloud cover, a common occurrence in savannas, deeper exploration of radar products in savannas is needed. In particular, we encourage greater use of LiDAR in calibrating and evaluating imaging radar data sets. Although LiDAR is a valuable tool in its own right, the fusion of LiDAR with data from other optical or radar sensors holds further advantages for the quantification of savanna fractional cover. Rango et al. (2000) elegantly utilized information derived from both LiDAR and high-resolution aerial photography to understand the relationship

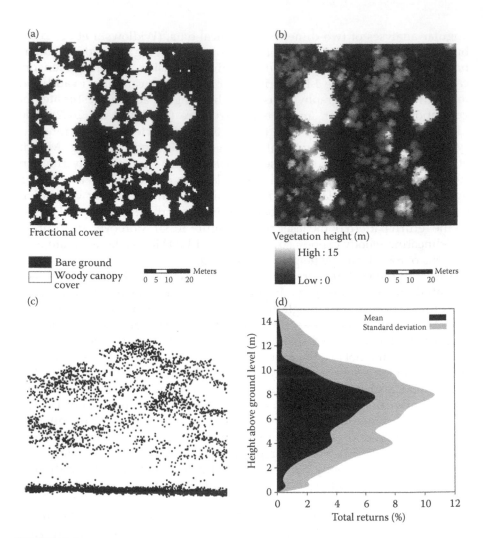

FIGURE 10.3
Small-footprint airborne LiDAR provides a means to measure the 3-D structure of woody canopies in a South African savanna: (a) fractional cover of woody vegetation and bare ground; (b) woody canopy height; (c) a cross-sectional view of canopy layering; and (d) spatially integrated vertical fractional cover of woody canopies.

between woody shrub encroachment and underlying sand dune development in New Mexico, the United States. Mundt et al. (2006) explored the use of hyperspectral imagery for mapping sagebrush in Idaho, the United States, and found that the inclusion of LiDAR data in the analysis boosted their classification accuracy from 74 to 89%. Similar results have been found in an African savanna, whereby the inclusion of LiDAR data increased classification accuracy of vegetation types from color aerial photography (Levick and Rogers, 2008).

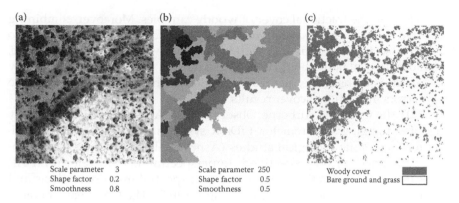

FIGURE 10.4
Object-based image analysis (OBIA) is well suited to fractional cover classification in heterogeneous savannas. (a) Fine-scale segmentation of aerial photograph ensuring image objects is smaller than the smallest tree in the scene; (b) Broad-scale segmentation of the same scene to provide context for the object-based classification rule set; and (c) classification of the woody cover fraction based on brightness differences between fine-scale and broad-scale image objects. (Reprinted from Levick, S. R. and K. H. Rogers. 2008. In *Object-Based Image Analysis: Spatial Concepts for Knowledge-Driven Remote Sensing Applications*, eds. T. Blaschke, S. Lang, and G. Hay, Springer, Berlin, pp. 477–492. © 2008. With kind permission of Springer Science+Business Media.)

An issue raised by Levick and Rogers (2008) was the value of adopting an object-based image analysis (OBIA) approach when fusing imagery and LiDAR in a savanna setting. The OBIA utilizes segmentation algorithms to divide remotely sensed images into discrete regions or objects that are homogenous with regard to spatial or spectral characteristics (Ryherd and Woodcock, 1996). In semiarid landscapes, OBIA outperforms pixel-based approaches to woody canopy cover classification (Laliberte et al., 2004), primarily through its ability to utilize spectral information from different scales and through the adoption of context-specific classification rules (Figure 10.4). Moreover, the inclusion of LiDAR data in the image segmentation phase of OBIA improves segmentation and classification results of aircraft and spaceborne imagery (Tiede et al., 2006). Therefore, the fusion of LiDAR with data from various passive optical sensors through OBIA is likely to greatly enhance fractional cover estimates in savannas.

A New Data Fusion Approach for Fractional Cover Mapping

As described earlier, imaging spectroscopy and, to some extent, multispectral imaging provide a way to map the fractional cover of photosynthetic and nonphotosynthetic vegetation as well as bare soil. The LiDAR provides a means to map the fractional cover of woody canopies. Moreover, combining these two technologies and their analytical approaches can provide a way to further dissect a landscape into woody canopies with and without foliage, live and senescent herbaceous cover, and bare soil. Combining spectroscopy and LiDAR is a challenge, as misregistration of the data types can lead to major errors in fractional cover results.

In 2007, the Carnegie Airborne Observatory (CAO) became the first fully integrated imaging spectrometer–LiDAR system made operational specifically for large-scale ecological studies (Asner et al., 2007). The system provides co-aligned, ortho-georectified, hyperspectral, and waveform-LiDAR imagery that facilitates the integrated use of spectral mixture analysis and LiDAR vegetation structural analysis (Figure 10.5). The CAO was operated over the Kruger National Park, South Africa, in 2008, yielding 3-D maps of terrain, vegetation structure, and species composition across large tracts of savanna. These data provided a new means to derive the cover fractions of woody and herbaceous vegetation along with their condition (live, senescent, and leafless) on an automated basis. The CAO mapped fractional cover in areas of Kruger with and without the presence of herbivores, showing that herbivore exclusion was correlated with up to 11-fold greater woody canopy cover and up to 80% less bare soil fractional cover (Asner et al., 2009). Differences in the woody canopy fraction were particularly striking on the northern basaltic soils of Kruger (Figure 10.6a and b), although only minor differences in the other fractions were observed at this site (Figure 10.6c and d).

FIGURE 10.5
(**See color insert following page 320.**) Detailed fusion of spectral and LiDAR data collected over the Kruger National Park, South Africa, by the Carnegie Airborne Observatory. (Adapted from Asner, G. P. et al. 2007. *Journal of Applied Remote Sensing* 1, DOI:10.1117/1111.2794018.) The imagery shows the 3-D partitioning of the landscape, including actual shadows cast by woody vegetation canopies onto the herbaceous and bare soil surface below. Co-alignment of imaging data opens new doors for quantitative 3-D fractional cover analysis of savannas.

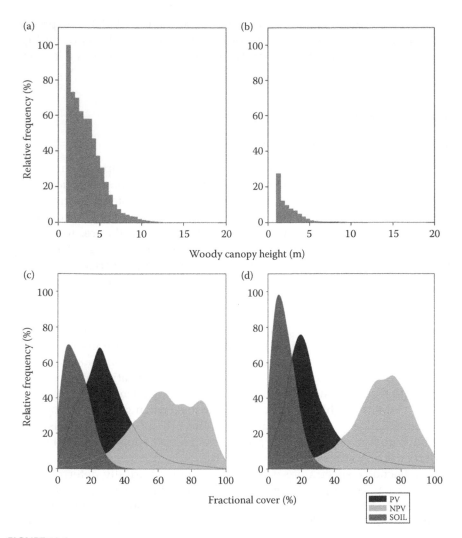

FIGURE 10.6
Differences in fractional cover between areas protected and accessible to herbivores on the northern basalts of the Kruger National Park, South Africa. (a) Height distribution of the woody cover fraction in the protected area; (b) height distribution of the woody cover fraction in the accessible area; (c) frequency distribution of the herbaceous and bare soil fractions in the protected area; and (d) frequency distribution of the herbaceous and bare soil fractions in the accessible area. (Adapted from Asner, G. P. et al. 2009. *Proceedings of the National Academy of Sciences* 106, 4947–4952.)

A follow-up study of herbivore-fire interactions indicated that long-term (34 year) herbivore exclusion was related to 36-fold greater woody canopy fractional cover, whereas short-term (7 year) fire suppression was related to less woody canopy cover in a restricted height class of 0–3 m, representing seedling to sapling stage individuals (Levick et al., 2009). In these early examples, the value of integrating hyperspectral and LiDAR data is expressed in the detailed mapping of the fractional cover and condition of multiple savanna lifeforms, with high certainty and immediate ecological relevance.

Mapping Canopy Chemistry

Compared with fractional cover mapping, studies of canopy chemistry are much less abundant in the savanna literature. Most canopy chemical mapping work has focused on forest and agricultural ecosystems. Nonetheless, novel progress has been made in a few savanna ecosystems. Four chemical groups have received attention, because these chemicals are, to varying degrees, expressed in the optical reflectance spectrum (400–2500 nm): pigments including chlorophylls, carotenoids, and tannins; nutrients including nitrogen and phosphorus; carbon fractions including primary lignin and cellulose; and canopy water content (Curran, 1989).

A number of studies have used multispectral imagery to derive vegetation indices, such as the Normalized Difference Vegetation Index (NDVI), which broadly correlates with vegetation greenness and herbaceous biomass in savannas (e.g., Ringrose and Matheson, 1991, Wessels et al., 2006, Skidmore and Ferwerda, 2008). Here we do not cover these vegetation index studies, but instead focus on quantitative analysis of canopy chemistry from remote sensing. Doing so highlights the unique role that imaging spectroscopy plays in savanna canopy chemistry studies, as the full spectrum analysis made possible with hyperspectral data provides a means to quantify leaf chemical properties (e.g., Ustin et al., 2004b).

The reflectance spectra of savanna vegetation is partially determined by canopy structural properties, especially when LAI is less than about 3 units (Asner et al., 2000). At higher LAI values, leaf chemistry can dominate the canopy reflectance spectrum. Since the LAI of many savanna canopies is less than three units (Asner et al., 2003b), remote sensing of canopy chemistry will necessarily be challenged by background variation in vegetation structure. Despite these limitations, important advances have been made in savanna canopy chemical mapping.

Photosynthetic and protective pigments, including chlorophylls, carotenoids, and tannins, have been remotely sensed at leaf and canopy scales in savanna and other semiarid regions. One particularly recent study represents the state of the art in pigment analysis: Cho et al. (2008) used hyperspectral data to estimate chlorophyll concentrations in semiarid vegetation in Italy. They found that the shape and position of the "red edge" (the transition from visible to near-infrared portions of the spectrum—690 to 800 nm) was most sensitive to chlorophyll variation. Ferwerda et al. (2006) used spectral data with a bootstrap analytical method to determine tannin concentrations ($r = 0.83$)

in South Africa savanna vegetation. They were able to explain the predictions using known spectral absorption features scattered throughout the shortwave-infrared range (1470–2175 nm). Their study also highlighted the advantage afforded by hyperspectral data in using the derivative of spectral reflectance, which produced the best prediction for tannin concentrations in foliage.

Foliar nutrients of savanna plants have also been estimated using hyperspectral data. At the leaf and canopy level using field spectrometers, nitrogen (N) concentration has been assessed in both herbaceous and woody canopies with a variety of spectroscopic methods. Mutanga et al. (2004) used field spectra with derivative and continuum-removal analysis, the latter of which isolates specific absorption features in the spectrum, to estimate N concentrations among five grass species. They developed regressions predicting N concentration with r^2-values reaching 0.60. They also highlighted the importance of measuring the full-range spectrum (0.4–2.5 µm), as spectral reflectance features spanning this range contributed to the successful prediction of foliar N (Figure 10.7). Ferwerda and Skidmore (2007) found that field spectra predicted N concentrations among four woody species, including the savanna tree *Colophospermum mopane* in Kruger National Park. Reflectance, derivative reflectance, and band-depth analysis produced regression coefficients relating predicted and measured N concentrations with $r^2 = 0.81$, 0.89, and 0.88, respectively.

Airborne imaging spectrometers have also been used to estimate leaf N concentrations in semiarid vegetation. Serrano et al. (2002) used airborne

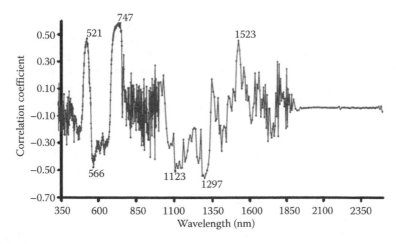

FIGURE 10.7
Correlation coefficients relating derivative spectral reflectance to leaf nitrogen concentrations in savanna grasses found in South Africa. Note that the spectral regions considered most important to the prediction of nitrogen concentrations are spread throughout the spectrum, suggesting that the entire wavelength range is important for hyperspectral studies of canopy chemistry. (Reprinted from Mutanga, O., A. Skidmore, and H. H. T. Prins. 2004. *Remote Sensing of Environment* 89, 393–408. © Elsevier. With permission.)

AVIRIS imagery to estimate N concentrations ($r^2 = 0.58$); they were able to use a combination of spectral regression analyses to decompose the relative contributions of canopy structure and biochemistry to the AVIRIS data. Mutanga and Skidmore (2004) used airborne HyMap imagery, combined with an artificial neural network approach, to highlight the coupled use of the visible and shortwave-infrared spectral ranges for canopy N estimation in African savanna. They demonstrated a tight correlation between predicted and measured N concentrations among training data ($r^2 = 0.92$) and moderate correlation ($r^2 = 0.60$) with test data collected from a different portion of the landscape (Figure 10.8a).

Nitrogen is not the only nutrient of interest to savanna ecological or remote sensing studies. Phosphorus (P) and several micronutrients including calcium (Ca), magnesium (Mg), potassium (K), and sodium (Na) regulate plant growth and herbivory. However, most of these elements are only indirectly expressed in the optical spectrum, as none directly contributes to the absorption or scattering features in foliage. Nonetheless, indirect correlations do exist, partially by way of plant chemical stoichiometry (McGroddy et al., 2004) or by way of the secondary role these elements play in the formation and maintenance of biochemical compounds that do have a major role in determining leaf and canopy spectra. Porder et al. (2005) capitalized on the correlation between leaf N and P to map canopy phosphorus concentrations using AVIRIS imagery and canopy radiative transfer model inversions. Mutanga and Kumar (2007) used HyMap imagery and a neural network algorithm to estimate leaf P concentrations among African savanna grasses. They achieved an r^2-value of 0.63 (Figure 10.8b), which was similar to that achieved for nitrogen (Figure 10.8a). Ferwerda and Skidmore (2007) tested a variety of transformations on field spectral data to estimate P, K, Ca, Mg, and Na concentrations in the foliage of four woody savanna canopy species in the Kruger National Park, South Africa (Figure 10.9). A close inspection of their chemical data suggested that inter-correlations between these nutrients and nitrogen may partially account for the apparent ability to estimate multiple nutrients.

Water is among the most important factors regulating plant growth and development in savannas and other semiarid ecosystems (Coughenour and Ellis, 1993). Remote sensing of leaf water concentration and canopy water content have rapidly progressed in recent years, allowing for routine mapping of vegetation water status using airborne and space-based imaging spectrometers. Gao and Goetz (1995) were among the first to quantitatively retrieve canopy water content from AVIRIS imagery using water absorption features located in the near-infrared portion of the reflectance spectrum (features centered near 0.94 and 1.2 µm). Ceccato et al. (2001) developed a unique quantitative approach for isolating leaf-level water concentration from hyperspectral imagery; the technique required a combination of near-infrared and shortwave-infrared measurements (1.0–2.5 µm). A variety of studies have applied spectroscopic techniques to estimate canopy water content in

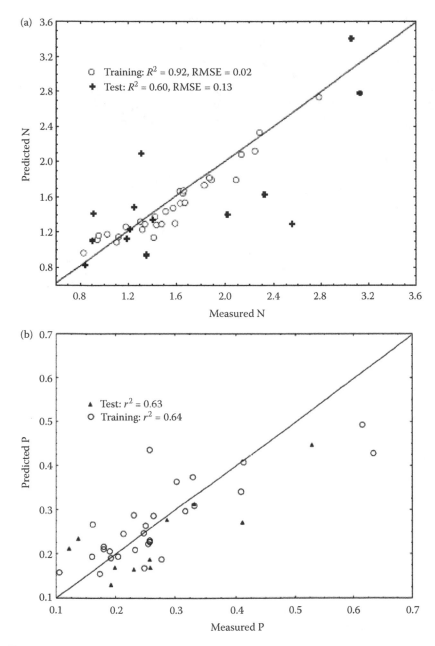

FIGURE 10.8
(a) Regression showing the relationship between measured and predicted nitrogen (N) concentrations in African savanna using airborne imaging spectroscopy and an artificial neural network (Reprinted from Mutanga, O. and A. Skidmore. 2004. *Remote Sensing of Environment* 90, 104–115. © 2004. With permission from Elsevier.) (b) Similar but for leaf phosphorus (P) concentration. (Reprinted from Mutanga, O. and L. Kumar. 2007. *International Journal of Remote Sensing* 28, 1–15. © 2007. With permission from Taylor and Francis.)

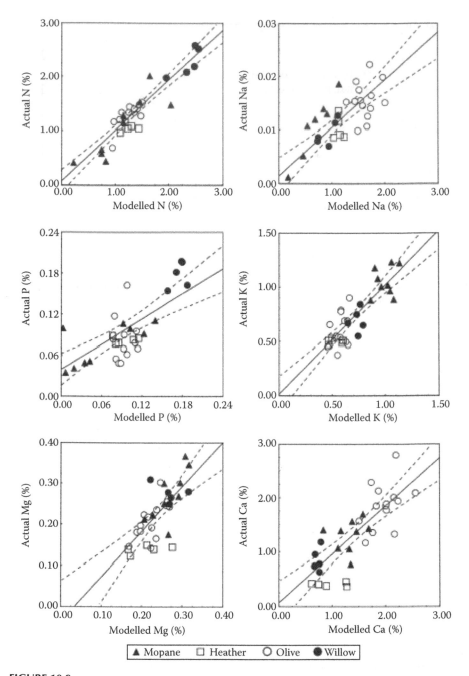

FIGURE 10.9

Estimation of multiple nutrient concentrations among four woody species using field spectroscopy. (Reprinted from Ferwerda, J. G., and A. Skidmore. 2007. *ISPRS Journal of Photogrammetry and Remote Sensing* 62, 406–414. © 2007. With permission from Elsevier.) Mopane is a savanna species found in northern Kruger National Park, South Africa.

semiarid regions: Ustin et al. (1998) and Serrano et al. (2000) used AVIRIS imagery to map water content in chaparral species. Asner et al. (in preparation) mapped canopy water content among a range of savanna vegetation types in the Cerrado region of Brazil using the EO-1 Hyperion satellite sensor (Figure 10.10). They showed that water content estimation can be used to quantify spatial variability in the timing of senescence during the annual dry season. Water content is another operational algorithm that will likely be part of the planned NASA HyspIRI satellite mission.

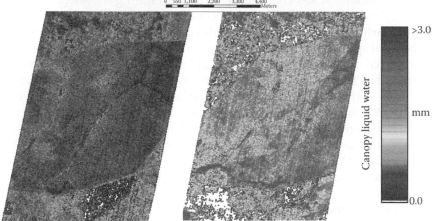

End of wet season   End of dry season

FIGURE 10.10
(See color insert following page 320.) Canopy water content in the Brazilian Cerrado savannas using Earth Observing-1 Hyperion imaging spectrometer data. Notice how canopy water content decreased variably across this savanna region based on differences in plant physiognomy: Campo Limpo, Campo Sujo, and Campo Ralo ecosystems are dominated by herbaceous species; Cerradao, Cerrado Tipico, and Cerrado Denso are areas dominated by drought-tolerant woody species.

Carbon fractions including lignin and cellulose are also of interest to ecological and remote sensing studies of savannas. The concentrations of lignin
and cellulose are predictors of a variety of ecosystem processes including fire
behavior and decomposition rates. The ligno-cellulose absorption features
centered at 1.75, 2.1, and 2.3 μm (Figure 10.2b) have proved particularly useful
in detecting not only the presence of dry vegetation (dead and senescent) but
also the leaf-level concentrations of these high molecular-weight chemicals.
Serrano et al. (2002) used AVIRIS imagery to estimate lignin concentrations
among chaparral species using the 1.75 μm absorption feature. Nagler et al.
(2000) used cellulose absorption features in the 2.0–2.3 mm wavelength region
to estimate surface residues from decomposing plants. Despite these promising results, to our knowledge, no studies aimed at estimating lignin or cellulose concentrations in savanna vegetation have been reported.

Conclusions and Research Needs

Fractional canopy cover and canopy chemistry continue as expanding foci in
remote sensing and ecological research on savannas and many other ecosystems. In this chapter, we emphasized the particularly strong role that hyperspectral and LiDAR remote sensing plays in studies of savanna cover and
chemistry. This apparent bias simply mirrors the literature, which presents
these technologies as key contributors to understanding ecosystem structure
and function. Although the fractional cover mapping approaches described
here are fairly mature and could be made operational in savanna studies, the
biochemical approaches have mostly come from forest and agricultural studies. This has resulted in a relatively limited understanding of how well
hyperspectral (and other) remote sensing can be used to estimate canopy
chemicals. Early savanna studies summarized here suggest that chemical
constituents can be quantified, even among a backdrop of highly varying
canopy structure. New integrated systems that combine technologies which
measure complementary aspects of vegetation—such as imaging spectrometers and LiDAR—will improve the estimation of fractional cover (structure) and canopy chemistry by providing orthogonal information to constrain
the interpretation of the individual data sets. In this context, systems such as
the Carnegie Airborne Observatory should be replicated and applied to a
range of savanna and other ecosystems.

Although the most accurate fractional cover and canopy chemical mapping has been demonstrated using imaging spectrometers and LiDAR, the
science has been limited in scope to a relatively few research groups.
Commercial LiDAR data sources are expanding much faster than hyperspectral sources. This likely reflects the higher state of readiness in LiDAR
measurements and, reciprocally, the continued research and development
needed to advance the hyperspectral approaches for canopy chemistry.

Imaging spectrometers face the issue of sensor fidelity, which determines the quality and applicability of spectroscopic reflectance signatures. Although many imaging spectrometers have been built, only a very small number provide high-fidelity data required for true spectroscopic remote sensing studies. The AVIRIS and HyMap are among the very few airborne sensors to provide the requisite data; no spaceborne spectrometers [including EO-1 Hyperion; Ungar et al. (2003)] have done so. With that in mind, the new NASA HyspIRI satellite project and the European Union's EnMap mission will eventually provide global access to hyperspectral data, with cascading effects on the information that can be provided for societal benefit (Hill et al., 2005). Current literature focuses primarily on one-time mapping of fractional cover and canopy chemistry, with a noticeable gap in studies exploring repeatable mapping with the aim of spatiotemporal monitoring. This void between snapshot mapping and multi-temporal monitoring needs to be crossed in order for remote sensing technologies to improve our understanding of ecological change in savannas.

Acknowledgments

We thank A. Skidmore for providing input on canopy chemistry mapping. We also thank M. J. Hill and N. P. Hanan for providing editorial guidance and two anonymous reviewers of the chapter. This work was supported by the Andrew Mellon Foundation.

References

Asner, G. P., S. Archer, R. F. Hughes, R. J. Ansley, and C. A. Wessman. 2003a. Net changes in regional woody vegetation cover and carbon storage in Texas drylands, 1937–1999. *Global Change Biology* 9, 316–335.

Asner, G. P., C. A. Wessman, C. A. Bateson, and J. L. Privette. 2000. Impact of tissue, canopy and landscape properties on the reflectance of arid ecosystems. *Remote Sensing of Environment* 74, 69–84.

Asner, G. P. and K. B. Heidebrecht. 2002. Spectral unmixing of vegetation, soil and dry carbon cover in arid regions: Comparing multispectral and hyperspectral observations. *International Journal of Remote Sensing* 23, 3939–3958.

Asner, G. P. and K. B. Heidebrecht. 2005. Desertification alters regional ecosystem–climate interactions. *Global Change Biology* 11, 182–194.

Asner, G. P., D. E. Knapp, T. Kennedy-Bowdoin, et al. 2007. Carnegie Airborne Observatory: In-flight fusion of hyperspectral imaging and waveform light detection and ranging (LiDAR) for three-dimensional studies of ecosystems. *Journal of Applied Remote Sensing* 1, DOI:10.1117/1111.2794018.

Asner, G. P., S. R. Levick, T. Kennedy-Bowdoin, et al. 2009. Large-scale impacts of herbivores on the structural diversity of African savannas. *Proceedings of the National Academy of Sciences* 106, 4947–4952.

Asner, G. P. and D. B. Lobell. 2000. A biogeophysical approach for automated SWIR unmixing of soils and vegetation. *Remote Sensing of Environment* 74, 99–112.

Asner, G. P., J. M. O. Scurlock, and J. A. Hicke. 2003b. Global synthesis of leaf area index observations: Implications for ecological and remote sensing studies. *Global Ecology and Biogeography* 12, 191–205.

Asner, G. P., C. A. Wessman, C. A. Bateson, and J. L. Privette. 2000. Impact of tissue, canopy and landscape factors on reflectance variability in spectral mixture analysis. *IEEE Transactions on Geoscience and Remote Sensing* 38, 1083–1094.

Asner, G. P., C. A. Wessman, and D. S. Schimel. 1998. Heterogeneity of savanna canopy structure and function from imaging spectrometry and inverse modeling. *Ecological Applications* 8, 1022–1036.

Babey, S. K. and C. D. Anger. 1989. A Compact Airborne Imaging Spectrographic Imager (CASI). *IGARSS '89: 12th Canadien Symposium on Remote Sensing* 2, 1028–1031.

Bateson, C. A., G. P. Asner, and C. A. Wessman. 2000. Endmember bundles: a new approach to incorporating endmember variability into spectral mixture analysis. *IEEE Transactions on Geoscience and Remote Sensing* 38, 1083–1094.

Borel, C. C. and S. A. W. Gerstl. 1994. Nonlinear spectral mixing models for vegetative and soil surfaces. *Remote Sensing of Environment* 47, 403–416.

Ceccato, P., S. Flasse, S. Tarantola, S. Jacquemoud, and J. M. Gregoire. 2001. Detecting vegetation leaf water content using reflectance in the optical domain. *Remote Sensing of Environment* 77, 22–33.

Cho, M. A., A. K. Skidmore, and C. Atzberger. 2008. Towards red-edge positions less sensitive to canopy biophysical parameters for leaf chlorophyll estimation using properties optique spectrales des feuilles (PROSPECT) and scattering by arbitrarily inclined leaves (SAILH) simulated data. *International Journal of Remote Sensing* 28, 2241–2255.

Cocks, T., R. Jenssen, A. Stewart, I. Wilson, and T. Shields. 1998. The HyMap airborne hyperspectral sensor: The system, calibration and performance. In *1st EARSEL Workshop on Imaging Spectroscopy*, Zurich.

Coughenour, M. B. and J. E. Ellis. 1993. Landscape and climatic control of woody vegetation in a dry tropical ecosystem: Turkana district, Kenya. *Journal of Biogeography* 20, 383–398.

Curran, P. J. 1989. Remote sensing of foliar chemistry. *Remote Sensing of Environment* 30, 271–278.

Defries, R. S., M. C. Hansen, J. R. G. Townshend, A. C. Janetos, and T. R. Loveland. 2000. A new global 1-km dataset of percentage tree cover derived from remote sensing. *Global Change Biology* 6, 247–254.

Ferwerda, J. G. and A. Skidmore. 2007. Can nutrient status of four woody plant species be predicted using field spectrometry? *ISPRS Journal of Photogrammetry and Remote Sensing* 62, 406–414.

Ferwerda, J. G., A. K. Skidmore, and A. Stein. 2006. A bootstrap procedure to select hyperspectral wavebands related to tannin content. *International Journal of Remote Sensing* 27, 1413–1424.

Gao, B.-C. and A. F. H. Goetz. 1995. Retrieval of equivalent water thickness and information related to biochemical components of vegetation canopies from AVIRIS data. *Remote Sensing of Environment* 52, 155–162.

Green, R. O., M. L. Eastwood, C. M. Sarture, et al. 1998. Imaging spectroscopy and the Airborne Visible Infrared Imaging Spectrometer (AVIRIS). *Remote Sensing of Environment* 65, 227–248.

Guerschman, J., M. J. Hill, L. Renzullo, D. Barrett, A. Marks, and E. Botha. 2009. Estimating fractional cover of photosynthetic vegetation, non-photosynthetic vegetation and bare soil in the Australian tropical savanna region upscaling the EO-1 Hyperion and MODIS sensors. *Remote Sensing of Environment* 113, 928–945.

Hill, M. J., G. P. Asner, and A. A. Held. 2005. A biogeophysical approach to remote sensing of vegetation in coupled human-environment systems - Societal benefits and global context. *Journal of Spatial Science* 53, 49–66.

Huang, C., G. P. Asner, R. E. Martin, N. N. Barger, and J. C. Neff. 2009. Multiscale analysis of tree cover and aboveground carbon stocks in pinyon-juniper woodlands. *Ecological Applications* 19, 668–681.

Huntley, B. and B. Walker, editors. 1982. *Ecology of Tropical Savannas*. Springer Verlag, Berlin.

Kokaly, R. F., G. P. Asner, S. V. Ollinger, M. E. Martin, and C. A. Wessman. 2009. Characterizing canopy biochemistry from imaging spectroscopy and its application to ecosystem studies. *Remote Sensing of Environment* 113, S78-S91.

Kumar, L., K. Schmidt, S. Dury, and A. Skidmore. 2001. Imaging spectrometry and vegetation science. In, *Imaging Spectrometry: Basic Principles and Prospective Applications*, ed. F. D. van der Meer and S. M. de Jong, Kluwer Academic Publishers, Dordrecht, pp. 111–155.

Laliberte, A. S., A. Rango, K. M. Havstad, et al. 2004. Object-oriented image analysis for mapping shrub encroachment from 1937 to 2003 in southern New Mexico. *Remote Sensing of Environment* 93, 198–210.

Lefsky, M. A., W. B. Cohen, G. G. Parker, and D. J. Harding. 2002. LiDAR remote sensing for ecosystem studies. *BioScience* 52, 19–30.

Levick, S. R., G. P. Asner, T. Kennedy-Bowdoin, and D. E. Knapp. 2009. The relative influence of fire and herbivory on savanna three-dimensional vegetation structure. *Biological Conservation* 142, 1693–1700.

Levick, S. R. and K. H. Rogers. 2008. Structural biodiversity monitoring in savanna ecosystems: Integrating LiDAR and high resolution imagery through object-based image analysis. In *Object-Based Image Analysis: Spatial Concepts for Knowledge-Driven Remote Sensing Applications*, ed. T. Blaschke, S. Lang, and G. Hay, Springer, Berlin, pp. 477–492.

Martin, R. E., G. P. Asner, R. J. Ansley, and A. R. Mosier. 2003. Effects of woody encroachment on soil nitrogen oxide emissions in a temperate Savanna. *Ecological Applications* 13, 897–910.

McGroddy, M. E., T. Daufresne, and L. O. Hedin. 2004. Scaling of C:N:P stoichiometry in forests worldwide: Implications of terrestrial Redfield-type ratios. *Ecology* 85, 2390–2401.

Mundt, J., D. Streutker, and N. Glenn. 2006. Mapping sagebrush distribution using fusion of hyperspectral and lidar classifications. *Photogrammetric Engineering and Remote Sensing* 72, 47.

Mutanga, O. and L. Kumar. 2007. Estimating and mapping grass phosphorus concentration in an African savanna using hyperspectral image data. *International Journal of Remote Sensing* 28, 1–15.

Mutanga, O. and A. Skidmore. 2004. Integrating imaging spectroscopy and neural networks to map grass quality in the Kruger National Park, South Africa. *Remote Sensing of Environment* 90, 104–115.

Mutanga, O., A. Skidmore, and H. H. T. Prins. 2004. Predicting *in situ* pasture quality in the Kruger National Park, South Africa, using continuum-removed absorption features. *Remote Sensing of Environment* 89, 393–408.

Nagler, P. L., C. S. T. Daughtry, and S. N. Goward. 2000. Plant litter and soil reflectance. *Remote Sensing of Environment* 71, 207–215.

Porder, S., G. P. Asner, and P. M. Vitousek. 2005. Ground-based and remotely sensed nutrient availability across a tropical landscape. *Proceedings of the National Academy of Sciences* 102, 10909–10912.

Rango, A., M. Chopping, J. Ritchie, et al. 2000. Morphological characteristics of shrub coppice dunes in desert grasslands of southern New Mexico derived from scanning LIDAR. *Remote Sensing of Environment* 74, 26–44.

Ringrose, S. and W. Matheson. 1991. A Landsat analysis of range conditions in the Botswana Kalahari drought. *International Journal of Remote Sensing* 12, 1023–1051.

Roberts, D. A., M. Gardner, R. Church, et al. 1998. Mapping chaparral in the Santa Monica Mountains using multiple endmember spectral mixture models. *Remote Sensing of Environment* 65, 267–279.

Roberts, D. A., M. O. Smith, and J. B. Adams. 1993. Green vegetation, nonphotosynthetic vegetation, and soils in AVIRIS data. *Remote Sensing of Environment* 44, 255–269.

Ryherd, S. and C. Woodcock. 1996. Combining spectral and texture data in the segmentation of remotely sensed images. *Photogrammetric Engineering & Remote Sensing* 62, 181–194.

Serrano, L., J. Penuelas, and S. L. Ustin. 2002. Remote sensing of nitrogen and lignin in Mediterranean vegetation from AVIRIS data: Decomposing biochemical from structural signals. *Remote Sensing of Environment* 81, 355–364.

Serrano, L., S. L. Ustin, D. A. Roberts, J. A. Gamon, and J. Peñuelas. 2000. Deriving water content of chaparral vegetation from AVIRIS data. *Remote Sensing of Environment* 74, 570–581.

Skidmore, A. and J. G. Ferwerda. 2008. Resource distribution and dynamics. In H. H. T. Prins and F. van Langevelde, editors. *Resource Ecology: Spatial and Temporal Dynamics of Foraging*. Springer, Berlin, pp. 57–77.

Theseira, M. A., G. Thomas, and C. A. D. Sannier. 2002. An evaluation of spectral mixture modelling applied to a semi-arid environment. *International Journal of Remote Sensing* 23, 687–700.

Tiede, D., S. Lang, and C. Hoffmann. 2006. Supervised and forest type-specific multiscale segmentation for a one-level-representation of single trees. *International Archives of Photogrammetry, Remote Sensing and Spatial Information Sciences*, XXXVI-4/C 42.

Ungar, S. G., J. S. Pearlman, J. A. Mendenhall, et al. 2003. Overview of the earth observing one (EO-1) mission. *IEEE Transactions on Geoscience and Remote Sensing* 41, 1149–1159.

Ustin, S. L., A. A. Gitelson, S. Jacquemoud, et al. 2009. Retrieval of foliar information about plant pigment systems from high resolution spectroscopy. *Remote Sensing of Environment* 113, Supplement 1, S67–S77.

Ustin, S. L., S. Jacquemoud, P. J. Zarco-Tejada, and G. P. Asner. 2004a. Remote sensing of environmental processes: State of the science and new directions. In S. L. Ustin, ed. *Manual of Remote Sensing Vol. 4. Remote Sensing of Natural Resource Management and Environmental Monitoring.* ASPRS, New York, pp. 679–730.

Ustin, S. L., D. A. Roberts, J. A. Gamon, G. P. Asner, and R. O. Green. 2004b. Using imaging spectroscopy to study ecosystem processes and properties. *Bioscience* 54, 523–534.

Ustin, S. L., D. A. Roberts, J. Pinzon, et al. 1998. Estimating canopy water content of chaparral shrubs using optical methods. *Remote Sensing of Environment* 65, 280–291.

Viergever, K. M., I. H. Woodhouse, and N. Stuart. 2008. Monitoring the world's savanna biomass by earth obervation. *Scottish Geographical Journal* 124, 218–225.

Waring, R. H., J. Way, E. R. Hunt, et al. 1995. Imaging radar for ecosystem studies. *BioScience* 45, 715–723.

Wessels, K. J., S. D. Prins, N. Zambatis, S. MacFadyen, P. E. Frost, and D. van Zyl. 2006. Relationship between herbaceous biomass and 1-km^2 Advanced Very High Resolution Radiometer (AVHRR) NDVI in Kruger National Park, South Africa. *International Journal of Remote Sensing* 27, 951–973.

Wessman, C. A., C. A. Bateson, and T. L. Benning. 1997. Detecting fire and grazing aptterns in tallgrass prairie using spectral mixture analysis. *Ecological Applications* 7, 493–512.

Widlowski, J., B. Pinty, N. Gobron, M. Verstraete, D. Diner, and A. Davis. 2004. Canopy structure parameters derived from multi-angular remote sensing data for terrestrial carbon studies. *Climatic Change* 67, 403–415.

11

Woody Fractional Cover in Kruger National Park, South Africa: Remote Sensing–Based Maps and Ecological Insights

Gabriela Bucini, Niall P. Hanan, Randall B. Boone, Izak P. J. Smit,
Sassan S. Saatchi, Michael A. Lefsky, and Gregory P. Asner

CONTENTS

Introduction

The variability in woody vegetation structure characterizes different types of savannas and is usually described by quantitative variables such as fractional cover, height, and biomass. Woody cover has been shown to directly affect important ecological processes of savanna ecosystems. It influences biomass production, fire regimes, herbivory, nutrient cycling, hydrology, and soil erosion (Rietkerk et al., 1997; Scholes and Archer, 1997; Archibald et al., 2009). Savannas are characterized by varied spatial combinations and relative proportions of woody and herbaceous vegetation. This heterogeneity results in diversified habitats and resources determining the large number of species that can establish and survive in these ecosystems. The ability to monitor woody fractional cover is therefore fundamental to understand savanna dynamics.

Mean annual precipitation (MAP) determines a potential maximum cover for woody plants, which however is rarely reached, because other regional and local factors intervene (Sankaran et al., 2005; Bucini and Hanan, 2007). The variability in observed woody cover at the regional scale can be the result of inter-annual variability in rainfall as well as local variability in soil properties and geomorphology, disturbance history, and patch dynamics, all of which impact tree-grass interactions and woody plant survival and recruitment (Ogle and Reynolds, 2004; Wiegand et al., 2006; Groen et al., 2008; Sankaran et al., 2008).

Kruger National Park (KNP) has a tradition of science-informed management that has recently focused on maintenance of heterogeneity as a means to create resilience. Woody cover assessments have direct impacts on the Park management decisions in relation to the extent they reflect and control fire effects, herbivore utilization, elephant damage, animal habitat suitability, and hydrology. Eckhardt et al. (2000), in a most recent study based on aerial photography covering the KNP southern-central part, found that between 1940 and 1998, woody cover slowly increased on granite bedrocks and decreased on basalt bedrocks. The most compelling result was a significant decline in trees >5 m in all sampled transects. Elephant numbers and interactions between elephant damage and fire were found to be the principal drivers of this decrease. The scale of woody cover dynamics in KNP has a combined local and regional nature, but the Park still lacks high-resolution wall-to-wall woody cover maps that allow a thorough cover assessment.

Current techniques to estimate woody canopy cover at the regional level are based on remote-sensing scaling-up approaches. At the fine scale, field instruments, aerial photographs, and high-resolution images provide cover estimates that can be used to calibrate remote-sensing images with coarser resolution and full coverage of the study area. The main difficulty in woody cover detection in arid and semiarid systems is to separate woody vegetation from the background of grass and soil. Research on woody cover mapping has tended to concentrate on either optical or radar-based techniques. Although optical reflectances and derived vegetation indices are very sensitive to photosynthetically active vegetation (e.g., green leaf area), they are not optimal to differentiate between woody and herbaceous attributes (Glenn et al., 2008). Synthetic aperture radar sensors, operating in the microwave C, L, and P bands, are instead sensitive to woody structure and biomass, including in low-density savanna situations (Santos et al., 2002; Lucas et al., 2006; Saatchi et al., 2007). The radar backscatter signal, however, is affected by spatiotemporal variability in soil and canopy moisture as well as surface roughness. The potential for combined optical and SAR systems should be explored, given the limitations and strengths in discriminating unique woody vegetation signatures using each technology separately.

In this case study, optical and radar imagery are combined to produce a woody cover map for the whole KNP at medium resolution (~90 m) and a map that quantifies woody-cover heterogeneity as a potential metric for biodiversity studies (Rogers, 2003). Then the mapped woody cover patterns are analyzed in relation to environmental and ecological factors using a classification and regression tree (CART) approach to construct an explanatory model of woody cover distribution (e.g., Breiman et al., 1984; De'Ath and Fabricius, 2000).

Study Area

The KNP is located in the lowveld semiarid savanna of northeastern South Africa (22.3–25.5° S, 30.8–32° E; Figure 11.1). The northern part of the park receives 300–500 mm MAP and is classified as arid bushveld (Venter et al.,

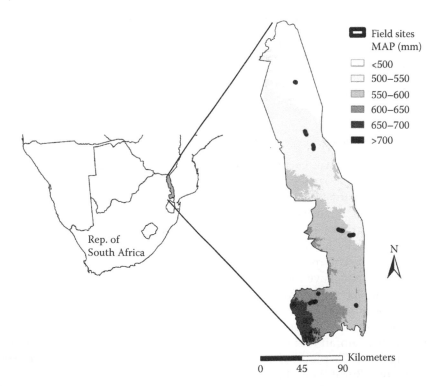

FIGURE 11.1
KNP and the distribution of the field plots across the MAP gradient.

2003). The southern part receives 500–700 mm MAP and is classified as lowveld bushveld. The long-term MAP is 506.6 ± 144 (mean ± standard deviation), and rainfall distribution has a monomodal pattern where the rainy season occurs mainly between October and March (austral summer). The KNP exhibits a longitudinal geomorphological division between underlying basalt bedrock (east) that forms soils rich in clay minerals and granite bedrock (west) that produces coarser sandy soils. The Park's terrain is fairly level (average slope of 1.6 ± 2.5°) except for the presence of few geological formations, varying from rugged outcrops and gorges to floodplains, in the north, in the southwestern corner, and along the eastern border.

The Park is subjected to natural disturbance, including drought cycles, fire, and herbivory. The physiognomy of the woody plants varies and reflects species genetic characteristics as well as phenotypic adaptations to disturbance. Woody plant heights fall in the 2–5 m range and can assume the forms of trees and shrubs with single or multi-stemmed physiognomy. Based on 1955–2004 data, fires occur with an overall average return of 4.4 years (frequency ~0.22). Fires are mostly surface fires, and they can suppress seedlings and saplings. Adult plants taller than 3 m generally survive and have the capacity to resprout from roots after a dieback. Large browsers (elephants, giraffes, and rhinos) have high potential to alter structure and compositions of plant communities (Shannon et al., 2008; Asner et al., 2009).

Data and Methods

Data

Field Measurements

The field campaign was carried out during April to May 2006. The 2005–2006 rainy season was above long-term average, and the trees still had their green canopy. We measured canopy cover using a handheld spherical densiometer (Lemmon, 1956), a simple and quick method comprising a convex mirror etched with a grid of 24 squares within which the observer estimates cover. Our measurements included woody plants taller than 1.3 m, and we subtracted canopy gaps present inside the crown. Due to the 60° angle of view, densiometer measurements have been shown to overestimate cover (Korhonen et al., 2006; Ko et al., 2009). However, savanna woody plants are generally scattered, and we conjectured that the bias would not be significant.

Field plots were distributed to sample the woody cover range in KNP as well as the rainfall and soil gradients but did not include riparian areas. We designated 73 plots of 250 × 250 m, with 63 plots laid out in the Experimental Burnt Plots (EBPs) and 10 plots in other areas. The EBPs have been maintained at different fire season and frequency treatments since 1954 (Biggs et al., 2003) and provide representative cover range levels in adjacent plots. In each plot, we collected 30 densiometer measurements on a regular 6 × 5 point grid with 50-m pace and averaged them to obtain the plot woody cover estimate. The sampled data ranged from 0.4% to 58% cover with an average of 23.8% ±13.7 percentage point (pp).

Remote Sensing Data and Processing

We sought medium resolution imagery with good quality and no cloud cover. We also wanted dry-season images to minimize the moisture contamination in the radar imagery and to maximize the contrast between photosynthetically active woody vegetation and senescent grass. When we started working (2006), the imagery available for the KNP with these characteristics included Landsat ETM+ and JERS-1 SAR.

The three orthorectified ETM+ images (source: Global Land Cover Facility, http://glcfapp.umiacs.umd.edu/index.shtml) were acquired between May 21 and 30, 2001 and covered the entire Park except for a small area (1%), for which we found an image from June 12, 2000. The images were mosaicked into a single scene at 28.5-m resolution using ENVI (http://www.ittvis.com) color adjustment, a method that uses a least-squares adjustment to ensure radiometric consistency among the scenes. Using the visible, near-infrared, and shortwave infrared bands, we calculated the NDVI, the Soil Adjusted Vegetation Index (SAVI; Huete, 1988), the Modified Soil Adjusted Vegetation Index (MSAVI; Qi et al., 1994), and the first three principal components (PCA). The 25-m resolution JERS-1 SAR imagery (Alaska Satellite Facility http://www.asf.alaska.edu/) included 11 georeferenced scenes of backscatter amplitude acquired in the L-band (frequency = 1275 MHz, wavelength = 23.5 cm) with HH polarization and at 35-degree off-nadir angle. Eight scenes were acquired between April 12 and August 20, 1996 and three scenes on March 12, 1995. In the two weeks before image acquisition, the average gauge rainfall was 20 ± 19 mm in 1996 and 7 ± 8 mm in 1995, which should not significantly contaminate the backscatter. We applied a Lee filter (standard deviation based filter) to smooth speckles in the radar scenes, but no radiometric corrections were made. Interaction with surface roughness produced some high backscatter areas unrelated to vegetation information. We corrected this effect by creating a mask for pixels with backscatter amplitude values >45 DN (Digital Number), which were generally facing east (aspect 45–135°) on a slope >2.5° and assigning them a reduced amplitude value of 45 DN. We mosaicked the radar images and coregistered them to the georectified Landsat mosaic (rms < 1 pixel, UTM 36 S projection), resampling at a common pixel resolution of 28.5 m. We then ran a focal analysis over all the remote-sensing layers to smooth image coregistration errors and worked at 85.5-m nominal resolution.

In between data acquisition dates, local woody cover changes likely occurred due to elephants and other large browsers' plant utilization, also in combination with fire. We decided to run the analysis while recognizing that the data could record local changes and give rise to some prediction instability.

We ultimately validated our model predictions using an independent woody cover data set obtained from the Carnegie Airborne Observatory (CAO, http://cao.stanford.edu) that integrates high-fidelity imaging spectrometers and waveform LiDAR sensors (Asner et al., 2007). The measurements were collected at 50 kHz laser pulse repetition frequency from an average flying altitude of 2000 m above ground level, with a ± 17 degree scan angle (after 2 degree cut-off) and 50% overlap between adjacent flight lines. This resulted in LiDAR data with 1.0 m laser spot spacing and 1.12 m beam diameter.

Geospatial Input Data for CART

The geospatial data included climate, geomorphology, soils, vegetation, herbivory, and fire data (Table 11.1). Most of the original data were provided without error estimates. We estimated errors for the variables derived from the original data through modeling (see notes in Table 11.1). Data with a temporal dimension (rainfall, fire, and herbivore census) were compiled into long-term

TABLE 11.1

Environmental and Ecological Variables Used in the Explanatory Model

Variable Name and Statistic	Notes
Climate	KNP database; 26 rain gauges, 1970–2004 Daily rainfall
Mean annual precipitation (mm)	Calculated: generalized regression model with x, y coordinates and elevation predictors ($R^2 = 0.75$, $p < 0.0001$, RSE = 38.8)
Variability of mean annual precipitation variability—coefficient of variation (CV)	Calculated: generalized regression model with x, y coordinates and elevation predictors ($R^2 = 0.31$, $p = 0.01$, RSE = 0.06)
Mean annual precipitation in the dry season (mm)	Calculated: generalized regression model with x, y coordinates predictors ($R^2 = 0.63$, $p < 0.0001$, RSE = 10.9)
Variability of mean annual precipitation in the dry season (CV)	Calculated: generalized regression model with x, y coordinates predictors ($R^2 = 0.3$, $p = 0.004$, RSE = 0.08)
Average length of dry spells in the wet season (no. of days)	Calculated: generalized regression model with x, y coordinates, elevation, and slope predictors ($R^2 = 0.8$, $p < 0.0001$, RSE = 0.65)
Number of dry days in the wet season	Calculated: Interpolation with regularized spline method (ESRI, http://www.esri.com/, ArcGIS 9.1)
Fire	KNP database; annual burnt areas (polygon file), 1955–2004
Fire frequency (number/year)	Calculated: sum of annual burnt area layers divided by 50-year time period
Herbivory	KNP database; annual dry-season aerial census, GPS points of herbivore counts, 1981–2005

TABLE 11.1 (continued)

Environmental and Ecological Variables Used in the Explanatory Model

Variable Name and Statistic	Notes
Elephant biomass density (mean kg/km²)	Calculated: For each census point and for each year, species counts were scaled to biomass using elephant average weight. An inverse-distance density function (ESRI, ArcGIS 9.1) with a radius of 40 km was applied to distribute point observations and create continuous elephant biomass maps. The annual mean of biomass maps gave the final map
Browsers and mixed-feeders biomass density (mean kg/km²)	Calculated: For each census point and for each year, species counts were scaled to biomass using browser and mixed-feeder species-specific average weights. Half of the mixed-feeders biomass was kept with the browser group, and half was put with the grazer group. An inverse-distance density function with a radius of 20 km was applied to distribute point observations and create continuous browser biomass maps. The annual mean of biomass maps gave the final map
Grazers and mixed-feeders biomass density (mean kg/km²)	Calculated: same as browser and mixed-feeders map but with gazer and mixed-feeder species-specific average weights
Geomorphology and Soils	KNP database; vectorized from the Chief Directorate Survey's and Mapping (CDSM) 20 m contour lines
Elevation (m asl)	Statistics: min = 88 m, max = 1172.3 m, mean = 376.1 m, sd = 124.3 m
Slope (degree)	Calculated (ESRI, ArcGIS 9.1); Statistics:min = 0°, max = 59.8°, mean = 2.1°, sd = 3.2°
Aspect (categorical)	Calculated (ESRI, ArcGIS 9.1): originally in degrees and then assigned to classes: north: (0–44° and 315–359°), east (45–134°), south (135–224°) and west (225–314°)
Basalt/granite bedrocks	
Soil texture (g/kg)	Subset on KNP
Total soil nitrogen (categorical)	Subset on KNP
Water	KNP database; rivers and water point maps
Distance to water (m)	Calculated (ESRI, ArcGIS 9.1): Euclidean distance function
Soil/Vegetation Complex	Polygon vegetation map (Gertenbach, 1983) and polygon bedrock map (Venter, 1990)
Soil vegetation	Calculated: Vegetation map was reclassified into three classes: mopane (*Colophospermum mopane*), non-mopane species, and sandveld communities. The intersection between the reclassified vegetation map and the bedrock map produced a five-class final map: mopane on basalt, mopane on granite, non-mopane on basalt, non-mopane on granite, and sandveld communities on granite

Note: The KNP source reference is the South African National Parks (SANParks) database (http://www.sanparks.org/parks/kruger/conservation/scientific/gis/). Most of the rainfall gauges had more than 20 years of data except for 3 gauges with 12 years of data. We kept elephants as a separate group, because they can affect woody vegetation not only as mixed feeders but also through direct mortality of trees. RSE = residual standard error; sd = standard deviation; " = same as above.

averages, discarding years with missing records. All data coverages were reprojected to a common format (UTM 36 S) and 1-km pixel spacing.

Analysis and Modeling

We carried out two modeling tasks based on two independent datasets: we first predicted woody cover using field measurements and remote-sensing data and then we explained woody cover by building a CART model (Brieman et al., 1984) with the environmental and ecological database (Table 11.1). The regression tree (CART) can be used to explain variation in a response variable such as woody cover, by inputting a range of explanatory variables to split the response variable into more homogeneous groups. Models may be selected by (a) iterative comparison based on a hypothesis test or (b) on the basis of the minimum error that can be assessed by a penalty for complexity using Akaike's Information Criteria (AIC; Akaike, 1973) or a cross validation or bootstrapping approach (De'Ath and Frabricius, 2000). All analysis and modeling were conducted using the software package R (version 8.0.2; http://www.r-project.org).

Woody Cover and Heterogeneity Maps

A relationship to calibrate the optical and radar data (predictors) was developed using the measured woody cover at our 73 field sites (response variable). The initial remote-sensing data set was reduced to eight covariates after eliminating one of a variable pair with Pearson correlations coefficient >0.8. Then, multiple linear regressions were run, and the AIC was used to select an optimal model. A jack-knife procedure was used to check the stability of the model parameter estimates. The best model was extrapolated to estimate woody cover percent over the entire KNP. We also quantified the relative contribution of the predictive variables to the total explanatory value of the model using the LMG metric (Gromping, 2006). From the map, the woody cover standard deviation was calculated as a metric for heterogeneity on a 1-km scale to identify where woody cover presents high variability hot spots. This is a scale at which many birds and mammals might perceive heterogeneity in the landscape.

Explanatory Ecological Model

We studied the mapped woody cover in association with ecological and environmental conditions (Table 11.1). We randomly sampled 1000 points (1-km square grid cells) with a minimum 2-km reciprocal distance (training set) and applied the regression tree analysis (CART, Breiman et al., 1984). Minimum cross-validation error was used to retrieve the optimal tree size. To verify the stability of the optimal tree, we created 10 independent random samples and grew 10 separate regression trees. After pruning, we compared them with the regression tree grown on the training set to check whether there was no, or minimal, deviation in tree structure and variable selection. A random forest analysis (Breiman, 2001; a combination of bootstrap aggregation and random selection of variables) was run to retrieve variable relative importance.

In order to validate the model, we applied the regression tree to 10 independent test samples of 1000 points and compared their predictions with the observed (mapped) woody cover. We evaluated the relative weight of the ecological or environmental versus spatial component in the woody cover patterns following the procedure described by Boone and Krohn (2000) based on Borcard et al. (1992). The total variance is portioned into independent components: pure environmental and ecological variation v_e; pure spatial variation v_s; environmental or ecological variation with spatial structuring v_{es}; and unexplained variation v_u. Accordingly, we created three types of woody cover models (CART):

1. Ecological (with ecological and environmental independent variables) explaining the variation

$$VE = v_e + v_{es} \tag{11.1}$$

2. Spatial (with UTM x and y coordinates as independent variables) explaining the variation

$$VS = v_s + v_{es} \tag{11.2}$$

3. Ecological spatial (with both ecological/environmental and spatial independent variables) explaining the variation

$$VES = v_e + v_s + v_{es}. \tag{11.3}$$

The single variation components can then be retrieved knowing that

$$v_{es} = VE + VS - VES \tag{11.4}$$

and

$$v_u = 1 - (v_e + v_s + v_{es}) \tag{11.5}$$

Result

Woody Cover Percent and Heterogeneity Maps

The optimal regression model (Table 11.2) that links field measurements to the remote-sensing variables explains 61% of the variability ($p < 0.0001$) with a residual standard error (RSE) of 8.9%. The residuals were near-normally distributed with a slight positive skew. The JERS-1 backscatter intensity and the Landsat green band together contributed 0.58 in the total R^2. According to the LMG metric for relative variable importance (normalized to sum 1), the JERS-1 backscatter intensity (LMG = 0.43 ± 0.14) and the Landsat green band (LMG = 0.38 ± 0.11) were the two most important variables to predict woody cover. The jack-knife analysis (Table 11.2) indicated model stability and absence of points in the data set with high leverage that could create biased estimates. The model performance (Figure 11.2) is good, but underestimations occur for some points having relatively dense woody cover (>50%). The validation of the model predictions against the LiDAR cover estimates gives strong agreement (Figure 11.2, $p < 0.0001$), and the RSE (=8.9%) was in the order of the model error.

The KNP woody cover map generated from the model is shown in Figure 11.3. The park is characterized by low to medium cover (WC = 34.7% ± 14.6 pp) with the granite landscapes (west) supporting higher woody cover (WC = 40.2% ± 11.8 pp) than basalt landscapes (WC = 24.4% ± 14.4 pp). Two emerging features are the sedimentary rocks (WC = 41.5% ± 15.3 pp) separating the granitic and basaltic bedrocks in the south and the intrusive gabbro sills in the granites with relatively low cover (WC = 19.7% ± 9.5 pp). The northern mountainous areas have the highest average cover (WC = 45.5% ± 13.4 pp). The heterogeneity map (Figure 11.2) brings to light the transitions between the savanna domain and the riparian areas (std. dev. > 14 %) and the outcrops (std. dev. 8–14%). Fairly large homogeneous patches occur on the basalt bedrocks and generally correspond to relatively flat upland areas far from drainage lines.

Woody Cover Explanatory Model

The best explanatory model (Figure 11.4) derived from the CART analysis explained 59.4 % of the variability in the mapped woody cover. The regression

TABLE 11.2

Woody Cover Regression Model

	Woody Cover Predictive Model			Jack-Knife Estimates		
	Estimate	Std. Error	p-value	Mean	Std. Error	Bias
Intercept	2.97	39.7	0.9404	3.00	33.88	1.62
JERS-1	1.35	0.4	0.0003	1.36	0.29	0.03
Landsat band2	−6.55	1.9	0.0013	−6.55	1.71	0.06
Landsat PCA3	−5.39	2.4	0.0255	−5.39	2.08	0.09
Landsat band5	3.01	1.3	0.0222	3.02	1.11	−0.06
SAVI	322.21	141.5	0.0259	322.12	123.57	−6.54

Source: Adapted from Efron, B. and R. J. Tibshirani 1993. *An Introduction to the Bootstrap.* Chapman & Hall/CRC, New York and London.

Note: Woody cover regression model (columns 2, 3, 4). Coefficient estimates, standard error, and p-value for the selected variables. The multiple R2 = 0.61, residual standard error = 8.9, and $p < 0.0001$. Jack-knife estimates of the model coefficients for the predictors (columns 5, 6, 7): mean, standard error, and bias.

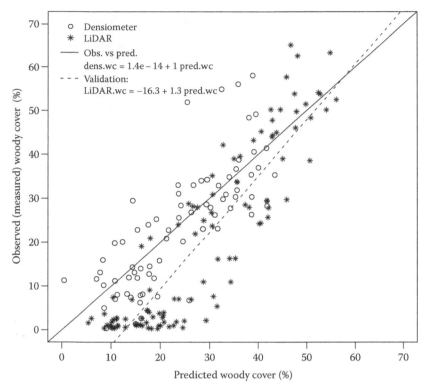

FIGURE 11.2
Model evaluation: observed (densiometer) versus predicted woody cover percent points (circles) and fitted line (solid, $R^2 = 0.61$ and RSE = 8.6, $p < 0.0001$). Model validation: LiDAR estimates versus predicted woody cover percent points (stars) and fitted line (dashed, $R^2 = 0.79$, RSE = 8.9, $p < 0.0001$).

tree structure was stable, and the predictions calculated for 10 test samples matched the mapped woody cover with an averaged $R2 = 0.54$ ($p < 0.0001$) and root mean squared error of 8%. The random forest function assigned the highest relative importance to the fire frequency, slope, MAP, and "basalt-granite" variables. The first splits in the regression tree occurred for the binary geological variable basalt granite. The two branches were not characterized by differences in woody cover value ranges but rather by two separate ecological behaviors occurring on the two different parent material layers. On the basalt rocks, the areas with lowest woody cover (WC = 17%) are located where fire recurs with frequency greater than once every 5 years (0.2), slope < 1.1°, and MAP < 567 mm. Less than 5-year fire return appears necessary for covers > 32%. The granite branch is first split in the northern and the eastern blocks. The CART did not further resolve the cover variability in the northern zone, predicting overall WC = 49 %. In the east, areas with MAP < 518 mm present less woody cover with crests (DEM > 391 m a.s.l.)

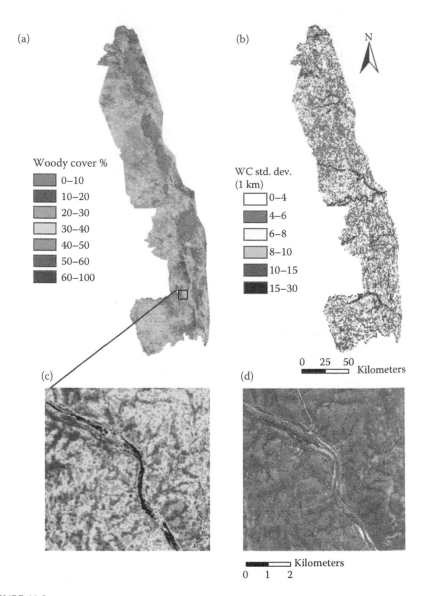

FIGURE 11.3
(See color insert following page 320.) (a) KNP woody cover percent map (90-m nominal resolution). (b) Heterogeneity map for Kruger National Park with focal resolution of 1 km. (c) Woody cover map zoom on Sabie river shows false color infrared-composite of a 4-m resolution IKONOS image (4–28–2001) (photosynthetically active vegetation in red shades), and (d) woody cover percent estimates.

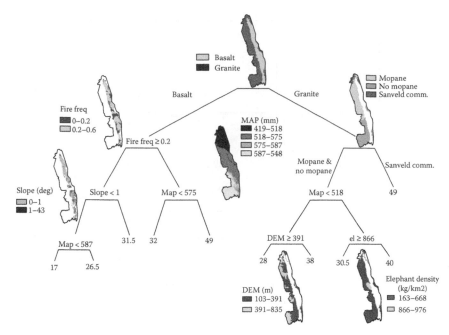

FIGURE 11.4
Ecological model (regression tree) with end node values being woody cover percent. Each split refers to the condition for the left child branch. The values at the end nodes are the mean response for that stratum. Node variables basalt/granite bedrocks, fire frequency (fire freq), elevation (DEM), slope, mean annual precipitation (map), mopane (*Colophospermum Mopane* species), sanveld community species (sanveld comm.), and elephant biomass density (el).

having the lowest cover (WC = 28%). If MAP > 518 mm, elephants density >866 kg/km^2 correlates with reduced woody cover.

Partitioning of variation among ecological, spatial, and combined models (Figure 11.5) showed that the variation in woody cover distribution was predominantly explained by ecological and environmental variables, because the pure spatial component explained only 2% of the variability. Two thirds (~40%) of the variation explained by the ecological model was associated with the shared ecological or spatial component. The other third (20%) has a pure ecological or environmental origin coming from variable interactions.

Discussion

Woody fractional cover and woody cover heterogeneity are important quantitative descriptors of savanna structure that capture biophysical processes driven by geomorphology, climate, and disturbances such as fire and

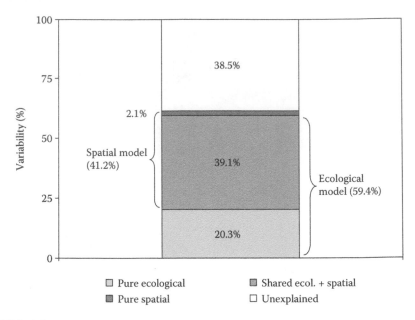

FIGURE 11.5
Variation partitioning for the CART models. Variability explained by the ecological-spatial model: 61.5 = 39.1 + 20.3 + 2.1%. Unexplained variability: 38.5%.

herbivory. We produced the first KNP woody cover map at medium-high (90 m) resolution. We also derived a woody cover heterogeneity map and a woody cover ecological model.

Woody Cover Map

It is helpful to combine optical and radar imagery for woody cover up-scaling and mapping: The JERS-1 L band and the Landsat ETM+ green band were the most important variables correlated with woody fractional cover. The strong correlation ($r = 0.67$) between woody cover percent and radar L-band backscatter is consistent with the results found in the savannas of Queensland, Australia (Lucas et al., 2006). In general, high backscattering is expected from tree boles due to their size that is similar to that of the L-band wavelength (23.5 cm). The signal return results from both direct backscatter and from stem-ground double bounce interaction (Durden et al., 1989). Topography contributes to woody plant establishment by creating local favorable conditions related to energy balance, higher soil moisture, or protection from fire and/or herbivory. However, woody cover overestimates could arise from enhanced backscatter due to aggregated structural information from both vegetation and topography (van Zyl, 1993). Our corrective approach, based on applying a mask (with a fixed intensity value) on topographic areas, was a rudimentary effort that can be substituted with robust radiometric corrections. The negative correlation ($r = -0.63$) between the ETM+ green band and woody cover is in agreement with research showing that in arid and semiarid systems, increasing vegetation cover corresponds to an increase in visible light absorption by the canopy leaves (mostly in blue and red bands) and a decrease in the background brightness from dry soil and senescent grasses (Yang and Prince, 1997; Xu et al., 2003). This behavior, however, can be non-linear, especially when accounting for shadows and complex backgrounds. We initially tested simple polynomial relationships, but the increased model complexity penalized model performance gains.

The overall good agreement with the woody cover estimates retrieved from LiDAR (Figure 11.2) means that our map gives an accurate woody cover representation at the KNP scale. Our approach, however, has limitations that influenced both the amount of model unexplained variability and model error. They can be related to woody cover temporal variations that occurred during the 11-year (1995–2006) span of data acquisition dates, field measurement errors, and remote-sensing signal variability (shadows and the background contamination effects; Spanner et al., 1990). Disturbances and vegetation demographic changes caused cover fluctuations during the data acquisition years. Our mapped woody cover is a model-combined value of different year signals and, therefore, it is not a point-in-time representation. Local spatial variability carries also a temporal component. The expected cover bias (overestimate) due to the densiometer design did not emerge in the validation comparison (Figure 11.2). The exclusion of canopy gaps in the densiometer measurements could have compensated the bias introduced by the densiometer angular view and lead to a good match with LiDAR estimates that are closer to crown cover (gaps included in cover at the operated resolution). Savanna sparse vegetation also probably reduced the densiometer angular effect. Given the densiometer problematic aspects, other field methods such as the line intercept or even a modified spherical densiometer would be better suited for their unbiased sampling and accuracy (Korhonen et al., 2006; Ko et al., 2009). The LiDAR cove deviations from the map predictions in the low woody cover range (Figure 11.2) are instead caused by the LiDAR sampling approach used for this data set. This approach tends to miss plants ≤2 m and, hence, underestimates cover in the more arid areas dominated by short woody plants (Bucini et al., unpublished manuscript).

Ecological Model

Both the regression tree and the variance partitioning analyses highlighted a strong spatial character of the woody plant patterns that arises from interactions with spatially structured environmental factors. The analysis was in accord with pieces of previous KNP research and combined them into an integrated model. Soil substrates and water availability are strong factors in determining woody cover. Excluding the mountainous areas, woody plants can reach fractional covers > 30% with MAP > 575 mm on basalt and with

MAP > 518 mm on granite where moisture is more available due to clay soil types. Water is more easily found in valleys and depressions. The two tree branches "slope > 1°" and "DEM < 391m" mainly correspond to this location type. In particular, "slope > 1°" separates both low elevation lands and some outcrops on the basalt side. Despite the long-standing concern on elephant impact (Eckhardt et al., 2000; Guldemond and Van Aarde, 2008), our model highlights only an area of the granites where elephant density is relatively high and woody cover is 30%, although MAP > 518 mm. Contrary to our expectations, the variables representing seasonal or annual rainfall variability only appeared at lower unstable nodes, indicating either an effect at smaller scales or sensitivity to data set idiosyncrasies. The 40% unexplained variability is likely to be linked to stochasticity in spatiotemporal drivers removed in the temporal means and spatially gridded data, and resulting population dynamics (Wiegand et al., 2006) that the ecological and environmental predictors do not capture. To resolve some finer-scale variability, we also run a similar model analysis maintaining the layers "distance to water," "elevation," "slope," and "aspect" at 90-m resolution, but no improvement was reached.

Research and Management Implications

The procedures used in this work are of interest for implementing a Park monitoring program that will collect woody-cover data for long-term monitoring and ecological assessments. In particular, heterogeneity has emerged in KNP as a management goal in association with maintenance of biodiversity (Rogers, 2003; Venter et al., 2008). The derived woody cover and heterogeneity maps can provide information at a scale equivalent to the watershed and landscape scales at which the Park is managed. Yet, given the KNP low-to-medium woody cover, the current 8.9% model error needs to be reduced. Very high-resolution imagery and LiDAR have the potential to provide a better spatiotemporal sampling of the KNP complex vegetation structure and, hence, to improve upscaling remote-sensing calibration models.

Acknowledgments

The authors thank Dr. R. Scholes for the valuable discussions on remote sensing and savanna ecology and Sandra MacFadyen, Navashni Govender, and Johan Baloyi from the Kruger Park Scientific Services for their help with the data collection. This research was funded by support from the National Science Foundation (NSF) on the project *Biocomplexity* in African Savannas (EAR 0120630) and the NASA Terrestrial Ecology Program on the project *African Carbon Exchange*. The Carnegie Airborne Observatory is funded by the W. M. Keck Foundation, William Hearst III, the Andrew Mellon Foundation, and the Gordon and Betty Moore Foundation.

References

Akaike, H. 1973. Information theory as an extension of the maximum likelihood principle. *Second International Symposium on Information Theory*, Akademiai Kiado, Budapest.

Archibald, S., D. P. Roy, B. W. van Wilgen, and R. J. Scholes. 2009. What limits fire? An examination of drivers of burnt area in Southern Africa. *Global Change Biology* 15, 613–630.

Asner, G. P., D. E. Knapp, T. Kennedy-Bowdoin, et al. 2007. Carnegie Airborne Observatory: In-flight fusion of hyperspectral imaging and waveform light detection and ranging (wLiDAR) for three-dimensional studies of ecosystems. *Journal of Applied Remote Sensing* 1, 21

Asner, G. P., S. R. Levick, T. Kennedy-Bowdoin, et al. 2009. Large-scale impacts of herbivores on the structural diversity of African savannas. *Proceedings of the National Academy of Sciences of the United States of America* 106, 4947–4952.

Biggs, R., H. C. Biggs, T. T. Dunne, N. Govender, and A. L. F. Potgieter. 2003. Experimental burn plot trial in the Kruger National Park: History, experimental design and suggestions for data analysis. *Koedoe* 46, 1–15.

Boone, R. B. and W. B. Krohn. 2000. Partitioning sources of variation in vertebrate species richness. *Journal of Biogeography* 27, 457–470.

Borcard, D., P. Legendre, and P. Drapeau. 1992. Partialling out the Spatial Component of Ecological Variation. *Ecology* 73, 1045–1055.

Breiman, L. 2001. Random forests. *Machine Learning* 45, 5–32.

Breiman, L., J. Friedman, R. Olshen, and C. Stone. 1984. *Classification and Regression Trees*. Wadsworth, Belmont, California.

Bucini, G. and N. P. Hanan. 2007. A continental-scale analysis of tree cover in African savannas. *Global Ecology and Biogeography* 16, 593–605.

De'Ath, G. and K. E. Frabricius. 2000. Classification and regression trees: A powerful yet simple technique for ecological data analysis. *Ecology* 81, 3178–3192.

Durden, S. L., J. J. van Zyl, and H. A. Zebker. 1989. Modeling and observation of the radar polarization signature of forested areas. *IEEE Transactions on Geoscience and Remote Sensing* 27, 290–301.

Eckhardt, H. C., B. W. van Wilgen, and H. C. Biggs. 2000. Trends in woody vegetation cover in the Kruger National Park, South Africa, between 1940 and 1998. *African Journal of Ecology* 38, 108–115.

Efron, B. and R. J. Tibshirani. 1993. *An Introduction to the Bootstrap*. Chapman & Hall/CRC, New York and London.

Gertenbach, W. P. D. 1983. Landscapes of the Kruger National Park. *Koedoe* 26, 9–121.

Glenn, E. P., A. R. Huete, P. L. Nagler, and S. G. Nelson. 2008. Relationship between remotely-sensed vegetation indices, canopy attributes and plant physiological processes: What vegetation indices can and cannot tell us about the landscape. *Sensors* 8, 2136–2160.

Groen, T. A., F. van Langevelde, C. de Vijver, N. Govender, and H. H. T. Prins. 2008. Soil clay content and fire frequency affect clustering in trees in South African savannas. *Journal of Tropical Ecology* 24, 269–279.

Gromping, U. 2006. Relative importance for linear regression in R: The package relaimpo. *Journal of Statistical Software* 17, 1–27.

Guldemond, R. and R. Van Aarde. 2008. A meta-analysis of the impact of African elephants on savanna vegetation. *Journal of Wildlife Management* 72, 892–899.

Huete, A. R. 1988. A Soil-Adjusted Vegetation Index (SAVI). *Remote Sensing of Environment* 25, 295–309.

Ko, D., N. Bristow, D. Greenwood, and P. Weisberg. 2009. Canopy cover estimation in semiarid woodlands: Comparison of field-based and remote sensing methods. *Forest Science* 55, 132–141.

Korhonen, L., K. T. Korhonen, M. Rautiainen, and P. Stenberg. 2006. Estimation of forest canopy cover: A comparison of field measurement techniques. *Silva Fennica* 40, 577–588.

Lemmon, P. E. 1956. A spherical densiometer for estimating forest overstory density. *Forest Science* 2, 314–320.

Lucas, R. M., A. C. Lee, and M. L. Williams. 2006. Enhanced simulation of radar backscatter from forests using LiDAR and optical data. *IEEE Transactions on Geoscience and Remote Sensing* 44, 2736–2754.

Ogle, K. and J. F. Reynolds. 2004. Plant responses to precipitation in desert ecosystems: Integrating functional types, pulses, thresholds, and delays. *Oecologia* 141, 282–294.

Qi, J., A. Chehbouni, A. R. Huete, Y. H. Kerr, and S. Sorooshian. 1994. A modified soil adjusted vegetation index. *Remote Sensing of Environment* 48, 119–126.

Rietkerk, M., F. vandenBosch, and J. vandeKoppel. 1997. Site-specific properties and irreversible vegetation changes in semi-arid grazing systems. *Oikos* 80, 241–252.

Rogers, K. H. 2003. Adopting a heterogeneity paradigm: Implications for management of protected savannas. In *The Kruger Experience: Ecology and Management of Savanna Heterogeneity*, eds. J. T. du Toit, K. H. Rogers, and H. C. Biggs, Island Press, Washington, pp. 41–58.

Saatchi, S., K. Halligan, D. G. Despain, and R. L. Crabtree. 2007. Estimation of forest fuel load from radar remote sensing. *IEEE Transactions on Geoscience and Remote Sensing* 45, 1726–1740.

Sankaran, M., N. P. Hanan, R. J. Scholes, et al. 2005. Determinants of woody cover in African savannas. *Nature* 438, 846–849.

Sankaran, M., J. Ratnam, and N. Hanan. 2008. Woody cover in African savannas: The role of resources, fire and herbivory. *Global Ecology and Biogeography* 17, 236–245.

Santos, J. R., M. S. P. Lacruz, L. S. Araujo, and M. Keil. 2002. Savanna and tropical rainforest biomass estimation and spatialization using JERS-1 data. *International Journal of Remote Sensing* 23, 1217–1229.

Scholes, R. J. and S. R. Archer. 1997. Tree–grass interactions in savannas. *Annual Review of Ecology and Systematics* 28, 517–544.

Shannon, G., D. J. Druce, B. R. Page, H. C. Eckhardt, R. Grant, and R. Slotow. 2008. The utilization of large savanna trees by elephant in southern Kruger National Park. *Journal of Tropical Ecology* 24, 281–289.

Spanner, M. A., L. L. Pierce, D. L. Peterson, and S. W. Running. 1990. Remote-sensing of temperate coniferous forest leaf-area index—the influence of canopy closure, understory vegetation and background reflectance. *International Journal of Remote Sensing* 11, 95–111.

van Zyl, J. J. 1993. The effect of topography on radar scattering from vegetated areas. *IEEE Transactions on Geoscience and Remote Sensing* 31, 153–160.

Venter, F. J. 1990. *A classification of land for management planning in the Kruger National Park.* PhD dissertation. University of South Africa.

Venter, F. J., R. J. Naiman, H. C. Biggs, and D. J. Pienaar. 2008. The evolution of conservation management philosophy: Science, environmental change and social adjustments in Kruger National Park. *Ecosystems* 11, 173–192.

Venter, F. J., R. J. Scholes, and H. C. Eckhardt. 2003. The abiotic template and its associated vegetation pattern. *The Kruger Experience: Ecology and Management of Savanna Heterogeneity.* eds. J. T. du Toit, Rogers, K. H. and Biggs H. C. Island Press, Washington, pp. 83–129.

Wiegand, K., D. Saitz, and D. Ward. 2006. A patch-dynamics approach to savanna dynamics and woody plant encroachment—Insights from an arid savanna. *Perspectives in Plant Ecology Evolution and Systematics* 7, 229–242.

Xu, B., P. Gong, and R. L. Pu. 2003. Crown closure estimation of oak savannah in a dry season with Landsat TM imagery: Comparison of various indices through correlation analysis. *International Journal of Remote Sensing* 24, 1811–1822.

Yang, J. L. and S. D. Prince. 1997. A theoretical assessment of the relation between woody canopy cover and red reflectance. *Remote Sensing of Environment* 59, 428–439.

12

Remote Sensing of Global Savanna Fire Occurrence, Extent, and Properties

David P. Roy, Luigi Boschetti, and Louis Giglio

CONTENTS

Introduction

Fire is a major determinant of the ecology and distribution of the world's savanna and grassland systems (Higgins et al., 2000; Bond et al., 2005). A prolonged annual dry season combined with relatively rapid rates of fuel accumulation create conditions conducive to frequent vegetation fires in the world's savannas. For a fire to occur, fine dead fuels such as grasses, twigs, and leaves must be present, and their moisture content must be low enough to sustain combustion. In addition, there must be a source of ignition. Humans have been using fire for the past 790,000 years, with remains in Kenya suggesting human exploitation of fire may be ancient (Alperson-Afil, 2008). Today, except in sparsely populated regions where late dry season and early wet season lightning fires predominate, most savanna fires are lit by people for numerous reasons including clearing land for cultivation, to provide nutrient rich ash for crops, to maintain and improve pasture and grasslands for livestock grazing and to attract game, to drive animals during hunting, to make fire breaks around fields and settlements, and due to accident or arson (Frost, 1999; Gill et al., 2009; Miranda et al., 2009).

Savanna fires are predominantly surface fires burning grasses and shrubs; representative fuel load and biomass burned amounts are 5.6 t ha^{-1} and

3.1 t ha^{-1}, respectively (median of 25 savanna fire experiments in Africa, Australia, and Brazil, Pereira 2003). Savanna fires can propagate rapidly under the hot, windy, low humidity conditions that occur in the dry season, with fire line intensities from 100 to 20,000 kW m^{-1} and rates of spread <0.1–2 m s^{-1} (Govender et al., 2006). Burned area patch sizes can be less than a hectare to several hundred square km, and fires occurring in the early dry season may be smaller and less intense than late dry season fires (Yates et al., 2008). In savanna systems, fine fuels can develop and cure within a year; and since grasses and shrubs can regrow rapidly post-fire, the same region can burn every year. Savanna fire return intervals are difficult to reliably define and are influenced by climatic factors and land use, typical return intervals are of the order of 1–3 years (van Wiglen et al., 2004).

Satellite sensors provide the only means to study and understand the global distribution and timing of fire. Figure 12.1 shows a satellite-derived annual global fire distribution map; the most extensive fires occur in the savannas of Africa, Australia, and South America. Such information is beginning to give scientists, managers, and policy makers the opportunity to understand fires in their environmental, economic, and social contexts.

Environmental satellite remote sensing systems have been sensing the Earth's land surface since NASA launched the first of a series of Landsat satellites in 1972 to provide observations over 185 × 170 km regions every 16 days (Williams et al., 2006). The capability to daily monitor the Earth's surface at a landscape to global scale was established in 1981 when the National Oceanic and Atmospheric Administration (NOAA) launched the first of a series of AVHRR satellites (El Saleous et al. 2000). These systems were not designed, however, for remote sensing of fire; and it was not until the late 1990s and this decade that space agencies have launched systems with specific fire-monitoring capabilities, including NASA's MODIS (Justice et al., 2002).

Satellite remote sensing of fire has predominantly used passive optical and thermal wavelength remote sensing techniques; active techniques, such as radar or lasers, have not been widely adopted for fire applications. Passive optical and thermal wavelength remote sensors measure the radiance of reflected and emitted surface electromagnetic radiation captured by wavelength calibrated sensors. The spectral radiance [units: W m^{-2} sr^{-1} m^{-1}] sensed by each detector is converted to a digital number for storage and digital transmission to receiving stations on the Earth. The digital numbers may be converted to spectral radiance using specified sensor calibration gain and bias coefficients. Radiance sensed at reflective wavelengths (0.3 μm to ~3.5 μm) may be converted to spectral reflectance [unitless, range 0–1], the ratio of the reflected radiation from the surface to the radiation incident on it; and at thermal wavelengths (~3.5 μm to ~12 μm) radiance may be converted to brightness temperature [units: K] to define the surface temperature assuming unit surface emissivity. These quantities, radiance (measured by the sensor), reflectance, and brightness temperature (both derived from radiance),

(a)

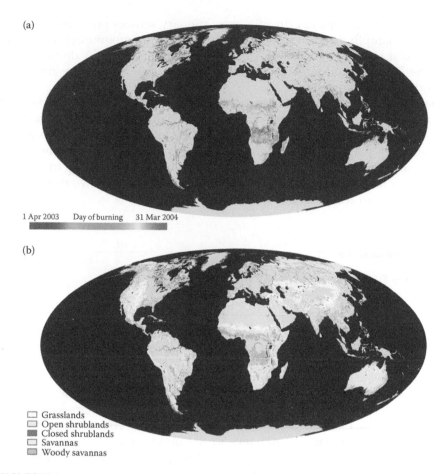

1 Apr 2003   Day of burning   31 Mar 2004

(b)

☐ Grasslands
☐ Open shrublands
■ Closed shrublands
☐ Savannas
▨ Woody savannas

FIGURE 12.1
(See color insert following page 320.) Satellite-derived global fire distribution: (a) global burned area distribution, derived from the MODIS burned area product for April 2003 to March 2004 shown superimposed over MODIS true color surface reflectance to provide geographic context. (Adapted from Roy, D. P. 2008. *Remote Sensing of Environment* 112, 3690–3707.) (b) Global distribution of savannas, grasslands, and shrublands defined by the MODIS land cover product (Adapted from Friedl, M. A., et al. 2002. *Remote Sensing of Environment* 83, 287–302.)

are the fundamental quantities used to retrieve surface properties from remotely sensed optical and thermal wavelength data.

The majority of environmental remote sensing sensors are onboard satellites in polar low Earth orbit (600–900 km altitude), where the surface is sensed intermittently depending on the sensor field of view and the satellite altitude and inclination; or they are in geostationary orbit, where the surface is sensed from the same position fixed above the Equator at an altitude of approximately 36,000 km. Polar orbiting sensors typically have higher spatial resolution but provide less frequent observation of the surface than geostationary sensors. In principle, geostationary sensors provide improved temporal sampling and so more opportunities for fire monitoring compared with polar orbiting systems. However, geostationary systems do not sense all the Earth, including high latitudes; and, to date, they have not been designed for fire mapping.

Satellite data have been used to monitor fire at regional to global scales for more than two decades, using computer algorithms that detect the location of active fires at the time of satellite overpass, and in the last decade using burned area mapping algorithms that map the spatial extent of the areas affected by fires. More recently, researchers have been attempting to use satellite data to characterize fire properties. This chapter reviews these three aspects with an emphasis on savanna systems.

Satellite Remote Sensing of Fire

Active Fire Detection

Satellite data have been used to monitoring fire using active fire detection algorithms developed for different sensors to take advantage of the elevated radiance of fires (Arino and Rosaz, 1999; Elvidge et al., 2001; Flasse and Ceccato, 1996; Giglio et al., 2003; Matson and Dozier, 1981; Prins and Menzel, 1992). The visible and near-infrared wavelengths are not usually used for fire detection, as natural surfaces can also have high reflectance and because smoke aerosols scatter reflective radiation and so obscure fires. Fire detection algorithms predominantly use the emitted radiance in the middle infrared (~4.0 μm) and thermal infrared (10–12 μm), which are in atmospheric windows where atmospheric absorption by water vapor, CO_2, O_3, and other gases is less prevalent and which are much less sensitive to smoke aerosol scattering. The temperatures of the flaming and smoldering phases of a fire range from about 500 K to 1800 K (Robinson, 1991). The radiance emitted from a smoldering fire is typically more than two orders of magnitude greater than the radiance emitted from a land surface (300 K) in the middle infrared (4.0 μm) and is about an order of magnitude greater in the thermal infrared (10 μm). Consequently, active fire detection algorithms nearly always take advantage of this middle-infrared and/or thermal-infrared elevated radiance.

The simplest algorithms for active fire detection apply a threshold to the middle-infrared brightness temperature. This type of algorithm has been applied to the 3.75 μm band of the NOAA AVHRR satellite series since the mid-1980s (Muirhead and Cracknell, 1985; Setzer and M. Pereira, 1991). The 3.75 μm AVHRR band becomes saturated, however, for surfaces hotter than about 320 K, which precludes reliable active fire detection over hot soil and rock surfaces (Csiszar and Sullivan, 2002). Algorithms that use the middle infrared must also account for the reflected radiation component that can cause false detections due to specular reflection of the Sun's radiation off water bodies (Justice et al., 2002). Single wavelength threshold algorithms have been applied using nighttime satellite data to reduce false detections due to high temperature surfaces and sensor saturation (Arino and Rosatz, 1999). The sole use of nighttime observations, however, leads to systematic underestimation of fire activity, because globally most fires occur in the daytime (Giglio, 2007). More refined algorithms that apply thresholds using both the middle infrared and thermal infrared have been developed, often with regional or land cover specific thresholds (Martin et al., 1999).

The emitted radiance, and hence the probability of active fire detection, varies with the areal proportions and temperatures of the smoldering and flaming fire and the nonburning components (Robinson, 1991; Giglio et al. 1999; Giglio and Justice, 2003). To provide more reliable active fire detection and to enable the detection of smaller and/or cooler fires, a variety of contextual detection algorithms have been developed (Prins and Menzel, 1992; Flasse and Ceccato, 1996; Giglio et al., 1999). Contextual detection algorithms first apply threshold(s) to the middle infrared and/or thermal infrared brightness temperature and then reject false detections by examining the brightness temperature relative to the temperature of the neighboring pixels. Contextual algorithms are the state of the practice for active fire detection; and when applied to sensors with specific fire detection capabilities, they provide reliable and consistent fire monitoring (Giglio et al., 2003).

A number of AVHRR-based fire monitoring systems have been developed by international initiatives such as the World Fire Web (Grégoire et al., 2001) and by national or regional agencies such as the Canadian Fire M3 system (Li et al., 2000). The NOAA Geostationary Operational Environmental Satellite (GOES) Wildfire Automated Biomass Burning Algorithm (ABBA) is providing half-hourly fire observations for the Western Hemisphere (Prins et al., 1998). The European satellites of the Meteosat Second Generation series (MSG) are used for active fire observation with 15 min temporal frequency and coverage including Africa and Europe. The middle-infrared channels of the European Space Agency Along-Track Scanning Radiometer (ATSR) and Advanced ATSR series have been used to create the World Fire Atlas (Arino and Rosatz, 1999). The Visible and Infrared Scanner (VIRS) on board the NASA Tropical Rainfall Measuring Mission (TRMM) satellites provides fire observations and, similar to GOES, allows sampling of the diurnal fire cycle (Giglio, 2007). None of the above sensing systems were designed for active fire detection. A major step forward has been the successful launch in 1999 and 2002 of the MODIS on board the morning descending Terra and afternoon ascending Aqua polar orbiting NASA satellites. The MODIS sensor design includes bands specifically selected for fire detection and is used to systematically generate the global 1km MODIS active fire product using a contextual algorithm (Giglio et al., 2003).

Fires in savanna environments are predominantly surface fires burning grasses and shrubs, and they move quickly relative to the temporal sampling

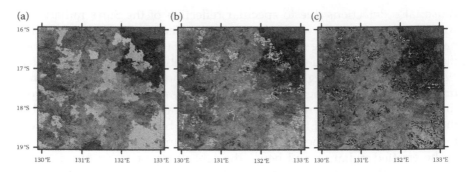

FIGURE 12.2

(See color insert following page 320.) Comparison of burned areas in Northern Australia in October 2002, as detected by the MODIS burned area product (Adapted from Roy, D. P. et al. 2008. *Remote Sensing of Environment* 112, 3690–3707) and the MODIS active fire product (Adapted from Giglio, L. et al. 2003. *Remote Sensing of Environment* 87, 273–282), superimposed on MODIS surface reflectance to provide geographic context. An area of 4° of longitude by 4° of latitude is shown. (a) The approximate day of burning defined by the MODIS 500 m burned area product with day of burning shown in a rainbow color scale (first day blue, last day red); (b) the date of the first detection by the MODIS 1km active fire product (Terra and Aqua, day and night detections), in the same color scale; (c) the number of times each pixel is detected by the MODIS active fire product (red = 1, yellow = 2, green = 3, cyan = 4).

of polar orbiting satellites. The satellite may not overpass sufficiently frequently to capture the spatial details of how the fire propagates across the landscape (Eva and Lambin, 1998a, Giglio, 2007, Roy et al., 2008). This is illustrated in Figure 12.2, which shows day and night MODIS Aqua and Terra active fire detections (middle) for a 4° × 4° savanna region of Northern Australia. Active fire detections in savanna environments typically underestimate the area burned (Roy et al., 2008). Geostationary systems have higher temporal resolution (15–30 min), but their lower spatial resolution may cause severe omission errors when fires are too small relative to the pixel size to be detected. For example, Roberts and Wooster (2008) report omission errors of more than 50% when comparing Meteosat SEVIRI with MODIS active fire detections over Africa.

Burned Area Detection

Methods to map burned areas have been developed predominantly in the last decade using moderate and coarse spatial resolution satellite data. There is no consensus burned area mapping algorithm. However, given that fire is an agent of land cover alteration, burned area mapping algorithms that use multitemporal moderate spatial resolution data under a change detection scheme to take advantage of the temporal persistency of fire effects have received considerable attention.

Vegetation fires deposit charcoal and ash, remove vegetation, alter the vegetation structure, and sometimes reveal the vegetation understory and/or soil (Roy et al., 1999; Pereira et al., 2003; Trigg et al., 2005). These changes are all amenable to optical wavelength remote sensing, although the changes may vary in complex ways in space and time. The most usual spectral change induced by fire is a decrease in reflectance. Black charcoal and ash deposited by fire have significantly lower reflectance than dry vegetation with reflectance increasing slightly with wavelength (0.3–3.5 µm). In savannas, vegetation desiccation and senescence during the dry season provide a fire prone fuel that is considerably more reflective than in the wet season. The reflectance spectra of dry grass can become similar to certain dry soil spectra, with suppressed blue, red, approximately 1.5 µm, and approximately 2.0 µm wavelength absorption features, because photosynthesis and the leaf water content are reduced significantly. Consequently, conventional vegetation indices based on combinations of red and near-infrared reflectance, such as the NDVI, do not provide good burned-unburned discrimination in dry savanna systems (Roy et al., 2005). Better spectral discrimination between unburned and burned vegetation is found in the near (~0.8 µm) and long wave near infrared (~1.6 µm and ~1.25 µm), and most burned area mapping algorithms are based on detecting decreased reflectance at these wavelengths (Pereira et al., 1997). Some surface changes not associated with fire, like cloud shadows, may induce similar spectral changes and depending on the algorithm and wavelengths used may cause false detections (Roy et al., 2005). Conversely, burned areas may not be detected using methods that expect a drop in reflectance if the fire is sufficiently hot and/or long lasting to deposit highly reflective white ash (Roy and Landman, 2005; Smith et al., 2005) or if the action of the fire is to reveal highly reflective soil (Roy et al., 2005).

The effects of fire on the spectral signature of vegetation tend to be non-permanent, especially in savanna systems where grasses regrow after the fire regardless of the degree of tree and shrub mortality. The persistence of the charcoal and ash signal depends on rates of dissipation by wind and rain and may be controlled by the unburned fuel load and structure (Pereira et al., 1997; Trigg and Flasse, 2000). Consequently, burned area detection accuracy may change temporally as the spectral characteristics of vegetation and burned areas change (Roy and Landmann, 2005). In all burned area mapping approaches, the definition of the magnitude of spectral change associated with the conversion of vegetation to burned vegetation is critical. In savanna ecosystems, the change in reflectance caused by fire can be small. For example, in southern Africa, the magnitude of variations in MODIS reflectance due to angular sampling may be greater than the reflectance change induced by biomass burning (Roy et al., 2002); and similar observations were made in Australia savannas using SPOT-VEGETATION data (Stroppiana et al., 2003). Early burned area mapping approaches included Fredericksen et al. (1990), who mapped burned areas in Senegal by analysis of time trajectories of AVHRR reflectance; Eva and Lambin (1998b) labeled burned ATSR pixels in Central Africa as those whose shortwave infrared reflectance fell with a simultaneous increase in brightness temperature; and

Barbosa et al. (1999) made continental burned area maps in different functional vegetation strata of Africa from AVHRR data by labeling burned pixels where a vegetation index value decreased and the brightness temperature increased. In these early approaches, the low geometric and spectral quality of the satellite data available constrained the sophistication of the detection algorithms.

A number of global multiannual burned area data sets have been developed recently using improved sensors, namely the L3JRC, GLOBCARBON, and MODIS products. The L3JRC burned area product (Tansey et al., 2008) is generated from 1 km SPOT-VEGETATION data. After preprocessing to remove clouds and shadows, a reference composite image is created moving forward in time in daily steps, each step keeping the most recent noncontaminated observation. At each step, an index based on the near infrared reflectance is computed, comparing the new observation with the previous composite; and the mean and standard deviation of the index is computed over a 200 km × 200 km window. A pixel is flagged as burned if its index value is lower than the mean minus two times the standard deviation, with additional spectral tests based on the 0.83 μm and 1.66 μm reflective bands. Each L3JRC pixel defines the day of the first burned area detection within a fire year starting on the 1st of April; if a pixel burns more than once per year, only the first day of burning is stored. The GLOBCARBON project, promoted by the ESA, has generated a number of land products, including a global monthly 1 km burned area product, for assimilation into global carbon models (Plummer et al., 2006). The GLOBCARBON burned area product is generated using three different change detection and spectral threshold algorithms applied to 1 km SPOT-VEGETATION and ERS2-ATSR2 or ENVISAT AATSR data, respectively; and all are reported as a monthly 1 km product defining the day of the first burned area detection within the month and which algorithm(s) detected it. The NASA MODIS burned area product uses a bidirectional reflectance model-based change detection approach applied independently to each 500 m MODIS pixel to detect the spectral, temporal, and structural changes that characterize vegetation fire (Roy et al., 2005). The MODIS reflectances sensed within a temporal window of a fixed number of days are used to model the directional dependence of reflectance, commonly defined by the BRDF, and to predict the reflectance on a subsequent day. A statistical measure is used to determine whether the difference between the predicted and observed reflectance in the near and middle-infrared bands indicates a significant change of interest. A temporal constraint is used to differentiate between impermanent changes, such as shadows, that are spectrally similar to more persistent fire-induced changes. The product is released as a monthly 500 m product, with an indication of the day of burning for each pixel.

Burned area products provide a more complete assessment of the area burned than active fire detections, particularly in savanna systems, where ground and surface fires are less likely to be obscured by overstory vegetation

(Pereira, 2003; Roy et al., 2008). This is illustrated in Figure 12.2, where the 1 km MODIS day and night active fire detections (middle) capture about one third of the burned area detected by the 500 m MODIS burned area product (left). The propagation of the fire and the multiple fire fronts moving in different directions is evident from the day of detection in both products, which are spatiotemporally coherent.

Fire Characterization

Despite several decades of research, there is no way to reliably characterize from satellite data whether a fire is a ground, surface, or crown (tree canopy) fire, let alone categorize the fire cause. Instead, fire type and cause have been inferred based on geographic context and proximity of satellite fire detections relative to thematic land cover classes and features such as roads and forest edges (Schroeder et al., 2005; Morton et al., 2008). Arguably, as a consequence, over large areas questions regarding the incidence, drivers, and impacts of different fire types have not been definitively addressed. In addition to type and cause, the prevailing characteristics of fire in a region are of great interest to scientists and ecosystem managers. The *frequency, seasonality, intensity, severity, fuel consumption, and spread patterns* of fires that prevail are referred to as the fire regime (Bond and Keeley, 2005; Gill, 1975). How fire regimes will change as human population, their land use practices, and the climate change is unclear. Climate and land cover or land use changes are likely to affect fire regimes directly, through changes in the ways fire is used, and, indirectly, by modifying the environmental conditions and the amount of fuel available for fire (Frost, 1999; Archibald et al., 2009).

Of the different variables defining a fire regime, only the frequency and seasonality of fire can be derived in a demonstrably reliable manner from satellite data, usually as summary statistics of burned area or active fire counts, with results depending on the scale of the analysis and on the completeness and reliability of the satellite time series (Giglio et al., 2006; Boschetti and Roy, 2008). Similarly, fire spread patterns may be derived from fire products; but the satellite spatial resolution relative to the burned area size distribution and spatial fragmentation may cause significant reporting biases (Eva and Lambin, 1998a; Laris, 2005; Roy and Landmann, 2005; Silva et al., 2005).

The fire intensity refers to the rate of energy release, usually the energy released per length of fire front [units: W m^{-1}], and is correlated strongly with the aboveground impacts of fire (Byram 1959). It cannot be reliably extracted from satellite data due to the spatiotemporal mismatch between satellite observations and fire fronts. Satellite retrievals of fire spread rates have been attempted but must use restrictive, and largely unverifiable, assumptions (Loboda and Csiszar, 2007). Fire severity is a qualitative indicator used to assess fire effects within burned areas; and definitions vary but are usually used to relate how fire changes ecosystems differentially or results in different biological responses (Lentile et al., 2006). Parameters used to estimate severity in the field include the condition and color of the soil, amount of fuel consumed, resprouting from burned plants, blackening or scorching of trees, depth of burn in the soil, and changes in fuel moisture (Key and Benson, 2005). Although several of these parameters may not be amenable directly to optical wavelength remote sensing, or may not be related in a linear way to reflectance (Roy et al., 2006), field-based measures of fire severity have been used to parameterize and assess fire severity maps created using optical wavelength satellite data (French et al., 2008). One of the main difficulties for assessment of fire severity from remotely sensed data is the empirical approach that has guided most studies. Empirical models are simple to calibrate and provide severity estimates, but they provide little confidence on whether they are applicable to other ecosystems or fuel characteristics. To date, for large savanna regions, fire severity has usually been inferred, for example, from the prevailing fire seasonality (Russell-Smith and Edwards, 2006).

Fuel consumption, that is, the total biomass consumed by a fire, has been conventionally estimated by multiplying the area burned by the fuel load and a combustion completeness parameter (Seiler and Crutzen, 1980), although the fuel load and combustion completeness are not reliably defined at landscape to global scales (Korontzi et al., 2004). A recent alternative has been developed using time-integrated instantaneous satellite measurements of the fire radiative power (FRP) (Kaufman et al., 1996). The FRP [units: W] may be retrieved from middle-infrared wavelength remotely sensed data and provides reasonably accurate (±15%) estimates of the rate of fuel consumed (Wooster et al., 2005). The time-integrated FRP provides the fire radiative energy (FRE) [units: J] and is linearly related to the total biomass burned (Wooster et al., 2005; Freeborn et al., 2008). Thus, fuel consumption may be estimated directly without the need for fuel load and combustion completeness information. The FRP retrieved from satellite active fire detections is only available under relatively cloud free conditions at the time of satellite overpass. As observed in Section 2.1, and illustrated in Figure 12.2, active fire detections from polar orbiting satellites typically under sample the temporal dynamic of fires due to the infrequent satellite overpass time and clouds (Giglio et al., 2007) and under sample the spatial extent of burned areas where the fire progresses rapidly across the landscape (Roy et al., 2008). Geostationary satellite active fire detections provide much improved temporal sampling (e.g., every 15 min for Meteosat Second Generation) over polar orbiting systems (e.g., four times a day for MODIS Terra/Aqua); but reliable geostationary active fire detection of small and cool fires is reduced due to the large pixel size (Roberts and Wooster, 2008; Schroeder et al., 2008). Despite this limitation, geostationary FRP estimates have been used to derive fuel consumption amounts for African savanna grasslands that are in broad agreement with literature values (Roberts et al., 2008).

Summary

Satellite data provide the only way to monitor fire at regional to global scales. Local fire information exists for some national parks, forests, and other conservation areas; but these lands are not usually representative of the region as a whole, because they are subject to specific fire management policies and are largely protected from the influence of ordinary people (Frost, 1999). In the last two decades, satellite fire products have been developed: first, active fire detection products and then burned area products, complemented more recently by research on the remote characterization of fire properties. Fires in savanna environments are predominantly surface fires burning grasses and shrubs that move quickly relative to the temporal sampling (overpass frequency) of satellites. Satellite active fire detection algorithms may fail to detect savanna fires, primarily because they do not overpass when the fire is burning (Giglio, 2007) or because the fires are too cool or are too small relative to the pixel size to be detected (Roberts and Wooster, 2008). Satellite burned area products provide a more complete fire record than active fire detections in savanna systems, as fire-induced changes in reflectance tend not to be obscured by overstory vegetation and are sufficiently persistent for detection (Roy et al., 2008). Integration of fire products from geostationary and polar-orbiting satellites to take advantage of their different sensing properties provides potential for improved fire monitoring and, with research on characterization of fire properties, is an area of ongoing research.

Despite the evident benefits and needs for satellite fire monitoring, fire remote-sensing scientists have not definitively demonstrated the accuracy and consistency of their products. Limited comparison studies have revealed large discrepancies in the areal estimates, timing, and location among satellite fire products and highlight the need for systematic product accuracy assessment (Boschetti et al., 2004; Korontzi et al., 2004; Roy and Boschetti, 2009). This need is expected to increase as the ease of producing satellite products increases, driven by factors including increasing computer processing and storage capabilities, decreasing computer costs, space agency support for low-cost satellite data, and the proliferation of satellite direct broadcast reception systems (Roy and Boschetti, 2009). As is the case with all the applications of remote-sensing research, fire algorithms and products are strongly constrained by the specifics of the sensors and satellite orbits. Until recently, existing and planned operational satellites have had limited fire monitoring capabilities (Townshend and Justice, 2002) but, significantly, several of the forthcoming operational systems (e.g., VIIRS, GOES-R, MTG) will have a mandate for fire detection. The challenge now is to transfer the technology developed in the research domain into the operational domain to provide long-term fire information with known accuracy and consistency. This is particularly important as satellite fire products are starting to be used in strategic ways in support of national environmental policy making (e.g., Russell-Smith et al., 2009).

References

Alperson-Afil, N. 2008. Continual fire-making by Hominins at Gesher Benot Ya'aqov, Israel. *Quaternary Science Reviews* 27, 1733–1739.

Archibald, S., D. P. Roy, B. W. Van Wilgen, and R. J. Scholes. 2009. What Limits Fire?: An examination of drivers of burnt area in sub-equatorial Africa. *Global Change Biology* 15, 613–630.

Arino, O. and J. M. Rosatz. 1999. 1997 and 1998 World ATSR Fire Atlas using ERS-2 ATSR-2 data. In *Proc. of the Joint Fire Science Conference, Boise, Idaho, 15–17 June 1999*, eds. L. F. Neuenschwander, K. C. Ryan and G. E. Gouberg, University of Idaho and the International Association of Wildland Fire, Boise, pp. 177–182.

Barbosa, P., J. M. Grégoire, and J. M. C. Pereira. 1999. An algorithm for extracting burned areas from time series of AVHRR GAC data applied at a continental scale. *Remote Sensing of Environment* 69, 253–263.

Bond W. J., F. I. Woodward, and G. F. Midgley. 2005. The global distribution of ecosystems in a world without fire. *New Phytologist* 165, 525–538.

Bond W. J. and J. E. Keeley. 2005. Fire as a global 'herbivore': the ecology and evolution of flammable ecosystems, *Trends in Ecology and Evolution* 20, 387–394.

Boschetti, L., H. Eva, P. A. Brivio, and J. M. Grégoire. 2004. Lessons to be learned from the intercalibration of three satellite-derived biomass burning products. *Geophysical Research Letters* 31, L21501.

Boschetti, L. and D. P. Roy. 2008. Defining a fire year for reporting and analysis of global interannual fire variability. *Journal of Geophysical Research* 113, G03020.

Byram, G. M. 1959. Combustion of forest fuels. In *Forest Fire: Control and Use*. ed. K. P. Davis. McGraw Hill, New York, 544pp.

Csiszar, I. and J. Sullivan. 2002. Recalculated pre-launch saturation temperatures of the AVHRR 3.7 micrometer sensors on board the TIROS-N to NOAA-14 satellites. *International Journal of Remote Sensing* 23, 5271–5276.

El Saleous N. Z., E. F. Vermote, C. O. Justice, J. R. G. Townshend, C. J. Tucker, and S. N. Goward. 2000. Improvements in the global biospheric record from the Advanced Very High Resolution Radiometer (AVHRR), *International Journal of Remote Sensing* 21, 1251–1277.

Elvidge, C. D., I. Nelson, V. R. Hobson, J. Safran, and K. E. Baugh. 2001. Detection of fires at night using DMSP-OLS data. In *Global and Regional Wildland Fire Monitoring from Space: Planning a Coordinated International Effort*. eds. F. J. Ahern, J. Goldammer and C.O. Justice, SPB Academic Publishing, The Hague, pp. 125–144.

Eva, H. and E. F. Lambin. 1998a. Remote sensing of biomass burning in tropical regions: Sampling issues and multisensor approach. *Remote Sensing of Environment* 64, 292–315.

Eva, H. and E. F. Lambin. 1998b. Burnt area mapping in Central Africa using ATSR data. *International Journal of Remote Sensing* 19, 3473–3497.

Flasse, S. P. and P. Ceccato. 1996. A contextual algorithm for AVHRR fire detection. *International Journal of Remote Sensing* 17, 419–424.

Freeborn, P. H., M. J. Wooster, W. M. Hao et al. 2008. Relationships between energy release, fuel mass loss, and trace gas and aerosol emissions during laboratory biomass fires. *Journal of Geophysical Research* 113, D01102.

French, N. H. F, J. L. Allen, R. J. Hall et al. 2008. Using Landsat data to assess fire and burn severity in the North American boreal forest region: An overview and summary of results. *International Journal of Wildland Fire* 17, 443–464.

Friedl, M. A., D. K. McIver, J. C. F. Hodges et al. 2002. Global land cover mapping from MODIS: Algorithms and early results. *Remote Sensing of Environment* 83, 287–302.

Fredericksen, P. S., S. Langaas, and M. Mbaye. 1990. NOAA-AVHRR and GIS-based monitoring of fire activity in Senegal—a provisional methodology and potential applications. In, *Fire in the Tropical Biota: Ecosystem Processes and Global Challenges*. ed. J. G. Goldammer. Springer-Verlag, Berlin, pp. 400–417.

Frost, P. G. H. 1999. Fire in southern African woodlands: Origins, impacts, effects, and control. In Proceedings of an FAO meeting on public policies affecting forest fires. *FAO Forestry Paper*, 138.

Giglio, L., J. D. Kendall, and C. O. Justice 1999. Evaluation of global fire detection algorithms using simulated AVHRR infrared data. *International Journal of Remote Sensing* 20, 1947–1985.

Giglio, L., J. Descloitres, C. O. Justice, and Y. J. Kaufman. 2003. An Enhanced Contextual Fire Detection Algorithm for MODIS. *Remote Sensing of Environment* 87, 273–282.

Giglio, L. and C. O. Justice. 2003. Effect of wavelength selection on wildfire characterization using the Dozier retrieval. *International Journal of Remote Sensing* 24, 3515–3520.

Giglio, L., I. Csiszar, and C. O. Justice. 2006. Global distribution and seasonality of active fires as observed with the Terra and Aqua Moderate Resolution Imaging Spectroradiometer (MODIS) sensors, *Journal of Geophysical Research* 111, G02016, doi:10.1029/2005JG000142.

Giglio, L. 2007. Characterization of the tropical diurnal fire cycle using VIRS and MODIS observations. *Remote Sensing of Environment* 108, 407–421.

Gill, A. M. 1975. Fire and the Australian flora: A review. *Australian Forestry* 38, 4–25.

Gill, A. M, R. J. Williams, and J. C. Z. Woinarski. 2009. Fires in Australia's tropical savannas: Interactions with biodiversity, global warming, and exotic biota. In M. A. Cochrane, ed. *Tropical Fire Ecology: Climate Change, Land Use and Ecosystem Dynamics*. Springer-Praxis, Heidelberg, Germany, pp. 113–141.

Govender, N., W. S. Trollope, and B. Van Wilgen, 2006. The effect of fire season, fire frequency, rainfall and management on fire intensity in savanna vegetation in south africa. *Journal of Applied Ecology* 43, 748–758.

Grégoire, J-M., D. R. Cahoon, D. Stroppiana et al. 2001. Forest fire monitoring and mapping for GOFC: Current products and information networks based on NOAA-AVHRR, ERS-ATSR, and SPOT-VGT systems. In *Global and Regional Vegetation Monitoring from Space: Planning a Coordinated International Effort*, ed. F. J. Ahern, J. G. Goldammer and C. O. Justice, SPB Academic Publishing, The Hague, Netherlands, pp. 105–124.

Higgins S. I., W. J. Bond, and W. S. W Trollope. 2000. Fire, resprouting and variability: A recipe for grass–tree coexistence in savanna. *Journal of Ecology* 88, 213–229.

Justice, C., L. Giglio, S. Korontzi et al. 2002. The MODIS fire products, *Remote Sensing of Environment* 83, 244–262.

Kaufman, Y., L. Remer, R. Ottmar et al. 1996. Relationship between remotely sensed fire intensity and rate of emission of smoke: SCAR-C experiment. In *Global Biomass Burning*. ed. J. Levine, MIT Press, Massachusetts, pp. 685–696.

Key, C. H. and N. C. Benson. 2005. Landscape Assessment: Ground measure of severity, the Composite Burn Index. CD-ROM. In *FIREMON: Fire Effects Monitoring and Inventory System*, ed. D. C. Lutes, R. E. Keane, J. F. Caratti, C. H. Key, N. C. Benson, and L. J. Gangi, Ogden, UT: USDA Forest Service, Rocky Mountain Research Station, General Technical Report, RMRS-GTR-164-CD:LA1-LA51.

Korontzi, S., D. P. Roy, C. O. Justice, and D. E. Ward. 2004. Modeling and sensitivity analysis of fire emissions in southern African during SAFARI 2000. *Remote Sensing of Environment* 92, 255–275.

Lentile, L. B., Z. A. Holden, A. M. S. Smith et al. 2006. Remote sensing techniques to assess active fire characteristics and post-fire effects. *International Journal of Wildland Fire* 15, 319–345.

Laris, P. 2005. Spatiotemporal problems with detecting and mapping mosaic fire regimes with coarse-resolution satellite data in savanna environments. *Remote Sensing of Environment* 99, 412–424.

Li, Z., S. Nadon, and J. Cihlar. 2000. Satellite-based detection of Canadian boreal forest fires: Development and application of the algorithm. *International Journal of Remote Sensing* 21, 3057–3069.

Loboda, T. and I. Csiszar. 2007. Reconstruction of fire spread within wildland fire events in Northern Eurasia from the MODIS Active Fire Product. *Global and Planetary Change* 56, 258–273.

Martin, P, P. Ceccato, S. Flasse, and I. Downey. 1999. Fire detection and fire growth monitoring using satellite data. In: *Remote Sensing of Large Wildfires in the European Mediterranean Basin*, ed. E. Chuvieco. Springer-Verlag, Berlin/Heidelberg.

Matson, M. and J. Dozier. 1981. Identification of subresolution high temperature sources using a thermal IR sensor. *Photogrammetric Engineering and Remote Sensing* 47, 1311–1318.

Miranda, H. S., M. N. Sato, W. N. Neto, and F. S. Aires. 2009. Fires in the cerrado, the Brazilian savanna. In M. A. Cochrane, ed. *Tropical Fire Ecology: Climate Change, Land Use and Ecosystem Dynamics*. Springer-Praxis, Heidelberg, Germany, pp. 427–450.

Morton, D. C., R. S. Defries, T. J. Randerson, L. Giglio, W. Schroeder, and G.R. van der Werf. 2008. Agricultural intensification increases deforestation fire activity in Amazonia. *Global Change Biology* 14, 2262–2275.

Muirhead, K. and A. P. Cracknell. 1985. Straw burning over Great Britain detected by AVHRR. *International Journal of Remote Sensing* 6, 827–833.

Pereira, J. M. C., E. Chuvieco, A. Beaudoin, and N. Desbois. 1997. Remote sensing of burned areas: A review. In: *A Review of Remote Sensing Methods for the Study of Large Wildland Fire Report of the Megafires Project ENV-CT96-0256*, ed. E. Chuvieco. Universidad de Alcala, Alcala de Henares, Spain, pp. 127–183.

Pereira, J. M. C. 2003. Remote sensing of burned areas in tropical savannas, *International Journal of Wildland Fire* 12, 259–270.

Plummer S., O. Arino, M. Simon, and W. Steffen. 2006. Establishing a Earth observation product service for the terrestrial carbon community: The GlobCarbon initiative. *Mitigation and Adaptation Strategies for Global Change* 11, 97–111.

Prins, E. M. and W. P. Menzel. 1992. Geostationary satellite detection of biomass burning in South America. *International Journal of Remote Sensing* 13, 2783–2799.

Prins, E. M., W. P. Menzel, and J. M. Feltz. 1998. Characterizing spatial and temporal distributions of biomass burning using multi-spectral geostationary satellite data. In *Proceedings of the Ninth Conference on Satellite Meteorology and Ocean-ography* 94–97.

Roberts, G. and M. J. Wooster. 2008. Fire detection and fire characterization over Africa using Meteosat SEVIRI. *IEEE Transactions on Geoscience and Remote Sensing* 46, 1200–1218.

Roberts, G., M. J. Wooster, and E. Lagoudakis. 2008. Annual and diurnal African biomass burning temporal dynamics, *Biogeosciences Discussions* 5, 3623–3663.

Robinson, J. M. 1991. Fire from space. Global fire evaluation using infrared remote sensing. *International Journal of Remote Sensing* 12, 3–24.

Roy, D. P., L. Giglio, J. Kendall, and C. Justice. 1999. Multitemporal active-fire based burn scar detection algorithm, *International Journal of Remote Sensing* 20, 1031–1038.

Roy, D. P., P. E. Lewis, and C. O. Justice. 2002. Burned area mapping using multi-temporal moderate spatial resolution data—a bi-directional reflectance model-based expectation approach. *Remote Sensing of Environment* 83, 264–287.

Roy, D. P. and T. Landmann. 2005. Characterizing the surface heterogeneity of fire effects using multi-temporal reflective wavelength data. *International Journal of Remote Sensing* 26, 4197–4218.

Roy, D. P., Y. Jin, P. E. Lewis, and C.O. Justice. 2005. Prototyping a global algorithm for systematic fire-affected area mapping using MODIS time series data. *Remote Sensing of Environment* 97, 137–162.

Roy, D. P., L. Boschetti, and S. N. Trigg. 2006. Remote Sensing of Fire Severity: Assessing the performance of the normalized burn ratio. *IEEE Geoscience and Remote Sensing Letters* 3, 112–116.

Roy, D. P., L. Boschetti, C. O. Justice, and J. Ju. 2008. The collection 5 MODIS burned area product—global evaluation by comparison with the MODIS active fire product. *Remote Sensing of Environment* 112, 3690–3707.

Roy, D. P. and L. Boschetti. 2009. Southern Africa validation of the MODIS, L3JRC and GLOBCARBON burned area products. *IEEE Transactions on Geoscience and Remote Sensing* 47, 1032–1044.

Russell-Smith, J. and A. C. Edwards. 2006. Seasonality and fire severity in savanna landscapes of monsoonal northern Australia. *International Journal of Wildland Fire* 215, 541–550.

Russell-Smith, J., B. P. Murphy, M. Meyer, G. D. Cook, S. Maier, A. C., Edwards, J. Schatz, and P. Brocklehurst. 2009. Improving estimates of savanna burning emissions for greenhouse accounting in northern Australia: Limitations, challenges, applications. *International Journal of Wildland Fire* 18, 1–18.

Schroeder, W., J. T. Morisette, I. Csiszar, L. Giglio D. Morton, and C. O. Justice. 2005. Characterizing vegetation fire dynamics in Brazil through multisatellite data: Common trends and practical issues. *Earth Interactions* 9, 1–26.

Schroeder, W., E. Prins, L. Giglio et al. 2008. Validation of GOES and MODIS active fire detection products using ASTER and ETM+ data. *Remote Sensing of Environment*, 112, 2711–2726.

Seiler, W. and P. J. Crutzen. 1980. Estimates of gross and net fluxes of carbon between the biosphere and the atmosphere from biomass burning. *Climatic Change* 2, 207–247.

Setzer, A. W. and M. Pereira. 1991. Operational detection of fires in Brazil with NOAA-AVHRR. *Proceedings of the 24th International Symposium on Remote Sensing of the Environment*, Rio de Janeiro, Brazil, Environmental Research, Fredericksen Institute of Michigan (ERIM), pp. 469–482.

Silva, J. M. N., A. C. L. Sá, and J. M. C. Pereira. 2005. Comparison of burned area estimates derived from SPOT-VEGETATION and Landsat ETM+ data in Africa: influence of spatial pattern and vegetation type. *Remote Sensing of Environment* 96, 188–201.

Smith, A. M. S., M. J. Wooster, N. A. Drake et al. 2005. Testing the potential of multispectral remote sensing for retrospectively estimating fire severity in African Savannahs. *Remote Sensing of Environment* 97, 92–115.

Stroppiana, D., J. M. Grégoire, and J. M. C. Pereira. 2003. The use of SPOT VEGETATION data in a classification tree approach for burnt area mapping in Australian savanna. *International Journal of Remote Sensing* 24, 2131–2151.

Tansey, K, J.-M. Grégoire, P. Defourny, et al. 2008. A new, global, multi-annual (2000–2007) burned area product at 1 km resolution and daily intervals, *Geophysical Research Letters* 35, L01 401, 2008.

Townshend, J. R. G. and C. O. Justice. 2002. Towards operational monitoring of terrestrial systems by moderate-resolution remote sensing. *Remote Sensing of Environment* 83, 351–359.

Trigg, S. N. and S. P. Flasse. 2000. Characterizing the spectral–temporal response of burned savannah using *in situ* spectroradiometry and infrared thermometry. *International Journal of Remote Sensing* 21, 3161–3168.

Trigg, S. N., D. P. Roy, and S. P. Flasse. 2005. An *in situ* study of the effects of surface anisotropy on the remote sensing of burned savannah, *International Journal of Remote Sensing* 26, 4869–4876.

Van Wilgen, B. W., N. Govender, H. C. Biggs, D. Ntsala, and X. N. Funda. 2004. Response of savanna fire regimes to changing re-management policies in a large African national park. *Conservation Biology* 18, 1533–1540.

Williams, D. L., S. Goward, and T. Arvidson. 2006. Landsat: yesterday, today, and tomorrow. *Photogrammetric Engineering and Remote Sensing* 72, 1171–1178.

Wooster, M. J., G. Roberts, G. Perry, and Y. J. Kaufman. 2005. Retrieval of biomass combustion rates and totals from fire radiative power observations: Calibration relationships between biomass consumption and fire radiative energy release. *Journal of Geophysical Research* 110, D21111: doi: 10.1029/2005JD006318.

Yates, C. P., A. C. Edwards, and J. Russell-Smith. 2008. Big fires and their ecological impacts in Australian savannas: Size and frequency matters. *International Journal of Wildland Fire* 17, 768–781.

Section IV

Perspectives on Patch- to Landscape-Scale Savanna Processes and Modeling

13

Understanding Tree–Grass Coexistence and Impacts of Disturbance and Resource Variability in Savannas

Frank van Langevelde, Kyle Tomlinson, Eduardo R. M. Barbosa,
Steven de Bie, Herbert H. T. Prins, and Steven I. Higgins

CONTENTS

Introduction

Savannas exhibit enormous spatiotemporal variability in woody and herbaceous biomass, structure, and plant functional forms; and the determinants of these are poorly understood (Bond, 2008; Lehmann et al., 2009). This is not surprising given that savannas are subject to an array of environmental drivers. Savannas occur over a huge range of climate conditions: mean annual precipitation (MAP) alone ranges from 150 mm to more than 2800 mm (Solbrig, 1996; Sankaran et al., 2005). Soil properties (fertility, texture, and depth) and irradiance also vary widely across savannas (Solbrig, 1996). Additionally, large areas of savanna are subject to defoliation by herbivores and fire, two quite different forms of defoliation on plants with consequent effects on their (re)growth, which also vary substantially over the biome (Olff et al., 2002; Bond and Keeley, 2005).

Ecologists have sought to gain predictive understanding of the responses of savanna communities to these multiple environmental drivers, to aid our understanding of these systems, and to guide management strategies in the

257

face of present human pressures and the future spectre of climate change. Ecologists have attempted to deal with the complexity of savannas by concentrating on two fundamental questions (Sarmiento, 1984; Scholes and Archer, 1997; Sankaran et al., 2004, 2005): how do the two very different life-forms coexist without one dominating the other? And, what environmental drivers determine the cover and biomass of each life-form? In this chapter, we focus on the first question, as the capacity to answer this question would allow answering the second one.

To date, several explanations have been proposed for the coexistence of trees and grasses in savannas. Several of these explanations have been translated into models designed to evaluate under which conditions this tree–grass coexistence occurs. The limitations of these explanations have led us to believe that an overly simple explanation cannot explain the sheer diversity of environments where savannas are encountered (Scholes and Archer, 1997; Sankaran et al., 2004; Bond, 2008). In this chapter, we will review nonspatial explanations for tree–grass coexistence in savannas. We avoid spatially motivated explanations for tree–grass coexistence, which are addressed in a separate chapter (Meyer et al., Chapter 14). We show that the existing models are only able to explain tree–grass coexistence under a subset of environmental conditions. By contextualising these explanations appropriately across the MAP gradient, we propose a general framework to better understand tree–grass coexistence across the broader spectrum of environments where savannas are encountered.

There are different types of general explanations proposed for the co-occurrence of organisms in communities. Below, we will discuss several of these general explanations for tree–grass coexistence in savannas: niche separation, disturbances, root–shoot partitioning, and facilitation (Figure 13.1). The first three are captured in existing models.

Explaining Coexistence of Trees and Grasses in Savannas

Niche Separation

The mechanism of niche separation proposes that plants segregate along various environmental niche axes, including gradients of light, soil moisture, and rooting depth (Sydes and Grime, 1984; Silvertown, 2004). The classical explanation for long-term coexistence of trees and grasses in savannas, the two-layer hypothesis of Walter (1971), is based on niche separation by rooting depth of trees and grasses. Assuming that water is the most limiting factor for plant growth in savannas, this hypothesis states that trees and grasses have different access to water due to differences in their rooting depth. Trees and grasses both have access to the upper soil layers, where grasses are thought to be competitively superior and trees have sole access to deeper soil

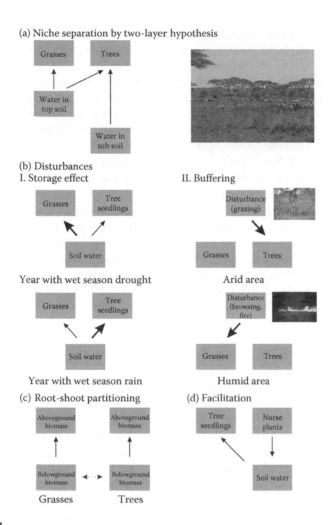

FIGURE 13.1
Schematic representation of models that explain coexistence of savanna trees and grasses:
(a) niche separation proposed by the two-layer hypothesis (Adapted from Walter, H. 1971.
Ecology of tropical and Subtropical Vegetation. Oliver and Boyd, Edinburgh, UK.); (b) disturbances
separated in (I) storage effect favoring grasses in years with wet season droughts (above) and
favoring tree seedlings in years with wet season rain (below) (Adapted from Higgins, S. I., W. J.
Bond, and W. S. W. Trollope. 2000. *Journal of Ecology* 88, 213–229.) and (II) buffering mechanism
where disturbances favor tree seedlings to prevent savannas developing into grassland in drier
areas (above) and disturbances favoring grass to prevent savannas developing into woodland in
humid areas (below) (Adapted from Jeltsch, F., G. E. Weber, and V. Grimm. 2000. *Plant Ecology*
161, 161–171.); (c) root–shoot partitioning where two life-forms coexist when root interactions are
balanced (Adapted from Scheiter, S. and S. I. Higgins. 2007. *American Naturalist* 170, 587–601.);
and (d) facilitation by plants enhance water availability and reduce stress favoring tree seedlings (Adapted from Anthelme, F. and R. Michalet. 2009. *Basic and Applied Ecology* 10, 437–446.).
The solid arrows represent positive effects (more resources promote plant growth, disturbances
stimulate growth of inferior competitor), whereas the dashed arrows represent negative effects
(competition). The thickness of the arrows indicates their relative importance.

layers (Figure 13.1a). The two-layer hypothesis has a long history, and several subsequent models have employed it as a basis to explain shifts in trees and grasses in savannas (Walker et al., 1981; Van Langevelde et al., 2003).

Over the last decade, this two-layer hypothesis has come under increasing criticism (Scholes and Archer, 1997; Higgins et al., 2000; Jeltsch et al., 2000) due to an accumulation of contrary empirical evidence. Some field experiments did not find clear niche separation of roots (Seghieri, 1995; Mordelet et al., 1997), whereas others appear to have done so (Weltzin and McPherson, 1997; Schenk and Jackson, 2002). Specifically, in drier areas with deep, sandy, free-draining soils where water penetration from rainfall does reach lower layers of the soil, allocation of root growth by trees to these layers has been observed (Schenk and Jackson, 2002). However, savannas can also occur on shallow soils with impermeable lower layers where grasses and trees are forced to share the same rooting space (Kim and Eltahir, 2004; Wiegand et al., 2005). Further, in certain environments, it appears that trees concentrate their rooting at shallow depths even though the underlying substrate is not necessarily impermeable to deeper root penetration (Mlambo et al., 2005; Mordelet and Le Roux, 2006). Presumably, advantages of competition with grasses for soil resources outweigh advantages of deeper rooting under these conditions.

A deeper problem with this two-layer hypothesis is the assumption that grasses are superior competitors to trees in the rooting zone that they both occupy and what this means for establishment of tree seedlings. Even if grasses are able to collect soil resources more efficiently than trees, this ability does not necessarily ensure their competitive dominance over trees, because of fundamental differences in the growth allocation of the two life-forms that is somewhat akin to the parable of the tortoise and the hare: grasses rapidly grow multiple short-lived shoots that they are forced to renew annually or biennially but which individually are cheap to build per unit carbohydrate spent, whereas trees grow single or few hard perennial stems that are more expensive to build but on which they are able to accumulate future resources. Thus, although tree seedlings may grow much more slowly than competing grasses, their sure accumulation of resources across seasons which they allocate to building a taller shoot ensures that in the long term they might overcome limitation by grass competitors (but see Riginos, 2009) and proceed to monopolize the light resources; slow, but surely wins the race. The only possible chance for the grasses is if their competitive effects are so severe that they reduce tree seedling survival. Models incorporating the two-layer hypothesis ignore this limited seedling recruitment and only consider adult trees able to reach deeper soil layers.

Disturbance

As a response to criticisms of the two-layer hypothesis, new explanations that recognised the fundamental role of disturbances in maintaining coexistence of trees and grasses were proposed (Menaut et al., 1990; Jeltsch et al., 1996, 1998, 2000; Higgins et al., 2000; Van Wijk and Rodriguez-Iturbe, 2002; Gardner, 2006). These explanations are premised on the assumption that climate variation (drought) and defoliation (fire and herbivory) are disturbances that differ in their impact on different life-history stages, causing demographic bottlenecks on tree recruitment. Repeated negative impacts of these disturbances on the superior life-form create opportunities for establishment and persistence of the competitively inferior life-form (Warner and Chesson, 1985). These explanations differ from the explanation based on niche separation in that they explicitly focus on population processes of trees.

The relative importance of these disturbances differs over the MAP gradient. Arid and semiarid savannas have relatively high interannual variation in MAP, and consequently these systems frequently experience drought events that can kill adult plants (Fensham and Holman, 1999), thereby opening resource space that may be reoccupied by grasses or tree seedlings (Jeltsch et al., 1996, 1998; Van Wijk and Rodriguez-Iturbe, 2002). In humid savannas, empirical evidence has accumulated which indicates that fire and herbivory are important elements that structure the vegetation, including the balance between trees and grasses, which suggested that these disturbances might play a fundamental role in ensuring coexistence (Sankaran et al., 2007).

The formulated explanations based on disturbances are underlain by two different arguments for the dynamics (Sankaran et al., 2004), namely nonequilibrium (Higgins et al., 2000; Van Wijk and Rodriguez-Iturbe, 2002; Gardner, 2006) and disequilibrium (Menaut et al., 1990; Jeltsch et al., 2000) dynamics. These two types of dynamics assume that savannas are systems in transition, which are maintained in their transitional state by disturbance events or temporal resource variability. The distinction between the two is that disequilibrium dynamics assume that disturbances prevent systems from reaching their equilibrium position, whereas nonequilibrium dynamics ignore the existence of such equilibrium positions. The nonequilibrium dynamics assume that tree recruitment is not limited by competition between tree seedlings and grasses, and it is only limited by defoliation and climatic events such as drought. On the other hand, the disequilibrium dynamics assume that competitive interactions between trees and grasses are occurring at all life stages such that local conditions determine the competitive advantage of each life-form and that disturbance prevents movement to dominance by one life-form or the other.

Higgins et al. (2000) built a model that focussed on the limitations of tree seedling establishment and transitions to adults (Figure 13.1b I, Storage effect). They assume that establishment of tree seedlings in arid savannas is not limited by competition with grasses but rather by rainfall variability which determines the opportunities for seedlings to germinate and establish. In these savannas, however, the critical assumption of the model is that in low-rainfall years, adult mortality exceeds recruitment else these savannas would be woody covered due to the presence of high-rainfall years. For humid savannas, it is assumed that small trees are unable to escape to taller size classes, because frequent fires prevent them from doing so, thus retaining space for grasses. In low-rainfall years, grass biomass accumulation is lower than for high-rainfall years, which reduces subsequent fire intensities. Low-intensity fires damage tree seedlings less severely, allowing them to recover more quickly and then to escape this fire trap. It is assumed here that tree seedlings establish in the herbaceous layer, which might, however, be problematic in humid savannas (Harrington, 1991; Van der Waal et al., 2009).

The ability of the nonequilibrium model of Higgins et al. (2000) to demonstrate tree–grass coexistence through disturbance and demography without explicit competition between the species suggests that both disturbance and demography are critical components describing tree–grass coexistence and that as long as disturbance is sufficiently frequent in the given system, competition between life-forms will never move to complete exclusion of one life-form. Unfortunately, since the authors ignore competition processes, they were not able to evaluate how the balance between the rates of plant growth in savannas due to resource competition and disturbances affects savanna structure.

Jeltsch et al. (2000) proposed that savannas only exist, because disturbances act as ecological buffering mechanisms that prevent local systems from moving to their true equilibria, either grassland or woodland depending on the local environmental conditions (Figure 13.1b II, Buffering). In arid savannas, disturbances provide favorable conditions for tree seedling establishment and survival by opening up the grass sward, thereby reducing competition with grasses. In contrast to the nonequilibrium models, tree recruitment in these disequilibrium dynamics is limited by competition with grasses and can only occur when this competition is reduced due to disturbances. Under humid conditions, similar processes as in the nonequilibrium models can be found: fire in combination with browsing is regarded as the buffering mechanism that impedes the transition to woodland by destroying the juvenile trees and shrubs and preventing their progression to taller, fire-resistant size classes. Unfortunately, the authors did not develop a model to explore the validity of their theory. This might have led them to reevaluate the assumption that competitive exclusion can work in both directions. In savannas, we know of no conclusive evidence that grasses can competitively exclude trees at any life stage, even if they are able to reduce their growth rates (Riginos, 2009), and our discussion on niche separation suggests that the differences in their growth strategy makes this likelihood improbable. The only areas within the savanna biome where trees may be completely absent are depressions subjected to seasonal soil saturation, and it is believed that this is because trees cannot survive these conditions (Joly and Crawford, 1982; Tinley, 1982).

Root–Shoot Partitioning

Scheiter and Higgins (2007) built a model with separate aboveground and belowground biomass components for both trees and grasses (Figure 13.1c). The division into aboveground and belowground parts allowed consideration of the consequences of competition for multiple resources: aboveground for light and belowground for water, rooting space, and nutrients. They included herbivory and fire as factors that only directly impacted aboveground biomass. The authors demonstrated that tree–grass coexistence could be achieved with balanced root competition without the two-layer niche separation, but only when root competition from both components remained below certain thresholds and only when there was a low level of competition for light aboveground. If the latter increased, then asymmetric light competition would favor tree growth and move the system to a tree-dominated state. The authors claim that this provides an explanation for coexistence in water-limited environments where light competition is restricted by the low levels of standing biomass. They further demonstrate that coexistence under high levels of light competition is only possible when browsing or fire reduces biomass of trees, thus reducing their competitive pressure on grasses. The difficulty with this model is that it is a phenomenological representation of the processes that shape tree–grass interactions; the prediction that root interactions should be balanced is hard to test. The parameters in their model are also difficult to directly estimate (Scheiter and Higgins, 2007), although Higgins et al. (2010) show that the parameters can be indirectly estimated and that the parameterised model can explain the patterns of tree abundance in Africa.

Facilitation

Although high grass biomass can suppress tree seedlings, other evidence suggests that tree seedling establishment can be facilitated by the presence of other trees (Holmgren et al., 1997; Ludwig et al., 2004) or grasses (Brown and Archer, 1989; Davis et al., 1998; Anthelme and Michalet, 2009). This facilitation is seen as a key structuring process for plant communities in stressful environments (Brooker et al., 2008). In arid conditions, the reductions in air and soil temperatures under nurse plant's canopies result in a decrease in evopotranspiration of shaded plants. This often translates into higher soil water content that promotes the recruitment of tree seedlings (Holmgren et al., 1997). Such facilitative interactions could explain coexistence of trees and grasses in arid areas when grass facilitates the recruitment of tree seedlings (Figure 13.1d, Anthelme and Michalet, 2009). The crucial assumption for this coexistence through facilitative effects of grasses on trees is that mortality of both tree seedlings and adult trees should exceed tree recruitment in the absence of the facilitator. So far, this possible explanation for coexistence of savanna trees and grasses has been overlooked.

An Environmental Framework for Tree–Grass Coexistence

It is clear that savanna tree–grass coexistence is mediated by different environmental processes across the broad range of environmental conditions where savannas are found and that none of the existing models are able to provide a general explanation for the existence of all savannas. To integrate and explore the importance of these different processes, should we jump to complex process models? Some complex models of savanna systems already exist (e.g., Coughenour, 1993; Gignoux et al., 1998), but invariably they are used for on-site prediction of dynamics and they need site-specific parameters to produce meaningful results. Critically, their results are not general and their complexity predisposes them to equifinality (Beven and Freer, 2001). Hence, they are not suitable for distinguishing between essential factors and contributory factors. If we wish to understand the underlying dynamics, we need to start with models that are as simple as possible and include the minimum number of contributing factors.

As a first step in defining such a model, we need a framework that summarises the environmental space of savannas, where we can contextualise the existing explanations and justify relevant system drivers (Figure 13.2). Most savanna ecologists agree that at least four environmental components are necessary to define this space occupied by savannas, namely water, fire, herbivory, and soil nutrients (Frost et al., 1986; Solbrig et al., 1996; Scholes and Archer, 1997). Of these, the first three have received much attention in savanna models, whereas the role of soil nutrients has been largely ignored. This latter omission is surprising given the well-described associations between species distributions and edaphic patterns (Cole, 1986; Solbrig, 1996). We will attempt to describe the environmental space covered by these four components and locate the proposed explanations for coexistence within this space (Figure 13.2). We use MAP as the main environmental gradient, because the change in savanna structure has been well described for this gradient: Sankaran et al. (2005) found that maximum observed tree cover in a large sample of African savannas increased linearly along the MAP gradient until approximately 650 mm per year, above which closed woodlands (>80% cover) could develop. This bound on woody cover is seen as a fundamental descriptor of the potential savanna physiognomy that could be achieved at a particular MAP. Notably, very few of the sites included in the study reached this maximum, suggesting that the other drivers of savanna dynamics (soil fertility, herbivory, and fire) reduce woody cover below the line. Recently, it has also been suggested that the potential cover achieved against MAP depends on the functional traits of the tree species involved (Lehmann et al., 2009), suggesting that models predicting savanna structure will necessarily need to include differences in functional types.

Bucini and Hanan (2007) separated the effects of disturbances (fire, domestic livestock, human population density, and cultivation intensity) over the rainfall

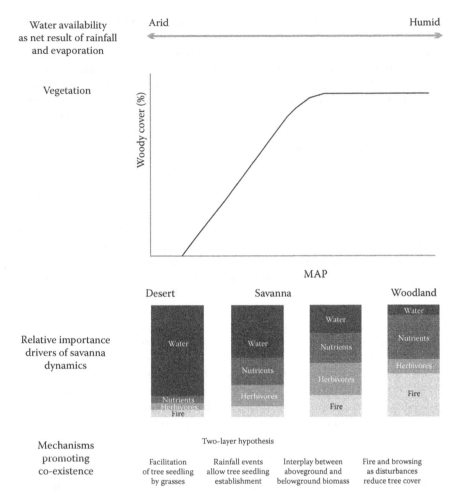

FIGURE 13.2
Change in woody cover (Adapted from Sankaran, M. et al. 2005. *Nature* 438, 846–849.), the relative importance of drivers of savanna dynamics (water and nutrient availability, herbivory, and fire), and the proposed explanations for coexistence of savanna trees and grasses over a rainfall gradient. The importance of these explanations changes over environmental gradients, for example, disturbances such as fire are more important at high rainfall levels, whereas water availability is more important at low rainfall sites. These explanations assume that tree seedlings struggle to establish under low rainfall conditions, whereas trees might be competitively superior under high rainfall conditions.

gradient and found that their absolute effect increased with MAP. This outcome agrees with the existing disturbance models where water supply limits seedling recruitment (Jeltsch et al., 1996, 1998; Higgins et al., 2000; Van Wijk and Rodriguez-Iturbe, 2002) and precipitates adult mortality (Fensham and Holman, 1999) in drier areas, whereas defoliation disturbances (herbivory and fire) limit adult recruitment in humid areas (Van Langevelde et al., 2003; Bond, 2008;

Scheiter and Higgins, 2008) (Figure 13.2). Interannual rainfall variability contributes to explaining tree cover: Environments with high MAP variation are
likely to have increased tree mortality or dieback events, leading to lower tree
cover. The tree cover curve found by Sankaran et al. (2005) suggests that tree
cover is limited by absolute rainfall amount as well and that tree–grass coexistence at the low MAP end may, thus, be mediated by total rainfall amount
regardless of its variability. This distinction between the responses of tree cover
to MAP and interannual variation of annual rainfall needs investigation.

The relative importance of fire versus herbivory is probably a function of
plant shoot quality, because low-quality vegetation can sustain much lower
large mammalian herbivore biomasses than high-quality vegetation, and thus
herbivory will increase in importance with increased forage quality whereas
fire will increase in importance in areas with low forage quality (Bond and
Keeley, 2005). These, in turn, are likely to have major implications for savanna
dynamics due to the different defoliation patterns associated with burning
(nonselective defoliation, destroying leaves, stems, and apical buds) and herbivory (selective defoliation, mainly removing leaves). Forage quality is positively correlated with soil fertility and negatively correlated with MAP (Olff
et al. 2002), indicating that both MAP and soil fertility are major drivers of
this disturbance regime. Hence, high rainfall savannas (>1000 mm) found on
nutrient-poor soils might be mediated by fire. At intermediate MAP, forage
quality could be sufficient to maintain large herbivores, meaning that these
environments are often mediated by herbivory (Olff et al., 2002).

Since rates of plant growth in savannas must depend on the productivity
of each environment, mainly determined by water and nutrient availability,
resource competition is likely to be more restricted by nutrient availability in
high MAP environments (Mordelet and Le Roux, 2006) and by water in low
MAP environments. In contrast to the increase of woody cover with MAP
(Sankaran et al., 2005), woody cover is negatively related to nutrient availability in African savannas (Sankaran et al., 2007), apparently because performance of tree seedlings in herbaceous swards is adversely affected by soil
fertility (Kraaij and Ward, 2006; Van der Waal et al., 2009). Since water is limiting at low MAP and soil nutrients are limiting at high MAP, the greatest
rates of plant growth are likely to be found in systems with intermediate
MAP, where rainfall and nutrients are both more abundant than at their
lower extremes. These differences in environmental productivity that cause
different rates of plant growth mean that disturbance regimes will lead to
substantially different savanna physiognomies across these gradients.

Future Avenues for Research

The next step would be to quantify the importance of the coexistence mechanisms over this proposed MAP gradient, including nutrient availability,

herbivory, and fire (Figure 13.2). This step should aim at evaluating under what conditions savannas are persistent and whether there are conditions where disturbance is not necessary to mediate coexistence. Therefore, a model should be developed that includes resource supply and capture (water, nutrients, and light), disturbance (fire, herbivory), and demographics as minimum components. The demographic part should be explicit about the size classes, as disturbance impacts size classes differently (Hanan et al., 2008). To separate the role of competition for soil resources and light, partitioning in aboveground versus belowground parts might be necessary.

Due to the inclusion of demographic structure, temporal resource variability, and disturbances, such a model is not likely to be analytically tractable. It is the interannual and intra-annual variation in water supply that determines dieback and death of plant individuals, and thus sub-annual time steps in such a model are essential to elucidate the relationship between demographic patterns and tree or grass dominance. Asymmetries in light competition can be imposed by considering plant heights, which means that grasses can outshade tree seedlings, but grasses themselves will be dominated by taller trees. Using resource capture and loss explicitly, such a model should allow evaluating whether disturbance is essential for coexistence or whether the differences in resource demand-supply rates over time affect the resulting savanna structure.

We might end up with a relatively complex model, especially because of the need to incorporate demographic processes; but inclusion of the demographic structure is a minimum requirement to simulate savanna dynamics meaningfully (Hanan et al., 2008). We stress that all processes in such a model should be based on measurable parameters. We can use estimates for plant nutrient requirements and growth rates to parameterise plant demands and growth rates. Effects of water supply can be treated using established relationships between soil water potential and plant wilting points as well as soil dry-down relationships with soil depth. Soil nutrient supply rates are still poorly understood, but estimates can be made of the supply rates across soil types in different environments by measuring growth rates in plants growing without competitors and supplied continuously with water. Disturbance events can be imposed as external events via fire and herbivory frequencies and rainfall data.

So far, several issues are hardly addressed in savanna ecology, which may contribute to our understanding of tree–grass coexistence and the determinants of tree cover. Including these issues in the type of models we propose can help us elucidate responses of savanna communities to water and nutrient availability, fire, and herbivory. First, the role of nutrient availability under different MAP, as highlighted in this chapter, should be the subject of future research. Second, there is a need for comparative studies on plant traits. The explanations discussed in this chapter ignore the diversity of growth strategies employed by different members of the same life-form. For example, Hoffmann and Franco (2003) compared the growth and physiological responses of pairs of tree species from forests and savannas in Brazil. Savanna seedlings had higher root-shoot ratios and higher root total nonstructural carbohydrate concentration than forest species. We not only do not know how the demography or physiology of trees in savannas differs from adjacent forests (Bond, 2008), but we also lack systematic overview of the traits exhibited by trees that occur in savannas worldwide. For the African continent, some comparisons have been made, for example, about saplings with different architecture occurring in heavily grazed savannas and in frequently burned savannas (Archibald and Bond, 2003); but a broader understanding is required. Third, demography of trees needs more attention. As shown in this chapter, some studies include demography of trees in their explanation of the coexistence between trees and grasses, and they highlight demographic bottlenecks as major constraints on adult tree density in savannas. These studies suggest that the establishment phase is probably the critical demographic bottleneck in drier areas (Jeltsch et al., 1996, 1998; Higgins et al., 2000; Wiegand et al., 2006), but it can also not be ignored in humid areas. Indeed, experimental studies found increased survival and growth of tree seedlings with grass removal, depending on water and nutrient availability (Van der Waal et al., 2009; but see Anthelme and Michalet, 2009). To date, comparative studies on plant traits, which allow seedlings to establish in a grass sward and to overcome periods of low rainfall, herbivory, or fire, are lacking (Bond, 2008).

In short, there is still much to be done to grasp the complexity of environmental conditions underlying the savannas we see. The models to date are mostly relevant, and their information can be synthesised with these future areas of expansion into an overall predictive framework designed to better understand savannas worldwide.

References

Anthelme, F. and R. Michalet. 2009. Grass-to-tree facilitation in an arid grazing environment (Air Mountains, Sahara). *Basic and Applied Ecology* 10, 437–446.

Archibald, S. and W. J. Bond. 2003. Growing tall vs growing wide: Tree architecture and allometry of *Acacia karroo* in forest, savanna, and arid environments. *Oikos* 102, 3–14.

Beven, K. J. and J. Free. 2001. Equifinality, data assimilation, and uncertainty estimation in mechanistic modelling of complex environmental systems. *Journal of Hydrology* 249, 11–29.

Bond, W. J. 2008. What limits trees in C-4 grasslands and savannas? *Annual Review of Ecology Evolution and Systematics* 39, 641–659.

Bond, W. J. and J. E. Keeley. 2005. Fire as a global 'herbivore': The ecology and evolution of flammable ecosystems. *Trends in Ecology and Evolution* 20, 387–394.

Brooker, R. W., F. T. Maestre, R. M. Callaway et al. 2008. Facilitation in plant communities: The past, the present, and the future. *Journal of Ecology* 96, 18–34.

Brown, J. R. and S. Archer. 1989. Woody plant invasion of grasslands: Establishment of honey mesquite (*Prosopis glandulosa var. glandulosa*) on sites differing in herbaceous biomass and grazing history. *Oecologia* 80, 9–26.

Bucini, G. and N. Hanan. 2007. A continental-scale analysis of tree cover in African savannas. *Global Ecology and Biogeography* 16, 593–605.

Cole, M. M. 1986. *The Savannas: Biogeography and Geobotany*. Academic Press, London, UK.

Coughenour, M. B. 1993. *Savanna—Landscape and Regional Ecosystem Model: Model Description*. Natural Resource Ecology Laboratory, Fort Collins, CO, USA.

Davis, M. A., K. J. Wrage, and P. B. Reich. 1998. Competition between tree seedlings and herbaceous vegetation: Support for a theory of resource supply and demand. *Journal of Ecology* 86, 652–661.

Fensham, R. J. and R. E. Holman. 1999. Temporal and spatial patterns in drought-related tree dieback in Australian savanna. *Journal of Applied Ecology* 36, 1035–1050.

Frost, P, J. C. Menaut, O. T. Solbrig, M. Swift, and B. H. Walker. 1986. *Responses of Savannas to Stress and Disturbance: A Proposal for a Collaborative Programme of Research*. Biology International Special Issue, No. 10. International Union of Biological Sciences, Paris, France.

Gardner, T. A. 2006. Tree–grass coexistence in the Brazilian cerrado: Demographic consequences of environmental instability. *Journal of Biogeography* 33, 448–463.

Gignoux, J., J. C. Menaut, I. R. Noble, and I. D. Davies. 1998. A spatial model of savanna function and dynamics: Model description and preliminary results. In *Dynamics of Tropical Communities*, eds. D. M. Newbery, H. H. T. Prins and N. D. Brown, Oxford, UK, Blackwell Scientific Publications, pp. 361–383.

Harrington, G. N. 1991. Effects of soil–moisture on shrub seedling survival in a semi-arid grassland. *Ecology* 72, 1138–1149.

Higgins, S. I., W. J. Bond, and W. S. W. Trollope. 2000. Fire, resprouting and variability: A recipe for grass–tree coexistence in savanna. *Journal of Ecology* 88, 213–229.

Higgins, S.I., S. Scheiter, and M. Sankaran. 2010. The stability of African savannas: insights from the indirect estimation of the parameters of a dynamic model. *Ecology* 91, 1682–1692.

Hoffmann, W. A. and A. C. Franco. 2003. Comparative growth analysis of tropical forest and savanna woody plants using phylogenetically independent contrasts. *Journal of Ecology* 91, 475–484.

Holmgren, M., M. Scheffer, and M. A. Huston. 1997. The interplay of facilitation and competition in plant communities. *Ecology* 78, 1966–1975.

Hanan, N. P., W. B. Sea, G. Dangelmayr, and N. Govender. 2008. Do fires in savanna consume woody biomass? A comment on approaches to modeling savanna dynamics. *American Naturalist* 171, 851–856.

Jeltsch, F., S. J. Milton, W. R. J. Dean, and N. Van Royen. 1996. Tree spacing and co-existence in semi-arid savannas. *Journal of Ecology* 84, 583–595.

Jeltsch, F., S. J. Milton, W. R. J. Dean, N. Van Rooyen, and K. A. Moloney. 1998. Modelling the impact of small-scale heterogeneities on tree–grass coexistence in semi-arid savannas. *Journal of Ecology* 86, 780–793.

Jeltsch, F., G. E. Weber, and V. Grimm. 2000. Ecological buffering mechanisms in savannas: A unifying theory of long-term tree–grass coexistence. *Plant Ecology* 161, 161–171.

Joly, C. A. and R. M. M. Crawford. 1982. Variation in tolerance and metabolic responses to flooding in some tropical trees. *Journal of Experimental Botany* 33, 799–809.

Kim, Y. and E. A. B. Eltahir. 2004. Role of topography in facilitating coexistence of trees and grasses within savannas. *Water Resources Research* 40, W07505.

Kraaij, T. and D. Ward. 2006. Effects of rain, nitrogen, fire and grazing on tree recruitment and early survival in bush-encroached savanna, South Africa. *Plant Ecology* 186, 235–246.

Lehmann, C. E. R., J. Ratman, and L. B. Hetley. 2009. Which of these continents is not like the other? Comparisons of tropical savanna systems: Key questions and challenges. *New Phytologist* 181, 508–511.

Ludwig, F., H. De Kroon, F. Berendse, and H. H. T. Prins. 2004. The influence of savanna trees on nutrient, water and light availability and the understorey vegetation. *Plant Ecology* 170, 93–105.

Menaut, J. C., J. Gignoux, C. Prado, and J. Clobert. 1990. Tree community dynamics in a humid savanna of the Cote-d'Ivoire: Modeling the effects of fire and competition with grass and neighbours. *Journal of Biogeography* 17, 471–481.

Mordelet, P. and X. Le Roux. 2006. Tree/grass interactions. In *Lamto, Structure, Functioning, and Dynamics of a Savanna Ecosystem*, eds. L. Abbadie, J. Gignoux, X. Le Roux and M. Lepage, Springer-Verlag, Ecological Studies, New York, 139–161.

Mordelet, P., J. C. Menaut, and A. Mariotti. 1997. Tree and grass rooting patterns in an African humid savanna. *Journal of Vegetation Science* 8, 65–70.

Olff, H., M. E. Ritchie, and H. H. T. Prins. 2002. Global environmental controls of diversity in large herbivores. *Nature* 415, 901–904.

Riginos, C. 2009. Grass competition suppresses savanna tree growth across multiple demographic stages. *Ecology* 90, 335–340.

Sankaran, M., N. P. Hanan, R. J. Scholes et al. 2005. Determinants of woody cover in African savannas. *Nature* 438, 846–849.

Sankaran, M., J. Ratnam, and N. P. Hanan. 2004. Tree–grass coexistence in savannas revisited—insights from an examination of assumptions and mechanisms invoked in existing models. *Ecology Letters* 7, 480–490.

Sankaran, M., J. Ratnam, and N. Hanan. 2007. Woody cover in African savannas: The role of resources, fire and herbivory. *Global Ecology and Biogeography* 17, 236–245.

Sarmiento, G. 1984. *The Ecology of Neotropical Savannas*. Harvard University Press, Cambridge, USA.

Scheiter, S. and S. I. Higgins. 2007. Partitioning of root and shoot competition and the stability of savannas. *American Naturalist* 170, 587–601.

Schenk, H. J. and R. B. Jackson. 2002. Rooting depths, lateral root spreads, and belowground/aboveground allometries of plants in water-limited ecosystems. *Journal of Ecology* 90, 480–494.

Scholes, R. J. and S. R. Archer. 1997. Tree–grass interactions in savannas. *Annual Review of Ecology and Systematics* 28, 517–544.

Seghieri, J. 1995. The rooting patterns of woody and herbaceous plants in a savanna: Are they complementary or in competition? *African Journal of Ecology* 33, 358–365.

Silvertown, J. 2004. Plant coexistence and the niche. *Trends in Ecology and Evolution* 19, 605–611.

Solbrig, O. T. 1996. The diversity of the savanna ecosystem. In *Biodiversity and Savanna Ecosystem Processes: A Global Perspective*, eds. O. T. Solbrig, E. Medina and J. F. Silva, Ecological Studies 121, Springer-Verlag, Berlin, pp. 1–27.

Solbrig, O. T., E. Medina, and J. F. Silva. 1996. Determinants of tropical savannas. In *Biodiversity and Savanna Ecosystem Processes: A Global Perspective*, eds. O. T. Solbrig, E. Medina, and J. F. Silva, Ecological Studies 121, Berlin, Springer-Verlag, pp. 31–41.

Sydes, C. L. and J. P. Grime. 1984. A comparative study of root development using a simulated rock crevice. *Journal of Ecology* 72, 937–946.

Tinley, K. L. 1982. The influence of soil moisture balance on ecosystem patterns in southern Africa. In *Ecology of Tropical Grasslands*, ed. B.J. Huntley and B. H. Walker, Springer, Ecological Studies, Berlin, pp. 193–216.

Van der Waal, C., H. De Kroon, F. De Boer et al. 2009. Water and nutrients alter herbaceous competitive effects on tree seedlings in a semi-arid savanna. *Journal of Ecology* 97, 430–439.

Van Langevelde, F., C. Van de Vijver, L. Kumar et al. 2003. Effects of fire and herbivory on the stability of savanna ecosystems. *Ecology* 84, 337–350.

Van Wijk, M. T. and I. Rodriguez-Iturbe. 2002. Tree–grass competition in space and time: insights from a simple cellular automata model based on ecohydrological dynamics. *Water Resources Research* 38, 18.11–18.15.

Walker, B. H., D. Ludwig, C. S. Holling, and R. M. Peterman. 1981. Stability of semi-arid savannah grazing systems. *Journal of Ecology* 69, 473–498.

Walter, H. 1971. *Ecology of Tropical and Subtropical Vegetation*. Oliver and Boyd, Edinburgh, UK.

Warner, R. R. and P. L. Chesson. 1985. Coexistence mediated by recruitment fluctuations: a field guide to the storage effect. *American Naturalist* 125, 769–787.

Weltzin, J. F. and G. R. McPherson. 1997. Spatial and temporal soil moisture resource partitioning by trees and grasses in a temperate savanna, Arizona, USA. *Oecologia* 112, 156–164.

Wiegand, K., D Saltz, and D. Ward. 2006. A patch-dynamics approach to savanna dynamics and woody plant encroachment – insights from an arid savanna. *Perspectives in Plant Ecology, Evolution and Systematics* 7, 229–242.

Wiegand, K., D. Ward, and D. Saltz. 2005. Multi-scale patterns and bush encroachment in an arid savanna with a shallow soil layer. *Journal of Vegetation Science* 16, 311–320.

14

Spatially Explicit Modeling of Savanna Processes

Katrin M. Meyer, Kerstin Wiegand, and David Ward

CONTENTS

Introduction

Savannas are inherently spatially variable regardless of where they occur (Sinclair and Fryxell, 1985; Jeltsch et al., 1997; Wiegand et al., 2000a; Ward 2009; Figure 14.1). Therefore, models of savanna processes and patterns often explicitly address spatial variation. In this chapter, we will first give an overview of the features of savannas that show spatial variation. We will then present evidence for the importance of spatial variation in savannas, because presence of spatial variation does not necessarily imply importance for savanna dynamics. In the main part of the chapter, we will explain the spatial modeling techniques that have been applied to savannas, give examples of their application, and discuss their strengths and limitations.

Small-scale features that affect spatial variation in savanna structure include geology and soil, rainfall, elevation and topography, fire, intra- and interspecific competition, propagule pressure, and herbivory (Table 14.1). Regarding geology and soil features, Britz and Ward (2007) found that there were significant differences in the densities of *Acacia mellifera* on adjacent rocky and sandy soil substrates less than 100 m apart. McNaughton (1983)

FIGURE 14.1
(See color insert following page 320.) Spatial structure in a semiarid savanna.

found that lawn grass patches occur in the Serengeti savanna (Tanzania) and may be associated with sodic soils or fertile locations. Scholes and Walker (1993) have shown that termite mounds and areas where people have lived and/or livestock corralled are more fertile (due to higher levels of nitrogen there), which has led to changes in savanna structure, shifting to *Acacia*-dominated savanna in an area dominated by broad-leaved trees.

TABLE 14.1

Major Spatial Processes in Savannas and Their Effect on Savanna Plant Mortality, Reproduction, and Growth; Their Direction; and the Scales They May Cover

| | Effect[a] on Plant ... | | | | |
Spatial Processes	Mortality	Reproduction	Growth	Direction[b]	Scale[c]
Abiotic Processes					
Rain events	–	+	+	h	100 m–10 km
Fire	+	–/+	0	h (+v)	10 m–10 km
Nutrient availability	–	+	+	h (+v)	1 mm–10 km
Biotic Processes					
Grazing	+	–	–/+	h	1 m–10 km
Termite mound degradation	0	0	+	h	10 cm–100 m
Belowground competition	+	–	–	v + h	1 cm–10 m
Facilitation via hydraulic lift	–	+	+	v	1 cm–10 m
Dispersal	–	0	0	h	1 cm–10 km

[a] Plants include trees, shrubs, and grasses; 0: neutral effect, –: negative effect, +: positive effect, –/+: negative and positive effects.
[b] Predominant direction of spatial variability; h: horizontal, v: vertical.
[c] Approximate scales covered by the process.

Reid and Ellis (1995) found that livestock increased seed densities of *Acacia tortilis* trees by 85 times inside corrals compared with similar areas outside of corrals. In the Kruger Park of South Africa, the dominant savanna vegetation patterns follow variation in climate, fire, geology, and soils (Scholes et al., 2003). In the western Llanos of Venezuela, Silva and Sarmiento (1976) (quoted by Sarmiento, 1996) have shown that species richness per 1000 m² can vary between 53 and 84 species in a single soil catena and is strongly related to water availability.

With regard to spatial variation in rainfall, there is a negative correlation between mean annual rainfall and its coefficient of variation (see e.g., Sharon, 1981; Ward, 2009), leading to large spatial variation in savanna structure and function as mean annual rainfall declines. Although rainfall is often considered as a large-scale feature, rainfall events can be highly localized, particularly as mean annual rainfall declines (Sharon, 1981).

Fire is considered a major determinant of spatial variation in savanna structure, particularly in mesic savannas (Higgins et al., 2000), although Australian studies indicate that fire can be important at rainfall values as low as 250–300 mm per annum (McKeon et al., 2004). Fires can result in increased space for trees to germinate en masse (Bond, 2008) or they can cause trees to be removed, resulting in a more open savanna (McKeon et al., 2004; Ward, 2005). Due to the large variation in fuel loads caused by differences in rainfall and soil type, fire intensity and duration vary considerably, resulting in considerable variation in spatial structure (Bond, 2008).

With regard to interspecific competition, Smith and Goodman (1986) have shown that *Acacia tortilis* and *Euclea divinorum* compete for space in the mesic Mkhuze reserve (South Africa); whereas Schleicher et al. (unpublished data) have shown that two species (*Tarchonanthus camphoratus* and *Acacia mellifera*) may compete for space in the semiarid area near Kimberley (South Africa). Propagule pressure may also show spatial structure. A spatially explicit simulation model by Jeltsch et al. (1996) showed that rooting niche separation might be insufficient to allow coexistence under a range of climatic situations. However, Jeltsch et al. (1998) found that introducing safe sites for seedling establishment by simulating the effects of various small-scale heterogeneities allowed coexistence. Grazing is widely purported to be the main factor affecting increases in tree density (Walker et al., 1981; Ward, 2005). This is largely ascribed to heavy stocking by domestic livestock, creating gaps for mass recruitment of trees. Piosphere formation (see e.g., Smet and Ward, 2006) and fenceline effects (e.g., Hendricks et al., 2005) are also a major source of the spatial variation in savannas, caused largely by heavy grazing near water holes and differential effects of animal densities on either side of a fence.

Even though savannas are often thought of as a homogeneous grass layer interspersed with trees, the grass component is also expected to be spatially heterogeneous (e.g., O'Connor, 1991; Fowler, 2002; Augustine, 2003; Wiegand et al., 2005). Jurena and Archer (2003) provided evidence that, in

Texas, there are always gaps in the grass for some *Prosopis* plants to germinate and establish. Spatial heterogeneity of the grass layer may thus facilitate woody invasion in grasslands. Such patterns are generally modified by grazing and fire, both of which are known to be highly patchy and to interact in spatially complex ways (e.g., Archibald et al., 2005; Getzin, 2007).

Large-scale spatial variation in savannas may occur too. In some cases, vast areas are affected. For example, Sinclair and Fryxell (1985) consider the Sahel, which covers 10 countries immediately south of the Sahara, as being subject to large changes due to a change from migratory livestock to fixed patterns of livestock control. Migratory ungulates use nutritious but seasonal food and probably maintain larger populations as a consequence. In the 1950s and 1960s, pastoralists were forced to settle with their livestock. This caused heavy grazing near these wells (leading to piosphere formation) and has resulted in frequent famines since this time (Sinclair and Fryxell, 1985). In North America, Schmitz et al. (2002) found that substrate heterogeneity in pine savannas in Florida, measured as variation in elevation, may influence species richness at larger spatial scales.

Evidence for the Importance of Space in Savannas

Patterns are the outcome of processes that have acted over time and space. Consequently, patterns are indicators of the processes that have shaped them, and spatial patterns indicate the presence of spatial processes. Thus, as a simple rule of thumb for model development, if there is at least one spatial pattern observed at the range of scales relevant to the question being studied, one should consider developing a spatial model. Arguably, this means that one should virtually always consider developing spatial models, because spatial patterns abound in our spatial world. The omnipresence of spatial patterns also makes it often easier to think spatially than nonspatially and to translate field knowledge into spatial rather than nonspatial models. Notwithstanding this, once spatial models have been thoroughly analyzed and have improved our understanding, it may well be possible to simplify the model and to remove the spatial components (e.g., Adler and Mosquera, 2000; Wiegand et al., 2004a). Most spatial savanna models presume that space is important without having explicitly tested the importance of space. A few exceptions exist and will be reviewed in the remainder of this section, but future research will need to clarify the level of spatial detail needed.

A common observation in savannas is the nonrandom spatial distribution of trees. However, whether the distribution is aggregated, random, or regular may change with age of savanna trees. This emphasizes the importance of space for explaining age-related phenomena. For example, small trees may be clumped whereas large trees are randomly or even regularly spaced (e.g., Skarpe et al., 1991; Meyer et al., 2008). This transition is readily explained by intraspecific competition leading to increased mortality of trees with many neighbors. However, other explanations may be possible such as in the case of *Acacia raddiana* trees in the Negev desert. At the seedling stage, these trees were clumped, changing to a random distribution thereafter (Wiegand et al., 2000a). In a spatially explicit simulation model, the breakup of the clumped pattern was explained by both density-dependent mortality (competition) or relatively high, yet density-independent, seedling mortality (Wiegand et al., 2000a). The lesson that one pattern may have more than one explanation is also true for the example of the clumping of fleshy-fruited *Grewia flava* relative to *A. erioloba* in the Kalahari and Kimberley Thorn Bushveld (Dean et al., 1999; Figure 14.2). Both facilitation and directed seed dispersal by birds and mammals are equally likely explanations. Spatial heterogeneity, for example, in soil resources, may also play a role, and further modeling is required to sort out these possibilities. These examples show that spatial patterns are key to better understand spatiotemporal processes (Table 14.1). The intricate mixture of space and time can be captured only by using spatial models.

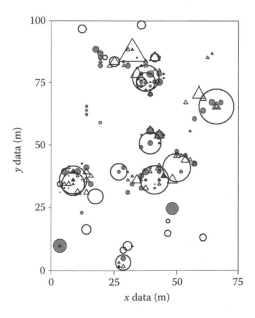

FIGURE 14.2

Example of spatial pattern of savanna plants. Spatial distribution of *Acacia erioloba* (open circles), *Grewia flava* (solid circles), and *Ziziphus mucronata* (triangles) at Pniel Estates (S 28°35′, E 24°29′), 30 km north of Kimberley, South Africa. Symbol size is proportional to canopy diameter. *G. flava* is significantly more frequently found beneath *A. erioloba* than in the open (Jana Schleicher, unpublished data and analyses).

Recently, empirical and modeling evidence for explaining savanna tree–grass coexistence with hierarchical patch dynamics has accumulated (e.g., Wiegand et al., 2005; Meyer et al., 2007b; Moustakas et al., 2009). Wiegand et al. (2005) investigated patterns of local tree size distributions across an *Acacia reficiens* savanna landscape in Namibia. Local tree size frequency distributions indicated that one or two age cohorts were present; whereas, at the landscape scale, a multitude of cohorts were detected after a negative exponential tree size–frequency distribution and reflecting an overall birth-death equilibrium (Wiegand et al., 2005). Using paleoecological techniques, Gillson (2004) found that patterns of vegetation change differed among micro, local, and landscape spatial scales, with the most rapid changes at the micro scale. In patch dynamic savannas, infrequent recruitment events generate even-aged stands at the local level. Competition at the micro level leads via self-thinning (Wiegand et al., 2008) from dense localized tree patches to open savanna with a few large trees. If the local recruitment events are not correlated in space and time, then a stable savanna emerges (Gillson, 2004; Wiegand et al., 2005, 2006). Other predictions of patch dynamics are supported by model results. Models show that mass recruitment driven by overlapping localized rain events in combination with self-thinning are the main drivers in arid savannas (Figure 14.3). In humid savannas, recruitment bottlenecks are more often mediated by fire (Higgins et al., 2000), whose spatial patchiness can drive patch dynamics as well. Since patch dynamics is a spatial scale-explicit mechanism, its validity in savannas supports the importance of spatial scales in savannas. Overall, these examples highlight the importance of including spatial structure in models of savanna dynamics.

Spatial Modeling Approaches

Models can incorporate space in various ways (Table 14.2). Descriptive approaches such as conceptual and statistical models or applications of remote sensing and GIS contribute to spatial savanna modeling by formalizing and quantifying spatial relationships. Equation-based modeling approaches require mathematical extensions to incorporate space explicitly, such as in the case of moment equations and matrix models. Grid-based approaches including cellular automata are specially designed to capture spatial processes. Therefore, grid-based approaches offer probably the greatest flexibility in describing, understanding, and predicting the impact of spatial relationships on savanna patterns and dynamics. Since grid-based modeling has been the most common approach in *spatial* savanna modeling thus far, we will present this approach in more detail a little later. Each modeling approach has strengths and limitations and should be applied accordingly to the savanna processes involved (Table 14.2).

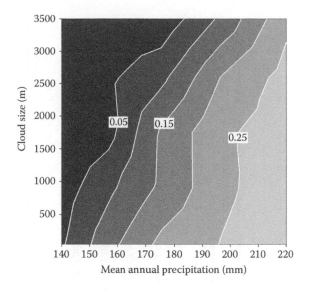

FIGURE 14.3

Tree–grass ratios as a function of precipitation cloud size (length) and mean annual precipitation. To test the hypothesis that local recruitment driven by localized rainfall and self-thinning may be major drivers of patch dynamic arid savannas, Eisinger, Wiegand, and Ward (in prep.) are developing a simulation model combining self-thinning of local shrub cohorts (Wiegand et al., 2008) with localized rainfall. (Adapted from Eisinger, D., and K. Wiegand 2008. *South African Journal of Science* 104, 37–42.) The shading indicates tree–grass ratio in terms of tree cover. Values below 0.05 indicate grassland, savanna extends from 0.05 to 0.25, and values beyond 0.25 indicate open woodland. This preliminary result shows that, given a fixed amount of mean annual precipitation, tree–grass ratios shift toward greater tree cover if rainfall is more patchily distributed in space (smaller precipitation cloud size). Especially in very dry areas, savannas occur under patchy rainfall only. Further preliminary results show that the spatial distribution of trees of different sizes generated by the model is in agreement with the patch-dynamics hypothesis (results not shown). Thus, tree–grass coexistence is indeed possible in a framework of hierarchical patch dynamics.

Conceptual Models

Conceptual models organize information, represent hypotheses and theories, and are particularly useful for capturing spatial relationships. For example, Walter's two-layer hypothesis explained tree–grass coexistence in savannas with a spatial segregation of rooting depths to access water (Walter, 1971), which was originally formulated as a conceptual model. Subsequent empirical and theoretical tests of Walter's hypothesis were based on this spatial conceptual model, yielding equivocal results (e.g., Knoop and Walker, 1981; Walker and Noy-Meir, 1982; Wiegand et al., 2005). In another spatial conceptual model, patch dynamics were mooted as a spatially explicit mechanism underlying tree–grass coexistence (Wiegand et al., 2006; Meyer et al., 2009). The explicit formulation of a conceptual model facilitated the derivation of specific predictions from the patch-dynamics mechanism such as the

TABLE 14.2

Major Spatial Processes in Savannas and Examples of How They Can Be Addressed in Spatial Modeling Approaches[a]

Spatial Processes	Conceptual Models	Remote Sensing	Spatial Statistics[b]	Moment Equations	Matrix Models	Cellular Automata	Grid-Based Models
				Modeling Approaches			
Abiotic Processes							
Rain events	Box	Possible	CV vs. average, autocorr	Forcing or input	Perturb or sub	Trans prob = f(rain)	Input + algorithm[e]
Fire	Box	Possible	Autocorr	Forcing or input	Perturb or sub	As cell state that can change trans probs	Input + rules
Nutrient availability	Box	Only indirectly[c]	Autocorr	Forcing or input	Perturb or sub	Trans prob = f(nutrient availability)	Input + rules
Biotic Processes							
Grazing	Arrow	Only indirectly[c]	Gradient analysis	Forcing or input	Perturb or sub	As cell state that can change trans probs	Input + rules
Termite mound degradation	Arrow	Only aerial photos	Autocorr	Forcing or input	Perturb or sub	As cell state that can change trans probs	Input + rules
Belowground competition	Arrow	Impossible	Point, NN	NN effect via equ[d]	Sub (if at all)	Trans probs between cells	Rules (+ input)
Facilitation via hydraulic lift	Arrow	Impossible	Gradient analysis	NN effect via equ[d]	Sub (if at all)	Trans probs between cells	Rules (+ input)
Dispersal	Arrow	Impossible	Cluster processes	Distr	Sub	Trans probs between cells	Rules+/distr (+ input)

Note: CV—coefficient of variation—refers to amount of rain in different with locations as at one point in time; autocorr—spatiotemporal autocorrelation analysis; point—point pattern analysis; NN—nearest-neighbor statistics; forcing—forcing function; input—input parameter; distr—distribution (e.g., Gaussian, Cauchy); perturb—via perturbation matrix approach; sub—via submatrix approach; trans prob—transition probability; $f(x)$—function of x.

  [a]   See text for more details on the modeling approaches.
  [b]   Examples of possible statistics, not conclusive.
  [c]   Via Normalized Differenced Vegetation Indices (NDVI).
  [d]   Effect of nearest-neighbor via additional equation.
  [e]   For example, SERGE. (Adapted from Eisinger, D. and K. Wiegand 2008. *South African Journal of Science* 104, 37–42.)

occurrence of cyclical successions of grass and tree dominance at the patch level (Meyer et al., 2009). State-and-transition models comprise a range of conceptual models that classify vegetation into multiple stable states and formulate probabilities for transitions between states. Transitions can be triggered by natural disturbances such as fire or by management actions such as grazing or wood harvesting. State-and-transition models have, for example, been used for the conceptual modeling of range management (Westoby et al., 1989) and bush encroachment (Dougill and Trodd, 1999), albeit not dealing with space explicitly. These examples illustrate the usefulness of conceptual models as a first step in the investigation of a new theory and for improving the communication between scientists, range managers, and the public. However, these models have the disadvantage that the applicability of conceptual models is limited to this early stage, because they are either purely descriptive or make only qualitative and not quantitative predictions.

Remote Sensing, GIS, and Spatial Statistics

After the formulation of conceptual models, remote sensing techniques such as satellite imagery or aerial photography can provide large-scale information on spatial patterns required to parameterize and validate subsequent modeling procedures (see also Section 3). Before this information can serve as input or validation for models, it needs to be processed to quantify the spatial patterns. Therefore, remote sensing is frequently used in combination with other descriptive techniques such as GIS and spatial statistics. The GIS capture spatial data in layers that can be superimposed and related to each other (Wiegand et al., 2000b). Spatial statistics cover a broad spectrum of techniques ranging from neighborhood analyses to spatial point pattern methods. These techniques are mainly used to describe spatial patterns of savanna vegetation and to infer the underlying processes.

Combining series of remotely sensed images, GIS, and spatial statistics can give insights into spatiotemporal patterns of savanna tree demography and help identifying their underlying drivers such as competitive or facilitative interactions. By applying nearest-neighbor and spatial point pattern analyses to remotely sensed data processed by a GIS to identify individual trees, Moustakas et al. (2008) found cyclical transitions between clumped, random, and regular tree patterns over time. Smaller nearest-neighbor distances and less clumped spatial point patterns coincided with high mortality, thus indicating the predominance of competitive interactions (Moustakas et al., 2008). Couteron (2002) found remote sensing followed by spectral analysis with Fourier transforms suitable for the description of patterned semiarid vegetation. The suitability of remotely sensed data for spatial savanna studies depends on the resolution of the images. To create maps of burned areas in savannas, the combined use of high- and low-resolution images seems more appropriate than using either of the two (Maggi and Stropiana, 2002).

Remote sensing, GIS, and spatial statistics have all, either alone or in combination, fruitfully been coupled with equation- or grid-based modeling approaches. To reconstruct land cover distributions in savannas, GIS have been combined with predictive land cover models for the analysis of pollen distributions (Flantua et al., 2007). In a Zambian savanna, land cover change has been predicted by linking remotely sensed data and a matrix model, indicating the continuation of a trend toward increasing proportions of bare and cultivated soils under unchanged environmental conditions (Petit et al., 2001). In contrast, a rule-based expert system fed with remotely sensed and GIS-processed data predicted canopy closure between woody clusters for a mesquite savanna in Texas, in the absence of management (Loh and Hsieh, 1995). Applying spatial point–pattern analysis to the output of various grid-based simulations of major savanna processes, Jeltsch et al. (1999) determined general relationships between patterns and processes that were subsequently used to infer the most likely processes generating the spatial tree patterns on remotely sensed images of the Kalahari in South Africa and Botswana. There, clumped tree patterns were promoted by grass fires, whereas random tree patterns arose from randomly distributed establishment patches (Jeltsch et al., 1999). However, the statistical analyses of the model output also showed that the spatial pattern observed in the aerial photographs from the Kalahari, that is, regular distributions at small scales and random or clumped distributions at intermediate scales, was diagnostic both for persistent open savannas and open savannas in transition toward woodlands. These combined approaches are particularly strong, because they benefit from the quantification strength of descriptive techniques and the predictive capabilities of equation- or grid-based models. The disadvantage of remote sensing, GIS, and spatial statistics approaches is their static nature. This could be overcome by combining them with dynamic approaches or by investigating temporal sequences of data (e.g., Moustakas et al., 2008).

Moment Equations

Incorporating spatial relationships explicitly into equation-based models is not an easy task, because it always requires an extension of the model with additional equations or matrices. In plant population modeling, deriving moment equations has been found to be a useful approach to capture spatial processes (Bolker et al., 2003). Moment equations or pair approximations consist of a pair of equations where one equation represents average population dynamics and the other represents the spatial covariance of the population dynamics, that is, the spatial interactions between an individual and its direct neighbors (Bolker and Pacala, 1997). This technique may be powerful but has only very recently been applied to model spatial savanna dynamics (Calabrese et al. 2010). A problem of moment equations is moment closure, that is the fact that usually only the effect of the direct neighbors is incorporated but not the effect of the neighbors' neighbors, to reduce mathematical complexity. Another disadvantage is that equation-based models are often less strongly linked to empirical data, because they often contain aggregated parameters that are difficult to parameterize. On the other hand, applications of moment equations benefit from the small data requirements, their analytical tractability, and their high potential for generalization. Therefore, their careful application to the spatially explicit modeling of savanna processes, probably in combination with other approaches, should be given more consideration. An example of a question suitable for moment equation modeling may be the extent to which local interactions matter for overall savanna dynamics.

Spatial Matrix Models

Another mathematical approach to spatially explicit savanna modeling is to add a spatial extension to matrix models. Matrix models often capture age- or stage-structured population dynamics and can be applied to implement state-and-transition models. Rows and columns of a matrix may refer to population states such as age or stage class and matrix entries to transition probabilities between the respective states. Nonhomogeneous matrix models can include perturbations to mean states and transitions due to spatial structure by adding an extra perturbation matrix. Another approach would be to create a matrix whose entries are submatrices, representing the transitions in each cell of a spatially discrete landscape. In savannas, the perturbation matrix approach has been used to predict land cover change (Petit et al., 2001), to evaluate individual plant architecture of fire-adapted species (Raventos et al., 2004), and to validate a new stability analysis method for nonhomogeneous Markov models (Li, 1995). The submatrix approach has been applied very rarely to savanna dynamics thus far (but see Miriti et al., 2001). In addition to the analytical tractability of an equation-based model, spatial matrix models have the additional advantage of capturing population structure when based on populations. Therefore, spatial matrix models can be used to investigate questions that focus on the demography of savanna plants as suggested by Sankaran et al. (2004). However, all transition probabilities need parameterization and are thus "data hungry"; and there is a trade-off between the ease with which an analytical solution of a matrix model can be found and the number of influencing factors such as spatial or temporal variation that can be concurrently analyzed.

Cellular Automata

In cellular automata, space is discretized into grid cells with a finite number of discrete states, in the narrow definition, only deterministic and synchronous transition rules between the states (Childress et al., 1996; Hogeweg, 1988). Transition rules depend on the direct neighborhood of a cell and represent local interactions only. The *direct neighborhood* is defined as the four cells sharing an edge with the focal cell (von Neumann neighborhood) or all eight cells touching the focal cell (Moore neighborhood). In the wider definition of cellular automata, stochastic and asynchronous transitions are also possible. The patterns that cellular automata produce can subsequently be compared with empirical data for validation. Spatial patterns emerging from cellular automata that have been successfully validated include the spread of prescribed fires (Berjak and Hearne, 2002); the distribution of crown heights and widths (Drake and Weishampel, 2001); and the spread of tree species depending on fire, demography, and seed dispersal (Hochberg et al., 1994). Cellular automata can be used to determine the importance and quality of spatial effects for the simulated patterns. In a sensitivity analysis of their cellular automaton, Drake and Weishampel (2001) found that spatial crown distributions were highly sensitive to spatially interactive parameters such as fire ignition, spread, and competition. A limitation of cellular automaton models is that they can quickly become highly complex, because all transitions between all possible combinations of states need to be described and parameterized from empirical data or expert guesses. To reduce this complexity, aggregation techniques are available, such as the voting system. In contrast to a unique neighbor system, in a voting system, only the number of cells in each state in a neighborhood is determined and not their exact location within the neighborhood (Childress et al., 1996). On the other hand, narrowly defined cellular automaton models are analytically tractable and very flexible due to the great range of possible cell identities from individual plants to pixels of satellite images (Childress et al., 1996).

Grid-Based Models

Grid-based models consist of a grid of cells similar to the grid on which cellular automata are based, but they can capture interactions between any two or more cells. These interactions are not confined to a specific neighborhood such as in cellular automata. Cells of cellular automata can only be in one of a defined set of states, whereas grid-based models can have complex and structured dynamics within each cell. Thus, grid-based models can capture spatial structure of almost any complexity and are, therefore, probably the most flexible spatially explicit modeling approach. The cells of a grid-based model can represent individual organisms or aggregated information such as biomass or presence or absence of a species. When cells contain individual organisms, agent-based modeling is a common approach in which the modeled dynamics emerges from individual interactions and depends on individual properties such as canopy width or spatial location (Grimm and Railsback, 2005). In the grid-, agent-, and rule-based savanna patch model SATCHMO, shrub agents, whose seed output depended on the individual property of canopy size, competed with other shrub or grass agents (Figure 14.4, Meyer et al., 2007a). A 10 × 10 cm cell contained either a shrub seedling, a part of a shrub or grass canopy, a part of the water uptake zone of one of eight roots per shrub, or bare ground and a certain amount of daily updated

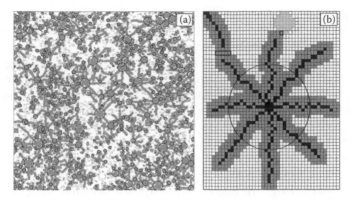

FIGURE 14.4
Overview of a typical model grid of an agent- and grid-based savanna model (a) and an enlarged section of the same grid (b). Grid-based models can be used to represent and analyse spatial interactions such as, in this example, water uptake and competition for soil moisture. The grid covers about 500 × 500 m in reality and consists of square cells with a side length of 10 cm. The interacting agents are shrubs and grass tufts. Each shrub is represented by its canopy extent aboveground (circles) and eight roots [black cells in (b)] below ground. Each root is surrounded by an uptake zone (dark gray) to model water uptake and competition for water in overlapping uptake zones [cells with bold black edges in (b)]. Grass tufts are represented by an uptake zone (light gray) and interact with shrubs when their uptake zones overlap. (Reprinted from K. M. Meyer et al., 2007a. *Ecological Modelling* 209, 377–391. © 2007. With permission from Elsevier.)

soil moisture. Spatial processes such as seed dispersal and shrub-shrub and shrub-grass competition were modeled by distributing seeds according to an empirical distribution around their parent and distributing soil moisture according to the number of individual shrubs and grass tufts overlapping in a certain cell, respectively (Meyer et al., 2007a). Since simulations were initialized with one shrub individual and 10 randomly distributed grass tufts, shrub and grass population and cover dynamics and spatial patterns were truly emergent from the simulations. These patterns were then used in a validation procedure and analyzed to yield conclusions about the question at hand. Parameterized with field data and a few expert guesses and successfully validated against several observed patterns, SATCHMO showed that cyclical successions between grassy and woody dominance can emerge from a setting typical of a semiarid savanna (Meyer et al., 2007b), corroborating a central prediction of patch-dynamics theory (Wiegand et al., 2006; Meyer et al., 2009, Chapter 16).

Grid-based modeling is one of the most common approaches in spatially explicit savanna modeling, addressing a large spectrum of questions such as the prediction of NPP, water fluxes, radiation absorption (Simioni et al., 2003), the impact of shrub encroachment on small mammals (Popp et al., 2007), the simulation of spatial rain cloud patterns (Eisinger and Wiegand, 2008), the relationship between the fragmentation of a grazed area and herbivore stocking rate in the SAVANNA model (Boone and Hobbs, 2004), or the ecological-socioeconomic assessment of management strategies in pastoral systems (Müller et al., 2007). Grid-based approaches are particularly useful in modeling spatial processes that range further than to the nearest plant neighbor, such as in the case of seed dispersal, fire, or grazing. Special emphasis has been placed on the role of grazing and herbivory for generating savanna dynamics and patterns (e.g., de Knegt et al., 2008). With a grid-based model whose grid cells were linked by seed dispersal and fire, Baxter and Getz (2005) showed that woody plant persistence was promoted by high tolerance to browsing by elephants, browsing-induced increase in germination success, and faster growth of woody plants. On the other hand, Jeltsch et al. (1996) found that persistence of tree clumps was promoted by intermediate values of grazing, fire, and moisture in their savanna modeling system. In an extension of this model, the effect of grazing on shrub encroachment depended on the time frame and became apparent at a century scale only (Jeltsch et al., 1997). Moreover, Weber et al. (2000) identified a threshold behavior of shrub cover response to grazing in their grid-based model, complicating predictions of grazing effects. However, predicting the effects of grazing is of major importance. For instance, grazing had a greater impact on shrub cover than climate change in a grid-based model of a southern African savanna (Tews et al., 2006, Chapter 15).

When compared with equation-based models, grid-based approaches do not have to make compromises with regard to complexity to ensure mathematical feasibility and are driven by and tightly linked to empirical data. Therefore, grid-based models are often structurally realistic (Wiegand et al., 2004b), which facilitates model parameterization and validation using qualitative expert knowledge that can sometimes be the only type of knowledge available in savannas. On the downside, they are very data hungry, which is alleviated by their great flexibility to adapt the level of modeled detail to data availability. Grid-based modeling is resource intensive regarding developing and computing time, and predictions are often difficult to generalize due to the lack of analytical solutions. First attempts to approximate the output of grid-based simulation models with analytical expressions (Calabrese et al., 2010) benefit from the full flexibility of one approach and the analytical tractability and general conclusions of the second.

Conclusions

Spatially explicit savanna modeling has focused on processes with strong spatial structure such as rainfall, fire spread, moisture fluxes, seed dispersal, competition, and grazing. This pool of spatial processes also provides the drivers of savanna tree–grass coexistence. Therefore, spatially explicit approaches, be they empirical or theoretical, are crucial to elucidate the underlying mechanisms of tree–grass coexistence. The few existing spatially explicit theories of tree–grass coexistence emphasize the importance of spatial structure such as small-scale heterogeneities (Jeltsch et al., 1998) and the scale explicitness of patch dynamics (Wiegand et al., 2006; Meyer et al., 2009). These examples share another characteristic, that is, the combination of different approaches such as grid-based modeling and spatial statistics. Future studies will greatly benefit from placing more emphasis on combining different approaches. Moreover, since savannas span a great range of spatial scales, more scale-explicit approaches are needed in future spatial savanna modeling. For example, among the grid-based approaches, agent-based models can cover several spatial scales by predicting large-scale population dynamics from local interactions between plant and animal individuals. Given that many people depend on savannas for a living, ultimately, the challenge will be to apply the modeling results to develop management strategies that sustain savanna ecosystems and improve people's living. For example, management may gain from applying the principles of patch dynamics, which indicates that there is no single appropriate scale and that changes are "normal" within any given savanna (Wiegand et al., 2006).

This will further enlighten the importance of spatially explicit savanna modeling for describing, understanding, and predicting savanna processes, patterns, dynamics, and management.

References

Adler, F. R. and J. Mosquera. 2000. Is space necessary? Interference competition and limits to biodiversity. *Ecology* 81, 3226–3232.

Augustine, D. J. 2003. Spatial heterogeneity in the herbaceous layer of a semi-arid savanna ecosystem. *Plant Ecology* 167, 319–332.

Archibald, S., W. J. Bond, W. D. Stock, and H. K. Fairbanks. 2005. Shaping the landscape: Fire–grazer interactions in an African savanna. *Ecological Applications* 15, 96–109.

Baxter, P. W. J. and W. M. Getz. 2005. A model-framed evaluation of elephant effects on tree and fire dynamics in African savannas. *Ecological Applications* 15, 1331–1341.

Berjak, S. G. and J. W. Hearne. 2002. An improved cellular automaton model for simulating fire in a spatially heterogeneous savanna system. *Ecological Modelling* 148, 133–151.

Bolker, B. M. and S. W. Pacala. 1997. Using moment equations to understand stochastically driven spatial pattern formation in ecological systems. *Theoretical Population Biology* 52, 179–197.

Bolker, B. M., S. W. Pacala, and C. Neuhauser. 2003. Spatial dynamics in model plant communities: What do we really know? *American Naturalist* 162, 135–148.

Bond, W. J. 2008. What limits trees in C_4 grasslands and savannas? *Annual Review of Ecology, Evolution, and Systematics* 39, 641–659.

Boone, R. B. and N. T. Hobbs. 2004. Lines around fragments: Effects of fencing on large herbivores. *African Journal of Range & Forage Science* 21, 147–158.

Britz, M. L. and D. Ward. 2007. Dynamics of woody vegetation in a semi-arid savanna, with a focus on bush encroachment. *African Journal of Range & Forage Science* 24, 131–140.

Calabrese, J. M., F. Vazquez, C. Lopez, M. San Miguel, and V. Grimm. 2010. The individual and interactive effects of tree–tree establishment competition and fire on savanna structure and dynamics. *American Naturalist* 175, E44–E65.

Childress, W. M., E. J. Rykiel, W. Forsythe, B. L. Li, and H. I. Wu. 1996. Transition rule complexity in grid-based automata models. *Landscape Ecology* 11, 257–266.

Couteron, P. 2002. Quantifying change in patterned semi-arid vegetation by Fourier analysis of digitized aerial photographs. *International Journal of Remote Sensing* 23, 3407–3425.

Dean, W. R. J., S. J. Milton, and F. Jeltsch. 1999. Large trees, fertile islands, and birds in arid savanna. *Journal of Arid Environments* 41, 61–78.

De Knegt, H. J., T. A. Groen, C. A. D. M. van de Vijver, H. H. T. Prins, and F. Van Langevelde. 2008. Herbivores as architects of savannas: Inducing and modifying spatial vegetation patterning. *Oikos* 117, 543–554.

Dougill, A. and N. Trodd. 1999. Monitoring and modelling open savannas using multisource information: Analyses of Kalahari studies. *Global Ecology and Biogeography* 8, 211–221.

Drake, J. B. and J. F. Weishampel. 2001. Simulating vertical and horizontal multifractal patterns of a longleaf pine savanna. *Ecological Modelling* 145, 129–142.

Eisinger, D. and K. Wiegand. 2008. SERGE: A spatially explicit generator of local rainfall in southern Africa. *South African Journal of Science* 104, 37–42.

Flantua, S. G. A., J. H. van Boxel, H. Hooghiernstra, and J. van Smaalen. 2007. Application of GIS and logistic regression to fossil pollen data in modelling present and past spatial distribution of the Colombian savanna. *Climate Dynamics* 29, 697–712.

Fowler, N. L. 2002. The joint effects of grazing, competition, and topographic position on six savanna grasses. *Ecology* 83, 2477–2488.

Getzin, S. 2007. *Structural Fire Effects in the World's Savannas. A Synthesis for Biodiversity and Land-Use Managers.* VDM Verlag, Saarbrücken.

Gillson, L. 2004. Evidence of hierarchical patch dynamics in an East African savanna? *Landscape Ecology* 19, 883–894.

Grimm, V. and S. F. Railsback. 2005. *Individual-based Modeling and Ecology.* Princeton University Press, Princeton.

Hendricks, H. H., W. J. Bond, J. Midgley, and P. A. Novellie. 2005. Plant species richness and composition along livestock grazing intensity gradients in a Namaqualand (South Africa) protected area. *Plant Ecology* 176, 19–33.

Higgins, S., W. J. Bond, and S. W. Trollope. 2000. Fire, resprouting and variability: A recipe for grass–tree coexistence in savannna. *Journal of Ecology* 88, 213–229.

Hochberg, M. E., J. C. Menaut, and J. Gignoux. 1994. Influences of tree biology and fire in the spatial structure of the West-African savanna. *Journal of Ecology* 82, 217–226.

Hogeweg, P. 1988. Cellular automata as a paradigm for ecological modeling. *Applications of Mathematics and Computing* 27, 81–11.

Jeltsch, F., S. J. Milton, W. R. J. Dean, and N. Van Rooyen. 1997. Analysing shrub encroachment in the southern Kalahari: A grid-based modelling approach. *Journal of Applied Ecology* 34, 1497–1508.

Jeltsch, F., S. J. Milton, W. R. J. Dean, N. Van Rooyen, and K. A. Moloney. 1998. Modelling the impact of small-scale heterogeneities on tree–grass coexistence in semi-arid savannas. *Journal of Ecology* 86, 780–793.

Jeltsch, F., S. J. Milton, W. R. J. Dean, and N. van Rooyen. 1996. Tree spacing and coexistence in semiarid savannas. *Journal of Ecology* 84, 583–595.

Jeltsch, F., K. A. Moloney, and S. J. Milton. 1999. Detecting process from snapshot pattern—lessons from tree spacing in the southern Kalahari. *Oikos* 85, 451–466.

Jurena, P. N. and S. Archer. 2003. Woody plant establishment and spatial heterogeneity in grasslands. *Ecology* 84, 907–919.

Knoop, W. T. and B. H. Walker. 1981. Interactions of woody and herbaceous vegetation in a southern African savanna. *Journal of Ecology* 73, 235–253.

Li, B. L. 1995. Stability analysis of a nonhomogeneous Markovian landscape model. *Ecological Modelling* 82, 247–256.

Loh, D. K. and Y. T. C. Hsieh. 1995. Incorporating rule-based reasoning in the spatial modeling of succession in a savanna landscape. *Ai Applications* 9, 29–40.

Maggi, M. and D. Stroppiana. 2002. Advantages and drawbacks of NOAA-AVHRR and SPOT-VGT for burnt area mapping in a tropical savanna ecosystem. *Canadian Journal of Remote Sensing* 28, 231–245.

McKeon, G., W. Hall, B. Henry, G. Stone, and I. Watson (eds). 2004. *Pasture Degradation and Recovery in Australia's Rangelands*. Queensland Department of Natural Resources, Mines and Energy. Indooroopilly, Australia, Department of Natural Resources, Mines and Energy, Queensland.

McNaughton, S. J. 1983. Serengeti grassland ecology—the role of composite environmental factors and contingency in community organization. *Ecological Monographs* 53, 291–320.

Meyer, K. M., D. Ward, K. Wiegand, and A. Moustakas. 2008. Multi-proxy evidence for competition between savanna woody species. *Perspectives in Plant Ecology, Evolution and Systematics* 10, 63–72.

Meyer, K. M., K. Wiegand, and D. Ward. 2009. Patch dynamics integrate mechanisms for savanna tree–grass coexistence. *Basic and Applied Ecology* 10, 491–499.

Meyer, K. M., K. Wiegand, D. Ward, and A. Moustakas. 2007a. SATCHMO: A spatial simulation model of growth, competition, and mortality in cycling savanna patches. *Ecological Modelling* 209, 377–391.

Meyer, K. M., K. Wiegand, D. Ward, and A. Moustakas. 2007b. The rhythm of savanna patch dynamics. *Journal of Ecology* 95, 1306–1315.

Miriti, M. N., S. J. Wright, and H. F. Howe. 2001. The effects of neighbors on the demography of a dominant desert shrub (*Ambrosia dumosa*). *Ecological Monographs* 71, 491–509.

Moustakas, A., K. Wiegand, S. Getzin, et al. 2008. Spacing patterns of an *Acacia* tree in the Kalahari over a 61-year period: How clumped becomes regular and vice versa. *Acta Oecologica* 33, 355–364.

Moustakas, A., K. Sakkos, K. Wiegand, D. Ward, K. M. Meyer, and D. Eisinger. 2009. Are savannas patch-dynamic systems? A landscape model. *Ecological Modelling* 220, 3576–3588.

Müller, B., K. Frank, and C. Wissel. 2007. Relevance of rest periods in non-equilibrium rangeland systems—A modelling analysis. *Agricultural Systems* 92, 295–317.

O'Connor, T. G. 1991. Local extinction in perennial grasslands: A life-history approach. *American Naturalist* 137, 753–773.

Petit, C., T. Scudder, and E. Lambin. 2001. Quantifying processes of land-cover change by remote sensing: Resettlement and rapid land-cover changes in south-eastern Zambia. *International Journal of Remote Sensing* 22, 3435–3456.

Popp, A., M. Schwager, N. Blaum, and F. Jeltsch. 2007. Simulating the impacts of vegetation structure on the occurrence of a small mammalian carnivore in semi-arid savanna rangelands. *Ecological Modelling* 209, 136–148.

Raventos, J., J. Segarra, and M. F. Acevedo. 2004. Growth dynamics of tropical savanna grass species using projection matrices. *Ecological Modelling* 174, 85–101.

Reid, R. S. and J. E. Ellis. 1995. Impacts of pastoralists on woodlands in South Turkana, Kenya: Livestock-mediated tree recruitment. *Ecological Applications* 5, 978–992.

Sankaran, M., J. Ratnam, and N. P. Hanan. 2004. Tree–grass coexistence in savannas revisited—insights from an examination of assumptions and mechanisms invoked in existing models. *Ecology Letters* 7, 480–490.

Sarmiento, G. 1996. Biodiversity and water relations in tropical savannas. In *Biodiversity and Savanna Ecosystem Processes*, ed. O. T. Solbrig, E. Medina and J. F. Silva. 61–75. Berlin, Springer.

Schmitz, M., W. Platt, and J. DeCoster. 2002. Substrate heterogeneity and number of plant species in Everglades savannas (Florida, USA). *Plant Ecology* 160, 137–148.

Scholes, R. J., W. J. Bond, and H. C. Eckhardt. 2003. Vegetation dynamics in the Kruger ecosystem. In *The Kruger Experience: Ecology and Management of Savanna Heterogeneity*, eds. J. T. Du Toit, K. H. Rogers and H. C. Biggs. Island Press, Washington, D.C., pp. 242–262.

Scholes, R. J. and B. H. Walker. 1993. *An African Savanna: Synthesis of the Nylsvley Study*. Cambridge University Press, Cambridge.

Sharon, D. 1981. The distribution in space of local rainfall in the Namib Desert. *International Journal of Climatology* 1, 69–75.

Simioni, G., J. Gignoux, and X. Le Roux. 2003. Tree layer spatial structure can affect savanna production and water budget: Results of a 3-D model. *Ecology* 84, 1879–1894.

Sinclair, A. R. E. and J. M. Fryxell. 1985. The Sahel of Africa—ecology of a disaster. *Canadian Journal of Zoology* 63, 987–994.

Skarpe, C. 1991. Spatial patterns and dynamics of woody vegetation in an arid savanna. *Journal of Vegetation Science* 2, 565–572.

Smet, M. and D. Ward. 2006. Soil quality gradients around water-points under different management systems in a semi-arid savanna, South Africa. *Journal of Arid Environments* 64, 251–269.

Smith, T. M. and P. S. Goodman. 1986. The effect of competition on the structure and dynamics of *Acacia* savannas in southern Africa. *Journal of Ecology* 74, 1013–1044.

Tews, J., A. Esther, S. J. Milton, and F. Jeltsch. 2006. Linking a population model with an ecosystem model: Assessing the impact of land use and climate change on savanna shrub cover dynamics. *Ecological Modelling* 195, 219–228.

Walker, B. H., D. Ludwig, C. S. Holling, and R. M. Peterman. 1981. Stability of semi-arid savanna grazing systems. *Journal of Ecology* 69, 473–498.

Walker, B. H. and I. Noy-Meir. 1982. Aspects of stability and resilience of savanna ecosystems. In *Ecology of Tropical Savannas*, ed. Huntley B. J. and B. H. Walker. Ecological Studies 42. Springer-Verlag, New York, pp. 556–590.

Walter, H. 1971. Ecology of tropical and subtropical vegetation. Oliver & Boyd, Edinburgh.

Ward, D. 2005. Do we understand the causes of bush encroachment in African savannas? *African Journal of Range and Forage Science* 22, 101–105.

Ward, D. 2009. *The Biology of Deserts*. Oxford University Press, Oxford, UK.

Weber, G. E., K. Moloney, and F. Jeltsch. 2000. Simulated long-term vegetation response to alternative stocking strategies in savanna rangelands. *Plant Ecology* 150, 77–96.

Westoby, M., B. Walker, and I. Noy-Meir. 1989. Range management on the basis of a model which does not seek to establish equilibrium. *Journal of Arid Environments* 17, 235–239.

Wiegand, K., F. Jeltsch, and D. Ward. 2000a. Do spatial effects play a role in the spatial distribution of desert-dwelling *Acacia raddiana*? *Journal of Vegetation Science* 11, 473–484.

Wiegand, K., F. Jeltsch, and D. Ward. 2004a. Minimum recruitment frequency in plants with episodic recruitment. *Oecologia* 141, 363–372.

Wiegand, K., D. Saltz, and D. Ward. 2006. A patch-dynamics approach to savanna dynamics and woody plant encroachment—Insights from an arid savanna. *Perspectives in Plant Ecology, Evolution and Systematics* 7, 229–242.

Wiegand, K., D. Saltz, D. Ward, and S. A. Levin. 2008. The role of size inequality in self-thinning: A pattern-oriented simulation model for arid savannas. *Ecological Modelling* 210, 431–445.

Wiegand, K., H. Schmidt, F. Jeltsch, and D. Ward. 2000b. Linking a spatially-explicit model of Acacias to GIS and remotely sensed data. *Folia Geobotanica* 35, 211–230.

Wiegand, K., D. Ward, and D. Saltz 2005. Multi-scale patterns in an arid savanna with a single soil layer. *Journal of Vegetation Science* 16, 311–320.

Wiegand, T., E. Revilla, and F. Knauer. 2004b. Dealing with uncertainty in spatially explicit population models. *Biodiversity and Conservation* 13, 53–78.

15

Interaction of Fire and Rainfall Variability on Tree Structure and Carbon Fluxes in Savannas: Application of the Flames Model

Adam Liedloff and Garry D. Cook

CONTENTS

Introduction

From an Australian perspective, the analysis by Sankaran et al. (2005) of the interactive effects of precipitation and disturbance on vegetation structure of African savannas raises a range of questions. These authors concluded on the basis of data from savannas across Africa that below 650 mm mean annual precipitation (MAP), African savannas were stable, in that rainfall constrained woody cover below about 70%, which allows the persistence of grass. For MAPs above this value, and in the absence of fire and other disturbances, woody cover could increase to levels that exclude grass, and thus savannas defined by the coexistence of trees and grass were not stable. Both above and below 650 mm mean annual precipitation (MAP), disturbance by fire and herbivory and stresses from edaphic conditions could and mostly did lead to woody cover being below the climatic optimum.

In Australia, savannas with a tree cover of about 30% and stable coexistence with grass dominate across hundreds of thousands of km² of northern

Australia with MAPs between 1000 and 2000 mm (Liedloff and Cook, 2007; Williams et al., 1996; Wilson et al., 1990). The first question is therefore: Why did Sankaran et al. not include sites with a MAP above 1185 mm in their study of the relationship between woody cover of savannas and MAP? Why are the wetter savannas apparently more uniform and stable in Australia than in Africa? The second major issue relates to the arid end of the relationship. Australian savanna researchers tend not to include the arid zone within their definition of a savanna. Is this merely pragmatic semantics or is there a functional difference between African and Australian arid systems, which is reflected in the research culture? Third, drought causes substantial fluctuations in tree density in savannas in Queensland, Australia (Fensham et al., 2009) in ways that are much less apparent elsewhere in Australia's north and in Africa. As an index of water availability at a particular site, does MAP hide drought as a significant driver of variation in vegetation structure?

Understanding the interactions between rainfall and disturbances is critical to predicting the effects of future changes in climate and management on savanna dynamics. The central thesis of Sankaran et al. is based on there being a physiological upper limit to woody cover determined by MAP, but disturbance-related demographic processes reduce cover below those limits. Demographic processes are key to tree dynamics in tropical savannas dominated by C4 grasses (Gardner, 2006; Bond, 2008; Hanan et al., 2008; Lehmann et al., 2009; Prior et al., 2009), because of the importance of direct competition between trees and grasses and secondary effects of fires and herbivores. These processes mean that tree populations in savannas are rarely at their climatic or "green world" optimum (Bond, 2005). Reliable modeling of the carbon stocks in savannas must take account of these demographic processes, and understanding the interaction between fire and rainfall is critical. Although of interest in itself, those dynamics are also critical to full-carbon accounting and the estimation of Net Biome Productivity as required under the 2006 IPCC guidelines for National Greenhouse Gas Inventories (Paustian et al., 2006). Given the extensive global coverage of savannas, such understanding is critical to understanding the dynamics of the terrestrial carbon stock.

In this chapter, we explore some of the questions raised by the work of Sankaran et al. from the context of Australian savanna vegetation and the associated carbon dynamics. First, we compare and contrast patterns of rainfall variability and seasonality in Australia and Africa as a background to understanding how water availability might limit tree growth in savannas. Then we describe Flames, which is a demographic model of Australian savanna vegetation dynamics that is used in this paper to explore the effects of fire as a landscape disturbance on a representative stand of trees. From this series of simulations, we also investigate the carbon budget of the stand in greater detail.

Rainfall Variability and Seasonality: How Different is the Australian Savanna Zone?

A comparison of rainfall records from northern Australia with records from sub-Saharan western African and southern Africa shows marked differences in inter-annual variability and intra-annual seasonality among regions. Compared with the African data, for MAPs above about 1000 mm, Australia maintains a high degree of summer dominance with more than 90% falling in six contiguous months (Figure 15.1a). Rainfall is less seasonally constrained at these levels of MAP in western Africa. For MAPs less than about 500 mm, rainfall in western Africa is highly summer dominant, but much less so in northern Australia. Southern Africa is intermediate. For MAPs between about 500 mm and 1000 mm, the seasonality is similar in all three zones (Figure 15.1a). However, the interannual coefficient of variation of rainfall in Australia between about 500 mm and 1500 mm is much greater than for western Africa and southern Africa (Figure 15.1b).

Differences in the seasonality of rainfall are probably a major contributor to the greater prevalence of shrub lands and perennial grasses in the northern Australian arid zone than in equivalent African arid zones. At high rainfall in Australia, the long dry season probably accounts for the dominance of savannas at levels of MAP that, in Africa, would support closed forests as the dominant vegetation type. The length of the seasonal drought has been shown to be the main mechanism by which rainfall limits tree density in Australian savannas rather than MAP per se (Cook and Heerdegen, 2001; Liedloff and Cook, 2007). The much higher coefficient of variation of rainfall in many of Australia's savanna lands than in Africa would also account for periodic drought-induced dieback being a more important issue for savanna dynamics in Australia than in Africa.

Modeling Savanna Dynamics in Australia: The Flames Model

For a model of savanna dynamics to contribute to an understanding of the applicability of the work of Sankaran et al. to Australia, it would need to incorporate both a mechanism for woody vegetation to reach the upper limit set by water availability and to take account of the demographic impact of disturbances. In Australia, such disturbances are dominated by fire, with browsing of the tree layer by megafauna as in Africa being negligible. The Flames model (Liedloff and Cook, 2007) was developed for northern Australian savannas and allows both these factors to be simulated. This model simulates the fate of a one-hectare stand of trees and grass with the ability to alter fire regime, fire weather, rainfall regime, soil properties, and vegetation properties. A water balance model allows daily rainfall to interact with stand and grass water use via soil water storage to create drought-induced death of individual trees (Cook et al., 2002) and curing in grasses. The structure and

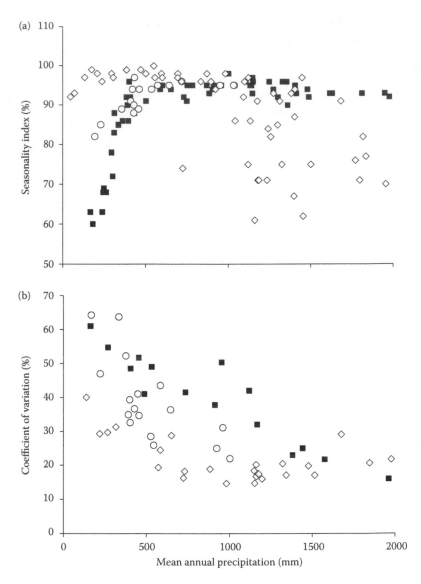

FIGURE 15.1
A comparison between northern Australia (■), southern Africa (○) and western Africa (◊) for the relationship between mean annual precipitation and (a) the proportion (%) of rainfall in the 6 wettest contiguous months and (b) the interannual coefficient of variation of rainfall.

density of tree stands is dynamic and depends on rainfall regime, soil properties, and tree water use. Fires can occur randomly within a fire frequency and seasonality set by the operator. Fire intensity is calculated from the relationship between the rate of fire spread, fire weather, fuel loads, fuel curing state, and the ignition style. Fuels comprise both grass litter and tree litter with tree litter typically making a substantial contribution to fuel loads. Unlike southern Australian eucalypts, those in northern Australia are rarely affected by canopy fires due to their sparser canopies and lower oil contents (Gill, 1997). Trees have a probability of being top killed by fires depending on the fire intensity and their species and stem diameter based on empirical relationships derived from field data for mild to intense fires where the smallest and oldest trees are most susceptible (Cook et al., 2005). Thus, the death of individual trees is stochastic with its probability determined by drought occurrences and fires. Tree death is accounted for monthly, because that is the resolution of fire occurrence in the model, whereas tree growth occurs annually. The fire weather is also stochastic through selection of a driving wind speed drawn from a population of wind speeds representing the site and fire timing in question. The model outputs a wide variety of factors including stand structure, fuel dynamics, pyrogenic emissions, and carbon stocks in biomass and litter.

Within the Flames model, water limits tree stand development through the death of individuals when the water demand exceeds that stored in the soil. This mechanism typically occurs at the end of the dry season, and depending on the stand size and the rainfall in relation to previous years it can lead to tree decline over several years. These processes allow the Flames model to simulate declines in tree populations along declining rainfall gradients from the coast (Williams et al., 1996), dieback and periods of recovery after droughts (Fensham et al., 2009), and differences between tree stands within the savanna (Liedloff and Cook, 2007; Lehmann et al., 2008).

The demographic modeling capabilities of the Flames model were combined with allometric relationships of Cook et al. (2005) and Werner and Murphy (2001), to track both the stand structure and biomass. Annual changes in biomass, and therefore carbon, provide measures of the carbon flux of the stand. The Flames model tracks the above ground carbon stored in trees, grasses, litter, and coarse woody debris as well as root biomass and below ground coarse woody debris (dead roots), but it does not currently account for soil carbon stores.

Unlike more physiological models where temperature and radiation are important factors, parameters such as soil depth, soil texture, tree mortality, growth, and resprouting rates are more important for our model, highlighting its demographic nature. To explore the predicted dynamics of this site, all these parameters were set to constant values representative of the site, and only fire frequency and rainfall pattern were altered. It is not the aim of this section to simulate an exact prediction for a given site but to provide a projection of general trends and understanding. In fact, exact site prediction at a given time is not the aim of the Flames model. This is because there are many stochastic influences on a stand that cannot be easily measured or predicted, and critical rainfall patterns are difficult to forecast. For example, the death of individual trees, wind speed at the time of fire, and the rate of fire spread are all dominant influences on the stand but are effectively unpredictable. We solve this problem by treating the model as providing one possible outcome of a given scenario. Running the model a number of times, or over extended periods (e.g., 200 years), can therefore provide valuable information in the form of means, expected variability, and population trends.

Methods

In this chapter, we compare the effects of two rainfall regimes with a range of fire regimes on total basal area, stand structure, and carbon flux in Australian savannas using the Flames model. To achieve this, the model was parameterized for a representative mesic savanna stand of trees based on a flux tower site located at Howards Springs, Northern Territory, Australia (12°29.712′ S, 131°09.003′ E), which was used by Beringer et al. (2007) and described in Chapter 4. The model was parameterized with the actual tree populations measured on the one hectare site surrounding the Howard Springs flux tower and using the properties of the 1.7 m deep sandy clay loam soil. We included any species that accounted for more than 2% of the total basal area of the stand. These were, from most to least abundant, *Eucalyptus miniata*, *Eucalyptus tetrodonta*, *Erythrophleum chlorostachys*, *Eucalyptus bleeseri*, *Terminalia ferdinandiana*, *Eucalyptus clavigera*, and *Eucalyptus porrecta*. A representative perennial grass species was used in the model and rooted to 30 cm with trees rooted to 1.7 m.

Simulations were performed for 200 years to capture long-term demographic processes. This required cycling through a second run of the actual daily rainfall records, which were about 120 years long. These records were preferred to synthetic rainfall data, because we wanted to capture the effects of real inter-annual and intra-annual variations, which have major demographic impacts. Since the initial population represents the actual stand of trees recorded at the flux tower site, and no other parameters in Flames require time to stabilize, no spin up period was performed. This allows examination of changes from the initial size structure that would be lost by undertaking a long-term spin up of the model.

We apply two sets of historic daily rainfall records in the simulations: that of Darwin, Northern Territory (12°27.73′ S, 130°50.62′ E, MAP = 1613.1 ± 337 (SD) mm year^{-1}) representing the monsoonal north of the Australian continent, and that of Charters Towers, Queensland (20°04.57′ S, 146°15.67′ E, MAP = 649.4 ± 246 (SD) mm year^{-1}), representing the El Niño influenced rainfall pattern of north-eastern Australia. These rainfall records represent one site above the level of MAP considered by Sankaran et al. and one at the boundary between their "stable" and "unstable" savannas.

We consider the effect of fire return interval on trees by running simulations with fires occurring with average frequencies of one fire every five years

(0.2 year^{-1}), two fires every three years (0.67 year^{-1}), and no fires. These fires represent the low and high fire return intervals experienced in contemporary fire management in the Australian savannas (Russell-Smith et al., 2007) and a system without fire. Fires were simulated as elliptical fires typical of the unconstrained landscape fires burning in northern Australia (Catchpole et al., 1992), where there is variation in the rate of spread at different angles to the prevailing wind direction. The fire weather was determined for the prevailing weather conditions representing afternoons in August in Darwin. The different fire return intervals determine the frequency with which trees are exposed to fire and influence fuel dynamics. Fuel loads under the no fire scenario are only subject to inputs and decomposition. The average fire return interval of five years provides the largest fuel loads in the presence of fire, with the frequent fires leading to minimum fuel accumulation.

A range of fire frequencies were simulated to provide the relationship between fire frequency and carbon fluxes of the stand. These included average fire frequencies of 0.05, 0.1, 0.5, and 1 in addition to the 0, 0.2, and 0.67 fire frequencies previously simulated. The carbon dynamics of the stand, including fluxes and carbon pools, were considered in greater detail using the Darwin rainfall simulations with no fires and an average two fires in three years (0.67) fire frequency.

Results

A 200-year simulation of the Howard Springs stand using Darwin's rainfall record in the absence of fire produced a relatively stable tree population with a mean total basal area of 12.6 m^2 ha^{-1} (Figure 15.2a). The modeled data shows the total basal area increasing above the initial population for about 20 years followed by sawtooth fluctuations of about 2–3 m^2 ha^{-1}. These fluctuations were brought about by periods of tree death when the stand's water requirements exceed the water supply followed by recovery. In contrast, modeling the tree population using the Charters Towers rainfall records in the absence of fire produced a mean total basal area of 6.5 m^2 ha^{-1} but with fluctuations of 8 m^2 ha^{-1} (Figure 15.2b).

The greater population fluctuations with the Charters Towers rainfall than with the Darwin rainfall led to substantial differences in size structure of the two populations. The simulated tree stand with the Charters Towers rainfall had more stems than with the Darwin rainfall and a much higher proportion of the total basal areas comprising small trees (Table 15.1). The greater impact of major droughts with the Charters Towers rainfall reduced the likelihood of trees reaching large size classes.

Compared with no fire, adding fire with an average frequency of one in five years had little effect on total basal area dynamics of either the Darwin or

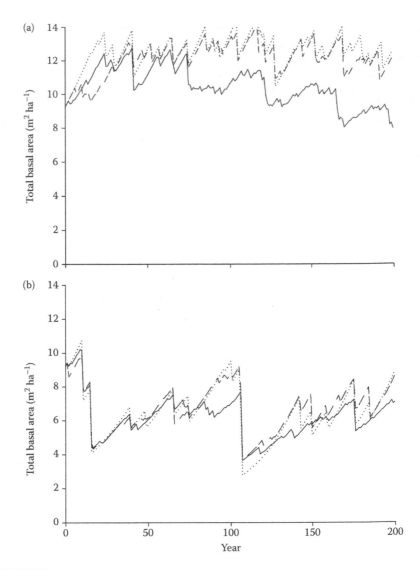

FIGURE 15.2
The simulated effect on stand total basal area (m² ha⁻¹) of average fire return intervals from no fires (dotted lines), to an average fire return interval of 5 years (dashed lines), and two fires every 3 years (solid lines) for a representative hectare stand of trees using (a) Darwin and (b) Charters Towers daily rainfall records.

Charters Towers rainfall simulations (Figure 15.2a and b). However, the highest fire frequency of two in three years resulted in greater tree death and longer recovery periods. Over time, this fire frequency reduced total basal area in Darwin, but its effect on the Charters Towers rainfall simulation was overshadowed by the impact of major tree death impacts driven by drought.

TABLE 15.1

The Initial and Simulated Stand Size Structure after 200 Years Based on 100 mm Diameter at Breast Height (DBH) Size Classes Using Darwin and Charters Towers Rainfall with No Fire ($F = 0$ year^{-1}) and Two Fires in 3 years ($F = 0.67$ year^{-1})

		Stem Density (ha^{-1})			
		Charters Towers		Darwin	
DBH (mm)	Initial Population	$F = 0$	$F = 0.67$	$F = 0$	$F = 0.67$
0–100	334	354	279	207	152
100–200	160	291	110	143	43
200–300	91	50	27	47	16
300–400	13	12	19	25	14
400–500	2		3	2	10
500–600	1	1	1	5	3
600–700				2	4
700–800		1	3	3	3
800–900				2	
>900				2	1
Total stems	601	709	442	438	246

For both the Charters Towers and Darwin rainfall simulations, the high fire frequency reduced the total number of stems compared with no fire (Table 15.1). Fire also resulted in a greater proportion of the tree population being in the small (<100 mm DBH) size class, as a result of post-fire resprouting. Fire had little effect on the abundance of mature (>400 mm DBH) trees, but Darwin had more than Charters Towers.

Across all fire frequencies, the overall mean annual carbon flux over 200 years of simulation was small, with a trend toward changing from a weak carbon sink to neutral or a weak source as fire frequency increased from no fires to annual fires (Figure 15.3a). However, when individual years are considered, the stands were a sink in most years (Figure 15.3b). The Charters Towers rainfall simulation was a sink in a lower proportion of years than the Darwin rainfall simulation for most fire frequencies. For the Darwin rainfall simulation with a fire frequency of one in 10 years, a sink was recorded in more than 80% of years (Figure 15.3b). The mean source strength in years when the stands were a source of carbon was generally greater in magnitude than the sink strength when the stands were a sink (Figure 15.3a). The annual carbon fluxes for the Darwin rainfall simulation at a range of fire frequencies varied from a sink of 1.75 t C ha^{-1} year^{-1} to a source of 7.5 t C ha^{-1} year^{-1}, whereas the respective values for the Charters Towers rainfall simulation were a sink of 1.5 t C ha^{-1} year^{-1} and a source of 6.22 t C ha^{-1} year^{-1}.

The track of carbon dynamics for the Darwin rainfall simulation over time indicates that disturbance from fire greatly increases the inter-annual

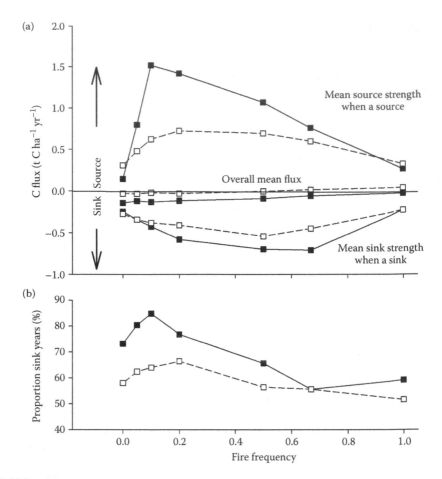

FIGURE 15.3
Analysis of simulated carbon fluxes over a 200-year simulation for the representative stand of trees exposed to a range of fire frequencies showing (a) the mean source strength for source years (upper line), overall mean flux (middle line), and mean sink strength during sink years (lower line) using Darwin (solid) and Charters Towers (dashed) rainfalls; and (b) the proportion of the 200 years defined as a sink.

fluctuations in sink and source strength, despite causing little difference in the running mean carbon flux (Figure 15.4a). In the no-fire simulation, the carbon stock was stable at about 55 t ha^{-1} after an initial period of increase (Figure 15.4a), whereas the carbon stock under a high fire frequency gradually declined over the second 100 years of the simulation (Figure 15.4b), consistent with trends in total basal area (Figure 15.2a). After the initial increase in carbon stock over the first few decades of the simulation, carbon flux in the no-fire treatment varied by less than 1 t C ha^{-1} year^{-1} interannually (Figure 15.4a) compared with variation of more than 3 t C ha^{-1} year^{-1} under frequent fires

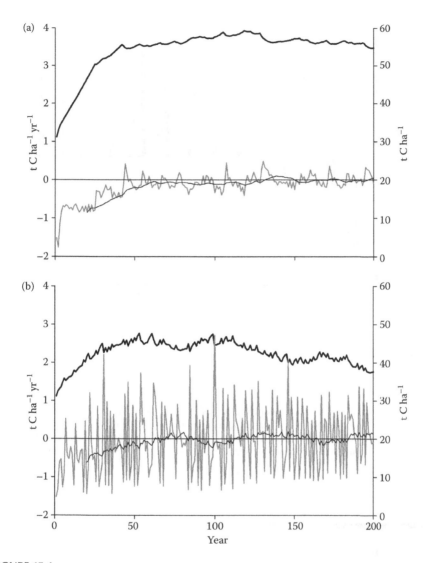

FIGURE 15.4
The simulated carbon stock (thick solid line, without soil carbon, t carbon) and annual carbon flux (gray line, t C ha⁻¹ year⁻¹) with the 20 year running mean (solid line) for a hectare stand of trees at Howard Springs, Northern Territory, experiencing (a) no fire and (b) two fires in 3 years for 200 years.

(Figure 15.4b). For both the no-fire and frequent fire simulation, leaf litter from trees and grass litter comprised most of the carbon sink (Figure 15.5). In the absence of fire, decomposition of litter was the source of carbon transfer to the atmosphere, but in the frequently burnt simulation, combustion accounted for more than half the source of carbon transfer (Figure 15.5).

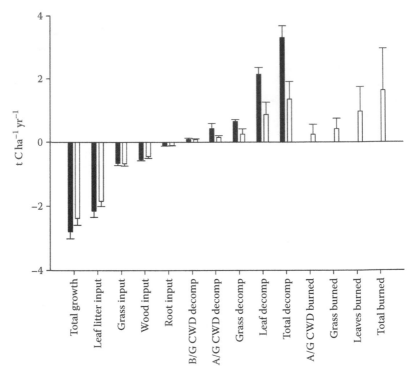

FIGURE 15.5
The average (±standard deviation) contribution of various components to the annual carbon flux at Howard Springs, Northern Territory, when unburned (black) and burned (white) for 200 years. B/G CWD and A/G CWD are belowground and aboveground coarse woody debris, respectively, decomp is decomposition, burned is consumed by fire, and the other fluxes are from growth.

Discussion

The simulation results presented here indicate that even at high rainfall, removal of disturbance by fire does not lead to the loss of grasses. Thus, the results indicate that, in contrast to the finding of Sankaran et al. for African savannas, Australian high rainfall savannas are stable in the sense that removal of disturbance does not lead inexorably to the loss of grasses. This simulation result is supported by experimental results where fire exclosure at decadal time spans did not lead to the development of closed forests or to the loss of grasses (Russell-Smith et al., 2003; Williams et al., 2003). The longer length of the dry season in high rainfall Australian savannas is at least one factor accounting for this difference. It has previously been shown that it is the length of the dry season rather than MAP per se that accounts for the strong relationship between rainfall and tree cover and tree basal area (Williams et al., 1996; Cook et al., 2002; Liedloff and Cook, 2007). The leaf architecture and high light transmission of eucalypt canopies also permit grass persistence at high tree densities.

The mechanism by which water limits tree stands implies that, where large interannual fluctuations in rainfall occur, MAP will be a poor index of water availability at a site. For Charters Towers, periodic droughts and recovery resulted in much greater variation in total basal area than did variations in fire frequency.

Compared with African savannas, tree stands in Australian savannas are more likely to be at or close to their climatic optimum due to their high fire tolerance and the absence of specialized browsing megafauna since their demise around 20,000 years ago (Johnson, 2006). The long dry season, even at high rainfall in Australian savannas, means that the climatic optimum will be lower than for African ecosystems at a similar MAP. The climatic optimum also is more likely to vary interannually in Australian savannas than in Africa due to the much higher coefficient of variation in rainfall. Drought death is more likely to be evident where other disturbances such as fire and browsing are either low or have little impact. In addition to drought, wind is another factor affecting Australian savannas. Of the world's wet dry tropics, Australia has the largest area exposed to tropical cyclones, which can cause massive destruction of trees (Cook and Goyens, 2008) akin to the long-term and localized influence of elephants in Africa (Shannon et al., 2008). Although cyclones are rare for a given location, the time taken for trees to recover from their stand altering impact results in the systems near the coast often being in a state of recovery from the cyclonic disturbance or other wind events (see Chapter 4). The increase in total basal area for the Darwin rainfall simulation may indicate that the stand at Howard Springs is still recovering from previous cyclones and is below its climatic optimum.

We found that despite the average carbon flux being close to zero for both Darwin and Charters Towers rainfall simulations regardless of fire frequency, the systems were a sink in the majority of years. When fire frequency was low (one in 10 years), the Darwin stand was a sink 85% of the time and a Charters Towers sink 64% of the time. These findings are consistent with the assertion of Gifford and Howden (2001) that most systems are sinks most of the time; disturbance can reduce NPP quickly, but recovery is slow. The interannual fluctuations in carbon flux were found to be far greater under frequent fire than under no fire, even where there was little difference in the overall carbon stock. This has important implications for interpretation of field studies. Many studies have found Australian savannas to be a sink (Burrows et al., 2002; Chen et al., 2003; Williams et al., 2004), but this is to be expected from relatively short-term studies. The results of long-term studies of carbon stocks in savannas are less likely to report high sink strength (Fensham et al., 2005; Lehmann et al., 2008; Murphy et al., 2009). With the move to full carbon accounting required under the Intergovernmental Panel on Climate Change's 2006 guidelines (IPCC, 2006), understanding such dynamics is critical.

References

Beringer, J., L. B. Hutley, N. J. Tapper, and L. A. Cernusak. 2007. Savanna fires and their impact on net ecosystem productivity in North Australia. *Global Change Biology* 13, 990–1004.

Bond, W. J. 2005. Large parts of the world are brown or black: A different view on the 'Green World' hypothesis. *Journal of Vegetation Science* 16, 261–266.

Bond, W. J. 2008. What limits trees in C4 grasslands and savannas? *Annual Review of Ecology, Evolution and Systematics* 39, 641–659.

Burrows, W. H., B. K. Henry, P. V. Back, et al. 2002. Growth and carbon stock change in eucalypt woodlands in northeast Australia: Ecological and greenhouse sink implications. *Global Change Biology* 8, 769–784.

Catchpole, W. R., M. E. Alexander, and A. M Gill. 1992. Elliptical-fire perimeter and area–intensity distributions. *Canadian Journal of Forest Research* 22, 968–972.

Chen, X., L. B. Hutley, and D. Eamus. 2003. Carbon balance of a tropical savanna of northern Australia. *Oecologia* 137, 405–416.

Cook, G. D. and R. G. Heerdegen. 2001. Spatial variation in the duration of the rainy season in monsoonal Australia. *International Journal of Climatology* 21, 1831–1840.

Cook, G. D., A. C. Liedloff, R. W. Eager, et al. 2005. The estimation of carbon budgets of frequently burnt tree stands in savannas of northern Australia using allometric analysis and isotopic discrimination. *Australian Journal of Botany* 53, 621–630.

Cook, G. D., R. J. Williams, L. B. Hutley, A. P. O'Grady, and A. C. Liedloff. 2002. Variation in vegetative water use in the savannas of the North Australian Tropical Transect. *Journal of Vegetation Science* 13, 413–418.

Cook, G. D. and C. M. A. C. Goyens. 2008. The impact of wind on trees in Australian tropical savannas: Lessons from the impact of Cyclone Monica. *Austral Ecology* 33, 462–470.

Fensham, R. J., R. J. Fairfax, and S. R. Archer. 2005. Rainfall, land use and woody vegetation cover change in semi-arid Australian savannas. *Journal of Ecology* 93, 596–606.

Fensham, R. J., R. J. Fairfax, and D. P. Ward. 2009. Drought-induced tree death in savanna. *Global Change Biology* 15, 380–387.

Gardner, T. A. 2006. Tree-grass coexistence in the Brazilian cerrado: Demographic consequences of environmental instability. *Journal of Biogeography* 33, 448–463.

Gifford, R. and M. Howden. 2001. Vegetation thickening in an ecological perspective: Significance to national greenhouse gas inventories. *Environmental Science and Policy* 4, 59–72.

Gill, A. M. 1997. Eucalypts and fires: interdependent or independent? In *Eucalypt Ecology: Individuals to Ecosystems*, eds. J. E. Williams and J. C. Z. Woinarski. Cambridge: Cambridge University Press, pp. 151–167.

Hanan, N. P., W. B. Sea, G. Dangelmayr, and N. Govender. 2008. Do fires in savannas consume woody biomass? A comment on approaches to modeling savanna dynamics. *The American Naturalist* 171, 851–856.

IPCC. 2006. *IPCC Guidelines for National Greenhouse Gas Inventories, Prepared by the National Greenhouse Gas Inventories Programme*, eds. H. S. Eggleston, L. Buendia, K. Miwa, T. Ngara and K. Tababe. Institute for Global Environmental Strategies (IGES), Hayama, Japan.

Johnson, C. 2006. *Australia's Mammal Extinctions: A 50 000 Year History*. Cambridge: Cambridge University Press.

Lehmann, C. E. R., L. D. Prior, R. J. Williams, and D. M. J. S. Bowman. 2008. Spatio-temporal trends in tree cover of a tropical mesic savanna are driven by landscape disturbance. *Journal of Applied Ecology* 45, 1304–1311.

Lehmann, C. E. R., L. D. Prior, and D. M. J. S. Bowman. 2009. Fire controls population structure in four dominant tree species in a tropical savanna. *Oecologia* 161, 505–515.

Liedloff, A. C. and G. D. Cook. 2007. Modelling the effects of rainfall variability and fire on tree populations in an Australian tropical savanna using the Flames simulation model. *Ecological Modelling* 201, 269–282.

Murphy, B., J. Russell-Smith, F. A. Watt, and G. D. Cook. 2009. Fire management and woody biomass carbon stocks in mesic savannas. In *Culture, Ecology, and Economy of Fire Management in North Australian Savannas: Rekindling the Wurrk Tradition*, eds. J. Russell-Smith, P. Whitehead and P. Cooke. Collingwood: CSIRO Publishing.

Paustian, K., N. H. Ravindranath, and A. van Amstel. 2006. Agriculture, forestry and other land use: Chapter 1 Introduction. In *2006 IPCC Guidelines for National Greenhouse Gas Inventories. Institute for Global Environmental Strategies (IGES)*, Hayama, Japan. 21 pp.

Prior, L. D., B. P. Murphy, and J. Russell-Smith. 2009. Environmental and demographic correlates of tree recruitment and mortality in north Australian savannas. *Forest Ecology and Management* 257, 66–74.

Russell-Smith, J., P. J. Whitehead, G. D. Cook, and J. L. Hoare. 2003. Response of Eucalyptus-dominated savanna to frequent fires: Lessons from Munmarlary, 1973–1996. *Ecological Monographs* 73, 349–375.

Russell-Smith, J., C. P. Yates, P. J. Whitehead, et al. 2007. Bushfires 'down under': Patterns and implications of contemporary Australian landscape burning. *International Journal of Wildland Fire* 16, 361–377.

Sankaran, M., N. P. Hanan, R. J. Scholes, et al. 2005. Determinants of woody cover in African savannas. *Nature* 438, 846–849.

Shannon, G., D. J. Druce, B. R. Page, H. C. Eckhardt, G. Grant, and R. Slotow. 2008. The utilization of large savanna trees by elephant in southern Kruger National Park. *Journal of Tropical Ecology* 24, 281–289.

Werner, P. A. and P. G. Murphy. 2001. Size-specific biomass allocation and water content of above- and below-ground components of three *Eucalyptus* species in a northern Australian savanna. *Australian Journal of Botany* 49, 155–167.

Williams, R. J., G. A. Duff, D. M. J. S. Bowman, and G. D. Cook. 1996. Variation in the composition and structure of tropical savannas as a function of rainfall and soil texture along a large-scale climatic gradient in the Northern Territory, Australia. *Journal of Biogeography* 23, 747–756.

Williams, R. J., L. B. Hutley, G. D. Cook, J. Russell-Smith, A. Edwards, and X. Chen. 2004. Assessing the carbon sequestration potential of mesic savannas in the Northern Territory, Australia: Approaches, uncertainties and potential impacts of fire. *Functional Plant Biology* 31, 415–422.

Williams, R. J., W. J. Muller, C. Wahren, S. A. Setterfield, and J. Cusack. 2003. Vegetation. In *Fire in Tropical Savannas*, eds. A. N. Andersen, G. D. Cook and R. J. Williams. 79–106. New York: Springer.

Wilson, B. A., P. S. Brocklehurst, M. J. Clark, and K. J. M. Dickinson. 1990. Vegetation Survey of the Northern Territory, Australia. In *Technical Report No. 49*. Palmerston: Conservation Commission of the Northern Territory.

16

Population and Ecosystem Modeling of Land Use and Climate Change Impacts on Arid and Semiarid Savanna Dynamics

Florian Jeltsch, Britta Tietjen, Niels Blaum, and Eva Rossmanith

CONTENTS

Introduction

Worldwide, drylands are under extreme threat due to predicted climate change and increasing land use pressure (Reynolds et al., 2007). This specifically holds for dry savanna rangelands, where livestock production is the dominant type of land use (Tietjen and Jeltsch, 2007). Already now, many savanna rangelands worldwide suffer from nonadapted land use in the form of overgrazing by livestock. Overgrazing, especially in combination with unfavorable rainfall conditions, is being held responsible either for a loss of vegetation cover and increasing risks of erosion or for an increase of less palatable woody vegetation at the cost of palatable herbaceous vegetation— the so-called phenomenon of *bush encroachment* (van de Koppel and Rietkerk, 2000; Weber and Jeltsch, 2000).

Unsustainable livestock production or unfavorable climatic conditions not only reduce rangeland productivity but also change the structural diversity of the dry savanna landscape formed by the spatiotemporal dynamics of the vegetation structure. This change in habitat structure and in the availability of key structural resources such as solitary nesting trees, foraging grounds, hiding places, and safe migration routes has severe consequences for animal species diversity (e.g., Dean et al., 1999; Tews et al., 2004; Blaum et al., 2007a, b).

Predicted climate change increases the probability of unfavorable climatic periods (i.e., increased temperatures with altered precipitation patterns, typically with a decrease in MAP for many regions and an increase in intra-annual variability, IPCC 2007). This further exacerbates environmental conditions for animal species in arid and semiarid savannas in two ways: (1) via alterations in precipitation-driven resource availability and (2) via additional, climate-driven changes in the structural diversity (Blaum and Wichmann, 2007).

Understanding the mechanisms through which climate and livestock production drive the interlinked dynamics of woody and herbaceous vegetation in dry savannas is, thus, a crucial prerequisite for sustainable savanna management and animal species conservation. The latter further needs a thorough understanding of how changes in the structural diversity in space and time drive animal species dynamics across a range of scales. Given the complexity of these tasks and the different spatiotemporal scales involved, empirical research needs to be complemented by ecological bottom-up models that simulate key population-level processes such as reproduction, mortality, and dispersal in relation to environmental conditions. These process-based models allow for systematic analyses of changes and mechanisms that drive savanna systems under different scenarios of future conditions. This is a challenging task, because savannas are often described as an inherently unstable mixture of woody and herbaceous vegetation that only persists owing to disturbances such as fire or other mechanisms buffering the transition to other biomes (Jeltsch et al., 2000; Sankaran et al., 2004). Fire plays only a minor role in arid and semiarid savannas (Sankaran et al., 2005), whereas precipitation and herbivory have been identified as key determinants of dry savanna vegetation structure and composition (e.g., Jeltsch et al., 1996, 1999; Sankaran et al., 2005). Nutrient limitations can also play an important role for semiarid savanna vegetation. In these systems, large herbivores often cause heterogenous nutrient distributions at a patch scale, that is, through dung deposition under large trees (Dean et al., 1999; Treydte at al., 2009).

In this chapter, we will briefly present recent process-driven, bottom-up modeling approaches focusing on the effects of climate change and (over) grazing on (1) the coupled dynamics of water and vegetation and (2) the response of selected animal species and populations in arid and semiarid savannas. Subsequently, we discuss modern approaches to spatial scaling from local to landscape scales as an emerging strategy that may offer a way to further increase the applicability of process-based bottom-up modeling approaches in the context of land management and biodiversity conservation.

Modeling the Vegetation Response of Dry Savannas to Grazing and Climate Change

Climate Change and Savanna Dynamics

One prime driver of vegetation dynamics in arid to semiarid savannas is the water availability to plants. Predicting the impacts of livestock production on savanna vegetation under future climate conditions, therefore, requires a full understanding of how different aspects of climate change are linked to soil moisture and, thus, to water availability. Climate change involves a variety of changes, among which the most important ones for savannas are (1) altered precipitation patterns with a decrease in mean annual precipitation for many regions and an increase in intra-annual variability; (2) increased temperature and closely related to climate change; and (3) elevated atmospheric CO_2 (IPCC, 2007). The predicted decrease in MAP naturally leads to a reduction in water availability. However, the increase in rainfall variability and, thus, in extreme events can have reverse effects: dependent on soil texture, topography, and vegetation cover, the same amount of rain can lead to complete infiltration (e.g., at sandy soils) or to high water losses by runoff (e.g., at clayey soils with low vegetation cover). If the maximum temperature of photosynthesis is not exceeded, temperature increases generally reduce water availability due to higher ET rates. However, this effect can be mitigated by an elevated CO_2 level, as the water use efficiency of plants increases with CO_2 concentration. Additionally, an increased atmospheric CO_2 level results in higher growth rates mainly for C_3 species but also for C_4 species (Körner, 2006).

An Ecohydrological Model

It is not a trivial mission to assess how grazing and climate change affect the state of the vegetation, because the coupled water-vegetation system is characterized by multiple feedbacks. If climate change modifies water availability, vegetation cover changes and results in altered infiltration, ET, and runoff. Livestock can have similar effects, as grazing can shift the vegetation state to a lower grass–shrub ratio (Roques et al., 2001) or to a higher proportion of bare ground (Rietkerk and van de Koppel, 1997), that is, altered forage availability. Understanding these feedback mechanisms and thus being able to predict the impacts of grazing under future climate conditions requires an ecohydrological approach with a similar resolution of hydrological and ecological processes. However, these approaches are widely missing (see Tietjen and Jeltsch, 2007). In particular, intra-annual rainfall patterns are seldom adequately resolved, and the effects of CO_2 are widely neglected in modeling. A recently developed ecohydrological model avoids these shortcomings by combining a grid-based hydrological model of two soil layers with a vegetation model that simulates the cover of grasses and shrubs with a spatial resolution of 5 m × 5 m (Figure 16.1; see also Tietjen et al., 2009a, 2009b).

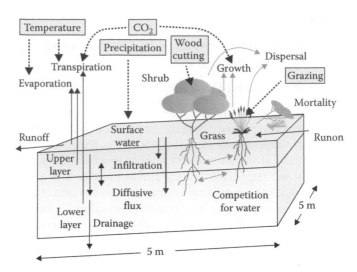

FIGURE 16.1
Overview of the ecohydrological model. (Adapted from Tietjen, B., E. Zehe, and F. Jeltsch. 2009a. *Water Resources Research* 45, W01418; and Tietjen, B. et al. 2009b. *Ecohydrology*. DOI: 10.1002/eco.70.) Main hydrological (black) and ecological (gray) processes and system components are shown for one grid cell. External drivers are indicated by gray boxes and dotted lines.

Soil moisture is simulated on an hourly basis to account for fast water redistribution by runoff. Additionally, infiltration, vertical water movement, and water losses by ET are calculated depending on vegetation cover. Although interception plays an important role for heterogeneity in soil moisture within one grid cell, it is less important on larger scales and, therefore, neglected in the model. Evapotranspiration is affected by changes in temperature, whereas an increased atmospheric CO_2 level leads to an increased water use efficiency of plants and increased growth rates. Plants compete for water in both layers (root fraction in the upper and lower soil layer: grasses 63% and 37%, shrubs 36% and 64%), and vegetation cover changes biweekly as a result of water-dependent growth, mortality, dispersal, and establishment. Simulated livestock grazing leads to a decrease in grass cover. Since the impact of fires on dry savannas is low (Sankaran et al., 2005), fire is not simulated in these model runs. This also applies to nutrient concentrations, which can act as limiting factors in many savannas but are of minor importance at the simulated site. Comparisons of model results with field data show that general patterns of soil moisture and vegetation cover can be reproduced well by the model (see Tietjen et al., 2009a, 2009b).

Scenario Analysis Using the Model

This model was used to examine the response of an African thornbush savanna with an MAP of 360 mm and a sandy loamy soil texture to various

TABLE 16.1

Parameters for Climate Change and Grazing Scenarios of the Ecohydrological Model

Climate Change Scenario	Precipitation (mm yr⁻¹)	Temperature Change (°C)	Extreme Events: Shifted Quantile (%)	Growth Rate of Grass (yr⁻¹)	Growth Rate of Shrubs (yr⁻¹)	Increase in Transpiration Efficiency (%)
0	360	0.0	0	0.8	0.2	0
1	345	0.75	10	0.85	0.25	10
2	330	1.5	20	0.9	0.3	20
3	315	2.25	30	0.95	0.35	30
4	300	3.0	40	1	0.4	40

Grazing Scenario	Reduction in Herbaceous Cover
no	0%
low	5%
high	15%

Source: Tietjen, B., E. Zehe, and F. Jeltsch 2009a. *Water Resources Research* 45, W01418; and Tietjen, B., et al. 2009b. *Ecohydrology*. DOI: 10.1002/eco.70.

climate change scenarios. The parameters used to define climate change and grazing scenarios of the ecohydrological model are shown in Table 16.1.

The combined effects of different climate change drivers lead to a clear decrease in soil moisture, especially in the upper soil layer (Figure 16.2a). Grass cover declines, as grasses cannot compensate for the lower MAP and higher temperatures by increased water use efficiency due to the elevated atmospheric CO_2 level (Figure 16.2b). In contrast, shrubs benefit from their higher water use efficiency and the reduced competition by grasses. These simulations show that future climate change can shift rangelands from a grass-dominated to a shrub-encroached state.

An increase in grazing intensity also caused grass cover to decline and shrub cover to increase. This shift in vegetation cover leads to a decrease in soil moisture under the same climatic conditions. Under future climate conditions combined with high grazing rates, soil moisture in the lower soil layers (Figure 16.2a and b) can even fall below a critical threshold. When a process of annual de-bushing (shrub clearing) was applied in the simulations, this resulted in an improvement in soil moisture in the lower layer due to reduced water losses by transpiration (Figure 16.2c and d). However, the upper soil layer is hardly affected by these changes, because shrub reduction at the same time increases water losses by evaporation due to reduced shading effects. As a result, rangeland productivity increases, because grasses are released from the high competition pressure by shrubs and can increase in cover. The given example shows that understanding the impact of climate

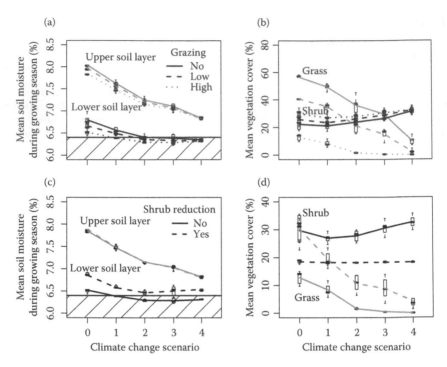

FIGURE 16.2
Impacts of climate change and grazing on soil moisture and vegetation cover. Resulting soil moisture (a, c) and vegetation cover (b, d) were evaluated for five climate change scenarios ranging from 0: no change to 4: strong simultaneous changes in precipitation patterns, temperature, and atmospheric CO_2. The upper panels show results for different grazing scenarios, and the lower panels show results for the high-impact grazing scenario with and without annual shrub manual to 20% cover. The striped rectangular panel in (c) indicates that the mean soil moisture is too low to allow for growth. For each scenario, mean results and quartiles of five model runs are shown.

change and grazing on vegetation dynamics requires explicit descriptions of the feedbacks between hydrological and vegetation processes at highly resolved spatiotemporal scales.

Impacts of Land Use and Climate Change on Population Dynamics

The previous example shows that both climate change and land use have a major impact on savanna vegetation. The resulting landscape changes are the main drivers of population dynamics and biodiversity in savannas worldwide (Sala et al., 2000). Plant and animal populations are affected by changes in basic life conditions, such as water and nutrient availability for germination and growth of plants and food and habitat availability for animals (Chesson et al., 2004; Schwinning and Sala, 2004; Wichmann et al., 2005). The consequences are temporal shifts, for example, flowering in plants with regard to emergence of insects, a reduction in savanna primary productivity, and an increase of the extinction risk of species that are neither able to shift their geographical range within the time scale of climatic changes nor able to adapt to the new climatic conditions (Erasmus et al., 2002; Thuiller, 2007).

Currently, two major modeling approaches for predicting the effects of climate change on species are applied:

1. Niche models based on the geographical distribution and preferred climatic conditions of species (e.g., Erasmus et al. 2002)

2. Process-based dynamic population models based on the relationship between population processes and environmental conditions (mostly triggered by climate) (e.g., Blaum and Wichmann, 2007; Schwager et al., 2008)

Process-based population models have the inherent advantage of considering a species' response to climate not as static but as resulting from distinct mechanisms. However, most of these models consider only short-term processes, as they occur under current climate in a restricted area, disregarding mechanisms of slow adaptation or other long-term mechanisms buffering negative climate change effects. The relevance of such mechanisms are indicated by Schwager et al. (2008), who showed that short-term responses of the reproductive output of the sociable weaver (*Philetairus socius*) to mean annual rainfall are buffered by long-term processes such as changes in habitat structure, plasticity in life-history traits, or shifts in interspecific interactions. Similarly, Wichmann et al. (2005) found the predicted increase in extinction risk of the tawny eagle (*Aquila rapax*) due to climate change to be much lower when including behavioral adaptations, that is, increase in territory size in relation to rainfall.

In addition to climate change, overgrazing has led to habitat loss and fragmentation with major implications for demographic and genetic structures of savanna plants and animals (Sankaran et al. 2005; Tews et al. 2006; Blaum et al. 2007b). The effects of climate change and land use impacts often occur simultaneously. Fragmentation in savannas can alter habitat quality not only in space (e.g., area wide shrub encroachment) but also in time (e.g., cropping rotation systems), and climate change effects occur not only in time (e.g., more rainfall extremes including extended droughts) but also in space (e.g., more localized rainfall). These effects may enhance or mitigate each other. For example, shrub-dominated savanna rangeland areas with little grass cover (caused by over-grazing) can act as a barrier and lead to separate rodent subpopulations living in scattered grass-dominated habitats (Blaum et al., 2007a). Clearly, this effect depends on the size of the shrub-encroached area between suitable habitats (see Figure 16.3a). However, when rare events of

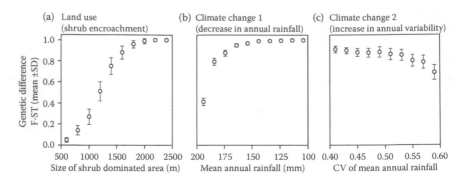

FIGURE 16.3

Effects of grazing (a) and climate change (b, c) on gene flow of the hairy-footed gerbil (*Gerbillurus paeba*) in the southern Kalahari (model simulations). (Adapted from Blaum, N. and M. C. Wichmann, 2007. *Journal of Animal Ecology* 76, 1116–1127.) Gene flow was measured as genetic difference (F_{ST}) between two neighboring gerbil populations. The model was validated with empirical data on population dynamics. (Adapted from Blaum, N., E. Rossmanith, and F. Jeltsch. 2007a. *African Journal of Ecology* 45, 189–195.) Genetic structure of the model species. (Adapted from Meyer, J., et al. 2009. *Mammalian Biology*. 74, 478–487.) Model validation and detailed model description are given. (Adapted from Blaum, N. and M. C. Wichmann, 2007. *Journal of Animal Ecology* 76, 1116–1127.)

heavy rainfall trigger mass germinations and growth of dormant annual grass seeds (Chesson et al., 2004), the degraded shrub areas can transform into hospitable habitats for a short period of time. This short-term habitat transformation may then allow gene flow between the otherwise-separated subpopulations, as a simulation study of gerbil dynamics shows (Blaum and Wichmann, 2007). This process can be affected by climate change: under decreased precipitation, general habitat quality and the frequency of high rainfall events will decrease and gene flow will be further reduced (Figure 16.3b). However, higher interannual variability in rainfall (i.e., an increased coefficient of variation of the annual precipitation) will have the opposite effect and gene flow will be enhanced through an increase of the frequency of extreme rain events (see Figure 16.3c). This example pinpoints the need to include possible synergistic or opposing effects of climate change and land use impacts in future savanna research.

The Road Ahead: Upscaling Ecological Processes in Space

The previous examples show that well-designed process-driven bottom-up models can be used to provide an improved understanding of complex factors and processes influencing savanna vegetation dynamics and population or community change under land use and climate change. Further, this dynamic modeling of key population processes allows the direct analysis of transient dynamics (e.g., gradual shift from savanna woodland to open savanna, Jeltsch et al. 1999) resulting from environmental change, thereby avoiding *ad hoc* equilibrium assumptions (Botkin et al., 2007; Jeltsch et al., 2008). However, given the necessary level of detail, the corresponding amount of input information, and the inherent model complexity, these type of process-driven models are typically restricted to small spatial scales (e.g., 100 × 100 grid cells with a size of 5 m × 5 m, Jeltsch et al., 1999). This restriction clearly limits the applicability of process-based models to large-scale land management and conservation assessments and, therefore, their relevance for challenges that are posed by land use and environmental change. To allow for sustainable decision making, the insights gained by process-based, bottom-up savanna modeling, therefore, have to be brought to a more general level, by moving the valuable mechanistic characterization of local dynamics to larger spatial scales.

When analyzing the effects of grazing or climate change on vegetation composition or population dynamics of plants and animals, one faces the problem that some processes involved act on regional spatial scales (e.g., dispersal, land use decisions, animal home ranges), whereas others act on local scales (e.g., plant establishment, water infiltration). Bringing such contrasting scales together, that is, understanding scale-crossing mechanisms and process interactions, is one of the biggest challenges in ecology (Urban, 2005). Different approaches to scaling up from local to regional scales or from short- to long-time scales can be found in a variety of disciplines. Wu and Li (2006) identified two general scaling approaches:

1. Similarity-based scaling methods are predicated on the concept of similarity and assume that two systems share some properties which can be related across the systems by a simple conversion factor (e.g., biological allometry). These methods are often characterized by relatively simple mathematical or statistical scaling functions despite the complexity of the system (Blöschl and Sivapalan, 1995).

2. Dynamic model-based scaling methods focus on processes and mechanisms. This latter approach explicitly simulates the processes of interest and the exchange of information across scales. In its simplest version, it is assumed that the same model is valid for both the local and the large scale. In this case, values for parameters and input variables are simply adapted to the large scale, for example, by averaging over the entire landscape ("extrapolating by lumping," King 1991) or by using representative parameters and inputs ("extrapolating by effective parameters," Wu and Li, 2006).

These relatively simple approaches are often used in soil physics or micrometeorology (L'Homme et al., 1996; Bierkens et al., 2000) and can only produce reliable results when processes do not change substantially between scales, a situation that rarely applies to ecological systems. In an alternative method ("direct extrapolation," Wu and Li, 2006), the local model is run separately for all patches of the landscape scale or—to reduce computational time—at least for all patches of the same type. In doing so, local differences between patches (e.g., elevation, soil type) can be accounted for. Through statistical analyses, quantitative or qualitative relationships can be found that serve as input for a large scale model (e.g., Zhang et al., 2007; Jeltsch et al., 2008; Köchy et al., 2008).

A similar approach was taken in a study of the Kalahari Thornveld savanna in Southern Africa (Rossmanith et al., unpubl.), where up-scaling methods were combined with a state and transition approach (Westoby et al., 1989). The landscape in this region is divided into a mosaic of farms for livestock production, with different livestock densities and different sizes (3000 ha to 20,000 ha). To predict the effects of intensification of livestock production and precipitation changes on land degradation at a regional scale, a grid-based landscape model was developed (grid-cell size 1 km^2, total cover: 70×100 km^2), which includes farm borders and their specific grazing intensities for all farms. To describe the condition of the vegetation, nine vegetation states were defined by the cover of shrubs and perennial grasses (see Figure 16.5). To determine transition probabilities between vegetation states, simulation experiments were conducted with a small-scale process-based vegetation model (5 m $\times$ 5 m grid-cell size, Jeltsch et al., 1997, 1999) on 1 km^2 for various grazing intensities and classes of annual precipitation. The frequency of occurrence of transitions between two states served as transition probabilities. For all combinations of grazing intensities and rain classes, transition probability matrixes were produced that were used to simulate vegetation change of basic 1 km^2 patches of the landscape model. Combining the dynamics at the different 1 km^2 patches, thus, allowed the vegetation

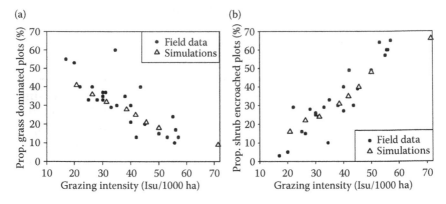

FIGURE 16.4
Validation of the upscaled vegetation model: field data on proportion of grass-dominated (a) and shrub-encroached (b) plots in relation to grazing intensity ($n = 25$ farms). (Adapted from Blaum, N., B. Tietjen, and E. Rossmanith 2009. *Journal of Wildlife Management* 73, 60–67.) Proportion of grass-dominated (a) and shrub-encroached (b) plots in model simulations (mean of 100 runs). For each grazing intensity, the model was run over 50 annual time steps (which equals the time since lifestock grazing started in the study area) and 100 runs under precipitation probabilities derived from weather data. (Adapted from Jeltsch, F., S. J. Milton, W. R. J. Dean, and N. VanRooyen 1997. *Journal of Applied Ecology* 34, 1497–1508.)

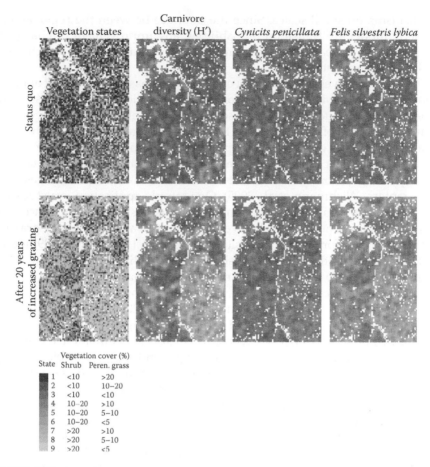

State	Vegetation cover (%) Shrub	Peren. grass
1	<10	>20
2	<10	10–20
3	<10	<10
4	10–20	>10
5	10–20	5–10
6	10–20	<5
7	>20	>10
8	>20	5–10
9	>20	<5

FIGURE 16.5

(See color insert following page 320.) Sample application of the upscaled regional vegetation model (70 km × 100 km) for a region in the southern Kalahari (South Africa, between 26°48′46 S, 20°44′56 E and 27°09′51 S, 21°07′20 E): 1. column: vegetation states, 2. column: carnivore diversity (Shannon Index) from dark gray = high (up to 0.70) to light gray = low diversity (up to 0.46), 3. & 4. column: relative abundance of yellow mongoose (*Cynictis penicillata*) and African wildcat (*Felis silvestris lybica*) from dark gray = high to light gray = low. The upper row shows the status quo (distribution of vegetation states derived by remote sensing) and the lower row shows predicted changes after 20 years under the scenario of 20% increase of current grazing intensity for all farms. The proportion of shrub-dominated vegetation states increases under this scenario, leading to a decrease in carnivore diversity and abundance of single species. However, the intensity of decrease is different between species: *Cynictis penicillata* on average decreases by only 2%, whereas *Felis silvestris* decreases by 18%.

dynamics to be scaled up to the entire rangeland area, explicitly considering varying grazing pressure and rainfall (see Figure 16.4 for model validation, Figure 16.5 for sample scenario).

This upscaling approach is a straightforward option to investigate rangeland dynamics on a regional scale by including relevant processes and mechanisms on small scales. Since the interface between the regional and the local model are transition probabilities, any improvement or change at the small scale can easily be transferred to the landscape model via an update of the transition matrix.

The large-scale vegetation model can serve as a basis for different applications. For example, it can be used for models intended to optimize sustainable livestock production on a large scale, for which predictions of vegetation change as a result of grazing or due to climate change is needed on the scale of a single farm (e.g., 10,000 ha, Popp et al., 2009) or even multiple farms. Further, the large-scale vegetation model can serve as a habitat map needed for population viability analyses of animals with large home ranges (e.g., large raptors, Wichmann et al., 2005). One sample application is a study on the effect of grazing intensification on carnivore species abundance (see Figure 16.5). It is known that shrub encroachment caused by heavy grazing is a major threat for species diversity, and Blaum et al. (2007b, 2009) found that carnivore abundance and diversity strongly decline in shrub-dominated rangeland areas. To test the impact of different scenarios of spatial changes of grazing intensities and, therefore, vegetation change, statistical models of empirical carnivore abundance data (Blaum et al., 2009) were linked to the regional vegetation model described earlier, including an algorithm accounting for neighborhood effects (see Figure 16.5). This combined large-scale analysis of grazing-induced vegetation changes and subsequent biodiversity effects clearly shows the differing sensitivity of animals of similar size within one guild (here: small carnivores) toward grazing effects.

The described up-scaling approach of the smaller scale vegetation model did not consider interactions between different 1 km² grid cells. This degree of complexity was not necessary to reflect vegetation dynamics correctly, which is shown by the model validation (see Figure 16.4). However, if horizontal interactions (e.g., hydrological flows) and feedbacks (e.g., caused by seed dispersal) between these spatial subunits turn out to be important in a system, more complex, spatially interactive modeling (Wu and Li, 2006) provides a suitable alternative, which explicitly integrates interactions among spatial subunits. In either case, the application of well-chosen approaches of spatial scaling from local to large scales allows combining the benefits of bottom-up process-based modeling with the relevance of ecosystem dynamics for management or nature conservation at regional and landscape scales.

Conclusion

Understanding the effects of land use and climate change on arid and semi-arid savanna systems and species dynamics requires thorough investigations of a complex network of interacting processes including soil–vegetation–hydrology feedbacks and the resulting spatiotemporal dynamics of vegetation structures and resources over a range of scales. Only on the basis of such a mechanistic understanding, we will be able to predict future changes; identify key threats; and develop sustainable management strategies of savanna biodiversity, ecosystem functioning, and ecosystem services, such as sustainable forage production or erosion prevention. Since savannas cover approximately 20% of the world's land surface including many hot spots of socioeconomic, climate, land use, and institutional change, the critical evaluation of existing strategies and the development of new management strategies are urgently needed. Process-based, bottom-up modeling approaches that integrate a variety of scales provide powerful tools in this process by systematically testing alternative scenarios of changes and comparing and optimizing related management options.

Acknowledgments

Parts of this work were funded by the German Ministry of Education and Research in the framework of BIOTA Southern Africa (01LC0024). F. Jeltsch acknowledges support from the European Union through Marie Curie Transfer of Knowledge Project FEMMES (MTKD-CT-2006-042261).

References

Bierkens, M. F. P., P. A. Finke, and P. de Willingen. 2000. *Upscaling and Downscaling Methods for Environmental Research*. Kluwer Academic Publishers, Dordrecht.

Blaum, N. and M. C. Wichmann. 2007. Short-term transformation of matrix into hospitable habitat facilitates gene flow and mitigates fragmentation. *Journal of Animal Ecology* 76, 1116–1127.

Blaum, N., E. Rossmanith, and F. Jeltsch. 2007a. Land use affects rodent communities in Kalahari savannah rangelands. *African Journal of Ecology* 45, 189–195.

Blaum, N., E. Rossmanith, M. Schwager, and F. Jeltsch. 2007b. Responses of mammalian carnivores to land use in arid savanna rangelands. *Basic and Applied Ecology* 8, 552–564.

Blaum, N., B. Tietjen, and E. Rossmanith. 2009. The impact of livestock husbandry on small and medium sized carnivores in Kalahari savannah rangelands. *Journal of Wildlife Management* 73, 60–67.

Blöschl, G. and M. Sivapalan. 1995. Scale issues in hydrological modeling—A review. *Hydrological Processes* 9, 251–290.

Botkin, D. B., H. Saxe, M. B. Araujo, et al. 2004. Resource pulses, species interactions, and diversity maintenance in arid and semi-arid environments. *Oecologia* 141, 236–253.

Dean, W. R. J., S. J. Milton, and F. Jeltsch. 1999. Large trees, fertile islands, and birds in arid savanna. *Journal of Arid Environments* 41, 61–78.

Erasmus, B. F. N., A. S. Van Jaarsveld, S. L. Chown, M. Kshatriya, and K. J. Wessels. 2002. Vulnerability of South African animal taxa to climate change. *Global Change Biology* 8, 679–693.

IPCC. 2007. Climate Change 2007: The Physical Science Basis. In *Fourth Assessment Report of the Intergovernmental Panel on Climate Change*, ed. S. Solomon, D. Qin, M. Manning, Z. Chen, M. Marquis, K. B. Averyt, M. Tignor, and H. L. Miller, 996pp., Cambridge: Cambridge University Press.

Jeltsch, F., S. J. Milton, W. R. J. Dean, and N. VanRooyen. 1996. Tree spacing and coexistence in semiarid savannas. *Journal of Ecology* 84, 583–595.

Jeltsch, F., S. J. Milton, W. R. J. Dean, and N. VanRooyen. 1997. Analysing shrub encroachment in the southern Kalahari: A grid-based modelling approach. *Journal of Applied Ecology* 34, 1497–1508.

Jeltsch, F., S. J. Milton, and K. Moloney. 1999. Detecting process from snap-shot pattern—lessons from tree spacing in the southern Kalahari. *Oikos* 85, 451–467.

Jeltsch, F., G. E. Weber, and V. Grimm. 2000. Ecological buffering mechanisms in savannas: A unifying theory of long-term tree–grass coexistence. *Plant Ecology* 150, 161–171.

Jeltsch, F., K. A. Moloney, F. M. Schurr, M. Kochy, and M. Schwager. 2008. The state of plant population modelling in light of environmental change. *Perspectives in Plant Ecology Evolution and Systematics* 9, 171–189.

King, A. W. 1991. Translating models across scales in the landscape. In *Quantitative Methods in Landscape Ecology*, ed. M. G. Turner and R. H. Gardner, 479–517. New York: Springer Verlag.

Köchy, M., M. Mathaj, F. Jeltsch, and D. Malkinson. 2008. Resilience of stocking capacity to changing climate in arid to Mediterranean landscapes. *Regional Environmental Change* 8, 73–87.

Körner, C. 2006. Plant CO_2 responses: An issue of definition, time and resource supply. *New Phytologist* 172, 393–411.

L'Homme, J. P., A. Chehbouni, and B. A. Monteny. 1996. Canopy to regional scale translation of surface fluxes. In *Boundary-Layer Meteorology*, ed. J. B. Stewart, E. T. Engman, R. A. Feddes and Y. Kerr, 161–182. Wiley, Chichester.

Meyer, J., A. Kohnen, W. Durka, et al. 2009. Genetic structure in the hairy-footed gerbil *Gerbillurus paeba*: Sex specific dispersal? *Mammalian Biology*. 74, 478–487.

Popp, A., S. Domptail, N. Blaum, and F. Jeltsch. 2009. Landuse experience does qualify for adaptation to climate change. *Ecological Modelling* 220, 694–702.

Reynolds, J. F., D. M. Stafford Smith, E. F .Lambin, et al. 2007. Global desertification: Building a science for dryland development. *Science* 316, 847–851.

Rietkerk, M. and J. van de Koppel. 1997. Alternate stable states and threshold effects in semi-arid grazing systems. *Oikos* 79, 69–76.

Roques, K. G., T. G. O'Connor, and A. R. Watkinson. 2001. Dynamics of shrub encroachment in an African savanna: Relative influences of fire, herbivory, rainfall and density dependence. *Journal of Applied Ecology* 38, 268–280.

Sala, O. E., F. S. Chapin, J. J. Armesto, et al. 2000. Biodiversity—Global biodiversity scenarios for the year 2100. *Science* 287, 1770–1774.

Sankaran, M., J. Ratnam, and N. P. Hanan. 2004. Tree-grass coexistence in savannas revisited—insights from an examination of assumptions and mechanisms invoked in existing models. *Ecology Letters* 7, 480–490.

Sankaran, M., N. P. Hanan, R. J. Scholes, et al. 2005. Determinants of woody cover in African savannas. *Nature* 438, 846–849.

Schwager, M., R. Covas, N. Blaum, and F. Jeltsch. 2008. Limitations of population models in predicting climate change effects: A simulation study of sociable weavers in southern Africa. *Oikos* 117, 1417–1427.

Schwinning, S. and O. E. Sala. 2004. Hierarchy of responses to resource pulses in and semi-arid ecosystems. *Oecologia* 141, 211–220.

Tews, J., N. Blaum, and F. Jeltsch. 2004. Structural and animal species diversity in arid and semi-arid savannas of the southern Kalahari. *Annals of Arid Zone* 43, 413–425.

Tews, J., S. J. Milton, A. Esther, and F. Jeltsch. 2006. Linking a population model with a landscape model: Assessing the impact of land use and climate change on savanna shrub cover dynamics. *Ecological Modeling* 195, 219–228.

Thuiller, W. 2007. Biodiversity—Climate change and the ecologist. *Nature* 448, 550–552.

Tietjen, B. and F. Jeltsch. 2007. Semi-arid grazing systems and climate change: A survey of present modelling potential and future needs. *Journal of Applied Ecology* 44, 425–434.

Tietjen, B., E. Zehe, and F. Jeltsch. 2009a. Simulating plant water availability in dry lands under climate change—a generic model of two soil layers. *Water Resources Research* 45, W01418.

Tietjen, B., F. Jeltsch, E. Zehe, et al. 2009b. Effects of climate change on the coupled dynamics of water and vegetation in drylands. *Ecohydrology*. DOI: 10.1002/eco.70.

Treydte, A. C., C. C. Grant, and F. Jeltsch. 2009. Tree size and herbivory determine the below-canopy grass quality and species composition in savannahs. *Biodiversity and Conservation* 18, 3989–4002.

Urban, D. L. 2005. Modeling ecological processes across scales. *Ecology* 86, 1996–2006.

van de Koppel, J., and M. Rietkerk. 2000. Herbivore regulation and irreversible vegetation change in semi-arid grazing systems. *Oikos* 90, 253–260.

Weber, G. E. and F. Jeltsch. 2000. Long-term impacts of livestock herbivory on herbaceous and woody vegetation in semiarid savannas. *Basic and Applied Ecology* 1, 13–23.

Westoby, M., B. Walker, and I. Noymeir. 1989. Opportunistic management for rangelands not at equilibrium. *Journal of Range Management* 42, 266–274.

Wichmann, M. C., J. Groeneveld, F. Jeltsch, and V. Grimm. 2005. Mitigation of climate change impacts on raptors by behavioural adaptation: Ecological buffering mechanisms. *Global and Planetary Change* 47, 273–281.

Wu, J. and H. Li. 2006. Perspectives and methods of scaling. In *Scaling and Uncertainty Analysis in Ecology: Methods and Application*, eds. J. Wu, B. Jones, H. Li and O.L. Loucks, pp. 17–44. Dordrecht: Springer-Verlag.

Zhang, N., Z. L. Yu, G. R. Yu, and J. G. Wu. 2007. Scaling up ecosystem productivity from patch to landscape: A case study of Changbai Mountain Nature Reserve, China. *Landscape Ecology* 22, 303–315.

Section V

Regional Carbon Dynamics in Savanna Systems: Remote Sensing and Modeling Applications

17

Integration of Remote Sensing and Modeling to Understand Carbon Fluxes and Climate Interactions in Africa

Christopher A. Williams

CONTENTS

Introduction

African carbon dynamics are of global significance, with as much as 40% of the world's fire emissions, about 20% of each of global net primary production, heterotrophic respiration, and land-use emissions, and acting as a major source of interannual variability in global net carbon exchange (Williams et al., 2007). Much of the continent is vulnerable to degradation from multiyear drought cycles (e.g., Nicholson, 2000), jeopardizing the livelihoods of millions who depend on local ecosystem services such as food, forage, fuel, and fiber (UNCTAD, 2002). The potential release of large amounts of carbon currently stored in African ecosystems is cause for concern (IPCC, 2001). Additionally, global environmental changes could trigger strong biophysical feedbacks to the climate system that may accelerate warming and prolong droughts (e.g., Wang and Eltahir, 2000a; Zeng and Neelin, 2000). Appraising these risks relies on detailed understanding of how ecosystems across the continent respond to climate and land use forcings. This motivates the current diagnostic analysis of carbon–climate connections in Africa.

Integration of remote sensing and land surface modeling lends unique insights about how carbon cycle processes are connected to climate variability and trends. For example, limits to plant productivity can be comprehensively attributed to specific ecophysiological and biophysical conditions by studying consistency between modeled plant stress and remotely sensed vegetation indices. This chapter illustrates aspects of what can be learned from such an approach.

Climate and Disturbance Controls

Most of Africa experiences significant water limitation, with ~94% of its surface area experiencing an annual deficit of precipitation relative to evaporative demand, particularly prevalent in the extensive seasonal and drier biomes (Figure 17.1). Climatological dryness is compounded by sizeable year-to-year variability in rainfall and even protracted drought (e.g., Nicholson, 2000, 2001; Nicholson and Kim, 1997; Tyson et al., 2002). Being a mostly tropical continent, coldness is a minor determinant of ecosystem function and composition. Thus, plant productivity across Africa is predominantly limited by water (Nemani et al., 2003), excepting the moist equatorial zone where maximal rates are realized. Correspondingly, most of the interannual variability in Africa's carbon cycle is caused by hydrologic fluctuations, mainly from rainfall, that modulate net ecosystem productivity (photosynthesis–respiration) as well as fire emissions (Patra et al., 2005; van der Werf et al., 2006; Williams et al., 2007, 2008).

Some of these hydrologic fluctuations are teleconnected to remote climate phenomena. Temporal variability of Africa's rainfall resonates with the coupled ocean–atmosphere phenomena of El Niño Southern Oscillation (ENSO) and the Indian Ocean Dipole Zonal Mode (abbreviated here as IOD) but the degree to which these oscillations excite carbon cycle anomalies across Africa remains unclear. Corresponding changes in African vegetation have been described for ENSO (e.g., Anyamba and Eastman, 1996; Anyamba and Tucker, 2005), but have only recently been examined for the IOD (Williams and Hanan, unpublished data). Furthermore, the regional-scale implications of sea surface temperature (SST) variability for Africa's carbon sources and sinks have received little focused attention, particularly in the case of IOD.

Although climate and soils are the dominant factors that determine potential vegetation in Africa, it is also important to appreciate the roles, especially at local scales, of disturbances, both natural (severe weather, fire, pests, pathogens) and anthropogenic (fuelwood extraction, grazing, fire, agriculture),

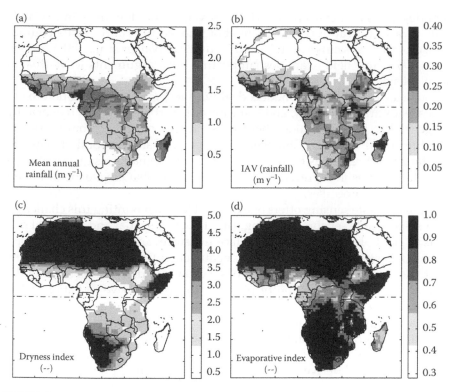

FIGURE 17.1
Continental patterns of (a) mean annual rainfall, (b) IAV of rainfall, (c) annual-scale dryness index from potential evapotranspiration divided by rainfall, and (d) annual-scale evaporative index from actual evapotranspiration divided by rainfall.

which act as important determinants of actual vegetation distributions. Disturbances modify how ecosystem productivity responds to weather and climate anomalies by altering resource use patterns and associated limitations, as well as by triggering endogenous growth and decay dynamics (e.g., Williams and Albertson, 2006). For example, disturbances can reduce vegetation abundance making water more available during ensuing regrowth and possibly dampening sensitivity to periods of low rainfall (i.e., drought), but may similarly inhibit ecosystem productivity from responding fully to periods of high resource (water) availability. Alternatively, disturbances may amplify sensitivity to some detrimental environmental conditions, for example, by increasing dieback and mortality induced by drought that follows a severe insect outbreak. Disturbances may even lock the system into one stable state as opposed to an alternate, if, for example, the vegetation–evaporation interactions shown to be particularly strong in parts of Africa (Charney et al., 1975; Koster et al., 2006; Wang and Eltahir, 2000b; Zeng and Neelin, 2000) do in fact contribute to long persistence of drought events. The relative importance of endogenous and exogenous drivers of vegetation dynamics in drylands is the heart of a captivating debate about nonequilibrium versus equilibrium ecological systems (e.g., Ellis and Swift, 1988; Walker et al., 1981; Westoby et al., 1989), and field examples supporting both arguments leave this an open debate.

Coupled carbon–climate models deal with responses of vegetation to climate trends and variability, and some even disturbance, but all rely heavily on generalized assumptions about ecosystem's structural, compositional, and functional changes. Characterizing actual spatiotemporal changes in vegetation as it responds to these various drivers remains a significant challenge to progress in representing continental to global biogeochemical and biogeophysical processes.

Satellite remote sensing offers one method for surmounting this challenge. Data from remote sensors are being used to detect spatial and temporal patterns of plant biomass, leaf area, surface moisture, and temperature, among other biophysical attributes. Spectral vegetation indices (SVIs) are among the most widely used remotely derived observations to date, characterizing the vegetation cover and fraction of photosynthetically active radiation (PAR) that is absorbed by plant canopies (f_{PAR}). Numerous production efficiency models utilize this attribute to represent the poorly characterized patterns of vegetation cover and vigor, providing one solution to the challenge of incorporating vegetation changes in biogeochemical and biogeophysical models.

Although model-derived estimates of global terrestrial photosynthesis are many (e.g., Ciais et al., 2001), relatively few attempts have been made to capture the influence of disturbance processes on productivity. Furthermore, few model-based reports on photosynthesis patterns provide detailed attribution to drivers such as precipitation, temperature, humidity, and vegetation with a clear display of their absolute effects and relative importance (Churkina and Running, 1998; Nemani et al., 2003; Schaefer et al., 2002). In addition, the global scope of such works enables only broad-brush descriptions of subcontinental-scale patterns. Thus much remains to be learned about how continental to regional productivity has responded to vegetation and climate conditions in recent decades, particularly throughout Africa.

Analytical Framework

The framework described here addresses several of the challenges described above. The basic idea is to combine remote sensing of vegetation cover with a model of plant ecophysiology to express their separate and combined influences on plant productivity. The method also enables attribution of variability in photosynthesis to specific weather and vegetation conditions. In addition, each factor that limits photosynthesis can be analyzed for its connection to remote climate indices to improve understanding of carbon cycle variability.

The method is based on simple biosphere (SiB), a land surface model originally designed (Sellers et al., 1986) for use with General Circulation Models but used here in an "offline" mode to represent ecosystem physiology as driven by weather and vegetation. With the current focus on carbon–climate connections across Africa, the model was used to simulate ecosystem water and carbon fluxes across Africa at 1×1 degree resolution for the period 1982–2003 with a 10-min time step, described further in Williams et al. (2008). The analysis here focuses on the model's representation of photosynthesis.

SiB (Sellers et al., 1996a, c) calculates photosynthetic rate from a physiologically defined leaf rate multiplied by a leaf-to-canopy effective area scaling based on remote sensing, thus separating physiological and biophysical controls. On the physiological side, the leaf-scale, gross photosynthetic rate (P) is reduced from a potential maximum (P_0) according to multiplicative stress modifiers for temperature and soil water (Sellers et al., 1996b, c) as well as regulation of intercellular CO_2 by stomatal closure (Collatz et al., 1991, 1992; Sellers et al., 1992, 1996b). The potential maximum "top-leaf" (full sun) photosynthetic rate is estimated with a modified version of the Farquhar et al. (1980) model (Collatz et al., 1991) coupled to the Ball–Berry stomatal conductance model (Ball et al., 1988; Collatz et al., 1991, 1992). The model's ecophysiological parameters that describe potentials and stresses vary across biomes (Los et al., 2000; Sellers et al., 1996c).

On the biophysical side, SiB scales these "top-leaf" photosynthetic rates to the canopy using the integration scheme of Sellers et al. (1992). The scheme defines a canopy-scale, PAR use parameter (Π) estimated from the fractional absorption of PAR (f_{PAR}) commonly derived from remotely sensed vegetation reflectances, here Normalized Difference Vegetation Index (NDVI) at ~8 km spatially averaged to 1-degree, and vegetation type. This canopy-scaling parameter (Π) has a potential maximum (Π_{max}) defined by solar geometry, leaf angle, and maximum leaf area density with parameters that vary by biome.

Combining the physiological and biophysical conditions, canopy-scale gross photosynthesis is obtained from $P\Pi$ (analogous to what many would define as gross primary productivity) and it has a potential maximum value of $P_0\Pi_{max}$. The scheme can be loosely represented in a general form as

$$P_c, g_c \approx P_0, g_0 \cdot f(W_f, h_s, T_l) \cdot \Pi, \tag{17.1}$$

where P_c and g_c are the canopy-integrated rates of photosynthesis and conductance solved simultaneously, P_0 and g_0 are the unstressed, fully illuminated "top-leaf" rates, and $f(W_f, h_s, T_l)$ represents the multiplicative stress modifiers for soil water W_f and temperature (T_l) plus influence from relative humidity at the leaf surface (h_s) (Sellers et al., 1992, 1996b).

With this formulation, the physiology is cleanly separated from the biophysical variable of vegetation cover and vigor, meaning chlorophyll abundance (Sellers et al., 1992). Separation enables quantitative attribution of photosynthetic fluctuations in time or patterns in space to particular drivers (vegetation, soil water, temperature, humidity), albeit through a complicated decomposition as described in Williams et al. (2008). Despite being separable in the model, the two core elements, biophysical and physiological, act as joint limits to canopy-scale photosynthesis according to the product $P\Pi$.

This multiplicative formulation has a major advantage of capturing the real suppression of photosynthesis where vegetation is sparse due to natural or human causes (e.g., fire, harvest, cultivation) despite conditions favorable for more extensive vegetation cover. Dynamic vegetation models, in contrast, would grow vegetation that is inherently consistent with a recent history of weather/climate and are thus typically devoid of information on disturbance or management effects. By integrating vegetation remote sensing with the model, the spatial and temporal patterns of photosynthesis can be separately attributed to acute physiological stress or low vegetation abundance, which can itself be a reflection of chronic physiological stress (e.g., structural/phenological responses to drought) but may also result from disturbance.

Geographic Variation in Emergent Limitations

It is instructive to begin by exploring where modeled productivity is low, as shown in Figure 17.2a, presenting the ratio of actual to potential canopy-scale gross photosynthesis, or $P\Pi/P_0\Pi_{max}$. Reassuringly, the broad pattern conforms to expectation, with high productivity in wet, humid climates (Congo Basin, coastal West Africa) and low productivity in drylands (Sahara, Sahel, Namib, Kalahari). Also evident is relatively high productivity over a wide expanse of southcentral to southeastern Africa, associated with dry, open woodlands and savannas (e.g., miombo).

While water availability is known to be the primary control on productivity across much of this dry continent, this ultimate cause can be expressed proximally in low vegetation, and for our purposes, low f_{PAR}. With this in mind, Figure 17.2b presents the relative importance of limitation by water stress as compared to low vegetation (i.e., f_{PAR}). Immediately evident (Figure 17.2b) is the shared importance of f_{PAR} and soil water limitation across the driest regions of the continent, where both factors exert severe reductions from the potential rate. However, sizeable limitation by acute soil water stress extends far into the equatorial zone of southern Africa, while f_{PAR} limitation is relatively strong (soil water limitation relatively weak) throughout most of West Africa as well as parts of East Africa. Heat and humidity stresses are both modest throughout the savanna regions in comparison with soil water and f_{PAR} limitations (Williams et al., 2008), though we cannot discount their

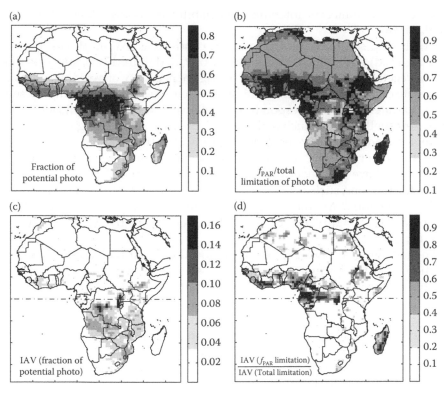

FIGURE 17.2
Continental patterns of (a) the ratio of actual to potential photosynthesis, (b) the proportion of limitation in photosynthesis due to low f_{PAR}, (c) IAV in the ratio of actual to potential photosynthesis, and (d) the proportion of IAV in photosynthetic limitation due to low f_{PAR}.

influence on vegetation cover and vigor at climate timescales and appearing here through f_{PAR}.

It is also instructive to inspect the magnitude and causes of interannual variation (IAV) in photosynthesis as shown in Figures 17.2c and d, respectively. Areas with high IAV of photosynthesis are concentrated south of the equator (Figure 17.2c) with notable absence of IAV for West Africa (Sudano-Sahelian zone) despite still high variability in precipitation. This may be partly explained by inspecting the relative importance of soil water compared to f_{PAR} controls, as presented in Figure 17.1d. Nearly all of the continent's interannual variability in photosynthesis is governed by variation in soil water availability (Figure 17.2d), with variation in f_{PAR} becoming important mainly in very wet locations (>1.5 m of rainfall per year) such as the moist tropical forests close to the equator and on the Guinea coast of West Africa. However, variation in f_{PAR} also assumes relatively high importance for the Ethiopian Highlands plus part of the Sudanian zone, relating to the region's unexpectedly low IAV in photosynthesis. Acute heat and humidity stresses induce almost no interannual variability in photosynthesis (Williams et al., 2008).

Combining soil water's dominance in controlling IAV in photosynthesis (Figure 17.2d) with prevailing limitation by f_{PAR} (Figure 17.2b), suggests that while vegetation density reflects the biological necessity of avoiding a persistent state of physiological stress, it is also extensive enough to be highly sensitive to water limitation, responding to relatively wet episodes and suffering from dry spells. While much of this simply reaffirms prevailing wisdom, the framework employed here enables deeper exploration of some unique questions that emerge. What consequences does uncommonly strong limitation by low f_{PAR} have for modeled photosynthesis? Is connection with hydroclimatic fluctuations, such as induced by ENSO or IOD, dampened by a general lack of vegetation? Are there unique ecoclimatic settings (rainfall, radiation, temperature, humidity) coincident with regions of relatively intense limitation by low f_{PAR}?

The following sections explore these questions, first with analysis of connection with the leading hydroclimate oscillations, and then by in-depth analysis centered on regions of particular interest. Both the Sudanian zone and the Ethiopian Highlands are chosen because of their aforementioned strong limitation by f_{PAR} and relatively low IAV of photosynthesis, while northern Tanzania, and southern Angola are chosen because of high IAV of photosynthesis. The Sahel and a similarly dry region of the Kalahari are chosen as semiarid extremes. In addition, a portion of southeastern Africa is highlighted owing to widely reported connection to ENSO.

Teleconnections to Climate Indices

Interannual variability in photosynthesis, as documented above, may derive partly from teleconnection to remote ocean–atmosphere oscillations. Therefore, we explore annual timescale sensitivities of photosynthesis and rainfall to climate indices using z-value normalized indices to facilitate quantitative comparison of the influence of ENSO and IOD signals on African terrestrial function. Additional details on methods are described in Williams and Hanan (unpublished data).

Beginning with ENSO, photosynthesis is strongly coupled to the Multivariate El Niño Index (i.e., ENSO) in regionally coherent spatial patterns across much of the African continent with sensitivities as strong as ± 200 gC m^{-2} year^{-1} z$_{ENSO}^{-1}$ (Figure 17.3a). In general, rainfall is coupled to ENSO with a similar pattern (Figure 17.3c) underscoring prevailing control by water availability across much of the continent. However, there are some notable exceptions; for example, rainfall increases observed during positive ENSO in most of Ethiopia are not reliably accompanied by increases in net

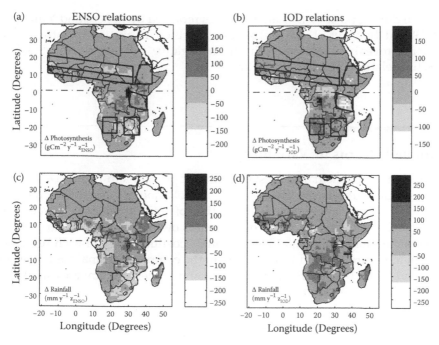

FIGURE 17.3

Spatial distribution across Africa of the slope of least-squares linear regressions of (a, b) annual net photosynthesis (P_{net}) or (c, d) annual rainfall on climate indices. Each column corresponds to analyses with a particular index with the left column (a, c) for the multivariate ENSO index, and the right column (b, d) for the IOD. Black polygons trace composite regions.

photosynthesis (as explored in Section 5.2). Relations between ENSO and rainfall shown here are broadly consistent with that documented by Nicholson and Kim (1997) and Nicholson (2001) analyzing rain gauge data, with the exception of negative, not positive, ENSO-related anomalies for the western coast of Africa just south of the equator (Gabon to northwestern Angola).

Relations with the IOD (Figure 17.3b and d) are of a similar intensity to those excited by ENSO, with anomalies as large as ±200 gC m⁻² year⁻¹ z_IOD⁻¹. Positive P_{net} correlations are found in most of the water-limited regions, particularly in parts of Namibia, Angola, and Zambia. The largest negatively correlated sensitivity occurs in Tanzania, northern Democratic Republic of the Congo, and along Africa's west coast within ~10° latitude of the equator (Guinean zone). IOD-driven photosynthesis anomalies are small in the Sahel and Sahara. Although often in agreement, net photosynthesis and rainfall sensitivities are not always of the same intensity or even sign (see Tanzania and Nigeria). These results are broadly consistent with Reason and Rouault (2002) who reports positive correlation between rain and IOD in southeastern Africa. Rainfall's negative relation with IOD in parts of East Africa appears to be at odds with the September to November (SON) elevation of rainfall during positive IOD anomalies as reported by Behera and Yamagata (2003) and Black et al. (2003). However, further inspection of intraannual patterns reveals a seasonal shift from negative relation during SON to positive relation the rest of the year.

Unique Patterns of Variability

Six regions of particular interest (see labeled polygons in Figure 17.4a) have been selected to serve as example of the spatiotemporal variability across the continent and to allow deeper inspection of unique patterns of seasonal variability that have important consequences owing to sensitivity to interannual fluctuations in climate. The seasonal ecoclimatology of each region is shown in Figure 17.4 and is referenced in the comparative analysis below. In addition, Table 17.1 highlights a variety of annualized ecoclimatic conditions relevant to the current discussion.

Contrasting the Sahelian and Namibian Regions

Beginning with the driest domains, we contrast the Sahel and Namibia regions, each receiving ~450 mm of rainfall annually but with a slightly higher IAV in Namibia. Evapotranspiration consumes nearly all of annual rainfall (Table 17.1) in both regions; however Sahelian photosynthesis exceeds Namibian photosynthesis partly because the Sahel has higher maximal rates of photosynthesis due to higher temperature that elevates biochemical reaction rates and hence P_0. In contrast, the Sahel has lower interannual variability in photosynthesis than that in Namibia, mainly owing to relatively modest water stress combined with less variable rainfall (Table 17.1).

The Sahel's more modest water stress is puzzling and explanation requires detailed inspection of the seasonal dynamics shown in Figure 17.4. Studying f_{PAR} we see that greening up with rains during the early growing season is lagged more in the Sahel than in Namibia. In addition, browning down toward the dry season occurs relatively rapidly in the Sahel compared to Namibia. Furthermore, when f_{PAR} is near its seasonal maximum, the Sahel has higher monthly rainfall and is hence wetter for a short period of time. These factors combine to decrease mean annual water stress as well as dampen ecosystem sensitivity to IAV in water availability in the Sahel compared to Namibia.

Contrasting the Sudanian and Tanzanian or Southeast African Regions

A similar story materializes in comparing the Sudanian and Tanzanian or southeast African regions. The Sudanian region has many months with a

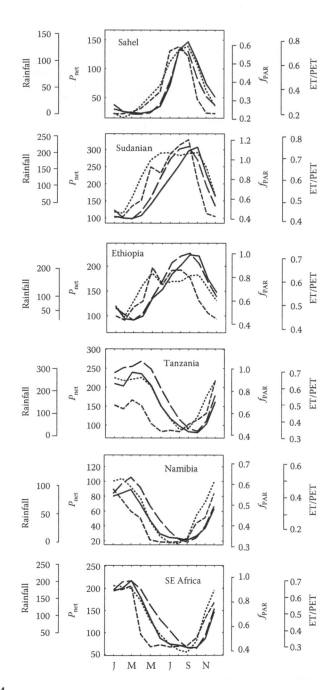

FIGURE 17.4
Seasonal climatology for selected regions of Africa, showing rainfall (mm month^{-1}) (short dash), net photosynthesis (g C m^{-2} month^{-1}) (solid line), f_{PAR} (long dash), and the ratio of actual to potential evapotranspiration (ET/PET) (dotted).

TABLE 17.1

Ecoclimatic Conditions of Select Regions Across Africa

Climatic Measure	Sahelian	Sudanian	Ethiopia	Tanzania	Southeast Africa	Namibia
MA PPT	0.46	1.19	1.24	1.10	0.90	0.44
IAV PPT	0.11	0.22	0.24	0.28	0.25	0.14
MA ET	0.43	0.81	0.75	0.93	0.76	0.44
ET/PPT	0.93	0.73	0.68	0.86	0.87	1.00
MA P_{net}	0.76	2.13	1.84	1.89	1.52	0.59
IAV P_{net}	0.17	0.14	0.15	0.30	0.30	0.24
MA water stress	0.22	0.09	0.13	0.15	0.19	0.31
IAV water stress	0.11	0.08	0.08	0.24	0.23	0.14
MA f_{PAR} limitation	0.51	0.42	0.39	0.38	0.38	0.41
IAV f_{PAR} limitation	0.01	0.03	0.03	0.01	0.01	0.01
IAV P_{net}/IAV PPT	1.60	0.63	0.61	1.11	1.24	1.70
MA water stress/ MA PPT	0.47	0.08	0.10	0.13	0.21	0.70
IAV water stress/ IAV PPT	1.03	0.38	0.34	0.89	0.94	1.02

MA = mean annual; IAV = interannual variability; PPT = precipitation (m year^{-1}); ET = evapotranspiration (m year^{-1}); P_{net} = net photosynthesis (kg C m^{-2} year^{-1}); "Water stress" refers to the ratio of photosynthetic limitation by water stress relative to potential photosynthesis (–); "f_{PAR} limitation" refers to the ratio of photosynthetic limitation by low f_{PAR} relative to potential photosynthesis (–).

sizeable hydrologic surplus, particularly during the late dry season to early growing season (March to May) when the only gradual rise in f_{PAR} restricts evapotranspiration and photosynthesis from responding fully to the high evaporative demand and seasonally high potential rates of photosynthesis. In addition, peak seasonal rainfall in the Sudanian zone is wetter than in the southern regions. Consistent with this picture, evapotranspiration consumes a smaller fraction of annual rainfall leaving about 27% to runoff compared to the ~14% that runs off in Tanzania or southeast Africa (Table 17.1). The lagged rise in f_{PAR} with the arrival of rains from late April to May (Figure 17.4) is likely linked to soaring air temperature and low humidity typical of these months in the Sudanian belt.

In contrast, f_{PAR} of the southern African regions rises more rapidly and declines more gradually with the evolution of wet and dry season transitions. This prolongs the period over which southern African vegetation is active during transition into the dry season, evident during April to July in southeast Africa (Figure 17.4). Correspondingly, water stress assumes greater importance in limiting photosynthesis and increases its year-to-year variation in these southern drylands compared to those of the Sudanian zone. In sum, the larger seasonal amplitude of wetness/dryness, combined with uniquely low f_{PAR}, dampens the Sudano-Sahelian variability in photosynthesis as it responds to interannual variability in rainfall.

Ethiopian Highlands

The final regional focus examines a domain centered over Ethiopia where mean annual rainfall is fairly high, as is its variability (Table 17.1). The region's evaporative index (ET/PPT) is relatively low, leaving about 32% of annual rainfall to runoff according to the model. While mean annual water stress is comparable to that experienced by vegetation in Tanzania, its interannual variability is one-fourth to one-half as large. Similar to that reported above for the Sudanian zone, this stems from relatively low vegetation cover and activity during the early wet season (March to May) (Figure 17.4). In addition, the Ethiopia region exhibits a long lag to peak f_{PAR}, occurring only in August or September with the second seasonal peak in rains, some three months after the year's first wet month of May. Thus, despite high interannual variability in rainfall for much of the Ethiopian region, little of this translates into year-to-year fluctuations in photosynthesis.

Ecoclimatic Controls on Coupling to Climate Oscillations

We now demonstrate how understanding regional ecoclimatological conditions provides insights into regional carbon-cycle connections to climate oscillations such as ENSO and IOD. Beginning with Sahelian teleconnections, we see positive responses of rainfall to ENSO matched by sizeable anomalies in photosynthesis (Figure 17.3, plus many meaningful responses missed by the coarse discretization imposed by a grayscale colorbar). In contrast, the negative association of rainfall with IOD is coincident with little to no response in photosynthesis. Examining the seasonal distribution of these anomalies (Figure 17.5), we find that the seasonal timing of rainfall stimulation is similar, occurring from July to August. So why then does the model simulate negative photosynthesis-ENSO correlation but not positive photosynthesis–IOD correlation?

Resolution appears in f_{PAR} and ET associations. For ENSO, f_{PAR} responds like rainfall with elevation in August to December and translating into elevated ET and photosynthesis. IOD fails to excite f_{PAR} and ET responses, indicating that additional rainfall at this time of year only enhances modeled photosynthesis if commensurate with increased f_{PAR}. Why does f_{PAR} respond to growing season rainfall anomalies stimulated by ENSO but not IOD? The answer may be linked to magnitude, with larger rainfall enhancement during ENSO. Additionally, in the months leading up to and during the growing season, ENSO but not IOD stimulates more favorable air temperature and relative humidity during the high rainfall phase, and the reverse during the low rainfall phase. Hence, ENSO may "prime the pump" for vegetation responses during the rainy season.

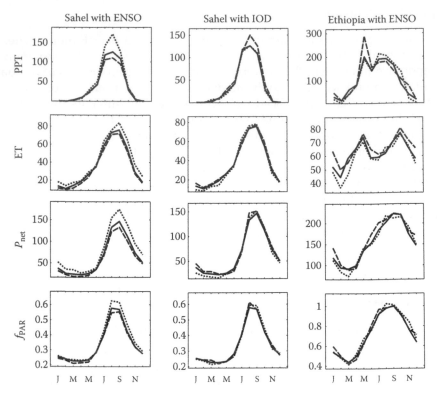

FIGURE 17.5
Monthly average, composite regional series averaged for all years (black), for the lowest three ENSO or IOD excursions (dotted) and highest three ENSO or IOD excursions (long dashed) shown for rainfall (PPT, mm month^{-1}), evapotranspiration (ET, mm month^{-1}), net photosynthesis (P_{net} g C m^{-2} month^{-1}), and fractional absorbance of PAR (f_{PAR}).

As a second example, we return to Figure 17.3 and note that the Ethiopian region experiences positive rainfall anomalies in association with the multivariate ENSO index but that photosynthesis lacks corresponding association. This disconnection is explained by inspection of seasonal distribution and prevailing conditions. Most of the additional rainfall coming with ENSO arrives in a single month (May), early in the growing season when vegetation is just beginning to become seasonally active. The month already experiences sizeable hydrologic surplus, with rainfall of 200 mm month^{-1} grossly exceeding evapotranspiration of only 75 mm month^{-1}. Hence, the sizeable ENSO-induced rainfall anomaly in May does little to enhance photosynthesis during the month, and only slightly enhances f_{PAR} and photosynthesis in the ensuing 2 months. This is consistent with the region's relatively low water stress and high runoff ratio of 0.32 (= 1 − ET/PPT, Table 17.1). The few pixels in this region that do show sizeable positive responses to ENSO are located in the polygon's dry margins to the north and south where water stress is more prevalent (Figure 17.3a).

Continuing Study and Emerging Opportunities

This final section mentions some particularly inviting opportunities for exploration. One of the first questions is to what degree is radiation limiting and how does this influence interannual variability. Although not explored here, the existing framework could easily be modified to also measure this aspect of plant ecophysiology that may be important in understanding global carbon cycle variability.

It would also be interesting to determine causes of seasonal lags in f_{PAR} responses to rainfall as noted for the Sudano-Sahelian and Ethiopian regions in comparison with those in southern Africa. For example, a satellite-derived rain use efficiency (f_{PAR}/rainfall) may be instructive, particularly when analyzed in the context of potential evapotranspiration, air temperature, and humidity. In addition to these potential climatological drivers of phenology, connection of f_{PAR} to land use, such as agricultural practices, should be examined but at a much finer resolution than that explored here. If successful, such analyses could help write new rules to be incorporated in dynamic global vegetation models to capture regional hydroclimatic controls on vegetation dynamics, as well as to better incorporate human dimensions of global environmental change.

For example, the lower PAR absorption in the Sudano-Sahelian zone compared to southern woodlands and savannas may be an expression of intense land use pressures (grazing, wood harvest, agriculture) well documented for the Sudano-Sahelian zone. Alternatively, the difference may be an ecoclimatic phenomenon; several recent studies have documented broad increases in satellite vegetation indices in the Sahel and Sudanian zones (Herrmann et al., 2005; Nicholson et al., 1998; Olsson et al., 2005; Prince et al., 2007) and increasing woody cover following the droughts of the 1970s and 1980s. Thus our inferred f_{PAR} limitation may reflect historical drought suppression of vegetation that has started to recover in later years. A third possibility is that mean climatic differences are partially or wholly responsible for regional differences in photosynthetic limitations, with the hot, dry April and May period in the Sudano-Sahelian zone potentially conditioning vegetation to delay activation of dormant buds until the wet season has fully developed, in order to avoid risks associated with growth in response to early sporadic rainfall.

Widespread observations are needed to determine if the meteorology or vegetation prescribed for the region is severely flawed. Process representations in the model may also be inadequate (or incorrect), calling for further model testing with land surface flux and state observations such as that provided with the eddy covariance technique. Prescription of absorbed PAR could be evaluated with new or existing vegetation surveys or with remotely sensed vegetation cover products that are independent of the AVHRR NDVI dataset used here. Leaf- and canopy-scale observations of ecophysiology throughout the region should provide the data needed to test regional application of the model's global, biome-scale parameterization of physiological stresses.

Up to now, this work has yet to mention fires, which are a critical aspect of Africa's carbon cycle, leading all other continents in annual fire emissions but with relatively small interannual variability (van der Werf et al., 2006). To what degree does seasonal burning respond to remote climate signals? Does it perhaps dampen anomalies of annual net biome productivity, by combusting more biomass in high-productivity years and less in low-productivity years, thus offsetting carbon sources/sinks from annual photosynthesis anomalies? Remotely sensed burned area and active fire detection (Barbosa et al., 1999; Giglio et al., 2000; Justice et al., 2002) could be readily explored for subcontinental analysis of unique regional patterns of variability in connection with climate and climate oscillations just as has been done here with f_{PAR} and photosynthesis.

Finally, some attention should be given to the study of interactions between ENSO and IOD forcings of African ecosystems. Analysis of these interactions (Williams and Hanan, unpublished data) has revealed the first indications of dominance by one or the other varying geographically across the continent. If famine and disease forecasts are to rely on such indices, these interactions must be better understood.

References

Anyamba, A. and J. R. Eastman. 1996. Interannual variability of NDVI over Africa and its relation to El Nino Southern Oscillation. *International Journal of Remote Sensing* 17, 2533–2548.

Anyamba. A. and C. J. Tucker. 2005. Analysis of Sahelian vegetation dynamics using NOAA-AVHRR NDVI data from 1981–2003. *Journal of Arid Environments* 63, 596–614.

Ball M. C., I. R. Cowan, and G. D. Farquhar. 1988. Maintenance of leaf temperature and the optimization of carbon gain in relation to water-loss in a tropical mangrove forest. *Australian Journal of Plant Physiology* 15, 263–276.

Barbosa P. M., D. Stroppiana, and J-M. Gregoire. 1999. An assessment of vegetation fire in Africa (1981–1991): Burned areas, burned biomass, and atmospheric emissions. *Global Biogeochemical Cycles* 13, 933–950.

Behera, S. K. and T. Yamagata. 2003. Influence of the Indian Ocean Dipole on the southern oscillation. *Journal of the Meteorological Society of Japan* 8, 169–177.

Black E, J. M. Slingo, and K. R. Sperber. 2003. An observational study of the relationship between excessively strong short rains in coastal East Africa and Indian Ocean SST. *Monthly Weather Review* 131, 74–94.

Charney J. G., P. H. Stone, and W. J. Quirk. 1975. Drought in the Sahara: A biogeophysical feedback mechanism. *Science* 187, 434–435.

Churkina. G. and S. W. Running. 1998. Contrasting climatic controls on the estimated productivity of global terrestrial biomes. *Ecosystems* 1, 206–215.

Ciais, P., P. Friedlingstein, A. Friend, and D. S. Schimel. 2001. Integrating global models of terrestrial primary productivity. In *Terrestrial Global Productivity*, ed. J. Roy, B. Saugier, and H. A. Mooney. San Diego, Academic Press, pp. 449–478.

Collatz, G. J., J. T. Ball, C. Grivet, and J. A. Berry. 1991. Physiological and environmental-regulation of stomatal conductance, photosynthesis and transpiration—A model that includes a laminar boundary-layer. *Agricultural and Forest Meteorology* 54, 107–136.

Collatz, G. J., M. Ribas-Carbo, and J. Berry. 1992. Coupled photosynthesis-stomatal conductance model for leaves of C4 plants. *Australian Journal of Botany* 19, 519–538.

Ellis, J. E. and D. M. Swift. 1988. Stability of African pastoral ecosystems: Alternate paradigms and implications for development. *Journal of Range Management* 41, 450–459.

Farquhar G. D., S. V. Caemmerer, and J. A. Berry. 1980. A biochemical-model of photosynthetic CO_2 assimilation in leaves of C-3 species. *Planta* 149, 78–90.

Giglio, L., J. D. Kendall, and C. J Tucker 2000. Remote sensing of fires with the TRMM VIRS. *International Journal of Remote Sensing* 21, 203–207.

Herrmann, S. M., A. Anyamba, and C. J. Tucker. 2005. Recent trends in vegetation dynamics in the African Sahel and their relationship to climate. *Global Environmental Change-Human and Policy Dimensions* 15, 394–404.

IPCC. 2001 Climate Change 2001: Working Group 2: Impacts, Adaptation, and Vulnerability. In *Third Assessment Report for the Intergovernmental Panel on Climate Change*, Ch 10: *Africa*. Cambridge University Press, New York.

Justice, C. O., L. Giglio, and S. Korontzi et al. 2002. The MODIS fire products. *Remote Sensing of Environment* 83, 244–262.

Koster, R. D., Z. Guo, P. A. Dirmeyer et al. 2006. GLACE: The Global land–atmosphere coupling experiment. Part I: Overview. *Journal of Hydrometeorology* 7, 590–610.

Los S. O., N. H. Pollack, M. T. Parris et al. 2000. A global 9-yr biophysical land surface dataset from NOAA AVHRR data. *Journal of Hydrometeorology* 1, 183–199.

Nemani, R. R., C. D. Keeling, H. Hashimoto et al. 2003. Climate-driven increases in global terrestrial net primary production from 1982 to 1999. *Science* 300, 1560–1563.

Nicholson, S. E. 2000. The nature of rainfall variability over Africa on time scales of decades to millenia. *Global and Planetary Change* 26, 137–158.

Nicholson, S. E. 2001. Climatic and environmental change in Africa during the last two centuries. *Climate Research* 17, 123–144.

Nicholson, S. E. and J. Kim. 1997. Relationship of ENSO to African rainfall. *International Journal of Climatology* 17, 117–135.

Nicholson, S. E., C. J. Tucker, and M. B. Ba. 1998. Desertification, drought, and surface vegetation: An example from the West African Sahel. *Bulletin of the American Meteorological Society* 79, 815–829.

Olsson, L., L. Eklundh, and J. Ardo. 2005. A recent greening of the Sahel—trends, patterns and potential causes. *Journal of Arid Environments* 63, 556–566.

Patra, P. K., M. Ishizawa, S. Maksyutov, T. Nakazawa, and G. Inoue. 2005. Role of biomass burning and climate anomalies for land-atmosphere carbon fluxes based on inverse modeling of atmospheric CO_2. *Global Biogeochemical Cycles* 19, GB3005, doi:10.1029/2004GB002258.

Prince, S. D., K. J. Wessels, C. J. Tucker, and S. E. Nicholson. 2007. Desertification in the Sahel: A reinterpretation of a reinterpretation. *Global Change Biology* 13, 1308–1313.

Reason, C. J. C. and M. Rouault. 2002. ENSO-like decadal variability and South African rainfall. *Geophysical Research Letters* 29, 1638–1642.

Schaefer, K., A. S. Denning, N. Suits, J. Kaduk, I. Baker, S. Los, and L. Prihodko. 2002. Effect of climate on interannual variability of terrestrial CO_2 fluxes. *Global Biogeochemical Cycles* 16, 1102, doi: 10.1029/2002GB001928.

Sellers, P. J., J. A. Berry, G. J. Collatz, C. B. Field, and F. G. Hall. 1992. Canopy reflectance, photosynthesis, and transpiration 3. A reanalysis using improved leaf models and a new canopy integration scheme. *Remote Sensing of Environment* 42, 187–216.

Sellers, P. J., L. Bounoua, G. J. Collatz et al. 1996a. Comparison of radiative and physiological effects of doubled atmospheric CO_2 on climate. *Science* 271, 1402–1406.

Sellers, P. J., S. O. Los, C. J. Tucker et al. 1996b. A revised land surface parameterization (SiB2) for atmospheric GCMs 1. Model formulation. *Journal of Climate* 9, 676–705.

Sellers, P. J., S. O. Los, C. J. Tucker et al. 1996c. A revised land surface parameterization (SiB2) for Atmospheric GCMs. Part II: The generation of global fields of terrestrial biophysical parameters from satellite data. *Journal of Climate* 9, 706–737.

Sellers, P. J., Y. Mintz, Y. C. Sud, and A. Dalcher. 1986. The design of a simple biosphere model (SiB) for use within general circulation models. *Journal of Atmospheric Science* 43, 505–531.

Tyson, P. D., G. R. J. Cooper, and T. S. McCarthy. 2002. Millennial to multi-decadal variability in the climate of southern Africa. *International Journal of Climatology* 22, 1105–1117.

UNCTAD. 2002. The Least Developed Countries Report 2002: Escaping the poverty trap. In *United Nations Conference on Trade and Development Secretariat*. United Nations, Geneva, Switzerland.

van der Werf, G. R., J. T. Randerson, L. Giglio, G. J. Collatz, P. S. Kasibhatla, and A. F. Arellano. 2006. Interannual variability in global biomass burning emissions from 1997 to 2004. *Atmospheric Chemistry and Physics* 6, 3423–3441.

Walker, B. H., D. Ludwig, C. S. Holling, and R. M. Peterman. 1981. Stability of semi-arid savanna grazing systems. *Journal of Ecology* 69, 473–498.

Wang, G. L. and E. A. B. Eltahir. 2000a. Biosphere–atmosphere interactions over West Africa. II: Multiple climate equilibria. *Quarterly Journal of the Royal Meteorological Society* 126, 1261–1280.

Wang, G. L. and E. A. B. Eltahir. 2000b. Ecosystem dynamics and the Sahel drought. *Geophysical Research Letters* 27, 795–798.

Westoby, M., B. Walker, and I. Noy-Meir. 1989. Range management on the basis of a model which does not seek to establish equilibrium. *Journal of Arid Environments* 17, 235–239.

Williams, C. A. and J. D. Albertson. 2006. Dynamical effects of the statistical structure of annual rainfall on dryland vegetation. *Global Change Biology* 12, 1–16.

Williams, C. A., N. P. Hanan, I. Baker, G. J. Collatz, J. Berry, and A. S. Denning. 2008. Interannual variability of photosynthesis across Africa and its attribution. *Journal of Geophysical Research* 113, G04015 doi:10.1029/2008JG000718.

Williams, C. A., N. P. Hanan, J. C. Neff et al. 2007. Africa and the global carbon cycle. *Carbon Balance and Management* 2, 1–13.

Zeng, N. and J. D. Neelin. 2000. The role of vegetation–climate interaction and interannual variability in shaping the African savanna. *Journal of Climate* 13, 2665–2670.

18

Timescales and Dynamics of Carbon in Australia's Savannas

Damian J. Barrett

CONTENTS

Introduction

Variations in the exchange of CO_2 between the land surface and atmosphere occur at all spatiotemporal scales. Routine methods now exist for the direct observation of hourly, daily, seasonal, and annual CO_2 exchange by photosynthesis, respiration, and decomposition using eddy covariance methods. Further, extensive flux tower networks exist that can provide time series of measurements for up to 10 years (Papale et al., 2006). However, variations in CO_2 exchange on multi-annual to decadal timescales over extensive regions, such as the 2×10^6 km^2 of Australian savannas, are much less well known. On these longer timescales, the response by the C cycle to disturbance (e.g., fire) or to slow climate cycles is poorly characterized, and their uncertainties are virtually unknown. Quantifying these long timescale variations in CO_2 exchange cannot be achieved through measurement alone. A combined approach, called model-data fusion (Barrett et al., 2005; Raupach et al., 2005), is required in which models are used to provide a theoretical framework for interpreting observations. These models need to be comprehensive, cover all significant pools and fluxes, and be simple enough to be constrained by limited available data. Over the last decade, the global change science community has turned its attention to the extensive use of model-data fusion (or model-data assimilation) techniques to better characterize the terrestrial carbon cycle. Rapid uptake of these methods has seen improved modeling capability in many disciplines, including hydrology, remote sensing, ecophysiology, and terrestrial biogeochemistry (Barrett et al., 2008; Renzullo et al., 2008; Barrett and Renzullo, 2009).

Global savannas occupy ~15% of total land area; contain 13% of global carbon stocks, account for ~30% of global NPP; and are highly susceptible to degradation by cultivation, grazing, mining, and bio-fuel harvesting (Grace et al., 2006). Little is known about the long-term impacts of climate change and altered disturbance regimes on savanna C fluxes, which, due to their extensive area and the dominant role of fire, have an important impact on global atmospheric CO_2 concentrations (Stephens et al., 2007). For these reasons, it is important to develop improved modeling approaches, such as model-data fusion methods, and use them to integrate disparate observations with land surface biophysical models. This will enable an improved understanding of the dominant ecosystem functions and environmental services they provide, better characterization of the variations in CO_2 exchange on all timescales, and more accurate prediction of future impacts of climate change and human management on this biome.

In a previous study, Barrett (2002) developed a continental scale model of carbon pools and fluxes for Australia (VAST1.1) that utilized a diverse array of ground-based observations of plant production, biomass, litter, and soil carbon pools and applied a multiple constraints model-data fusion scheme to estimate model parameters (Barrett et al., 2005). In this chapter, the original VAST model has been modified to better characterize the dynamics of carbon for the Australian continent. With this new model parameterization, the underlying relationships between climate forcing and plant production and decomposition are explored for Australian savannas to characterize the carbon dynamics, quantify the uncertainties in carbon pools and fluxes, and improve understanding of the key processes controlling the storage and release of carbon.

Model-Data Fusion

Model Structure

VAST1.2 is a continental scale dynamical model of net ecosystem production and carbon pools that spans all major biomes of the Australian continent (Barrett, 2002). In VAST, biosphere C is discretized into vegetation, litter, and soil carbon pools to 1.0 m depth (Figure 18.1). Each pool represents mean C-mass density (t_C ha^{-1}) of a single grid cell at 0.05° resolution for monthly time steps between January 1981 and December 2000. First-order rate equations control plant growth, mortality, litterfall, soil organic matter accumulation, decomposition, and emissions from biomass burning. Model complexity is commensurate with the type and quantity of ground data available for parameterization. For this work, the original model was improved in three ways: (1) bootstrap methods were used to quantify uncertainties in estimated parameters and state variables; (2) model equations were modified to calculate emissions due to biomass burning based on satellite observations of fire hotspots and fire scars; and (3) a dynamical version of the model was parameterized to account for non-linearities in the relationships between soil

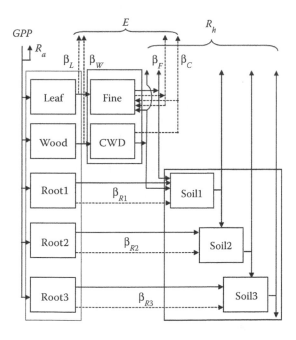

FIGURE 18.1
Schematic of VAST1.2 model showing primary fluxes of carbon determining net ecosystem productivity (NEP = NPP − R_h; solid lines) and net biome productivity (NBP = NPP − R_h − E; dotted lines). Model equations are given in Appendix and parameters are defined in Table 18.1.

moisture and plant production, canopy growth, and decomposition. The model contains 22 estimated parameters (Table 18.1).

Forcing Data: Normalized Difference Vegetation Index

The fraction of canopy-absorbed radiation was estimated from the Global Area Coverage NDVI at 8 km resolution obtained from the AVHRR on board the NOAA 7, 9, 11, and 14 polar orbiting satellites (Ichii et al., 2001). Systematic errors in these data were removed using an empirical correction procedure (Ichii et al., 2001; Lovell and Graetz, 2001) to account for sensor degradation, drift in equatorial crossing time, seasonal variation in solar zenith angle,

TABLE 18.1

Definitions of Parameters and Their Units in VAST1.2

Parameter Symbol	Units	Definition
Timescale Parameters of Carbon Pools		
τ_L	Years	Turnover time of C in live leaf pool
τ_W	Years	Turnover time of C in live wood pool
τ_{Rj}	Years	Turnover time of C in fine root pool for the jth soil layer
τ^*_k	Years	Moisture and temperature corrected turnover time of C in kth pool
		$k = F$ (fine littermass), C (coarse littermass), S_j (C in jth soil layer)
Partitioning Parameters of NPP Among Carbon Pools		
α_k	—	Allocation of net primary production (NPP) to the kth plant tissues:
		$k = L$ (leaf), W (stem), R (fine root)
ζ_j	—	Proportion of root allocated NPP to the jth soil layer:
		$j = 1$ (0–20 cm), 2 (20–50 cm), 3 (50–100 cm).
θ_F	—	Carbon partitioning coefficient of the fine litter pool
θ_C	—	Carbon partitioning coefficient of the coarse woody debris pool
θ_{Sj}	—	Partitioning coefficient of soil C in the jth soil layer
η	—	Fraction of C exiting the coarse woody debris pool by mechanical breakdown
Other Parameters		
ε^*	$gC\ MJ_{PAR}^{-1}$	Maximum light use efficiency
ρ	—	Ratio of total respiration to total photosynthesis per month
γ	—	Fraction of aboveground biomass as coarse woody debris

and variation in aerosols and water vapor concentration of the atmosphere. The NDVI data were not available for 10 months in 1994, and so all data for that year were omitted and replaced with the monthly mean NDVI from 1981 to 1993.

Forcing Data: Fire

Two satellite fire products were used to force the model. The ATSR-ERS2 World Fire Atlas Project product (Mota et al., 2006; http://dup.esrin.esa.it/ionia/wfa/index.asp) provided a 1 km resolution data set since 1995 to the present day using calibrated and georeferenced nighttime imagery. The ATSR fire algorithm 1 product was used, which detects fire hotspots when sensor thermal channels exceed a surface radiative temperature threshold of 312 K. The second product was the 0.01° resolution AVHRR fire scar product available from the FireWatch project (McMillan et al., 1997; http://www.landgate.wa.gov.au). This fire scar product is derived from NOAA-AVHRR data and provides information on burnt area based on a color detection algorithm using NOAA-12 and NOAA-14 data, which takes into account differences in reflectivity of different soil types. A fraction of burnt area at 0.05° resolution was generated from these two satellite fire products.

Forcing Data: Climate

Climate forcing data were obtained from a variety of sources at different resolutions and were interpolated to a common 0.05° resolution grid for use in VAST1.2. Soil water balance was determined by numerically solving a set of simple water balance equations using rainfall and temperature data, net radiation, and soil hydraulic characteristics, as described in Barrett (2002).

Model Parameterization and Forward Modeling

Model parameterization followed previous methods (Barrett, 2002; Barrett et al., 2005) but with an additional step that used bootstrap methods to quantify parameter uncertainty. First, multiple realizations of observations were sampled from an existing data base of plant, litter, and soil observations (Barrett, 2001). Then, parameters were estimated using a multiple constraints approach in which a multiobjective cost function was minimised using a genetic algorithm search function. This process was repeated for each realization of observations. In total, 20 realizations were generated for each of three vegetation classes (Tall Eucalypt Forest, Open Woodlands, and Arid Shrublands), resulting in a total of 60 parameter vectors being produced by the model-data fusion scheme. The uncertainty in estimated parameters was represented by the range of variation seen in each of the parameters. In addition, individual parameter vectors retained the correlation structure among parameters induced by data and the model.

Forward modeling of NBP, NPP, NEP, and C pools was conducted by drawing a single parameter vector at random from the set of 20 vectors per vegetation class and running the model forward for 20 years. This process was repeated 100 times to generate means and confidence intervals of predicted C pools and fluxes based on uncertainties in the parameters. Burning efficiency parameters were derived from published empirical studies (Table 18.2) and applied to each carbon pool in the model forward runs. Before the forward modeling, the model was spun-up to a stationary mean for all pools by repeating the 20-year simulation period 80 times, thus generating 1600 years of model run before the final simulations. For NBP, average monthly fire frequencies were used for model forcing until 1997, when actual monthly data were available. Only results between 1997 and 2000 for NBP are reported here.

Timescales of Carbon Fluxes

Mean age and transit time of carbon was calculated using impulse-response analysis to characterize the timescales of carbon in the biosphere. Mean age of C, τ_a, is defined as the average age of all C molecules in all pools of the biosphere. Transit time of C, τ_t, is the average time period between entry and exit of a molecule of C through the biosphere. These parameters were estimated by perturbing the model using a Dirac Delta function and then calculating the age distribution functions of the respired flux, $\phi(t)$, and of biosphere

TABLE 18.2

Burning Efficiency Parameters Used in VAST1.2 to Estimate Carbon Emission from Fire Adapted from Published Empirical Studies

Vegetation Class	Carbon Pool			
	Leaf	Wood	Fine Litter	Coarse Litter
Tall Forests	0.33[a,k]	0.27[a,b,c,f,g,j,l,n]	0.93[a,b,c,f,g,j,l,n]	0.37[a,b,c,f,g,j,l,n]
Open Woodlands	0.84[d,m]	0.33[d,i,m]	0.93[d,i,j,l,m,n]	0.40[a,b,c,f,g,j,l,n]
Arid Shrublands	0.92[d,m]	0.54[d,m]	0.98[o]	0.60[o]

[a] Araújo et al. (1999).
[b] Carvalho et al. (1998).
[c] Carvalho et al. (2001).
[d] De Castro et al. (1998).
[e] Fearnside et al. (2001).
[f] Fearnside et al. (1993).
[g] Fearnside et al. (1999).
[h] Guild et al. (1998).
[i] Graca et al. (1999).
[j] Hoffa et al. (1999).
[k] Prasad et al. (2001).
[l] Scholes (1996).
[m] Shea et al. (1996).
[n] Van der werf et al. (2003).
[o] Assumed value.

pools, $\psi(t)$, respectively. Transit time of C was then obtained from the first moment of $\phi(t)$,

$$\tau_t = \int_0^\infty t\phi(t)\mathrm{d}t \cdot (\mathrm{n})$$

and mean age of C from,

$$\tau_a = \int_0^\infty t\psi(t)\mathrm{d}t \cdot (\mathrm{m})$$

Model Parameters

Parameter Values and Uncertainties

Examination of the confidence intervals estimated from the coupled bootstrap resampling and multiple constraints model-data fusion method (Table 18.3) showed that most parameters were not significantly different among the three vegetation classes (based on a t-test at 95% confidence). Exceptions were turnover times of leaf tissues, fine and coarse litter, and soil carbon at 50–100 cm. Light use efficiency was significantly greater, and leaf and fine litter turnover times were significantly shorter for tall forests than for open woodlands and arid shrublands, consistent with greater leaf longevity and slower fine litter decomposition as aridity increased. In addition, coarse woody debris turnover times were significantly longer in tall forests, presumably due to larger log diameters slowing decomposition in these forests. Model parameterization corrects for moisture and temperature effects on C turnover by decomposition, and so significant differences in estimated turnover times in Table 18.3 reflect differences in plant tissue nutrient composition or soil texture effects on decomposition rather than climatic effects.

Australian savannas extend from high productivity coastal, high rainfall forests at low latitudes to inland sparse open forests intermingling with arid shrublands under low rainfall conditions. The majority of the savanna area in Australia coincided with the Open Woodland class of VAST1.2 (Table 18.3). This class is characterized by relatively high light use efficiency compared with most temperate and boreal forests having values <0.5 g_C MJ^{-1} absorbed photosynthetic active radiation (APAR). The majority of carbon fixed by photosynthesis was allocated between leaf tissues (grass and overstorey canopy) and relatively long-lived woody tissues (overstorey vegetation). Turnover of carbon in the fine and coarse litter pools was between two and three years, and soil carbon at depths between 50 and 100 cm is retained on century timescales.

TABLE 18.3

Summary Statistics of Mean and Standard Error for Estimated Parameters in VAST1.2 Derived Using the Multiple Constraints Method Repeatedly with Multiple Realizations of NPP, Plant, Litter, and Soil Carbon Observations ($n = 20$ per Vegetation Class)

Parameter	Vegetation Class		
	Tall Forest	Open Woodland	Arid Shrubland
ε^*	1.36[a] (0.06)	0.85[b] (0.05)	0.27[c] (0.15)
ρ	0.55	0.45	0.45
α_L	0.59[a] (0.02)	0.71[b] (0.02)	0.74[b] (0.10)
α_W	0.22[a] (0.01)	0.25[a] (0.02)	0.05[b] (0.02)
α_R	0.20[a] (0.03)	0.04[b] (0.01)	0.21[a] (0.07)
γ	0.30	0.25	0.1
ζ_1	0.50 (0.02)	0.48 (0.02)	0.50 (0.04)
ζ_2	0.42 (0.01)	0.43 (0.02)	0.39 (0.03)
ζ_3	0.08 (0.01)	0.09 (0.01)	0.12 (0.01)
θ_F	0.21 (0.04)	0.21 (0.03)	0.23 (0.04)
θ_C	0.24 (0.02)	0.24 (0.02)	0.25 (0.02)
η	0.05 (0.01)	0.05 (0.00)	0.05 (0.01)
τ_L	1.74[a] (0.20)	3.19[b] (0.42)	4.76[b] (0.56)
τ_W	162.27 (14.47)	215.23 (15.64)	224.02 (23.45)
τ_{R1}	5.79 (0.57)	4.98 (0.36)	4.98 (0.42)
τ_{R2}	11.42 (1.30)	10.02 (0.76)	10.48 (0.82)
τ_{R3}	31.63 (1.97)	25.02 (2.08)	24.48 (1.83)
τ_F^*	1.19[a] (0.08)	2.04[b] (0.24)	3.18[b] (1.00)
τ_C^*	10.05[a] (1.09)	2.59[b] (0.34)	4.20[b] (0.92)
τ_{S1}^*	39.58 (4.50)	40.57 (4.00)	34.30 (13.00)
τ_{S2}^*	94.48 (20.08)	111.42 (11.49)	101.57 (24.07)
τ_{S3}^*	123.08[a] (21.66)	259.76[b] (26.53)	221.29[b] (41.39)
θ_{S1}	0.05 (0.01)	0.05 (0.01)	0.05 (0.01)
θ_{S2}	0.05 (0.01)	0.04 (0.01)	0.05 (0.01)

Note: Values without standard errors in parentheses were fixed during model parameterizations. Superscripts show significantly different parameter estimates among vegetation classes at 95% confidence interval (*t*-test). Absence of superscripts indicates no significant difference.

Forward Modeling

Savanna Fires

Satellite observations of fire extent from the fire scar and hotspot products were compared with on-ground mapping of fire in Western Australia and Northern Territory for those years where data overlapped (Graetz, 2002

TABLE 18.4

Comparison of on-Ground Assessment of Burnt Area with Burnt Area Estimates from AVHRR Fire Scar and ATSR Hotspot Products

Year	Ground-Based Burnt Area (Mha)	AVHRR (Mha)	ATSR (Mha)	Total Satellite (Mha)	Ratio
1996–97	32.1	18.6	0.3	18.9	0.59
1997–98	42.9	28.6	0.6	29.2	0.68
1998–99	25.0	18.0	0.3	18.3	0.73
1999–00	50.5	39.5	0.6	40.1	0.79

Note: On-ground estimates were aggregated over Western Australia and Northern Territory from Graetz (2002) and Meyer (Personal Communication). Summations were performed for the year ending March 30, 1996 to 2000 to align with fire season presented in Graetz (2002). Ratio compares total satellite product with ground-based burnt area mapping.

and Meyer, Personal Communication). This comparison provides an overall measure of gross detection accuracy of fire for the two different products (Table 18.4). Comparisons showed that an underestimate in detected fire area relative to ground data was evident, ranging from 21% to 41%. The relative distribution of fire among vegetation classes (Table 18.5) showed that both underestimation and overestimation of fire area occurred. Burnt area was underestimated by both satellite methods in shrubland communities, particularly Tall Shrublands, but it was consistently overestimated in woodlands. Russell-Smith et al. (2009) showed that Landsat detection of fire scars in Western Arnhem Land, Northern Territory, was overestimated by 29% in early season fires and by 11% in late season fires due to burn patchiness. Underestimation of fire area in arid shrublands is likely due to poor detection by satellites. Overestimation of fire in woodlands is likely due to false classification by satellites, patchiness, and difficulty in mapping of fire in dissected and forested terrain.

An analysis of the coincidence of AVHRR fire scars and ATSR hotspots within 0.05° grid cells for Northern Territory and Western Australia indicated

TABLE 18.5

Proportional Distribution of Burnt Area among Australian Survey and Land Information Group (AUSLIG) Vegetation Types for On-Ground Observations, AVHRR Fire Scar, and ATSR Hotspot Products Summed over Western Australia and Northern Territory for the Years 1996–2000

Vegetation Class	On Ground (%)	AVHRR Scar (%)	ATSR Hotspot (%)
Low Open Woodland	25.9	36.0	33.5
Woodland	9.2	15.1	16.4
Low Woodland	5.3	13.6	10.4
Tall Open Shrubland	27.3	22.5	21.6
Tall Shrubland	20.6	5.3	6.8
Other	11.6	7.4	11.3

TABLE 18.6

Estimated Carbon Emission from Biomass Burning for Australian Savannas for 1996–2000 (Calendar Year)

Year	Emissions (Tg_C year^{-1})	NPP(%)	Bias Corrected Emissions (Tg_C year^{-1})
1997	96.9 (1.2)	48	130.8
1998	81.5 (1.6)	38	110.1
1999	148.5 (3.6)	58	200.5
2000	110.9 (3.3)	51	149.7
Mean	109.5	49	147.8

Note: Emissions based on model output. Bias corrected emissions are a pro-rata increase in emissions to account for the ~35% underestimate in burnt area relative to on-ground observations (Table 18.4).

that only a 4% overlap occurred between these two products (data not presented). In addition, 93% of fire scars were detected without hotspots, and 3% of hotspots were detected without fire scars. These differences in fire detection are due to differences between sensor type (visible versus thermal), the processing algorithm (temperature threshold versus color change), and overpass time (morning versus nighttime) and result in different types of fire being detected in different parts of the landscape. The AVHRR fire scar product will preferentially detect large, open, and fast-moving fires; whereas the ATSR hotspot product will detect hotter fires burning through the night where biomass and fuel loads are greater.

Model estimates of carbon emissions from fire between 1997 and 2000, based on the overlaid hotspot and fire scar data sets, varied by 61% between 81.5 and 148.5 Tg_C year^{-1} around a mean value of 109.5 Tg_C year^{-1} (Table 18.6) with a coefficient of variation (CV; the ratio of standard deviation to mean) of 2.1%; but this uncertainty does not incorporate the uncertainty in burning efficiency parameters that was estimated to lie between 6 and 49% for different fuel types and fire intensities (Russell-Smith et al., 2009). The emissions from fire accounted for between 38 and 58% of savanna NPP in any year and constituted ~96% of Australia's continental emissions from fire. This result is similar to the estimates of Graetz (2002), who calculated a continental burnt area of 37.8 Mha yr^{-1} releasing 109 Tg_C year^{-1} in emissions. However, this apparent good agreement may be coincidental, given that the satellite fire products appeared to underestimate on-ground mapping of fire. With pro rata adjustment for bias, fire emissions could be between 110 and 201 Tg_C year^{-1} (or 52–79% of NPP) with an average of 148 Tg_C year^{-1} (Table 18.4). Globally, an estimated 8.7 million tons of biomass is burned every year (De Groot, 1992), releasing up to 3 ~ 5 Pg_C year^{-1} (Levine, 1991) and consuming between 5% and 10% of global NPP. Based on these figures, burning in Australian savannas contributes between 1.6% and 5.0% of global carbon emissions from fire in any year.

Dynamics of Savanna Carbon Fluxes

The total savanna biosphere carbon pool averaged 27.1 Pg over the 20 years (CV = 4.6%), with plant, litter, and soil pools averaging 9.1, 2.4, and 15.6 Pg_C (CV = 9.5%, 5.1%, and 5.6%), respectively. Thus, just over half of biosphere C was found in the soil, one-third was found in plants, and the remaining carbon was in fine and coarse litter. Variation in the savanna C stock (maximum–minimum/mean) over the 20-year model run represented 0.6% or about 155 Tg_C, which, when distributed among plant, litter, and soil-C pools, represented variations of 3.2%, 10.4%, and 0.1%, respectively.

Variation in total annual NPP, NEP, and net biome production (NBP) of Australia's savannas is presented in Figure 18.2. Annual NPP for this biome over the 20 years varied between 73.8 Tg_C year^{-1} in 1992 and 255.1 Tg_C year^{-1} in 1999, with a coefficient of variation in NPP ranging from 5.0% to 5.4%.

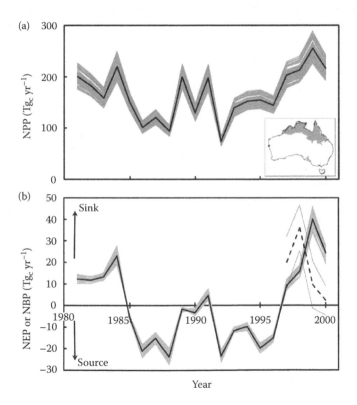

FIGURE 18.2

Variations in annual carbon fluxes for Australian savannas from 1981 to 2000 based on the VAST1.2 model and parameter uncertainties from the multiple constraints model-data fusion using *in situ* observations of plant, litter, and soil carbon pools. (a) shows mean NPP (solid line). (b) shows mean NEP (solid line) and mean NBP (dashed line). Light gray lines show individual model runs. Light gray lines associated with NBP show 95% confidence intervals.

Year-to-year variation in NPP resulted from variation in the duration and intensity of the monsoon. In tropical Australia (latitudes north of 18°S), the South-East Asian monsoon generates 95% of annual rainfall between December and March, producing an NPP maximum in February to March (Hutley et al., 2000). In some years, stronger monsoons penetrate further inland into more semiarid regions of the savannas, leading to peaks in annual NPP and episodes of high growth followed by mortality and decomposition.

Annual NEP ranged from a maximum sink of 39.8 Tg_C year^{-1} in 1999 to a maximum source to the atmosphere of -23.7 Tg_C year^{-1} in 1988 with a coefficient of variation on these estimates of up to 34% (positive values in Figure 18.2 indicate C accumulation by the biosphere). Years where savanna NEP was a net source of carbon generally followed periods of monsoon driven NPP, as the previously assimilated carbon was released back to the atmosphere. Fire in the 5 years till the year 2000 shifted peak C-sink activity back by one year from 1999 to 1998 and decreased the amplitude of NBP to 35.8 Tg_C year^{-1} (CV = 15%; Figure 18.2). The maximum carbon sink here equated to a spatial average of 0.2 t_C ha^{-1} year^{-1} over the 1.9×10^6 km^2 Australian savannas and represents an extended period of sink activity from before the start of 1997 to after the end of 2000 (Figure 18.2). This sink strength is about an order of magnitude less than the short-term maximum sink observed for northern Australian savannas of ~2 t_C ha^{-1} year^{-1} (Williams et al., 2009), because it included the lower productivity semiarid southern extent of the Australian savannas. However, it is of a similar magnitude to the estimated global savanna sink of 0.14 t_C ha^{-1} year^{-1} by Grace et al. (2006). Given a global NEP savanna sink of 0.39 Gt_C year^{-1} (Grace et al., 2006), the Australian savannas at their maximum sink strength contributed ~10%. However, Figure 18.2 shows the transient nature of this sink with 11 of the 20 years between 1981 and 2000 being a source of carbon to the atmosphere. The sum of NEP over the 20 years was a weak sink of 3.4 Tg_C or 0.02 t_C ha^{-1}.

Variation in the timing of seasonal rainfall, temperature, radiation, phenology, and fire altered the magnitude of C uptake and loss terms, and the combination of opposing terms determined whether the land was a net source or sink for carbon at any time. The dynamics of carbon fluxes were further complicated by the presence of time lags in the flow of C through pools that determined how long C took to traverse the biosphere before reappearing in the atmosphere. The combination of abiotic climate forcing and biotic responses to climate determined whether the land was a net source or sink for carbon at any time.

Lag Anomaly Correlations of NEP and NBP

The short-term response by NPP and R_h (and hence NEP and NBP) to external forcing is apparent in the monthly lag anomaly correlations of APAR (Φ_P), effective LUE ($\varepsilon^* \omega$), and the moisture scaling function of decomposition

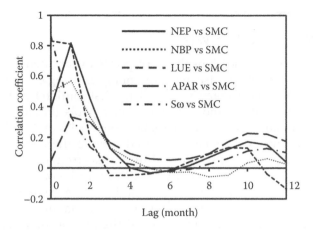

FIGURE 18.3
Lag anomaly correlations of various components of NEP and NBP against soil moisture content for a typical northern savanna woodland site averaging 1380 mm rainfall per year. The components are effective light use efficiency ($\varepsilon^* \omega$), absorbed photosynthetic active radiation (Φ_P), and the decomposition moisture scaling function ($\bar{S}_\omega$).

($\bar{S}_\omega$; Figure 18.3). These correlations have the seasonal signal removed and, therefore, isolate the proximal effect of variations in soil moisture on variations in the components of NPP and R_h (and hence NEP and NBP). They show that NEP lags soil moisture content by about 1 month with this correlation decreasing to zero by 4 months. The spread in the NEP signal over a 3-month period is due to contributions by both NPP and R_h. The NPP component can be further broken down into an immediate and direct response by effective LUE of up to 1 month, which declines to zero by 3 months; and a weaker, lagged correlation by APAR to SMC due to canopy growth, which peaks at 1 to 2 months and then slowly declines to zero at ~6 months. This strong dependence on soil moisture resembles other woodland and semiarid regions in the southern hemisphere, where variability in NDVI and NPP on annual to decadal timescales is closely correlated with rainfall variability and trends (Ichii et al., 2002). The R_h component of NEP is comprised of an immediate response through the moisture scaling function of decomposition ($\bar{S}_\omega$) that declines to zero by five months. The anomaly correlation of NBP to soil moisture shows a pattern similar to NEP but with a lower correlation coefficient due to randomness introduced by fire. Van der Werf et al. (2008) also demonstrated positive correlations between rainfall, APAR, NPP, and fire in northern Australian and African savannas. Thus, subannual variations in the net flux of carbon in the Australian savannas are controlled on monthly timescales by the combined rapid response of effective LUE on NPP and soil moisture impacts on decomposition. On timescales of 1–3 months, variations in net flux are controlled by the influence of slow variation in canopy leaf area and its impact on light interception and NPP.

Impulse–Response Analysis

On multiannual timescales, biospheric response to variations in C input are elucidated from model impulse–response analysis using estimated parameters from Table 18.3. This analysis yielded age distribution functions for both C flux to the atmosphere and C in biospheric pools (Figure 18.4). For the productive woodland site shown here, between 91.4% and 95.5% of carbon fixed by photosynthesis has been released back to the atmosphere within five years over the range of fractional burnt areas from 0% to 60% area per year (Figure 18.4a). In addition, between 10% and 50% of carbon in pools is more than 10 years old; and between 3% and 7% is more than 100 years old (Figure 18.4b). These distribution functions show that an increase in burnt area from 0 to 60% preferentially depleted the proportion of carbon efflux from decadal to century pools (Figure 18.4a), meaning that carbon was more quickly returned to the atmosphere, and shortened mean transit times. In addition, an increase in burnt area fraction depressed the proportion of carbon held in decadal pools (Figure 18.4b) but increased the small proportion of carbon contained within century pools. With burnt areas of 40 to 60%, the loss of carbon from biomass to the soil resulted in between 1 and 3% of carbon residing in soil pools for more than 400 years. In an analysis of Australian, Queensland, and Northern Territory savanna soils, Lehmann et al. (2008) estimated that between 5 and 7% of total organic carbon resided in the inert organic matter pool with mean residence times of between 718 and 9260 years. As a result, mean transit time of carbon decreased with an increase in fire from 49 years at 0% burnt area to 8 years at 60% burnt area (Table 18.7). However, the response by mean age was more complicated. First, mean age decreased from 128 to 119 years with an increase in burnt area from 0 to

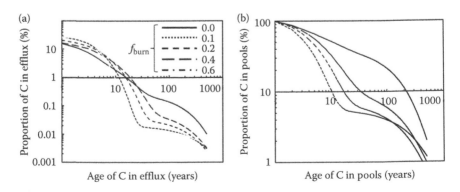

FIGURE 18.4
(a) Age distribution functions of carbon in biospheric efflux (fire emissions and autotrophic and heterotrophic respiration) and (b) age distribution functions of biospheric carbon pools (plant, litter, and soil) for a range of fractional burnt areas (f_{burn}) for the same site as Figure 18.3.

TABLE 18.7

Mean Transit Time of Carbon in Efflux and Mean Age of Carbon in Biosphere Pools Based on Dynamics of the VAST1.2 Model Determined using Impulse–Response Analysis from Figure 18.4

Fraction Burned	Mean Transit Time (years)	Mean Age (years)
0.0	49	128
0.1	21	119
0.2	15	122
0.4	11	148
0.6	8	178

10% but then increased to 178 years with an increase burnt area to 60% (Table 18.7). This occurred, because fire preferentially diverted burnt carbon that was not oxidized away from fast turnover surface pools and into slow turnover soil pools. In particular, since wood and coarse woody debris pools were consumed by fire, the mean transit time of carbon decreased whereas carbon diverted to soil pools increased mean age. The resulting estimated transit time and mean age was a result of the balance between these two fates of carbon as determined by fire.

Summary

This chapter has presented a multiple constraints model-data fusion scheme that has provided insights into the dynamics and uncertainties of net carbon fluxes for these tropical savannas including the impact of fire on net biome production. The total amount of carbon in the Australian savannas averaged 27 Pg with about half in the soil, about one-third in biomass, and the remainder in litter. The uncertainty in the estimated parameters, expressed as CV, was between 5% and 10%. Net primary productivity between years varied threefold depending on the monsoon strength and ranged from 74 to 255 Tg_C year^{-1}. The uncertainty of these estimates of NPP was ~5%. Net ecosystem productivity varied between a sink of 40 Tg_C year^{-1} and a source of 24 Tg_C year^{-1} depending on the NPP of preceding years and current soil moisture conditions. The uncertainty in these estimates of NEP was up to 34%. Fire consumed between one-third and more than one-half of NPP in any year, and emissions varied more than twofold depending on fuel loads. On monthly timescales, variations in NEP and NBP were initially due to soil moisture mediated variation in effective LUE and decomposition; but on 2–3 month timescales, these variations were also responding to slower variation in canopy phenology. On multi-annual timescales, more than 90% of carbon is respired within five years, increasing to more than 95% when annual fractional burnt area equaled 60%. Fire acts to divert carbon from multiyear and decadal pools into either the atmosphere or longer-lived (century) soil pools. As a result, mean transit times of carbon decreased from 49 to 8 years at high fire frequencies. However, mean age (which initially declined with an increase in burnt area) increased from 128 to 178 years at high fire frequency. This occurred, because a small amount of carbon was diverted away from wood and coarse woody debris to very long-lived soil pools. This work has demonstrated the utility of model-data fusion methods at integrating models and observations to yield important information on the timescales and dynamics of carbon in Australia's savannas. This information provides the scientific basis for improved understanding of the long-term impacts of climate change and human interventions on these systems.

Acknowledgments

This work was supported by prior funding from the Australian Government (Australian Greenhouse Office) to both the CSIRO Climate Change Research Program and the Cooperative Research Centre for Greenhouse Accounting. The AVHRR data used in this study includes the data produced through funding from the Earth Observation System Pathfinder Program of NASA's Mission to Planet Earth, in cooperation with the U.S. NOAA. These data were provided by the Earth Observing System Data and Information System, Distributed Active Archive Centre, at Goddard Space Flight Center. Thanks to Edward King, Jenny Lovell, and Alan Marks for assistance with processing these data. Thanks also to Johnny Xu for development of the VAST model and processing driver data; to Bevan McBeth for compiling satellite fire data for Queensland, Australia; and to Mick Meyer and Dean Graetz for providing information on fire mapping. The author is also grateful to many colleagues for important discussions over the last decade that have shaped the fundamental ideas presented in this paper. In particular, I thank Michael Raupach, Michael Hill, Dean Graetz, Ying Ping Wang, Peter Rayner, Ian Galbally, Chris Mitchell, Roger Francey, Michael Roderick, Stephen Roxburgh, and John Carter. I also thank Rowena Foster for her support over many years.

References

Araújo, T. M., J. A. Carvalho Jr, N. Higuchi, A. C. P. Brasil Jr, and A. L. A. Mesquita. 1999. A tropical rainforest clearing experiment by biomass burning in the state of Para, Brazil. *Atmospheric Environment* 33, 1991–1998.

Barrett, D. J. 2001. NPP Multi-Biome: VAST Calibration Data, 1965–1998, report, Oak Ridge Natl. Lab. Distrib. Active Arch. Cent., Oak Ridge, TN, available at http://www.daac.ornl.gov/).

Barrett, D. J. 2002. Steady state turnover time of carbon in the Australian terrestrial biosphere. *Global Biogeochemical Cycles* 16, doi:10.1029/2002GB001860.

Barrett, D. J., M. J. Hill, L. B. Hutley, et al. 2005. Prospects for improved savanna biophysical models by using multiple-constraints model-data assimilation methods. *Australian Journal Botany* 53, 689–714.

Barrett, D. J., V. A. Kuzmin, J. P. Walker, T. R. McVicar, and C. Draper. 2008. Improving stream flow forecasting by integrating satellite observations, *in situ* data and catchment models using model-data assimilation methods. eWater CRC Technical Report, Canberra, 54pp. http://ewatercrc.com.au/reports/Barrett_et_al-2008-Flow_Forecasting.pdf.

Barrett, D. J. and L. J. Renzullo. 2009. On the efficacy of combining thermal and microwave satellite data as observational constraints for root-zone soil moisture estimation. *Journal of Hydrometeorology* 10, 1109–1127.

Carvalho Jr., J. A., F. D. Costa, C. A. G. Veras, et al. 2001. Biomass fire consumption and carbon release rates of rainforest-clearing experiments conducted in northern Mato Grosso, Brazil. *Journal of Geophysical Research* 106, 17877–17887.

Carvalho Jr., J. A., N. Higuchi, T. M. Araújo, and J. C. Santos. 1998. Combustion completeness in a rainforest clearing experiment in Manaus, Brazil. *Journal of Geophysical Research* 103, 13195–13199.

De Castro, E. A. and J. B. Kauffman. 1998. Ecosystem structure in the Brazilian Cerrado: A vegetation gradient of aboveground biomass, root mass and consumption by fire. *Journal of Tropical Ecology* 14, 263–283.

De Groot, W. J. 1992. Global biomass burning: Atmospheric, climatic and biospheric implications. *New Scientist*, 12 September 1992.

Fearnside, P. M., P. M. L. A. Graca, N. L. Filho, F. J. A. Rodrigues, and J. M. Robinson. 1999. Tropical forest burning in Brazilian Amazonia: Measurement of biomass loading, burning efficiency and charcoal formation at Altamira, Pará. *Forest Ecology and Management* 123, 65–79.

Fearnside, P. M., P. M. L. A. Graca, and F. J. A. Rodrigues. 2001. Burning of Amazonian rainforests: Burning efficiency and charcoal formation in forest cleared for cattle pasture near Manaus, Brazil. *Forest Ecology and Management* 146, 115–128.

Fearnside, P. M., N. Leal Jr., and F. M. Fearnside. 1993. Rainforest burning and the global carbon budget: Biomass, combustion efficiency, and charcoal formation in the Brazilian Amazon. *Journal of Geophysical Research* 98, 16733–16743.

Graca, P. M. L. A., P. M. Fearnside, and C. C. Cerri. 1999. Burning of Amazonian forest in Ariquemes, Rondônia, Brazil: Biomass, charcoal formation and burning efficiency. *Forest Ecology and Management* 120, 179–191.

Grace, J., J. San Jose, P. Meir, H. S. Miranda, and R. A. Montes. 2006. Productivity and carbon fluxes of tropical savannas. *Journal of Biogeography* 33, 387–400.

Graetz, R. D. 2002. The net carbon dioxide flux from biomass burning on the Australia continent. CSIRO Atmospheric Research Technical Paper, No. 61. 68pp.

Guild, L. S., J. B. Kauffman, L. J. Ellingson, et al. 1998. Dynamics associated with total aboveground biomass C, nutrient pools, and biomass burning of primary forest and pasture in Rondônia, Brazil during SCAR-B. *Journal Geophysical Research* 103, 32091–32100.

Hoffa, E. A., D. E. Ward, W. M. Hao, et al. 1999. Seasonality of carbon emissions from biomass burning in a Zambian savanna. *Journal of Geophysical Research—Atmospheres* 104, 13841–13853.

Hutley, L. B., A. P. O'Grady, and D. Eamus. 2000. Monsoonal influences on evapotranspiration of savanna vegetation of northern Australia. *Oecologia* 126, 434–443.

Ichii, K., A. Kawabata, and Y. Yamaguchi. 2002. Global correlation analysis for NDVI and climatic variables and NDVI trends: 1982–1990. *International Journal of Remote Sensing* 23, 3873–3878.

Ichii, K., Y. Matusi, Y. Yamaguchi, and K. Ogawa. 2001. Comparison of global net primary production trends obtained from satellite-based normalized difference vegetation index and carbon cycle model. *Global Biogeochemical Cycles* 15, 351–363.

Lehmann, J., J. Skjemstad, S. Sohi, et al. 2008. Australian climate–carbon cycle feedback reduced by soil black carbon. *Nature Geosciences* 1, 832–835.

Levine, J. S. 1991. *Global Biomass Burning: Atmospheric, Climatic, and Biospheric Implications*. The MIT Press: Cambridge, MA.

Lovell J. L. and R. D. Graetz 2001. Filtering pathfinder AVHRR land NDVI data for Australia. *International Journal of Remote Sensing* 22, 2649–2654.

McMillan, C., R. Craig, M. Steber, J. Adams, F. Evans, and R. Smith. 1997. Fire history mapping in the Kimberley Region of Western Australia 1993–1996—Methodology and results. In *Bushfire'97 Processings*, eds. B. J. Kaige, R. J. Williams, and W. M. Waggitt. Cooperative Research Centre for the Sustainable Development of Tropical Savannas, Darwin, Australia, pp. 257–262.

Mota, B. W., J. M. C. Pereira, D. Oom, M. J. P. Vasconcelos, and M. Schultz. 2006. Screening the ESA ATSR-2 World Fire Atlas (1997–2002). *Atmospheric Chemistry and Physics* 6, 1409–1424.

Papale, D., M. Reichstein, M. Aubinet, et al. 2006. Towards a standardized processing of Net Ecosystem Exchange measured with eddy covariance technique: Algorithms and uncertainty estimation. *Biogeosciences* 3, 571–583.

Prasad, V. K., Y., Kant, P. K. Gupta, C. Sharma, A. P. Mitra, and K. V. S. Badarinath. 2001. Biomass and combustion characteristics of secondary mixed deciduous forests in Eastern Ghats of India. *Atmospheric Environment* 35, 3085–3095.

Raupach, M. R., P. J. Rayner, D. J. Barrett, et al. 2005. Model-data synthesis in terrestrial carbon observation: Methods, data requirements and data uncertainty specifications. *Global Change Biology* 11, 278–397.

Renzullo, L. J., D. J. Barrett, A. S. Marks, et al. 2008. Multi-sensor model-data fusion for estimation of hydrologic and energy flux parameters *Remote Sensing of Environment* 112, 1306–1319.

Russell-Smith, J., B. P. Murphy, C. P. Meyer, et al. 2009. Improving estimates of savanna burning emissions for greenhouse accounting in northern Australia: Limitations, challenges, applications. *International Journal of Wildland Fire* 18, 1–18.

Scholes R. J., J. Kendall and C. O. Justice. 1996. The quantity of biomass burned in southern Africa. *Journal of Geophysical Research-Atmospheres* 101, 23667–23676.

Shea, R. W., B. W. Shea, J. B. Kauffman, D. E. Ward, C. I. Haskins, and M. C. Scholes. 1996. Fuel biomass and combustion factors associated with fires in savanna ecosystem of South Africa and Zambia. *Journal of Geophysical Research* 101, 23551–23568.

Stephens, B. B., K. R. Gurney, P. P. Tans, et al. 2007. Weak northern and strong tropical land carbon uptake from vertical profiles of atmospheric CO_2. *Science* 316, 1732–1735.

Van der werf, G. R., J. T. Randerson, and L. Giglio. 2003, Carbon emissions from fires in tropical and subtropical ecosystems. *Global Change Biology* 9, 547–562

Van der Werf, G. R., J. T. Randerson, L. Giglio, N. Gobron, and A. J. Dolman. 2008. Climate controls on the variability of fires in the tropics and subtropics. *Global Biogeochemical Cycles* 22, GB3028, doi:10.1029/2007GB003122.

Williams R. J., D. J. Barrett, G. Cook, et al. 2009. Landscape-scale fire research in northern Australia: Delivering multiple benefits in a changing world. In *Culture, Ecology and Economy of Fire Management in North Australian Savannas*, eds. J. Russell-Smith, P. Whitehead and P. Cooke, Collingwood, Victoria, CSIRO Publishing, pp. 181–200.

Appendix

The equations of VAST1.2 including emissions from fire are

$$P_n = (1 - \rho)\varepsilon * \omega \, \Phi_F \tag{A1.1}$$

$$\frac{dq_L}{dt} = \alpha_L P_n - (1 - f_b)q_L - fb \, qL \tag{A1.2}$$

$$\frac{dq_W}{dt} = \alpha_W P_n - (1 - f_b)q_W/\tau_W - f_b \, q_W \tag{A1.3}$$

$$\frac{dq_{Rj}}{dt} = \alpha_R \xi_j \, P_n - (1 - m_j f_b)q_{Rj}/\tau_{Rj} - m_j \, f_b q_{Rj}, \quad j = 1, 2, 3 \tag{A1.4}$$

$$\frac{dq_F}{dt} = (1 - f_b)[q_L/\tau_L + \eta q_C/\tau_C - q_F/\tau_F] + f_b[(1 - \beta_L)q_L - \beta_F q_F + (1 - \beta_C)q_C] \tag{A1.5}$$

$$\frac{dq_c}{dt} = (1 - f_b)[q_W/\tau_W - q_C/\tau_C] + f_b[(1 - \beta_W)q_W - q_C] \tag{A1.6}$$

$$\frac{dq_{s1}}{dt} = (1 - f_b) \, [q_{R1}/\tau_{R1} + \Theta_F \, q_F/\tau_F + \Theta_C q_C/\tau_C - q_{s1}/\tau_{s1}] + f_b m_1 \, q_W \tag{A1.7}$$

$$\frac{dq_{sj}}{dt} = (1 - f_b) \, [q_{Rj}/\tau_{Rj} + \Theta_{Sj-1}q_{Sj-1}/\tau_{Sj-1} - q_{Sj}/\tau_{Sj}] + f_b m_j \, q_{Rj}, \quad j=2, 3 \tag{A1.8}$$

where $(\varepsilon^*\omega)$ is the effective (or moisture) limited light use efficiency, $\tau_k = \tau_k^* / \bar{S}_T\bar{S}_\omega$ is the turnover time of carbon in the kth pool, $f_b = f_s(1 - c)$ is the fraction of a 0.05° grid cell burnt, $\beta = \beta^* - (0.5 \, \omega\beta^*)$ is the combustion efficiency of fuel at field moisture content, ω is the relative volumetric soil moisture content, $\bar{S}_T$ is the temperature scaling function of decomposition, $\bar{S}_\omega$ is the soil moisture scaling function of decomposition, f_s is the fraction of a $0.05°$ grid cell that is a fire scar, c is the proportion of fire scar left unburnt after fire, m_j is the fraction of fine root mass mortality after fire in the jth soil layer, and β^* is the combustion efficiency of dry biomass. Other parameters are defined in Table 18.1, and variables are specified in Barrett (2002). Subscripts are leaf (L), live wood (W), root (R), fine litter (F), coarse litter (C), and soil (S). By rearrangement, heterotrophic respiration, R_h, is

$$R_h = (1 - \theta_F)q_F/\tau_F + (1 - \eta - \theta_C)q_C/\tau_C + \sum_{j=1}^3 (1 - \theta_{Sj})q_{Sj}/\tau_{Sj}, \quad \theta_{S3} = 0 \qquad (A1.9)$$

and emissions from biomass burning, E, are

$$E = f_b(q_L\beta_L + q_W\beta_W + q_F\beta_F + q_C\beta_C). \qquad (A1.10)$$

19

Land Use Change and the Carbon Budget in the Brazilian Cerrado

Mercedes M. C. Bustamante and L. G. Ferreira

CONTENTS

Introduction: The Cerrado Region

The Cerrado covers approximately 2 million km^2, mostly in central Brazil. It consists of a gradient from grasslands (locally called *campo limpo*) to sclerophylous forests (*cerradão*) (Eiten, 1972). Between these, there are intermediate physiognomies with increasing density of woody species (*campo sujo, campo cerrado*, and *cerrado stricto sensu*). The annual precipitation varies from 600 to 2200 mm and the dry season lasts from 4 to 7 months, whereas the mean annual temperature ranges from 22°C to 27°C (Adámoli et al., 1986). Oxisols and Entisols represent approximately 46% and 15% of the Cerrado soils, respectively (Reatto et al., 1998). In general, soils are acidic, with low nutrient concentrations and high aluminum content. Under such conditions, soil organic matter (SOM) is particularly important to processes related to nutrient cycling, soil aggregation, and plant available water (Resck, 1998). Gallery forests (closed canopy with 75–95% cover) occupy about only 5% of the Cerrado but contain approximately 32% of its biodiversity (Felfili et al., 2001). They are associated with a great variety of soils, ranging from dystrophic to mesotrophic and seasonal flooded hydromorphic soils with different levels of organic matter (Reatto et al., 1998).

Land Use Patterns and Processes

Between 1975 and 1996, the area cleared in the Cerrado for farming almost doubled, from 34.7 to 64.5 million hectares. During this period, pasturelands (mainly African *Brachiaria* species) increased from 16.0 to 49.2 million hectares, whereas the crop areas increased from 6.9 to 8.2 million hectares. Presently, the region is one of the most important beef producing regions in Brazil, with approximately 44% of the national herd. The major export crops are soybeans and maize. Crop farming in the region is characterized by large farm units, monocultures with high external inputs, and heavy mechanization (Cadavid-Garcia, 1995). Land use changes in the Cerrado also include forestry plantations for the pulp and paper sector (by 2000, planted forests, mostly *Eucalyptus* and *Pinus*, represented approximately 1.7 million hectares in the Cerrado region; Sociedade Brasileira de Silvacultura, 2009) and the deforestation and burning of native species for charcoal production for the steel industry.

Land use change mapping (Sano et al., 2009), based on Landsat ETM+ (for 2002), as well as on ancillary information and field campaigns, showed that approximately 80 millions of hectares (about 39.5% of the total Cerrado area), which showed that approximately 80 million of hectares (about 39.5% of the total Cerrado area) have been already converted into different land uses (accuracy ranging from 71% to 90%). Cultivated pastures and agriculture fields occupy 26.5 and 10.5% of the Cerrado, respectively. If cattle ranching over natural grasslands (about 23 million ha) is accounted for, the overall agro-pastoral occupation in the Cerrado biome increases to about 50%. Landscape fragmentation in the Cerrado, strongly dependent on topography and dominant land use type, is more prominent in the States of Mato Grosso, Bahia, and Goiás (Brannstrom et al., 2008; Carvalho et al., 2009).

Although heterogeneously distributed, the remnant vegetation is mostly found in the northern portion of the biome (Figure 19.1); whereas in the southern areas, with easier access to major cities and consumption centers (e.g., Sao Paulo) and predominance of more fertile soil and adequate topography, land conversions were much more intense and widespread (Miziara and Ferreira, 2008). In fact, 146 watersheds, located in the southern states of Goiás and Mato Grosso, are nearly 100% converted; whereas in another 3101 watersheds, located mainly in Goiás, Minas Gerais, Mato Grosso, and Mato Grosso do Sul, land conversion is already more than 70%, that is, beyond the minimum area required by the Brazilian Forest Code to be maintained as legal forest reserves (20% in the core of the Cerrado region) and permanent protected areas (mean value of 10% in every property/basin assuming a buffer zone of 100 m along the stream detected at the 1:250,000 scale) in the core Cerrado region (Bonnet et al., 2006). By contrast, in the northern regions, particularly in the States of Tocantins, Piauí, and Maranhão, there are 3386 watersheds with nearly 0% disturbance (Figure 19.1).

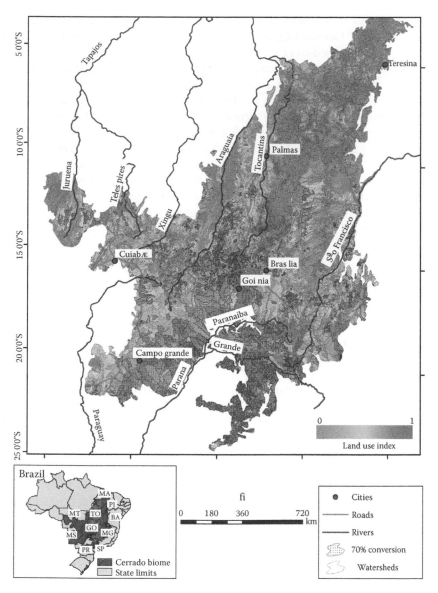

FIGURE 19.1
(See color insert following page 320.) Land use in the cerrado biome relative to 15,612 watersheds (0–100% range). Black dots correspond to watersheds with conversion over 70%.

Data from the Cerrado Warning Deforestation System (SIAD), based on the annual comparison of 250 m MODIS vegetation index images (Huete et al., 2002), indicate that the conversion of pristine areas continues (Ferreira et al., 2007a). From June 2002 to June 2009, about 37,227 km^2 of native Cerrado were deforested (maps available at: http://mapas.mma.gov.br/i3geo/), mainly along the Goiás–Bahia State border and in Mato Grosso, as part of the well-known Amazon Deforestation Arc (Ferreira et al., 2007b). These clearings, predominantly located over flat terrains (~74% in slopes up to 2°) and in the vicinity of major roads and existent pastures, also impacted protected areas and areas identified as high priority for biodiversity conservation (MMA, 2007). Areas were identified on the basis of biodiversity and landscape data (e.g., endangered species, physiognomic domains, geomorphological units, etc), according to concepts such as irreplaceability, vulnerability, and complementarity (maps available at: http://mapas.mma.gov.br/i3geo/). Among the 180 protected areas (amounting to about 161,757 km²), 1223 km² of changes were detected in 54 of them (i.e., 0.82% of their vegetative cover, according to the PROBIO map). The 250 priority areas (765,948 km² in area) were even more affected, as nearly 10,008 km² of changes were found in 172 of them, impacting around 1.8% of their remnant vegetation.

Carbon Budget and Land Use

The fast land cover changes in the Cerrado have already impacted water resources and significantly increased runoff and discharge at major river basins (Costa et al., 2003) as well as the regional CO_2 and energy balance (Potter et al., 2009). The natural and anthropogenic vegetation cover exhibits a clear contrast in the production of sensible heat (Figure 19.2). Among the impacts on ecosystem functioning, a major concern is the carbon cycle affected both by land conversions and by management after conversion.

Carbon Stocks and Fluxes

The native Cerrado, generally lower in carbon stocks than the humid forest, has larger stock in the belowground biomass in the different physiognomies, especially in grasslands due to the dense root systems of grasses (Table 19.1). In general, soil C stocks up to 100 cm depth, under different vegetation covers and soil types, range from approximately 97 to 210 Mg C ha^{-1} (see Bustamante et al., 2006 for a review), with organic C content increasing with clay content. Soil compartment is an important component of C budget in Cerrado systems.

Cerrado *stricto sensu* is a sink of CO_2 during the rainy season and a source during a brief period at the end of the dry season. Annual balances indicated a net uptake of about 0.1 (Rocha et al., 2002) to 2.5 Mg C ha^{-1} year^{-1} (Miranda et al., 1996). These differences could be related to heterogeneity of vegetation and locations. Grace et al. (2006) estimated for tropical savannas an uptake of 0.14 Mg C ha^{-1} year^{-1}, contributing to a total sink of 0.39 Pg C year^{-1}, almost 15% of all C fixed by world vegetation. Soil CO_2 flux is an important component of

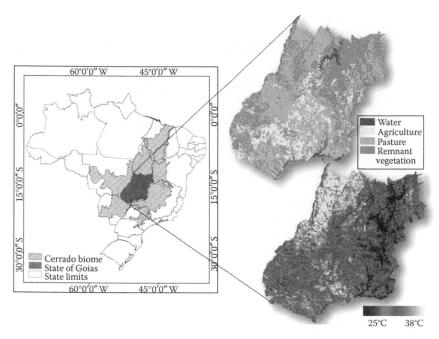

FIGURE 19.2
(See color insert following page 320.) Land surface temperature for the State of Goiás (core cerrado region) (1 km spatial resolution MOD11 daytime product for day-of-year 161).

ecosystem flux that is strongly influenced by temperature and humidity (Pinto et al., 2002). The annual fluxes in native areas were around 14 Mg C ha^{-1} year^{-1} (Pinto, 2003). Often, Cerrado soils are also a net sink for methane (CH_4) due to the dominance of well-drained soils (Bustamante et al., 2009a).

Impacts of Fire on C Budget

Fire is not only a natural determinant of Cerrado (Miranda et al., 2002) but it has also been used as a tool for land clearing and management of native pastures (Roscoe et al., 2000). Fire affects the dynamics of the vegetation, particularly the grass or woody biomass ratio. Surface fires, which consume the fine fuel of the herbaceous layer, are the most common. The fine fuel of the herbaceous layer ranges from 85 to 97% from woodland to open cerrado (Miranda et al., 2002). Losses of C, N, and S after combustion of the aboveground biomass have been observed to be greater in grasslands than in woodlands (Kauffman et al., 1994). The high mortality rate of woody vegetation (27–39%) suggests that frequent fires change vegetation structure to a more open form (higher grass biomass), which favors the occurrence of more intense fires (Miranda et al., 2002). On the other hand, fire suppression (at least 26 years) in a cerrado *sensu stricto* resulted in an increase of 19% in

TABLE 19.1

Above and Belowground Biomass in Different Cerrado Vegetation Types

Biomass Stocks	Compartments	Mg ha^{-1}	Reference
Campo limpo	Total aboveground biomass[a]	3.78–16.57	Ottmar et al. (2001)
	Herbaceous layer[b]	3.51–16.00	
	Woody material	0.27–0.57	
	Trees	0.00–0.00	
	Total belowground biomass	16.3	Castro and Kauffman (1998)
	Below/aboveground ratio	7.7	
Campo sujo	Total aboveground biomass	6.68–15.77	Ottmar et al. (2001)
	Herbaceous layer	4.27–10.28	
	Woody material	2.41–3.05	
	Trees	0.00–2.44	
	Total belowground biomass	30.1	Castro and Kauffman (1998)
	Below/aboveground ratio		
Open cerrado	Total aboveground biomass	12.55–39.05	Ottmar et al. (2001)
	Herbaceous layer	4.58–5.05	
	Woody material	4.66–4.06	
	Trees	3.31–29.94	
	Total belowground biomass	46.6	Castro and Kauffman (1998)
	Below/aboveground ratio	2.6	
Cerrado stricto sensu	Total aboveground biomass	20.9–58.01	Ottmar et al. (2001)
	Herbaceous layer	4.10–10.39	
	Woody material	2.25–4.66	
	Trees	14.28–42.96	
	Total belowground biomass		Castro and Kauffman (1998)
	Below/aboveground ratio		
Dense cerrado	Total aboveground biomass	29.90–71.89	Ottmar et al. (2001)
	Herbaceous layer	3.76–5.44	
	Woody material	3.18–11.78	
	Trees	22.90–54.67	
	Total belowground biomass	52.9	Castro and Kauffman (1998)
	Below/aboveground ratio		

[a] Includes live biomass (herbaceous and woody), litter, and woody material (dead and down woody with diameter ≤ 0.6–22.9 cm)

[b] Includes grasses, dicots, and litter.

density of the woody layer and of 15% in basal area over a period of 13 years (1991–2004) (Roitman et al., 2008).

Nardoto et al. (2006) detected a reduction of 80% in litterfall after four fire events in cerrado *stricto sensu*. However, Roscoe et al. (2000) did not observe significant changes in soil C stocks in cerrado *stricto sensu* burned 12 times during 21 years (compared to a protected area), even though the woody vegetation was significantly reduced.

Ecosystem CO_2 fluxes measured by eddy covariance were also lower in a cerrado frequently burned (1.4 Mg C ha^{-1}) than in an area protected from fire (2.6 Mg C ha^{-1}) (Breyer, 2001). Frequent burning also impacts soil respiration. After the first rains, soil respiration in burned areas became higher than in unburned areas (Pinto et al., 2002), probably due to root activity of grasses. In cerrado *stricto sensu*, approximately 60% of soil CO_2 fluxes came from C4 plants (grasses) in the wet and dry seasons (Aduan, 2003). Despite the high resilience of Cerrado vegetation with regard to fire, the increase in fire frequency due to human intervention is interfering in the recovery of the woody layer, changing the proportion of plant functional types (grass and woody plants) with consequences for ecosystem carbon fluxes.

Impacts of Conversion

After anthropgenic fire, native vegetation is usually converted to pastures or croplands. There is evidence of an effect of time since conversion on the carbon stocks and fluxes in the Cerrado, which is related to the management of the converted systems. Well-managed cultivated pastures may provide enough organic C to maintain or even increase native C contents (Roscoe et al., 2001) due to high ecosystem gas exchange rates (Santos et al., 2004). However, at least half of the pastures in the Cerrado are in different stages of degradation due to poor management (Oliveira et al., 2004), with carbon inputs too low to sustain the high soil C storage under native Cerrado. Therefore, the potential for soil C sequestration under pastures relies heavily on the recovery or restoration of degraded pastures through nutrient inputs (Oliveira et al., 2001; Silva et al., 2004) and good management of stocking rates. In Rondônia and Mato Grosso States, soil stocks decrease by about 0.28 Mg C ha^{-1} year^{-1} in degraded pastures and increase up to 0.72 Mg C ha^{-1} year^{-1} under improved management (Maia et al., 2009).

Root turnover might be an important factor regulating soil carbon stocks in deep soils. The root turnover time in a native cerrado was between 9 and 12 years from 40 to 400 cm depth; whereas in a planted pasture, it was between 5 and 10 years (Aduan, 2003). The low productivity systems (native Cerrado) maintain the stability of organic carbon in deep soil layers due to the low supply of fresh C (Fontaine et al., 2007). The greater supply of fresh C to deeper soil layers (mainly exudates) under pasture may stimulate the loss of older carbon. The effect of converting Cerrado to pasture on global warming potentials (GWPs) was assessed by compiling data from different sources over a 100-year period (Table 19.2). Estimated emission of methane by live-stock in Brazil for 1994 was 9.8 Tg CH_4 y^{-1} (enteric fermentation and animal wastes). Beef cattle contributed to 81% of this total, and the Center-West region (most of the Cerrado) represented 30% (Lima et al., 2001).

In croplands, conventional tillage was associated with soil erosion and car-bon losses in the Cerrado region (Lepsch et al., 1994; Resck et al., 1999). However, soil carbon losses are not always observed (Freitas et al., 2000; Roscoe and Burma, 2003), probably due to the high stability of SOM in Cerrado soils (Resende et al., 1997). Currently, no-tillage systems are used in about 70% of the soybean cultivated area (Landers, 2000). Bayer et al. (2006) indicated that soil C stocks in no-till sandy clay loam Oxisols increased by 2.4 Mg ha^{-1} (0.30 Mg ha^{-1} $year^{-1}$) and in the clayey Oxisols by 3.0 Mg ha^{-1} (0.60 Mg ha^{-1} $year^{-1}$).

By reducing tillage (or number of passes by cultivating equipment) tillage, farmers can sow the main commercial crop earlier in the season, allowing the planting of a second crop during the same growing season. Plant biomass production in no-tillage systems with cover crops can be up to 3 times higher than in the conventional soybean monocropping (Scopel et al., 2004), thus representing a significant increase in C input to the soil. Soil carbon storage potential of no-tillage in the Cerrado is, on average, 0.83 Mg C ha^{-1} $year^{-1}$ in the 0–20 cm topsoil, primarily due to the introduction of a second crop that leads to higher C inputs to the soil. The N input into the soil from green

TABLE 19.2

Net Effects on Global Warming Potentials (GWPs) of *Cerrado*-To-Pasture Conversion

Net Source (+)/Sink (–)	Range	Estimate	CO_2 Equivalents (Mg CO_2 ha^{-1})	Reference
Aboveground biomass (Mg C ha^{-1})	+2 to +16	+15	+55	Ottmar et al. (2001), Castro and Kauffman (1998)
Belowground biomass (Mg C ha^{-1})	+4 to +10	+10	+35	Grace et al. (2006)
Soil carbon (Mg C ha^{-1})	0 to +50	+30	+110	Bustamante et al. (2006)
Soil N_2O (kg N ha^{-1})	0 to +1	0	0	Pinto et al. (2006), Varella et al. (2004)
Soil CH_4 (kg CH_4 ha^{-1})	0 to +1	0	0	Bustamante et al. (2009a)
Fire N_2O (kg N ha^{-1})	0 to +5	0	+2	Ferek et al. (1998), Krug et al. (2002)
Fire CH_4 (kg CH_4 ha^{-1})	0 to +70	0	+2	Andreae and Merlet (2001), Krug et al. (2002)
Total			+220	

manure or fertilizers is also instrumental to SOM accumulation or conservation under no-tillage (Corbeels et al., 2006).

New production systems, such as integrated crop-livestock management systems (ICLS), have also been recently developed. Pasture is sown in association with the main crop to achieve a dense soil cover at crop harvest (Kluthcouski et al., 2000). Pasture is an excellent cover crop and can be maintained for a few years. The field can be returned to crop production before the pasture degrades by direct sowing of a crop into the mulch of pasture residue after a total herbicide application.

Relatively little information is available on the functioning of these new cropping systems in the Cerrado, and the first studies are not conclusive with regard to effects of soil carbon. Marchão et al. (2008) evaluated soil C storage under continuous pasture, continuous crop, and four years of crop followed by four years of pasture and vice versa, with two levels of soil tillage (conventional and no-tillage) and fertility (maintenance and corrective fertility). An adjacent native Cerrado was used as control. Carbon storage under Cerrado was 61 Mg ha^{-1} and ranged from 52 Mg ha^{-1} under the ICLS rotation to 60 Mg ha^{-1} with continuous cropping. The decrease in SOC stocks was approximately 8.5 and 7.5 Mg ha^{-1}, or 14% and 12% for continuous pasture and ICLS, respectively; whereas there was essentially no change in SOC under a no-till continuous cropping system. The larger decrease in SOC stocks under pasture (comparatively to crops) was probably due to low N inputs that limited pasture productivity (Silva et al., 2004). Fernandes (2008) found significantly lower soil C stocks (0–30 cm depth) under soybean followed by natural fallow (51.7 Mg C ha^{-1}) adopted for 31 years after conversion in comparison to an adjacent native Cerrado remnant (62.3 Mg C ha^{-1}). By contrast, 10 years of ICLS (rotation bean-sorghum-pasture), after 10 years of continuous crop and 13 years of continuous pasture, resulted in soil carbon values similar to the native area.

Occupation Trends and Future Scenarios

Diniz-Filho et al. (2009), in a study on the spatial patterns and causes behind habitat loss in the Cerrado, concluded that the proportion of remnant vegetation across the biome is highly structured and well explained (89%) by socioeconomic variables (e.g., road network size), land productivity of soybean, climatic-environmental variables (e.g., temperature, actual ET), and the inherent dependence on the geographic connectivity at multiple scales.

Assuming the current trends as the basis for future occupation (Ferreira, 2009), the PROBIO map as a reference baseline, SIAD data as an indicative of the ongoing clearings (based on which a constant annual deforestation rate of 0.4% was derived), six explanatory variables (i.e., terrain slope, precipitation, soil fertility, human development index, and the proximity to roads and streams), and a business-as-usual scenario to estimate land conversion rates. The scenario analysis predicted that about 14% of the current native Cerrado would be converted by 2050. This would represent 40,000 km² of land cover change every decade, affecting mainly the States of Bahia (in particular, its western portion, already a consolidated agricultural frontier), Piauí and Maranhão (presently, well preserved), and Tocantins (strengthening the Amazon deforestation arc in its southern portion).

Although a significant amount of these potential conversions are expected to be related to the expansion of the cattle ranching industry, the growing demand for biofuels, particularly ethanol from sugarcane and biodiesel from soybean, will certainly be an important driver of land use changes. Substantial increases in land use, due to the sugarcane expansion, are already occurring in core Cerrado States, such as Mato Grosso do Sul and Goiás, for which the increments between the 2005–2006 and 2008–2009 harvesting years were 126.21% and 99.93%, respectively (Silva et al., 2009). Ribeiro et al. (2009) estimated a threefold increase in the Cerrado area occupied with sugarcane above the area mapped up to 2007 [Canasat: Brazilian Space Research Institute (INPE) sugarcane mapping project; Sugawara et al., 2008; Data available at: http://www.dsr.inpe.br/mapdsr]. This predictive modeling was based on the main physical characteristics of areas suitable for sugarcane production, available infrastructure, and environmental constraints and assumed that expansion would be restricted to pasturelands. About 66% of the increase showed by more recent (2008) sugar cane mapping was in agreement with this deterministic model and involved replacement of existing pastures. The conversion of degraded pasturelands to sugarcane in the Cerrado has the potential to avoid soil degradation, while maintaining or building soil carbon with appropriate tillage practices and mechanical harvesting (Bustamante et al., 2009b). These land management practices, however, are not yet ubiquitous in the Cerrado.

Additionally, this expansion is also clearly compromising other crops and remnant native vegetation. Even if a best case scenario of sustainable sugarcane expansion on underutilized cattle pastures was completely effective, the pressure on the Cerrado would still be very high, as cattle are displaced, inducing both new clearings, and due to the intensification of ranching on well-managed pastures. Sugarcane expansion may also cause a major change in the Cerrado land tenure structure with the displacement of small landowners. Such impacts need to be further investigated and incorporated in the modeling of future landscape spatial patterns.

Concluding Remarks

Carbon dynamics with conversion of native Cerrado to pasture or cropland shows different patterns depending on differences in establishment, history, and stage of degradation. The variability of the data reported here indicates the urgent need to capture the spatial heterogeneity and temporal dynamics of land use in the Cerrado with additional data on stocks and fluxes of C as well as on the environmental controls (e.g., seasonality impacts on C budget along the region). This would provide a better understanding of the interactions among land use history, diversification of management systems, and physical environment, a fundamental input for regional scale modeling. The extension of conversion of native Cerrado, however, implies that this process can represent a substantial source of CO_2 to the atmosphere. New factors, such as the expansion of bioenergy crops, impose additional challenges for a sustainable land use in the region. Carbon and biodiversity costs of converting native Cerrado into new agricultural areas would be substantial.

In particular, analysis of new scenarios for the Cerrado should consider the overall landscape structure and functioning and the relationships with land cover change in adjacent biomes (such as the Amazon rain forest). In addition, predictive scenarios, based on deterministic and stochastic approaches, should encompass (a) a wider range of (socioeconomic and environmental) variables, (b) all the possible land cover—land use and transitions, (c) different levels of governance practices, (d) all the planned or under-construction infrastructure, (e) market forces inducing changes, and (f) potential economic mechanisms promoting conservation, such as REDD (reducing deforestation from deforestation and degradation). The current scope of REDD and its focus on the Amazon forest may exacerbate agricultural pressure on the Cerrado. By contrast, and diametrically in opposition to prevalent thinking, a Cerrado REDD, aiming at avoiding or minimizing deforestation in the agriculture frontiers, mainly in Mato Grosso and Bahia, as well as recovering degraded areas along the riparian environments, could complement and expand tropical forest focus to encompass the tropical woodlands and savanna lands.

References

Adámoli, J., J. Macêdo, L. G. Azevedo, and J. M. Neto. 1987. Caracterização da região dos cerrados. In *Solos dos cerrados: tecnologias e estratégias de manejo*, ed. W. J. Goedert, 33–98. Embrapa-CPAC/Nobel, Brasília/São Paulo.

Aduan, R. E. 2003. Respiração de solos e ciclagem de carbono em cerrado nativo e pastagem no Brasil central. PhD dissertation, Universidade de Brasília, Brasília.

Andreae, M. O. and P. Merlet. 2001. Emission of trace gases and aerosols from biomass burning. *Global Biogeochemical Cycles*, 15, 955–966.

Bayer, C., L. Martin-Neto, J. Mielniczuk, A. Pavinoto, and J. Dieckow. 2006. Carbon sequestration in two Brazilian Cerrado soils under no-till. *Soil and Tillage Research* 86, 237–245.

Bonnet, B. R. P., L. G. Ferreira, and F. C. Lobo. 2006. Uso de dados SRTM como suporte à implementação de um sistema de reserva legal extra-propriedade por bacia hidrográfica no Cerrado Brasileiro. *Revista Brasileira de Cartografia* 58, 129–137.

Brannstrom, C., W. Jepson, A. M. Filippi, D. Redo, Z. Xu, and S. Ganesh. 2008. Land change in the Brazilian Savanna (Cerrado), 1986–2002: Comparative analysis and implications for land-use policy. *Land Use Policy* 25, 579–595.

Breyer, L. M. 2001. Fluxos de energia, carbono e água em áreas de cerrado sensu stricto submetidas a diferentes regimes de queima. PhD Dissertation, Universidade de Brasília, Brasil.

Bustamante, M. M. C., M. Keller, and D. A. da Silva. 2009a. Sources and sinks of trace gases in Amazonia and the Cerrado. In *Amazonia and Global Change*, eds. J. Gash, M. Keller, M. M. C. Bustamante and P. L. S. Dias. American Geophysical Union. pp. 337–354.

Bustamante, M. M. C., J. Melillo, D. J. Connor, et al. 2009b. What are the final land limits? In *Biofuels: Environmental consequences and interactions with changing land use*, eds. R.W. Howarth and S. Bringezu. Proceedings of the Scientific Committee on Problems of the Environment (SCOPE) International Biofuels Project Rapid Assessment.

Bustamante, M. M. C., M. Corbells, and E. Scopel. 2006. Soil carbon storge and sequestration potential in the Cerrado region of Brazil. In *Carbon Sequestration in Soils of Latin America*, eds R. Lal, C.C. Cerri, M. Bernoux, J. Etchevers and C.E.P. Cerri, Harworth Press Inc., Binghamton, NY.

Cadavid-Garcia, E. A. 1995. Desenvolvimento econômico sustentável do Cerrado. *Pesquisa Agropecuária Brasileira* 30(6), 759–774.

Carvalho, F. M. V., P. De Marco, and L. G. Ferreira. 2009. The Cerrado into-pieces: Habitat fragmentation as a function of landscape use in the savannas of Central Brazil. *Biological Conservation* 142, 1392–1403.

Castro, E. A. and J. B. Kauffman. 1998. Ecosystem structure in Brazilian cerrado: A vegetation gradient of aboveground, root biomass and consumption by fire. *Journal of Tropical Ecology* 14, 263–283.

Corbeels, M., E. Scopel, A. Cardoso, M. Bernoux, J. M. Douzet, and M. S. Siqueira Neto. 2006. Soil carbon storage potential of direct seeding mulch-based cropping systems in the Cerrados of Brazil. *Global Change Biology* 12, 1773–1787.

Costa, M. H., A. Botta, and J. A. Cardille. 2003. Effects of large-scale changes in land cover on the discharge of the Tocantins River, Southeastern Amazonia. *Journal of Hydrology* 283(12): 206–217.

Diniz-Filho, J. A. F., G. Oliveira., F. Lobo, L. G. Ferreira, L. M. Bini, and T. F. V. L. B. Rangel. 2009. Agriculture, habitat loss and spatial patterns of human occupation in a biodiversity hotspot. *Scientia Agricola* 66, 764–771.

Eiten, G. 1972. The cerrado vegetation of Brazil. *The Botanical Review*, 38(2), 201–341.

Felfili, J. M., R. C. Mendonça, Walter, B. M. T. et al. 2001. Flora fanerogâmica das matas de galeria e ciliares do Brasil Central. In *Cerrado—caracterização e recuperação de matas de galeria*, eds. Ribeiro, J. F., C. E. L. Fonseca, J. C. Souza-Silva, Embrapa Cerrados, DF, Planaltina, pp. 195–209.

Ferek, R. J., J. S. Reid, P. V. Hobbs, D. R. Blake, and C. Liousse. 1998. Emission factors of hydrocarbons, halocarbons, trace gases and particles from biomass burning in Brazil. *Journal of Geophysical Research* (Atmosphere) 103, 32107–32118.

Fernandes, E. B. 2008. Emissões de CO_2, NOx e N_2O de solos sob diferentes sistemas de cultivo no Cerrado. PhD Dissertation, University of Brasília, Brasília, Brazil. 138p.

Ferreira, M. E. 2009. Modelagem da Dinâmica de Paisagem do Cerrado. PhD Dissertation, Federal University of Goiás, 92pp.

Ferreira, N. C., L. G. Ferreira, A. R. Huete, and M. E. Ferreira. 2007a. An operational deforestation mapping system using MODIS data and spatial context analysis. *International Journal of Remote Sensing* 28, 47–62.

Ferreira, N. C., L. G. Ferreira, and F. Miziara. 2007b. Deforestation hotspots in the Brazilian Amazon: Evidences and causes as assessed from remote sensing and census data. *Earth Interactions* 11, 1–16.

Fontaine, S., S. Barot, P. Barré, N. Bdioui, B. Mary, and C. Rumpel. 2007. Stability of organic carbon in deep soil layers controlled by fresh carbon supply. *Nature* 450, 277–281.

Freitas, P. L., P. Blancaneaux, E. Gavinelli, M. C. Larré-Larrouy, and C. Feller. 2000. Nível e natureza do estoque orgânico de latossolos sob diferentes sistemas de uso e manejo. *Pesquisa Agropecuária Brasileira* 35, 157–170.

Grace, J., J. San José, P. Meir, H. S. Miranda, and R. A. Montes. 2006. Productive and carbon fluxes of tropical savannas. *Journal of Biogeography* 33, 387–400.

Huete, A. R., T. Miura, K. Didan, E. P. Rodrigues, X. Gao, and L. G. Ferreira. 2002. Overview of the radiometric and biophysical performance of the MODIS vegetation indices. *Remote Sensing of Environment* 83, 195–213.

Kauffman, J. B., D. L. Cummings, and D. E. Ward. 1994. Relationships of fire, biomass and nutrient dynamics along a vegetation gradient in the Brazilian cerrado. *Journal of Ecology* 82, 519–531.

Kluthcouski, J., T. Cobucci, H. Aidar, et al. 2000. Integração lavoura/pecuária pelo consórcio de culturas anuais com forrageiras, em áreas de lavoura, nos sistemas direto e convencional. Embrapa Arroz e Feijão, Circular Técnica N° 38, 28pp.

Krug, T., H. B. Figueiredo, E. E. Sano, et al. 2002. Emissões de gases de efeito estufa da queima de biomassa no cerrado não-antrópico utilizando dados orbitais. Ministério da Ciência e Tecnologia, Brasília. 53pp.

Landers, J. N. 2001. *Zero tillage development in tropical Brazil—The story of a successful NGO activity*. Wageningen University and FAO, Wageningen, The Netherlands. 57pp.

Lepsch, I. F., J. R. F. Menk, and J. B. Oliveira. 1994. Carbon storage and other properties of soils under agriculture and natural vegetation in São Paulo State, Brazil. *Soil Use and Management* 10, 34–42.

Lima, M. A., R. C. Boeira, V. L. S. Castro, et al. 2001. Estimativa das emissões de gases de feito estufa provenientes de atividades agrícolas no Brasil. In *Mudanças climáticas Globais e a Agropecuária Brasileira*, eds. M. A. Lima, O. M. R. Cabral and J. D. G. Miguez, 169–189, Jaguariúna, Embrapa Meio Ambiente.

Maia, S. M. F., S. M. Ogle, C. E. P. Cerri, and C. C. Cerri. 2009. Effect of grassland management on soil carbon sequestration in Rondônia and Mato Grosso states, Brazil. *Geoderna* 149, 84–91.

Marchão, R. L., T. Becquer, D. Brunet, L. C. Balbino, L. Vilela, and M. Brossard. 2008. Carbon and nitrogen stocks in a Brazilian clayey Oxisols: 13-year effects of integrated crop-livestock management systems. *Soil and Tillage Research* 103, 442–450.

Miranda, A. C., H. S. Miranda, J. Lloyd, et al. 1996. Carbon dioxide fluxes over a cerrado sensu stricto in Central Brazil. In *Amazonian Deforestation and Climate*, eds. J. H. C. Gash, C. A. Nobre, J. M. Roberts and R. L. Victoria, 353–363. Chichester, John Wiley & Sons.

Miranda, H. S., M. Bustamante, and A. C. Miranda. 2002. The fire factor. In *The Cerrados of Brazil: Ecology and Natural History of a Neotropical Savanna*, eds. P. S. Oliveira and R. J. Marquis, 51–68. New York, Columbia University Press.

Miziara, F. and N. C. Ferreira. 2008. Expansão da fronteira agrícola e evolução da ocupação e uso do espaço no Estado de Goiás: subsídios à política ambiental. In *A encruzilhada socioambiental—biodiversidade, economia e sustentabilidade no cerrado*, ed. L. G. Ferreira. Goiânia, Editora Universidade Federal de Goiás, pp. 107–125.

MMA. 2007. Áreas Prioritárias para Conservação, Uso Sustentável e Repartição de Benefícios da Biodiversidade Brasileira—Portaria MMA n° 9, de 23 de janeiro de 2007.

Nardoto, G. B., M. M. C. Bustamante, A. S. Pinto, and C. A. Klink. 2006. Nutrient use efficiency at ecosystems and species level in savanna areas of Central Brazil and impacts of fire. *Journal of Tropical Ecology* 22, 1–11.

Oliveira, O. C., I. P. Oliveira, E. Ferreira, et al. 2001. Response of degraded pastures in the Brazilian Cerrado to chemical fertilization. *Pasturas Tropicales* 23, 14–18.

Oliveira, O. C., I. P. Oliveira, S. Urquiaga, B. J. R. Alves, and R. M. Boddey. 2004. Chemical and biological indicator of decline/degradation of *Brachiaria* pastures in the Brazilian Cerrado. *Agric. Ecosyst. Environ.* 103, 289–300.

Ottmar, R., R. E. Vihnanek, H. S. Miranda, M. N. Sato, and S. M. A. Andrade. 2001. *Stereo photo series for quantifying Cerrado fuels in Central Brazil.* Volume I. General Technical Report PNW-GRT-519. USDA-FS, Northwest Research Station, Portland.

Pinto, A. S. 2003. Contribuição dos solos de cerrado do Brasil—Central para as emissões de gases traço (CO_2, N_2O e NO): sazonalidade, queimadas prescritas e manejo de pastagens degradadas. PhD Dissertation, Universidade de Brasília, Brasília, Brazil, 114p.

Pinto, A. S., M. M. C. Bustamante, K. Kisselle, et al. 2002. Soil emissions of N_2O, NO, and CO_2 in Brazilian savannas: Effects of vegetation type, seasonality, and prescribe fires. *Journal of Geophysical Research* 107, 8089–8096.

Pinto, A. S., M. M. C. Bustamante, M. R. S. S. da Silva et al. 2006. Effects of different treatments of pasture restoration on soil trace gas emissions in the Cerrados of Central Brazil. *Earth Interactions* 10, 1–26.

Potter, C., S. Klooster, A. R. Huete, et al. 2009. Terrestrial carbon sinks in the Brazilian Amazon and Cerrado Region predicted from MODIS Satellite Data and ecosystem modeling. *Biogeosciences Discussions* 6, 1–23.

Reatto, A., J. R. Correia, and S. T. Spera. 1998. Solos do Bioma Cerrado: Aspectos pedológicos. In *Cerrado: Ambiente e flora*, eds. S. M. Sano and S. P. Almeida. Embrapa-CPAC, Planaltina, pp. 47–86.

Resck, D. V. S. 1998. Agricultural intensification systems and their impact on soil and water quality in the Cerrados of Brazil. In *Soil Quality and Agricultural Sustainability*. ed. R. Lal, 288–300. Ann Arbor Press, Chelsea.

Resck, D. V. S., C. A. Vasconcelos, L. Vilela, and M. C. M. Macedo. 1999. Impact of conversion of Brazilian Cerrados to cropland and pasture land on soil carbon pool and dynamics. In *Global Climate Change and Tropical Ecosystems*, eds. R. Lal, J. M. Kimble and B. A. Stewart, 169–196. Adv. Soil Sci., CRC Press, Boca Raton, FL.

Resende, M., N. Curi, S. B. Rezende, and G. F. Correa. 1997. *Pedologia: Base para a descrição de ambientes.* 2ed. Viçosa, NEPUT, 367pp.

Ribeiro, N. V., L. G. Ferreira, and N. C. Ferreira. 2009. Expansão da Cana-de-Açúcar no Bioma Cerrado: Uma análise a partir da modelagem perceptiva de dados cartográficos e orbitais. In *Proceedings of Simpósio Brasileiro de Sensoriamento Remoto (SBSR)* 14, 4287–4293. DVD, On-line. ISBN 978–85–17–00044–7. Available at: <http://urlib. net/dpi.inpe.br/sbsr@80/2008/11.14.00.00>. São José dos Campos: INPE.

Rocha, H. R., H. C. Freitas, R. Rosolem, R. I. N. Juárez, R. N. Tannus, M. A. Ligo, O. M. R. Cabral, and M. A. F. Silva Dias. 2002. Measurements of CO2 exchange over a woodland savanna (cerrado sensu stricto) in southeastern Brazil. *Biota Neotropica* 2, 1–11.

Roitman, I., J. M. Felfili, and A. V. Rezende. 2008. Tree dynamics of a fire-protected cerrado sensu strict surrounded by forest plantations, over a 13-year period (1991–2004) in Bahia, Brazil. *Plant Ecology* 197, 255–267.

Roscoe, R., P. Buurmann, E. J. Velthorst, and J. A. A. Pereira. 2000. Effects of fire on soil organic matter in a "cerrado sensu stricto" from southeast Brazil as revealed by changes in $\delta^{13}C$. *Geoderma* 95, 141–160.

Roscoe, R., P. Buurmann, E. J. Velthorst, and C. A. Vasconcellos. 2001. Soil organic matter dynamics in density and particle size fractions as revealed by the $^{13}C/^{12}C$ isotopic ratio in a Cerrado's Oxisol. *Geoderma* 104, 185–202.

Roscoe, R. and P. Buurman. 2003. Effect of tillage and no-tillage on soil organic matter dynamics in density fractions of a cerrado oxisol. *Soil and Tillage Research*, 70, 107–119.

Sano, E. E., R. Rosa, J. L. S. Brito, and L. G. Ferreira. 2009. Land cover mapping of the tropical savanna region in Brazil. *Environmental Monitoring and Assessment*, DOI: 10.1007/s10661-009-0988-4.

Santos, A. J. B., C. A. Quesada, G. T. Silva, et al. 2004. High rates of net ecosystem carbon assimilation by *Brachiaria* pasture in Brazilian cerrado. *Global Change Biology* 10, 877–885.

Scopel, E., B. Triomphe, M. F. dos Santos Ribeiro, L. Séguy, J. E. Denardin, and R. A. Kochhann. 2004. Direct seeding mulch-based cropping systems (DMC) in Latin America. In *Proceedings of the 4th International Crop Science Congress "New directions for a diverse planet"*, Brisbane, Australia.

Silva, J. E., D. V. S. Resck, E. J. Corazza, and L. Vivaldi. 2004. Carbon storage in clayey Oxisol cultivated pastures in the "Cerrado" region, Brazil. *Agriculture, Ecosystems and Environment* 103, 357–363.

Silva, W. F., D. A. Aguiar, B. F. T. Rudorff, L. M. Sugawara, and T. L. I. N. Aulicino. 2009. Análise da expansão da área cultivada com cana-de-açúcar na região Centro-Sul do Brasil: Safras 2005/2006 a 2008/2009. In, *Proceedings of Simpósio Brasileiro de Sensoriamento Remoto* (SBSR), 14, 467–474. DVD, On-line. ISBN 978–85–17–00044–7. Available at: <http://urlib.net/dpi.inpe.br/sbsr@80/2008/11.14.00.00>. São José dos Campos: INPE.

Sociedade Brasileira de Silvicultua. 2009. www.sbs.org.br, accessed in 04/14/09.

Sugawara, L. M., B. F. T. Rudorff, R. M. S. P. Vieira, et al. 2008. Imagens de satélite na estimativa de área plantada com cana na safra 2005/2006—Região Centro-Sul. São José dos Campos: Instituto Nacional de Pesquisas Espaciais, 74pp. (INPE-15254-RPQ/815).

Varella, R. F., M. M. C. Bustamante, A. S. Pinto, K. W. Kisselle, R. V. Santos, R. Burke, R. Zepp and L. T. Viana. 2004. Soil fluxes of CO2, CO, NO and N2O from an active old-pasture and from native savanna in Central Brazil. *Ecological Applications* 14, S221–S231.

20

Land Use Changes (1970–2020) and Carbon Emissions in the Colombian Llanos

Andrés Etter, Armando Sarmiento, and Milton H. Romero

CONTENTS

Introduction

Regional and local land use changes driven by the growing demands of the world population are considered prime factors contributing to the observed global climate change trends (Foley et al., 2005; Pielke et al., 2002). One major impact of such land use changes along the expanding agricultural fronts on climate is through carbon emissions from deforestation in forests and changing fire patterns, cattle densities, and vegetation biomass in savannas. Although forests have been considered to be the key land cover component of the terrestrial carbon cycle (Ramankutty et al., 2007; Houghton, 2005), the importance of savannas is being increasingly recognized as growing human impacts transform this biome (Grace, 2004). The global study by Goldewijk (2001) estimated that around half of the land clearing during the past three centuries has taken place in the savanna biomes. Grace et al. (2006) estimated that worldwide savannas are being transformed at an average rate of more than 1% per year, but reliable data on savanna transformation rates are currently not available (see Chapter 19 for cerrado data).

Land management practices are a major factor controlling the amounts of carbon within the terrestrial pool by influencing vegetation and soil characteristics (Grace et al., 2006; Ramankutty et al., 2007). Historically, savanna ecosystems have been extensively managed worldwide by humans for different production purposes, driving ecological processes such as fire frequency and biomass accumulation and consequently affecting the carbon cycle (Barbosa and Fearnside, 2005; Grace et al., 2006; Hoffmann et al., 2002). Beerling and Osborne (2006) contend that the intensification of land use and the anthropogenic forcing of climate will be the major determinants of the ecological future of savannas.

However, the spatiotemporal variability of carbon emissions from fire and other sources such as cattle in savannas is high due to the local and regional biophysical and land use heterogeneity (Wassmann and Vlek, 2004). Emissions from fires, for instance, vary significantly due to local and regional differences in ecosystem type, biomass, and combustion efficiency (Van Der Werf et al., 2003; Romero-Ruiz et al., 2010). Grace et al. (2006) point to important comprehension gaps in the knowledge and uncertainty about the ecological consequences of the transformation processes resulting from increasing land use changes in savanna ecosystems, with several contradictory views and unresolved aspects about their role in the global carbon emissions. An important aspect of such knowledge gaps concerns the intra and interregional biophysical and land use variability that makes extrapolations difficult.

Savannas are becoming increasingly important in the world food supply, especially in Latin America, and they are, therefore, prone to increasing human impacts (Ayarza et al., 2007; Brannstrom et al., 2008). In Venezuela, the extent of natural savannas decreased annually by 2.3% during the 1980s and 1990s from conversion to agriculture (San José and Montes, 2001); whereas in Brazil more than 700,000 km^2 (50%) of savannas had already been cleared for agriculture by the early 2000s (Brannstrom et al., 2008). The Llanos of Colombia and Venezuela are a significant part (18%) of savannas of northern South America that are subject to strong human pressures (Ayarza et al., 2007). These changes affect carbon emissions; a full understanding of these requires a spatially explicit analysis that has not yet been undertaken.

To address the above-mentioned knowledge gaps, the objectives of this study were to quantify the changes in carbon emissions from the ongoing and future land use changes in the Colombian Llanos savannas for the 1970–2020 period. The analysis relied on the construction of a scenario for 2010 and 2020, based on empirical data of savanna clearing for the 1970–2007 period. To address the effect of the spatial heterogeneity of carbon emission factors, a spatially explicit approach was applied using savanna ecosystems and land use or management systems data, along with data on burned area and frequency and cattle stocking rates.

Background and Study Area

The Llanos region is the northern component of the 2.6 million km² of South American savannas, spanning 450,000 km² east of the Andes across

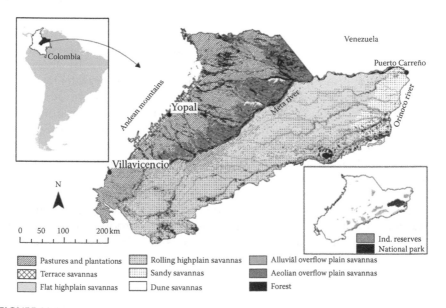

FIGURE 20.1
Location of the study area (black—Colombian Llanos), showing main ecosystem types and cleared areas for 2007 (Derived from Etter, A. 1998a. In *Informe Nacional sobre el Estado de la Biodiversidad Colombia 1997*, edited by M. E. Chaves and N. Arango. Bogotá, DC: Inst. Alexander von Humboldt y UNDP; Etter, A. 1998b. In *Informe Nacional sobre el Estado de la Biodiversidad en Colombia—1997 (National Biodiversity Report—1997)*, edited by M. E. Chaves and N. Arango. Bogotá, DC: Instituto Alexander von Humboldt. and Landsat images.), and areas with conservation status (Derived from Instituto Geografico Agustin Codazzi-IGAC-. 2005. *Mapas de Resguardos y AreasProtegidas*. Bogotá: IGAC.).

eastern Colombia and southern Venezuela in the Orinoco watershed below 250 m of altitude. In Colombia the Llanos region covers the western 170,000 km² (Figure 20.1), which corresponds to the wetter part of the region with rainfall varying from 2500 to 1500 mm in a west to north-east gradient (3 to 5 dry months). About 75% of the area was originally covered by savanna grassland ecosystems. These savannas are mostly chemically poor pedobiomes (sensu Walter, 1973), where Ultisol and Oxisol soil types predominate (Instituto Geográfico Agustin Codazzi—IGAC, 1983). The region is composed of two main physiographic units divided by the river Meta: the Andean piedmont to the north and the high plains to the south. The Andean piedmont has generally more fertile soils and includes large areas of wetlands and seasonally flooded savannas in the lower parts toward the river Meta. However, savanna ecosystems in the region are heterogeneous and can be divided into several types according to their geomorphology, soils, flooding, and vegetation type. The piedmont includes the Terrace savannas, the Alluvial overflow plain savannas, and the Aeolian overflow plain; whereas the high plains to the south include the Flat high plain savannas, the Rolling high plain savannas, and the Sandy high plain savannas (Figure 20.1).

Although fire is a natural phenomenon, historically burning has also been an important management practice in the Llanos; but there are no quantitative figures about their frequency and extent. Fire was used by aboriginal communities for hunting and savanna cleaning purposes and later by ranchers to provide fresh regrowth for cattle during the dry season. The region was occupied by aboriginal hunter-gatherer and farmer communities for several thousand years (Rausch, 1994). From the late 1500s, these savannas have been increasingly settled by Spanish missions and cattle ranches displacing aboriginal communities. The Llanos remained a frontier area dominated by extensive cattle grazing and extractive activities until a few decades ago. Only from the 1980s onward has the exploitation of oil fields led to the development of infrastructure, enhancing the regional economy and increasing the opportunities for more profitable agricultural production. As a result, the land use in large areas of natural savannas has changed from the predominant extensive grazing on natural grasslands to more intensive agricultural uses.

The Colombian Llanos are still sparsely populated, with a total population not exceeding 1.4 million equivalent to less than 5% of the national population, with a third living in rural areas (DANE—National Statistics Department, 2005). The population is concentrated within 50 km of the Andean mountain range where the main urban centers are located. The major economic centers are Villavicencio (380,000 inhabitants), which is well connected to the capital Bogotá and Yopal (85,000 inhabitants). There are another 20 minor towns, mostly municipal headquarters with up to 20,000 inhabitants. At present, the economy of the region depends mainly not only on the exploitation of oil fields, intensive agriculture (irrigated rice, soy, and

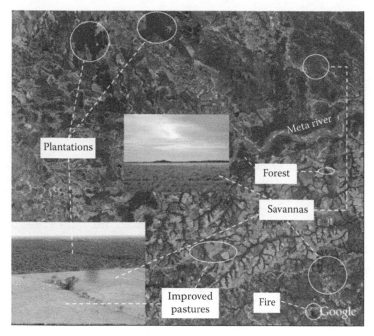

FIGURE 20.2
(See color insert following page 320.) Satellite image showing the main land covers of the Colombian Llanos and the savanna transformation process (Insert photos by A. Etter).

sorghum), and cattle ranching, but increasingly also on oil palm and forestry plantations (Figure 20.2). Extensive grazing of cattle is the dominant land use with a cattle herd of approximately 5.5 million head. Land tenure consists of a network of Indigenous Reserves covering around 1,340,000 ha, a National Park of 548,000 ha, whereas the remaining 90% is privately owned by ranchers and farmers, often through vacant possession without title (Figure 20.1).

With the agricultural frontier expanding into the region, the once-extensive land uses are rapidly changing (Etter, 1998a, b; San José et al., 2003; Ayarza et al., 2007). New land uses, such as intensive grazing (exotic pastures), plantations (oil palm, rubber, and timber), and intensive high input cropping (rice, soy, and corn), are changing several ecological processes including fire regimes and soil water retention (Amézquita et al., 2004; Romero-Ruiz et al., 2010). Some of the envisaged consequences of the land use changes in the Llanos include (i) reduction in fire frequency; (ii) increased tree cover from plantations; and (iii) increased cattle stocking rates. The possible effects of human transformation of the Llanos savannas on the carbon exchange with the atmosphere (Figure 20.3) are dependent on the diverse ecological systems, the biophysical processes that drove the evolution of the ecosystems and the differences in land tenure and land use regimes.

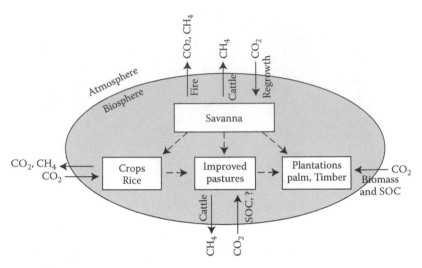

FIGURE 20.3
Diagram indicating the major land cover types (boxes) and transitions (broken lines) for the Colombian Llanos savannas and the implicated carbon fluxes to and from the atmosphere. Thick lines indicate the carbon fluxes addressed in this study.

Data and Methods

To analyze recent (1970–2007) and projected (2010, 2020) land use change and the effects on carbon emissions, a variety of data sources including remote sensing and GIS tools was used for three main purposes (Figure 20.4):

  i. Mapping land use change based on Landsat (1985 and 2000) and CBERS (China-Brazil Earth Resources Satellite: 2007) images

 ii. Prediction of future (2010 and 2020) savanna clearing using a logistic modeling framework

iii. Calculation of carbon emissions from past (1970–2007) and predicted (2010 and 2020) land use or cover change

Land Cover and Land Use Changes

Land cover patterns for 1985 and 2000 were mapped using remote sensing data from Landsat TM and ETM+ images, which have a 30 m ground resolution in 6 bands and 120 m resolution in one band. For 2007, CBERS data with 20 m resolution in 4 different bands were used. Information was downloaded from the University of Maryland (http://www.glcf.uniacs. umd.edu/) and Institute of Pesquisas Espaciales (INPE) (http://www.cbers.

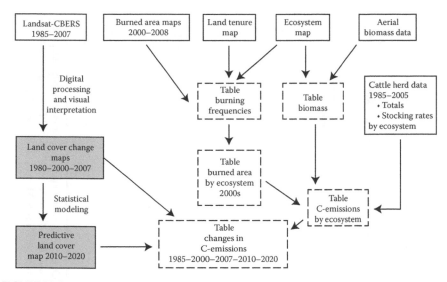

FIGURE 20.4
Methodological diagram indicating the inputs and intermediate and final results of the research process.

inpe.br) web sites. Each image was georeferenced using between 15 and 30 ground control points, with an RMSE between 0.4 and 0.7. All data were re-projected to UTM (WGS84 datum) coordinate system. No ancillary data on the atmospheric conditions were available for all information. Only enhanced using linear contrast stretching and histogram equalization was done (Jensen et al., 1996). A supervised processing and visual interpretation was used to obtain the land cover classification. Four general land cover types were examined: (i) forests, (ii) savannas, (iii) planted pastures and annual crops, and (iv) plantations. Supervised classification was done by selecting control areas in the form of training areas using an algorithm that considers the spectral characteristics of a group of pixels assigned to each one of the classes using a maximum likelihood algorithm. The accuracy of the 2000 and 2007 classified maps was checked with aerial photos and fieldwork by a stratified random sampling method distributed among small (369), medium (371), and large (351) polygons. A cross-tabulation table was produced and a change matrix was evaluated using Kappa analysis.

The savanna class was then re-classified into seven savanna ecosystem types by overlaying the pre-clearing ecosystem map from Etter et al. (1998; 2008) (Table 20.1): low terrace savannas; flat high plain savannas; rolling high plain savannas; sandy savannas; dune savannas; alluvial overflow plain savannas; and aeolian overflow plain savannas.

TABLE 20.1

Parameters Used in the Calculations of Equations 20.1 and 20.2

Ecosystem Code [1]	Vegetation Structure [1]	(b_f) Aboveground Biomass (ton.ha⁻¹) [1]	C in Biomass [2]	(a_f) Consumption by Fire [2]	(f_f) Fire Recurrence Factor [3]	(g_f) Stocking Rate (head/ha) [1]	(m) CO₂e (kg head⁻¹ y⁻¹) [4]
Terrace savannas	*Trachypogon, Andropogon, Curatella,* Dense tussock grassland with scattered trees	7	0.49	0.8	0.02	0.45	1840
Flat highplain savannas	*Trachypogon, Panicum, Leptocoryphium,* Dense tussock grassland	6	0.49	0.85	0.31	0.35	1840
Rolling highplain savannas	*Paspalum, Trachypogon, Bulbostylis,* Open tussock grassland	3.5	0.49	0.9	0.28	0.2	1840
Sandy savannas	*Bulbostylis, Paspalum, Curatella, Byrsonima,* Open grassland with scattered trees	3.5	0.49	0.95	0.39	0.2	1840
Dune savannas	*Paspalum, Byrsonima,* Open grassland with scattered trees	4	0.49	0.7	0.21	0.15	1840
Alluvial overflow plain savannas	*Andropogon, Axonopus, Leptocoryphium,* Dense tussock grassland	7	0.49	0.7	0.03	0.45	1840
Aeolian overflow plain savannas	*Trachypogon, Andropogon,* Dense tussock grassland	6.5	0.49	0.75	0.23	0.4	1840
Introduced pastures	*Brachiaria,* Dense grassland	8	n.a	n.a	n.a	1.2	2000

Source: [1] From Etter, A. 1985. A Landscape Ecological Approach for Grazing Development: A Case in the Colombian Llanos Orientales. MSc, Department of Land and Rural Surveys, ITC, Enschede, The Netherlands. [2] From Barbosa, R. I., and P. M. Fearnside. 2005. *Forest Ecology and Management* 204, 371–384. [3] From Romero, M. et al. 2010. *Global Change Biology* 16, 2013–2023. [4] From McGinn, S. M. et al. 2004. *Journal of Animal Science* 82, 3346–3356.

Predictive Model of the Expansion of the Agricultural Frontier in the Llanos, 2007–2020

The savanna clearing process began in the early 1970s, the reason for which the year 1970 was used as the pre-clearing reference in the region. With the land cover maps of 1985, 2000, and 2007, the cleared areas for the 1970–1985, 1985–2000, and 2000–2007 were mapped and calculated. A cross-tabulation was performed between the clearing map for each period and the ecosystem map (Etter, 1998a, b), to calculate the affected areas for each ecosystem type by period.

To estimate the area of savanna to be cleared by 2010 and 2020, a logistic curve was applied using the total cleared area data of 1970, 1985, 2000, and 2007:

$$CA_t = \frac{S}{1 + Ke^{bt}}$$

(20.1)

where CA_t is cleared area in year t, S is the expected savanna clearing saturation point, K is proportion of the saturation area cleared at the beginning of observed reference period (1985–2000), e is base logarithm constant, b is the clearing rate of the observed reference period, and t is time in years. The saturation (S) was estimated at 7.2 million ha based on the exclusion of areas not available for clearing including National Parks, Indigenous Reserves, and areas with no or very limited mechanization possibilities (70% of rolling high-plain savannas and 100% of dune savannas).

The future expansion of the savanna clearing for 2010 and 2020 was projected, assuming that conditions and clearing trends remain constant in time. The sum of all cleared savanna areas (pastures, oil palm, and forest plantations) was included within this expansion figure, 412,000 ha for the 2007–2010 period and 1,823,000 ha for the 2010–2020 period. To spatially allocate the projected area, we used a statistical modeling framework based on logistic regression. The logistic regression model was applied to the binary (1,0) 2007 cleared area map using distance to roads (*distROADS*), distance to towns (*dist TOWNS*), distance to rivers (*distRIVERS*), and soil fertility (*distSOIL*) as explanatory variables (Equation 20.2). Colinearity of explanatory variables was tested with Pearson's correlation coefficient, leading to retaining all explanatory variables for further analysis, as the colinearities were below 0.7 (Green, 1979). To run the regression, we locked the areas corresponding to the National Park and the Indigenous Reserves, assuming that land clearing and land development will mainly take place in private lands. The modeling was performed using the *Logisticreg* module in the IDRISI 15.01 software, using the following model:

$$Logit\ (CLEAR) = b1 + b2 * distROADS + b3 * distRIVERS + b4 * distTOWNS + b5 * fertSOIL$$

(20.2)

The model was assessed through the Receiver Operator Characteristics (ROC) curve (Metz, 1978), with the *Logisticreg* module from IDRISI. Statistical significance of variables was calculated with the software package R.

To plot the predicted savanna clearing for the years 2010 and 2020 (Equation 20.1) on a map, the values of the predictive model were ordered from the highest probability downward and were cut to match the number of expected cleared cells for the areas from the regression curve (Equation 20.1) for each of the 2010 and 2020 dates.

Calculations of Carbon Emissions

The calculation of the carbon emissions was done for the fluxes expected to change and for which there is reliable data (Figure 20.3). The analysis concentrated on the three major land use change processes expected in the coming years: (a) intensification of cattle grazing, (b) increase in intensive cropping (irrigated rice), and (c) increase in plantations (mainly oil palm).

The data used to calculate the changes of the carbon emissions of the past and predicted land use scenarios included the following (Figure 20.4):

a. Multitemporal land cover maps (1985, 2000, 2007; and predicted 2010, 2020)

b. Multitemporal burning maps (2000–2008) (Romero-Ruiz et al., 2010)

c. Ecosystem map (Etter, 1998b; Etter et al., 2008)

d. Land tenure map (Instituto Geografico Agustin Codazzi—IGAC, 2005)

e. Aboveground biomass data for each savanna ecosystem (Etter, 1985)

f. Cattle stocking rate data for each savanna ecosystem (Etter, 1985)

All land uses (extensive grazing in savannas, intensive grazing in improved pastures, irrigated rice, and oil palm plantations) were analyzed for all dates (1970, 1985, 2000, 2007, 2010, and 2020), for each savanna type and each land tenure category, based on total non cleared area, biomass data, burned area, fire frequency, and stocking rates, using the sources indicated in Table 20.1.

The calculation of carbon emissions was done in CO_2e (carbon dioxide equivalents) following the IPCC standards (IPCC, 2006). For biomass accumulation in oil palm plantations, an average annual carbon uptake of 5.1 ton CO_2e ha^{-1} was used (Germer and Sauerborn, 2008). For irrigated rice, a methane emission factor of 20 g CH_4 m^2 per crop cycle was used, taken from Yan et al. (2005). Recapture of emissions from fire through regrowth during the next year is assumed to be 85% of the CO_2e emissions, with 15% non-recaptured (methane, nitrous oxides, and nonmethane hydrocarbons) calculated from the warming potential using fire emission factor data from Delmas et al. (1995). We assumed that most grass regrowth in pastures is consumed by cattle, of which more than 90% is returned by respiration or oxidation in the form of CO_2 to the atmosphere following San José and Montes (2001); and that under current low input management conditions, soils are not a significant carbon (da Silva et al., 2004). Since savannas are almost exclusively grass savannas, the only stock of aboveground biomass to change during the current land use process is where plantations (basically oil palm) are grown.

To integrate these data into calculations, the following equations were used:

$$YE = \sum_{i,j} [\![X_{ji} \cdot (f_j \cdot [\![b_j \cdot a_j + g_j \cdot m) + \sum_i (P_i \cdot g_p \cdot m + R_i \cdot r) \quad (20.3)$$

$$YC = \sum_{i,j} X_{ij} \cdot b_j \cdot f_j \cdot c + \sum_i Z_i \cdot d_z \quad (20.4)$$

$$Y_{net} = YE - YC \quad (20.5)$$

where YE and YC are the annual emission and annual capture in CO_2e, X_{ij} is area of ecosystem j in year i, f_j is the average proportion of annually burned area for ecosystem j, b_j is the average aboveground biomass per hectare of ecosystem j, and a_j is the proportion of biomass consumption by fire for ecosystem j; g_j is the average cattle density per hectare for ecosystem j, and m is the CO_2e emission of enteric fermentation and manure per head per year; P_i is the total area under improved pastures in year i, and g_p is the average cattle density per hectare in improved pastures; R_i is the cultivated irrigated rice area in year i, and r is the annual CO_2e emission factor for irrigated rice; c is the CO_2 grass regrowth factor in savannas and pastures; and Z_i is the area of oil palm plantations in year i, and d_z is the annual CO_2 uptake (aboveground biomass and soil) per area for palm plantations.

Results

Observed 1970–2007 and Predicted 2010–2020 Land Cover Changes

In 1985, over 500,000 ha of savanna had already been cleared and converted to pastures, 85,000 ha to rice crops, and some 14,000 ha to oil palm and few timber plantations, mostly located in the vicinity of the major towns Villavicencio and Yopal. In general, the expansion of the agricultural frontier had concentrated in the south west and west of the region corresponding to the Meta Department but has increasingly moved to the north west into the

Departments of Casanare and Arauca. Based on the growth trends of pasture and plantation area from 1970, 1985, 2000, and 2007, the clearing of savannas has been increasing almost exponentially and is currently progressing at an annual rate of more than 100,000 ha for pastures and 5–10,000 ha for oil palm plantations.

The predictive model for 2007 had an ROC greater than 0.8 (ROC = 0.84), indicating a good model fit (Fielding and Bell, 1997; Table 20.2). The clearing pattern was well explained by variables related to accessibility, such as distance to roads and towns and to the fertility of soils and slope (Table 20.2). If the current trends hold, a projected additional 2,235,000 ha of savannas will be cleared for pastures (1,935,000), plantations (200,000), and rice (100,000) by 2020; which is an annual clearing rate of more than 200,000 ha by 2020. The clearing rate increased from 0.3% (1970–1985) to 0.9% (2000–2007) and is expected to reach 2% by 2020. Most of the predicted expansion of the agricultural frontier in the Llanos will occur in a westward direction along the Andean range where accessibility in the region is best (Figure 20.5).

The savanna ecosystems predicted to be most affected by the frontier expansion of pastures and plantations (oil palm and timber) are the alluvial overflow plain savannas, the flat high plain savannas, and the aeolian overflow plain savannas (Table 20.3). Areas limited for mechanization or that are relatively inaccessible such as the rolling high plain savannas and the sandy high plain savannas are expected to continue under natural vegetation covers and more extensive land uses.

Trends in C Emissions for 2020 from Land Use Changes in Cattle Grazing and Plantations

According to the land use model and the calculations performed, the net longer term effects of land use changes result in a regional increase in carbon emissions. The calculated net emissions of the region for 2007 are 13.91 Tg

TABLE 20.2

Results of the Logistic Regression Model for the 2007 Cleared Savanna Areas

Variable	Estimate	Std. Error	z Value	Pr(>|z|)
(Intercept)	−4.823	0.182	−26.427	<2e-16 ***
distTOWNS	0.309	0.022	14.082	<2e-16 ***
distRIVERS	−0.036	0.022	−1.61	0.107 ns
distROADS	−0.226	0.015	−15.042	<2e-16 ***
fertSOIL	0.637	0.026	24.424	<2e-16 ***
SLOPE	−0.294	0.05	−5.683	1.32e-08 ***

*** 0.001; ** 0.01; * 0.05; ns 0.1–1.

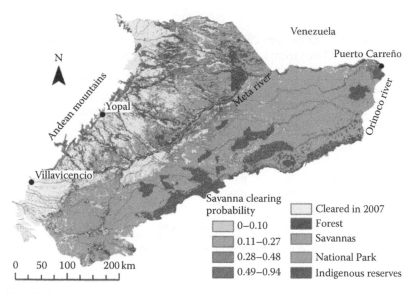

FIGURE 20.5
(**See color insert following page 320.**) Predicted future agricultural frontier expansion into the savanna ecosystems of the Colombian Llanos for the period 2007–2020 obtained by applying the logistic regression model. The progression of the expansion is shown by the red shadings from darker to lighter ones.

CO_2e. The expansion of agricultural land uses up until 2007 increased the overall net carbon emissions by 21.5% with reference to 1970 (approximate preclearing), representing some 2.46 Tg yr^{-1} CO_2e. The projected expansion from 2007 to 2020 is expected to add an additional 1.16 Tg yr^{-1} CO_2e, meaning an overall increase of 31.5% from the 1970 baseline (Table 20.4). Until 2007,

TABLE 20.3

Progression of the Savanna Clearing Process as Showed by the Proportion Area Cleared for Each Ecosystem

Savanna Type	1970	1985	2000	2007	2010 Projected	2020 Projected
Terrace savannas	0.1	0.62	0.89	0.92	0.94	0.98
Alluvial overflow plain savannas	0.00	0.06	0.34	0.50	0.71	0.86
Flat highplain savannas	0.00	0.08	0.14	0.19	0.21	0.32
Aeolian overflow plain savannas	0.00	0.01	0.11	0.19	0.23	0.41
Rolling highplain savannas	0.00	0.00	0.04	0.06	0.1	0.12
Sandy savannas	0.00	0.00	0.00	0.01	0.03	0.06
Dune savannas	0.00	0.00	0.00	0.01	0.01	0.01
All savannas	0.00	0.05	0.15	0.21	0.25	0.37

TABLE 20.4

Carbon Emissions in CO_2e from Past and Predicted Land Use Changes (1985–2020), Indicating Contributions from Land Cover/Land Use Sources, and Differences in Relation to the 1970 Reference

Year	Savannas ($Tg.yr^{-1}$) Fire[a]	Cattle	Pastures ($Tg.yr^{-1}$) Cattle	Irrigated Rice ($Tg.yr^{-1}$)	Plantations ($Tg.yr^{-1}$)	Net Emissions ($Tg.yr^{-1}$)	Difference from 1970 (%)
1970	3.90	7.24	0.00	0.35	−0.03	11.45	0.0
1985	3.70	6.82	1.00	0.49	−0.11	11.90	3.9
2000	3.40	5.89	3.76	0.69	−0.45	13.29	16.0
2007	3.24	5.37	5.19	0.86	−0.75	13.91	21.5
2010	3.14	5.08	6.00	1.04	−0.98	14.29	24.8
2020	2.75	4.05	8.71	1.44	−1.88	15.07	31.6

[a] Net fire emissions discounting the regrowth of the savannas.

the average annual increase in emissions from cattle was around 0.09 Tg CO_2e. An expansion of improved pastures replacing close to 2 million hectares by 2020 implies that the regional cattle herd will grow by more than 1.8 million head and trigger two major changes with reference to 2007: an increase in annual methane emissions from enteric fermentation (2.2 Tg CO_2e) and a reduction of the net annual fire emissions (0.49 Tg CO_2e) due to smaller burned area.

Calculations indicate that the net emissions from fires decreased by 16.9% by 2007 and are expected to decrease close to 29.5% by 2020 with current trends. Cattle emissions, in turn, increased by 36% by 2007 and are expected to exhibit a 77% increase by 2020 with respect to 1970. Rice emissions are predicted to increase fourfold over the 1970–2020 period, adding an additional 1.1 Tg CO_2e per year to the atmosphere. The main factor having a slowing effect on emissions is the growing area of plantations during the 2000s. Land use intensification shows two factors that are reducing the regional emissions: the carbon sequestration due to the increase in oil palm plantations and the decreasing emissions from savanna burning as a result from increased areas of planted pastures. These processes are partly compensating the increasing cattle and irrigated rice emissions.

Discussion

The extent of land transformation in savanna ecosystems in the Colombian Llanos is currently around 20%, which is low when compared with Brazilian (50%) and Venezuelan (35%) savannas. In Venezuela, the grassland savannas decreased annually by 2.3% (500,000 ha) during the 1980s and 1990s from conversion to pastures, plantations, and crops (San José et al., 2003); whereas for the Cerrado savannas of Brazil, Brannstrom et al. (2008) give figures between 1.3 and 2.6% for the period 1986–2002, depending on region. In Colombia, the rate increased from 0.5% during the 1980s–1990s to 1% in 2000s and is expected to reach 2% in the 2020s, equivalent to more than 200,000 ha per year. The likelihood of the predicted agricultural expansion into the Colombian Llanos is high, given the increasing integration of the region into the national economy. This integration arises from expansion of oil fields and national development policies that emphasize expanded food and bio-fuel production to match international demand.

Our study shows how the various ecosystems contribute differently to emissions (Table 20.1) and are differentially impacted by the expansion of the agricultural frontier (Table 20.3), so that under different scenarios the results are sensitive to both the amount and location of the future land use changes. The net carbon emissions from savanna fires in Colombia have been diminishing with the changing land use trends in both absolute terms and per unit area, because the more fertile areas with higher biomass are undergoing a faster conversion. However, this is largely compensated by the CH_4 emissions from increased cattle stocking rates in the improved pastures replacing the savannas. By contrast, in conservation areas such as National Park and Indigenous Reserves, fires remain high; but emissions will not increase due to the absence of cattle and/or changes in land use (Romero-Ruiz et al., 2010).

There are large discrepancies between the existing data on the carbon balance of the Colombian and Venezuelan savannas that cannot be con-veniently explained by differences in area or major land use trends. Our results suggest that in the year 2000 the Colombian Llanos accounted for an annual emission of close to 13.3 Tg CO_2e. In contrast, San José and Montes (2001) estimated for the 1982–1992 period an average sink of 17.53 Tg C per year (64.15 Tg CO_2e) for the Venezuelan Llanos. However, if the warming potential of the presented methane emissions from cattle is taken into account, the carbon sink would be reduced to only about 48 Tg CO_2e per year. A possible source of the conflict with the results presented here is the appar-ent contradiction in the data for Venezuela. These data show an important and increasing reduction in the natural vegetation area; however, most of the estimates of captured carbon are derived from fluxes associated with high native vegetation regrowth values. Although the data from the two studies are not strictly comparable, these discrepancies in carbon balances point to the need for additional comparative studies that help identify possible gaps and quantify uncertainties.

A controversial aspect that we did not address in this study is the effect of replacing savanna by improved pasture on the soil carbon stocks. The study in Venezuela by San José et al. (2003) showed that after 30 years both the SOC and the total nitrogen increased in fallows but remained in equilibrium in the

Brachiaria sp. pastures and in the native savanna. In contrast, Fisher et al. (1994) have claimed that there is an increase in SOC as a result of introduced pastures in Brazil of around 0.1–0.5 Gt C year⁻¹. However, da Silva et al. (2004) contend that although carbon accumulation does occur in the Brazilian Cerrado savannas under fertilized and well-managed pastures, most of the 50 million ha of cultivated pastures in the region are poorly fertilized and inadequately grazed and are not acting as a driver of soil carbon storage. The low level of inputs in most pastures in the Llanos would point to low accumulation levels, if any. However, Grace et al. (2006) caution that the behavior of soil carbon in savannas is not yet sufficiently well understood to draw general conclusions.

Our study did not include non-carbon emissions associated with irrigated rice and cattle such as N_2O. Also, land use changes in the Llanos will increasingly affect several other ecological processes such as fire activity, water availability, soil productivity, and the vulnerability to introduced species invasions and the persistence of native biodiversity, which have effects even on emissions themselves. Soils in South American savannas are prone to compaction and erosion under intensive mechanical farming; soils in Colombian savannas are particularly problematic, because they tend to be shallow and have high bulk density and low infiltration rates (Amézquita et al., 2004). Protection from fire and grazing in Venezuelan savannas during 25 years leads to annual carbon sequestration rates of 1 tC ha⁻¹ (San Jose and Montes, 1998). However, what the longer term response of ecosystem functioning will be to fire exclusion, such as the increasing fire risks and changes in nutrient cycling, is still unclear. Although globally savannas currently act as a C sink, careless management can lead to degradation of vegetation and soil, with losses of C to the atmosphere (Grace et al., 2006).

Other direct and indirect consequences from land use changes on emissions concern impacts from invasive processes of introduced pastures and plantations. Introduced grasses have been planted for decades in the Llanos savannas to improve the forage quality. Common alien grasses in the Llanos such as Braquiaria (*Brachiaria* spp.) and Gordura (*Melinis minutiflora*) have been reported in Brazilian savannas as changing fire patterns (Mistry and Berardi, 2005), inhibiting tree regeneration (Hoffmann and Haridasan, 2008), and posing threats to biodiversity by outcompeting natural grasses (Hoffmann et al., 2004; Pivello et al., 1999). Also, pastures tend to decline in productivity and enter phases of degradation 4–10 years after sowing, needing in general high maintenance fertilization (da Silva et al., 2004; Boddey et al., 2004).

Finally, an important aspect that was not addressed here is the impact that the mechanization process in intensive agriculture such as oil palm plantations has on carbon emissions. Some emissions from fuels may originate locally, whereas others may take place somewhere else, such as where fertilizers are being produced.

Concluding Remarks

This study demonstrates that the impact on carbon emissions from the land use intensification trends in the Colombian Llanos is sensitive to both the type of change in land use and where it will actually take place. Therefore, management oriented toward protecting and enhancing carbon stocks needs to take into account not only what land use transitions are occurring but also under what biophysical and management conditions they are taking place. Since water use efficiency in this type of biophysical context is constrained by climate seasonality and soil fertility, the carbon storage capacity is limited (da Silva et al., 2004), and global climate change is likely to have important impacts in the future.

Land use change and the increasing conversion of the natural savannas into pastures of introduced African grasses, crops, and plantations is changing the setting of a broad range of ecological processes in the Llanos savannas. We advanced some of the specific effects of the major land use changes so far, increasing the trend of carbon emissions in the Llanos. Additionally, a more thorough calculation of carbon storage or release in soils after land use change, based on deeper understanding of the SOC and total soil nitrogen, is needed to devise adequate management systems for the Llanos in the future. Especially important is the differential response of the soils of different ecosystem types to land use change and disturbance, such as well-drained poor soils and flood savanna soils.

Policies to manage or reduce carbon emissions and other impacts from savannas need to consider current land use change trends and their spatiotemporal patterns. A way of doing so would be to apply a model such as the "Carbon calculator" (Hill et al., 2003, 2006) adapted to the Llanos context. Such a system would provide a means to systematically include new data and interactively evaluate different scenarios.

How changes in processes such as the capture and release of carbon do interact with other ecological changes such as soil fertility, water availability, the expansion of invasive plants, and changes in composition and structure of vegetation in the savannas is not well known and needs to be addressed to anticipate more comprehensively the overall environmental impacts of current and prospective land uses.

Acknowledgments

Research was funded by the Javeriana University Research Fund. The Alban Program of the European Union, the Colfuturo Fund (Colombia), and The Nature Conservancy (Northern Andes Program) provided funding to MR.

We acknowledge valuable inputs from reviewers and editors that helped improve the manuscript.

References

Amézquita, E., R. J. Thomas, I. M. Rao, D. L. Molina, and P. Hoyos. 2004. Use of deep-rooted tropical pastures to build-up an arable layer through improved soil properties of an Oxisol in the Eastern Plains (Llanos Orientales) of Colombia. *Agriculture, Ecosystems & Environment* 103, 269–277.

Ayarza, M., E. Amézquita, I. Rao, et al. 2007. Advances in improving agricultural profitability and overcoming land degradation in savanna and hillside agroecosystems of tropical America. In *Advances in Integrated Soil Fertility Management in Sub-Saharan Africa: Challenges and Opportunities*, ed. A. Bationo. New York: Springer.

Barbosa, R. I. and P. M. Fearnside. 2005. Fire frequency and area burned in the Roraima savannas of Brazilian Amazonia. *Forest Ecology and Management* 204, 371–384.

Beerling, D. J. and C. P. Osborne. 2006. The origin of the savanna biome. *Global Change Biology* 12, 2023–2031.

Boddey, R. M., R. Macedo, R. M. Tarro, et al. 2004. Nitrogen cycling in Brachiaria pastures: The key to understanding the process of pasture decline. *Agriculture, Ecosystems & Environment* 103, 389–403.

Brannstrom, C., W. Jepson, A. M. Filippi, D. Redo, Z. Xu, and S. Ganesh. 2008. Land change in the Brazilian Savanna (Cerrado), 1986–2002: Comparative analysis and implications for land-use policy. *Land Use Policy* 25, 579–595.

da Silva, J. E., D. V. S. Resck, E. J. Corazza, and L. Vivaldi. 2004. Carbon storage in clayey Oxisol cultivated pastures in the "Cerrado" region, Brazil. *Agriculture, Ecosystems & Environment* 103, 357–363.

DANE—National Statistics Department. 2007. *Censo Nacional de Poblacion (National Population Census)*. http://www.dane.gov.co/inf_est/poblacion 2005 [cited 15.06.03 2007].

Delmas, R., J. P. Lacaux, and D. Brocard. 1995. Determination of biomass burning emission factors: Methods and results. *Environmental Monitoring and Assessment* 38, 181–204.

Etter, A. 1985. A Landscape Ecological Approach for Grazing Development: A Case in the Colombian Llanos Orientales. MSc, Department of Land and Rural Surveys, ITC, Enschede, The Netherlands.

Etter, A. 1998a. Ecosistemas de Sabanas. In *Informe Nacional sobre el Estado de la Biodiversidad Colombia 1997*, edited by M. E. Chaves and N. Arango. Bogotá: Inst. Alexander von Humboldt y UNDP.

Etter, A. 1998b. Mapa general de ecosistemas de Colombia—1:2 000 000 scale, (General ecosystem map of Colombia). In *Informe Nacional sobre el Estado de la Biodiversidad en Colombia—1997 (National Biodiversity Report—1997)*, eds. y M. E. Chaves and N. Arango. Bogotá, D.C.: Instituto Alexander von Humboldt.

Etter, A., C. McAlpine, and H. Possingham. 2008. A historical analysis of the spatial and temporal drivers of landscape change in Colombia since 1500. *Annals of the Association of American Geographers* 98, 1–27.

Fielding, A. H. and J. F. Bell. 1997. A review of methods fopr assessment of prediction errors in conservation presence/absence models. *Environmental Conservation* 24, 38–49.

Fisher, M. J., I. M. Rao, M. A. Ayarza, et al. 1994. Carbon storage by introduced deep-rooted grasses in the South American savannas. *Nature* 371, 236–238.

Foley, J. A., R. DeFries, G. P. Asner, et al. 2005. Global consequences of land use. *Science* 309, 570–574.

Germer, J. and Æ J. Sauerborn. 2008. Estimation of the impact of oil palm plantation establishment on greenhouse gas balance. *Environment Development and Sustainability* 10, 697–716.

Goldewijk, K. K. 2001. Estimating global land use change over the past 300 years: The HYDE database. *Global Biogeochemical Cycles* 15, 417–433.

Grace, J. 2004. Understanding and managing the global carbon cycle. *Journal of Ecology* 92, 189–202.

Grace, J., J. San Jose, P. Meir, H. S. Miranda, and R. A. Montes. 2006. Productivity and carbon fluxes of tropical savannas. *Journal of Biogeography* 33, 387–400.

Green, R. H. 1979. *Sampling Design and Statistical Methods for Environmental Biologists.* New York: Wiley.

Hill, M. J., R. Braaten, and G. M. McKeon. 2003. A scenario calculator for effects of grazing land management on carbon stocks in Australian rangelands. *Environmental Modelling & Software* 18, 627–644.

Hill, M. J., S. J. Roxburgh, G. M. McKeon, J. O. Carter, and D. J. Barrett. 2006. Analysis of soil carbon outcomes from interaction between climate and grazing pressure in Australian rangelands using Range-ASSESS. *Environmental Modelling and Software* 21, 779–801.

Hoffmann, W. A., W. Schroeder, and R. B. Jackson. 2002. Positive feedbacks of fire, climate, and vegetation and the conversion of tropical savanna. *Geophysical Research Letters* 29, 2052–2055.

Hoffmann, W. A. and M. Haridasan. 2008. The invasive grass, *Melinis minutiflora*, inhibits tree regeneration in a Neotropical savanna. *Austral Ecology* 33, 29–36.

Hoffmann, W. A., Verusca M. P. C. Lucatelli, F. J. Silva, et al. 2004. Impact of the invasive alien grass *Melinis minutiflora* at the savanna-forest ecotone in the Brazilian Cerrado. *Diversity and Distributions* 10, 99–103.

Houghton, R. A. 2005. Aboveground forest biomass and the global carbon balance. *Global Change Biology* 11, 945–958.

Instituto Geográfico Agustin Codazzi—IGAC. 1983. *Mapa general de suelos de Colombia - 1:1,500,000 (General soils map of Colombia).* Bogotá: IGAC.

Instituto Geografico Agustin Codazzi-IGAC. 2005. *Mapas de Resguardos y Areas Protegidas.* Bogotá: IGAC.

IPCC 2006. Agriculture, Forestry and Other Land Use (Chapter 4). In *2006 IPCC Guidelines for National Greenhouse Gas Inventories*, Prepared by the National Greenhouse Gas Inventories Programme, eds. B. L. Eggleston, H. S., Buendia, L., Miwa K., Ngara T., and Tanabe K., IGES, Japan.

McGinn, S. M., K. A. Beauchemin, T. Coates, and D. Colombatto. 2004. Methane emissions from beef cattle: Effects of monensin, sunflower oil, enzymes, yeast, and fumaric acid. *Journal of Animal Science* 82, 3346–3356.

Metz, C. E. 1978. Basic principles of ROC analysis. *Seminars in Nuclear Medicine* 8, 283–298.

Mistry, J. and A. Berardi. 2005. Assessing Fire Potential in a Brazilian Savanna Nature Reserve1. *Biotropica* 37, 439–451.

Pielke, R. A., G. Marland, R. Betts, et al. 2002. The influence of land use change and landscape dynamics on the climate system: Relevance to climate change policy beyond the radiative effect of greenhouse gases. *Philosophical Transactions of the Royal Society of London A* 360, 1705–1719.

Pivello, V. R., V. M. C. Carvalho, P. F. Lopes, A. A. Peccinini, and S. Rosso. 1999. Abundance and distribution of native and alien grasses in a "Cerrado" (Brazilian Savanna) Biological Reserve1. *Biotropica* 31, 71–82.

Ramankutty, N., H. K. Gibbs, F. Achard, R. Defries, J. A. Foley, and R. A. Houghton. 2007. Challenges to estimating carbon emissions from tropical deforestation. *Global Change Biology* 13, 51–66.

Rausch, J. M. 1994. *Una frontera de la sabana tropical los Llanos de Colombia 1531–1831, Colección Bibliografica Banco de la Republica—Historia Colombiana*. Bogotá: Banco de la Republica.

Romero-Ruiz, M., A. Etter, A. Sarmiento, and K. Tansey. 2009. Spatial and temporal variability of burned area in relation to ecosystems, land tenure and rainfall in the Colombian Llanos savannas. *Global Change Biology* 16, 2013–2023..

San José, J. J. and R. A. Montes. 2001. Management effects on carbon stocks and fluxes across the Orinoco savannas. *Forest Ecology and Management* 150, 293–311.

San José, J. J., R. A. Montes, and C. Rocha. 2003. Neotropical savanna converted to food cropping and cattle feeding systems: Soil carbon and nitrogen changes over 30 years. *Forest Ecology and Management* 184, 17–32.

Van Der Werf, G. R., J. T. Randerson, G. J. Collatz, and L. Giglio. 2003. Carbon emissions from fires in tropical and subtropical ecosystems. *Global Change Biology* 9, 547–562.

Walter, H. 1973. *Vegetation of the Earth in Relation to Climate and the Ecophysiological Conditions*. Universities Press, Heidelberg Science Library, London.

Wassmann, R. and P. L. G. Vlek. 2004. Mitigating greenhouse gas emissions from tropical agriculture: Scope and research priorities. *Environment, Development and Sustainability* 6, 1–9.

Yan, X. and K. Yagi. 2005. Statistical analysis of the major variables controlling methane emission from rice fields. *Global Change Biology* 11, 1131–1141.

21

Modeling Vegetation, Carbon, and Nutrient Dynamics in the Savanna Woodlands of Australia with the AussieGRASS Model

John O. Carter, Dorine Bruget, Beverley Henry,
Robert Hassett, Grant Stone, Ken Day, Neil Flood,
and Greg McKeon

CONTENTS

Introduction

Models of savanna and grassland systems in Australia are used for two distinct aims. The first aim is an applied operational role, where models are run in near real-time and results are used to address the information needs of public policy decisions, for example, drought situation analysis; State of Environment reporting; and provision of climate risk assessments of pasture resources for tactical management of grazing and fire. The second aim is to improve the scientific understanding of processes and to simulate the likely long-term outcomes and risks of current management practices for strategic policy and planning purposes.

AussieGRASS (Carter et al., 2000) provides a simulation of the soil water balance and pasture dry matter fluxes over the Australian continent. Simulations are spatial and are performed on individual grid cells (approximately 5 km × 5 km) across Australia. AussieGRASS uses a version of the GRASP model (Rickert et al., 2000) and simulates the hydrology and dry matter flow of pasture communities at a daily time step. It includes, as gridded inputs (5 km × 5 km), information layers of daily climate data, soil attributes (e.g., plant available water range), tree cover, pasture community, and densities of grazing animals (i.e., domestic livestock, feral and native herbivores). The information layers of soil type and pasture community allow specification of parameters for available water range of four soil layers and pasture attributes affecting plant growth, senescence, detachment, and decomposition. A key feature of GRASP is that hydrological processes such as pasture transpiration and runoff are driven by dynamic groundcover estimates that are updated daily. Thus, AussieGRASS is a dynamic model, responding to daily variation in climatic variables and moderated by management, fire, and herbivory.

Major applications of AussieGRASS since the mid-1990s include

- Ranking current conditions for drought assistance using 120 years of simulation outputs (Day et al., 2005)
- Generating products for near real-time assessment of impacts of climate variability (NAMS, 2009)
- Simulating production from grazing lands (Kokic et al., 2004)
- Operational climate risk assessment using conditional probabilities based on current conditions and the behavior of El Niño-Southern Oscillation (Henry, 2006)
- Simulating dynamic cover for regional catchment management groups and other modeling applications for assessment of carbon dynamics (Hill et al., 2005)
- Detecting climate change impacts on soil moisture, flow to stream, and pasture growth (McKeon et al., 2009)

- Generating time series for State of Environment reporting (e.g., Stone et al., 2008)

This chapter describes recent progress in the development of the AussieGRASS model and simulation studies to address current major issues, including

- Assimilation of model-driving variables and observations including remote sensing (i.e., model data fusion)
- Improvement of the modeling of surface cover (especially for use in seasonal cover forecasts, times before the satellite era and for those areas where tree density precludes use of remote sensing)
- Evaluation of the model in terms of the simulation of carbon and nitrogen fluxes and evaluation of the sensitivity of the model to changes in livestock numbers and fire

Assimilation of Observations for Model Calibration and Validation

The calibration and validation of a spatial model of continental coverage presents formidable computing and data assimilation challenges. Ideally, any observation in time or space (with error estimate if available) should be reflected in the simulation output of that variable from the model. Observational data may be available in a number of forms, such as

- Point observations that can be matched by specifying geographic coordinates with a date
- Grids of observations (e.g., satellite images)
- Average data over an area (e.g., daily runoff from small catchments)
- Area and time-averaged data (e.g., wool production for a statistical survey region over a 12-month period)

Integrating a diverse range of data types into the simulation framework assists with model calibration and validation but presents challenges for data management, as discussed below.

Model-Driving Data Variables

Climate Data

Climate data are essential to produce accurate outputs for all locations simulated. In Australia, daily gridded rainfall data surfaces are available for the continent back to 1890 (Jeffrey et al., 2001). Climate data are only consistently available back to about 1890 in the eastern half of the continent and 1907 in western regions (Rayner et al., 2004). Long-term daily averages are used where no actual measurements exist, though this data deficiency is likely to be addressed in the future by extracting information from Global Circulation Model (GCM) re-analysis data grids (Compo et al., 2006). If savanna or grassland models are to be useful to resource managers, they need to be run in near real-time and in forecast mode; and, therefore, meteorological observations need to be available on a near-daily basis. Seasonal and longer-term model simulations require gridded forecasts of future climate. Such forecasts are readily obtained from historical analogs (i.e., data selected from earlier years), where the analogs are provided either from statistical systems (e.g., Stone and Auliciems 1992), individual GCM forecasts (Syktus et al., 2003), or consensus forecasts from multiple forecast sources (Barnston et al., 2003). Downscaled daily data grids from numerical models producing seasonal forecasts are now becoming a reality, as is the availability of future climate change scenarios. However, significant challenges remain in the downscaling and bias removal processes. Even without seasonal climate forecasts, process models such as AussieGRASS can produce plant growth forecasts with skill, due to information about current conditions (i.e., ground cover, soil moisture status, and nitrogen availability; Henry, 2006).

Fire Scar Data

After climate, the most important variables for Australian savannas are fire, grazing animal density, and land use or land cover change. Location and timing of fire information are supplied by remote sensing. There are systems that use the older series of NOAA satellites (Craig et al., 2002) and the newer MODIS satellite data on a near real-time basis. Information regarding the extent of fire is usually available within a few days of any event, given constraints of satellite orbits, smoke, and cloud cover.

Fire scar data not only provide the opportunity to reset model state variables and provide users with realistic near real-time model output, but they also offer the potential for model validation and calibration. Model estimates of fuel mass and curing before the occurrence of fire scars are used to test whether the simulation is producing at least enough fuel for a fire to occur. In some regions where burns are nearly complete, direct observations of pasture growth are also possible in the post-fire growing season, as the pasture standing dry matter does not include dead material from the previous season's growth, thus ensuring calibration of model growth parameters is more robust.

Fire scar mapping was problematic before the advent of meteorological satellites, although some manual and air photograph-based detailed mapping were available (Zylstra, 2003). Development of a model of daily synthetic fire maps at a continental scale to apply to simulations before the satellite era remains a priority. However, some progress has been made by using regressions of pasture growth against late season fire extent (Hill et al., 2006).

Grazing Pressure

Large areas of Australian savannas are grazed by domestic livestock (Figure 21.1). The contribution of native and feral animals to total grazing pressure can be significant, not only on commercial grazing enterprises but also on lands designated for other purposes (e.g., national parks and military zones). Macropod numbers in some regions are highly variable over time, due to seasonal and climate conditions (Pople et al., 2007). Similarly, estimates of feral animal numbers should be incorporated when they are at significant densities. Historically, rabbit populations in some areas of Australia have become very high, and in the modern era they have varied with periodic control by induced disease epidemics (e.g., myxomatosis and calicivirus). Feral animal populations are rarely comprehensively surveyed or monitored on a spatiotemporal basis.

Historical data are sparse and require considerable effort to translate into density maps as a grazing pressure model input. Insect populations can also remove large quantities of biomass, but likewise they are poorly quantified in space and time. Models of population dynamics may provide useful histori-

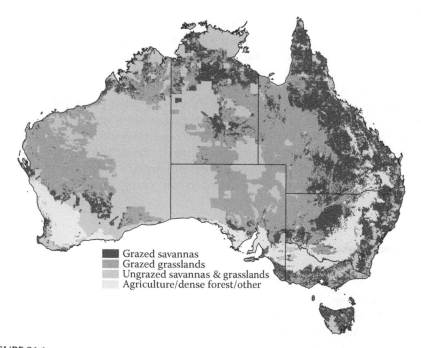

Grazed savannas
Grazed grasslands
Ungrazed savannas & grasslands
Agriculture/dense forest/other

FIGURE 21.1
(See color insert following page 320.) Map showing distribution of grazed and ungrazed savannas and grasslands in Australia.

cal regional reconstructions (e.g., for rabbits; Scanlan and Berman, 1999) and should be included in models that aim at calculating aboveground NPP based on observed temporal changes in pasture biomass. It remains a challenge to comprehensively address total grazing pressure at continental scales.

Land Cover Dynamics

In Queensland, satellite mapping shows about 7.3 million hectares of woodland, savanna, and forest clearing from 1988 to 2006 (NRW, 2008). Not all clearing of woody remnant vegetation is successful; up to 40% of clearing events revert to regrowth that may be re-cleared or slowly recover to mature woody systems. Land clearing, however, is not the only mechanism of loss of tree cover in savanna systems. Between 1991 and 1997, tree death due to prolonged drought was estimated to affect at least 60,000 ha (DNR, 1999, Fensham and Holman, 1999). Increases in the extent of woody cover also occur due to invasion of woody plants into natural grasslands, replanting of old cleared landscapes to plantation forest, or plantings for conservation or biodiversity purposes.

Woody density in existing savanna systems may also increase due to land management (e.g., reduced fire frequency, high pasture utilization), climate variability, and increasing atmospheric CO_2 concentration. The estimated rate of change in savannas in Queensland is about a 1% increase in live tree basal area per year, with increases in dead tree mass also occurring. *Woodland thickening* represents a significant carbon sink estimated to be equivalent to 65–125 Mt CO_2 per year (after Burrows et al., 2002), and there is evidence that this phenomenon is much more widespread than the area surveyed (Lewis, 2002). Historical dynamics of savanna systems remain a difficult issue for modeling of carbon stocks, due to the lack of detailed mapping and stand density measurements before field monitoring projects and remote sensing.

Two approaches for improving the representation of the dynamics of woody plant cover in continental models are (1) the inclusion of fire histories and their likely effects on foliage cover and woody plant density and/or (2) estimates of changes in perennial cover using remote sensing (e.g., Donohue et al., 2008).

Improving the Modeling of Surface Cover

The measurement and simulation of surface cover is of increasing importance in assessing resource condition (e.g., Stone et al., 2008). Process models such as AussieGRASS provide a quantitative approach to separating the impacts of climate and management (e.g., rainfall and livestock numbers). However, one major limitation in measurement and modeling has been the difficulty of assessing the contribution of tree leaf litter to ground cover.

Cover to Mass Relationships for Standing Biomass and Litter Pools

Remote sensing and field measurements can provide estimates of vegetation cover such as percentages of bare ground, tree foliage projective cover, crown cover, and total green cover. These cover estimates all have a corresponding mass component and in simulation models, it is necessary to convert mass estimates to cover estimates and vice versa for the purpose of model calibration and validation. Conversion algorithms need to be robust and operate over a range of scales and should be parameterized for a range of vegetation types and stages of maturity. Tree FPC is readily remotely sensed, as it is often possible to obtain imagery when the understorey is dead, providing good spectral discrimination between the green woody and dead understorey components.

Green cover of grasses is estimated from satellite imagery, but it is not always easy to separate the relative contributions of the tree and grass signals without recourse to time-series imagery and analyses. Perennial and annual greenness signals can be separated directly from satellite data (Donohue et al., 2008) or combined in a model simulation to produce a model estimate of total greenness as viewed by the satellite.

Ground cover estimates for simulating erosion risk can be remotely sensed (Chapter 11); but simulation studies, such as presented later in this chapter, require an ability to model ground cover. In operational assessments, remotely sensed estimates of ground cover are restricted to areas of less than 20% FPC, due to the effects of tree canopies and their shadows. To fill this data gap, methods are required to estimate ground cover in the approximately 25% of grazed savanna systems with FPC greater than 20%. It is possible to estimate bare ground from an empirical function relating field measurements of ground cover and tree FPC (Figure 21.2). Variability in this relationship arises due to grazing and nonuniform litter distribution patterns on the ground. Percentage bare ground can vary by 20% or more even with similar tree density (see points labeled "Mulga Litter").

Calibrating Mass to Cover Algorithms

Field measurement of ground cover is not a straightforward task. Cover estimates based on quadrats can differ from point-based measurements (Murphy and Lodge, 2002) and, hence, need to be corrected to a consistent basis before intercalibration of satellite, field measurement, and model estimates. To date there is no clearly best algebraic function to translate pasture mass to ground cover. Many possible functions have the key attributes of being asymptotic

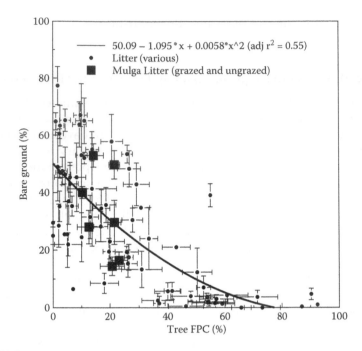

FIGURE 21.2
Estimated bare ground percentage from tree litter and grasses as a function of tree foliage projective cover for different vegetation communities largely in Queensland using point intercept
methods: X and Y bars represent standard errors. Highlighted data points show variability at a
single location in a Mulga woodland for grazed and ungrazed plots.

to 100% cover (at some mass) and having zero cover at zero mass. However,
functions where the parameters are intuitive and practical to estimate (e.g.,
mass needed to provide 50% cover) are less common. The two parameter
functions used in AussieGRASS are as follows:

$$\text{Proportion Cover} = \frac{\text{TSDM}^{0.95}}{\text{TSDM}^{0.95} + \text{yield\_cov}_{50}^{0.95}} \qquad (21.1)$$

and

$$\text{Proportion Cover} = 1 - \exp(\text{TSDM} \times -0.6931/\text{yield\_cov}_{50}) \qquad (21.2)$$

where the first equation is used for green cover, the second for ground
cover, TSDM = total standing dry matter (pasture biomass), and yield_
cov_{50} = pasture mass that gives 50% cover.

Cover functions need to be flexible enough to parameterize the degree
of variation resulting from management and pasture effects, for example,
100% cover at very low harvestable pasture biomass ("lawn-like" stoloniferous

grasses) to low cover at very high biomass (e.g., *Triodia* spp. in desert environments). In some pasture communities, composition can change from annual herbaceous species to erect perennial grasses; and, hence, different mass to cover relationships are required. The relationship between mass and visible green cover is further complicated in perennial systems where there is dead material overlaying green leaf.

In calibrating the AussieGRASS model, the pasture mass required to produce 50% ground cover was estimated by model inversion, where model parameters for all pasture communities (including the mass to cover ratio) were modified, so that the model simulated (1) the temporal pattern of NDVI across Australia; (2) estimates of total pasture biomass (Hassett et al., 2000); and (3) dry season ground cover derived from LANDSAT data (Queensland and NSW) (Figure 21.3).

Across a range of environments, the mass of pasture required to produce 50% green pasture cover was related to average annual vapor pressure deficit (VPD). For the 186 pasture communities covering the Australian continent and simulated in the AussieGRASS model, there was a strong linear relationship between the mass for 50% cover and long-term average annual VPD.

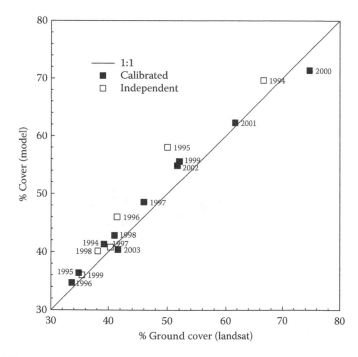

FIGURE 21.3

Comparison of remotely sensed dry season ground cover for Queensland (areas where tree canopy cover is less than 20%) and predicted cover (including estimates for years not used in calibration) from the AussieGRASS model from 1994 to 2003.

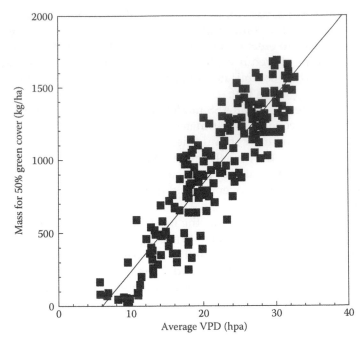

FIGURE 21.4
Relationship between annual average vapor pressure deficit and pasture mass required for 50% projected pasture cover for 186 pasture communities in Australia. Mass for 50% Cover = −363 + 60*VPD (adjusted r^2 = 0.81).

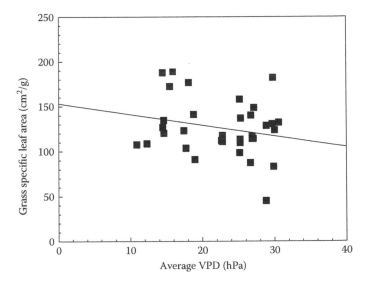

FIGURE 21.5
Pasture grass leaf specific area (±S.E.) and annual average VPD for a range of grass species from 34 sites across Queensland (trend line is not significant).

Average daytime VPD was estimated from temperature and vapor pressure (after Tanner and Sinclair, 1983):

$$VPD = VP_{\text{saturated}}(T_{\min}) + 0.75(VP_{\text{saturated}}(T_{\max}) - VP_{\text{saturated}}(T_{\min})) - VP. \quad (21.3)$$

The mass required to produce 50% cover increases with increasing VPD (Figure 21.4). This relationship exists possibly due to leaf angle and tussock structure with upright green elements in situations with high radiation loading and low rainfall, whereas leaves are flatter in more benign environments. Other factors that contribute to a higher mass to cover ratio in high VPD environments are (1) increased lignin concentration at high temperatures (Wilson et al., 1991) and (2) reductions in solar radiation interception due to leaf rolling (e.g., *Triodia* species) or leaf folding (e.g., *Heteropogon contortus*). Grass-specific leaf area is less well correlated with VPD (Figure 21.5).

Grazing pressure can also have a significant impact on sward structure, especially if species composition changes from tussock grasses to prostrate species under high rates of grazing utilization (Corfield and Nelson, 2008). Tree cover and high shade environments are also likely to change mass to cover relationships. Analysis of the data reported in Figure 21.5 suggests that the correlation is slightly improved for use in savanna systems if average tree basal area is included (mass per unit cover is reduced at the rate of 11 kg/m² tree basal area). These factors probably account for some of the scatter around the line of best fit. The above results indicate that substantial improvement can be made in model development by combining observations from a number of sources (e.g., field observations, remote sensing) and examining consistency of parameters across vastly different climatic zones.

Carbon and Nitrogen Stocks

There is a need to evaluate alternative management options in grazing lands, particularly with regard to livestock densities. Some management actions that may increase the size of the domestic herd are additional fencing, improved distribution of water points, and optimization of animal supplementation (e.g., nitrogen and phosphorus). However, reduction in livestock numbers would reduce herbivore methane emissions; possibly rebuild rangeland soil carbon; and modify fire regimes in northern Australia. Thus, choice of stocking rate and associated grazing and livestock management (e.g., use of fire) will be important management issues over the next century, which need to be evaluated in terms of their impact on C and N stocks. In addition, the understanding of the effects of interactions between grazing pressure and ecosystem processes (including woodland thickening, soil carbon, runoff, drainage, soil erosion, and potential fire frequency) requires the use of models such as AussieGRASS.

Case Study: Broad Scale Sensitivity of Natural Resource Attributes to Stocking Rates

In order to examine some of the above issues, three AussieGRASS simulation experiments were conducted:

1. Evaluation of recent (1998–2008) C and N fluxes resulting from the 2001 stocking density and actual fires that occurred in areas where remotely sensed fire scars were available
2. Evaluation of model sensitivity to a range of stocking rates (using 2001 Australian Bureau of Statistics livestock densities as a base) with tree density and the number or density of non-domestic herbivores held constant and with no fires
3. Same as Experiment 2 but with synthetic (modeled) to fires simulated with fuel loads modified by grazing

In the case of Experiment 3, the biomass generated by the model was allowed to burn completely on a pixel-by-pixel basis as soon as grassy fuel was more than 1400 kg DM/ha and curing greater than 70% (with the assumption that ignition sources were always available). Carbon (estimated at 0.46 * biomass) and nitrogen in burnt biomass (grassy fuels and tree litter) were calculated on an annual basis and averaged over the period 1890–2008. The production of charcoal was calculated using a conservative estimate of 2.5% of carbon burnt (Lehmann et al. 2008).

Tabular statistics were generated for three vegetation types of Australia (Figure 21.1):

- Grasslands (i.e., tree basal area generally < 1 m²/ha, 36% of continent)
- Wooded areas with a tree basal area averaging less than 25 m²/ha (i.e., savannas, 22% of continent) grazed by domestic livestock
- Ungrazed areas (i.e., mix of grassland and savanna, 33% of continent)

Croplands, dense forests, and irrigated lands were excluded from the analysis.

Syktus et al. (2003) used AussieGRASS to evaluate the impact of changing stocking rates in response to hindcasts derived from an RCM. These authors used two indicators of degradation risk: (1) the frequency of years that exceeded 30% utilization (% eaten of pasture growth) and (2) percentage of time surface cover was less than 40%. We have used similar indicators to evaluate the simulation studies just described.

Impacts of Current Fire and Stocking Rate Regimes

Fire removals of biomass (1998–2008) averaged over the entire area were 159 kg DM ha^{-1} year^{-1} (including tree litter) for grazed savannas and 95 kg DM/ha/year for grasslands (Table 21.1). This was between 8% and 9% of annual aboveground pasture growth for savannas and grazed grasslands, respectively. For 2001 stocking rates, the average amount eaten (144 kg DM ha^{-1} year^{-1}) was 15% of aboveground pasture growth for grazed savannas and grazed grasslands, consistent with estimates of property safe utilization rates (Hall et al., 1998). Fire and grazing combined removed 20–25% of annual pasture growth.

Simulated average nitrogen loss per hectare ranged from 7.5 to 8.6 kg N ha^{-1} year^{-1} in burnt areas with grazed savannas having the least loss and grazed grasslands and ungrazed areas having higher losses. The loss rate represents approximately 40% of the annual aboveground uptake in native pastures in northern Australia (about 20 kg N/ha/year). The mechanisms for replacement of N losses are poorly characterized with inputs from rainfall (~2 kgN/ha/1000 mm rainfall in northern Australia), dry deposition, and N fixation from plants and micro-organisms contributing to total inputs. Changes in the fire regimes are likely to affect carbon and nitrogen cycles if imposed over the long term. The extent to which fire regimes may change in response to a change in grazing pressure is uncertain, as is the subsequent impact on the N cycle and associated change in soil fertility. Fires provide a long-term charcoal sink (Lehmann et al., 2008) and buffer against climate change; in this simulation, charcoal production was estimated to be about 0.77 Mt C/year for the recent fire regime in grazed areas.

TABLE 21.1

Estimates of Biomass and Nitrogen Currently Burnt as Estimated from the AussieGRASS Model and Fire Scars Mapped from NOAA Satellite Imagery for Areas Depicted in Figure 21.1

System	Percentage of Area Burnt (average of 1998–2008)	Biomass Burnt (Mt DM/year) (rate kg DM/ha/event)	Nitrogen Volatilized (ktN/year) (rate kg N/ha/event)
Australian grazed savannas ~1.67 × 10^6 km^2	8.4	26.6 (1880)	106 (7.5)
Australian grazed grasslands ~2.76 × 10^6 km^2	4.5	26.2 (2096)	108 (8.6)
Australian ungrazed areas (grass and savanna) ~2.57 × 10^6 km^2	9.1	49.6 (2136)	199 (8.5)

TABLE 21.2

Sensitivity to a 30% Change in Grazing Pressure of Domestic Livestock across Australia on Key System Variables, Absolute Change, and Percentage Change from Current Base

Change in System Variable (Average 1890–2008)	Australian Savannas		Australian Grasslands	
Stocking rate change (%)	+30	−30	+30	−30
Pasture eaten (kg DM/ha/year)	+58 (+19%)	−70 (−23%)	+26 (+18%)	−32 (−22%)
Pasture growth (above ground) (kg DM/ha/year)	−28 (−1%)	+28 (+1%)	−5 (−0.6%)	+5 (+0.5%)
Root growth (below ground) change (kg DM/ha/year)	−33 (−2%)	+35 (+2%)	−6 (−0.5%)	+5 (+0.5%)
Standing pasture (kg DM/ha)	−67 (−5%)	+58 (+5%)	−46 (−5%)	+55 (6%)
Grass litter (kg DM/ha)	−16 (−4.4%)	18 (+5%)	−7 (−2.9%)	8 (3.5%)
Ground cover (%)	−1.5 (−2)	+1.6 (2)	−1.1 (−3)	+1.2 (3)
Change in area of ground cover <40% km^2	−24,000 (+10.3%)	+24,000 (−10.3%)	+47,000 (+3.6%)	−54,000 (−4.1%)
Soil loss (kg/ha/year)	+129 (18%)	−114 (−24%)	+76 (+11%)	−68 (−13%)
Change in area (km^2) where utilization exceeds 30%	111,000 (+34%)	−117,000 (−36%)	116,000 (+17%)	−142,000 (−21%)
Soil water (mm in top 1 m)	+0.11 (0.26%)	−0.14 (−0.32%)	−0.27 (−0.69%)	+0.32 (0.81%)
Runoff + drainage (mm/year)	+1.4 (1.43%)	−1.45 (−1.43%)	−0.38 (−0.57%)	+0.37 (0.56%)

Broad Scale System Sensitivity to Changing Livestock Numbers

As expected, the largest change resulting from changing livestock numbers were for pasture eaten (Table 21.2). At the low level of average utilization, pasture eaten was almost linearly related to livestock numbers (head/hectare). Changing stocking rate had a relatively small average impact for many variables; however, examination of spatial maps showed regions of both higher and lower change than the average. Simulated average pasture growth was not greatly affected by changes in stocking rate; and hence, average standing pasture, litter, and groundcover were not greatly changed. However, in terms of impact on the soil resource, a small increase in average ground cover was associated with a 13 to 24% average reduction in simulated erosion (based on Scanlan et al., 1996, Mckeon et al., 2000) for the grazed grasslands and grazed savannas of Australia, respectively, with the AussieGRASS baseline erosion (1.28 t ha^{-1} year^{-1}) for areas excluding cropping similar to published measurements of 1.34 t/ha/year (Lu et al., 2003). Similarly, the percentage change in the area where pasture utilization exceeded 30% (considered a safe threshold) was also quite sensitive to changes in stocking rate, which is consistent with simulated sensitivity to erosion hazard.

In terms of carbon storage, reducing domestic stocking rate to 70% of 2001 values in the model increased the mass of standing grass and grass litter by 63–76 kg DM ha^{-1} on average. This represents a nominal once off sink over the

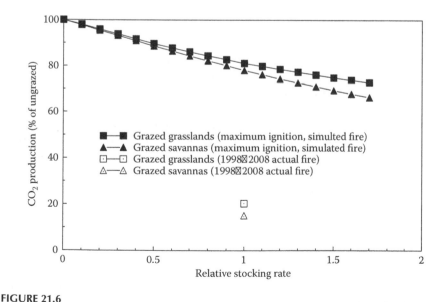

FIGURE 21.6
Relative CO$_2$ production from simulated burning (using a maximum possible ignition rule) and a range of stocking rate scenarios and CO$_2$ emissions based on actual fire with all results scaled relative to burning at maximum ignition with no livestock. Stocking rate is scaled relative to 2001 livestock numbers.

main grazing areas of Australia of about 14 Mt C (51 Mt CO_2), assuming no increased biomass losses due to increased fire and non-domestic herbivory.

Broad Scale Sensitivity of C and N Fluxes with Changing Grazing Pressure and Fire

Simulations with synthetic fire using a maximum ignition rule produced more than four times as much CO_2 from combustion than actual fires recorded for 1998–2008 (Figure 21.6). An analysis of the relative change in synthetic fire potential with a change in grazing pressure indicates that nitrogen losses associated with combustion would be about +6% for grassland and +8% for savanna systems for a 30% decrease in stocking rate from the 2001 baseline. This analysis does not take into account changes in enteric losses, change in the prevalence of N-fixing organisms, or changes due to modification of the water balance. In the long term, if the increases in nitrogen loss from additional fire were not compensated by an increase in N fixation, then a slow loss of plant productivity would result.

For savanna systems, complete removal as opposed to a 30% reduction of domestic livestock increased the simulated area burnt and fire-induced fluxes, for example, fluxes of CO_2 (+28%), area burnt (+21%), and associated N loss (+26%). Removal of livestock in grasslands had a smaller impact on fluxes of CO_2 (+23%), area burnt (+16%), and associated N loss (+21%) relative to savanna systems.

Case Study Discussion

The results from broad scale sensitivity analysis Experiment 2 (Table 21.2) indicated that varying livestock densities in the AussieGRASS model did not produce substantial long-term changes in pasture productivity. However, historically, Australia's rangelands have been subjected to regional episodes of degradation (i.e., substantial loss of livestock carrying capacity) due to overgrazing (McKeon et al., 2004). The processes of resource degradation include decreasing grass basal cover, loss of desirable perennial grass species, reduced surface soil infiltration capacity, increased runoff, increased soil erosion, increases in unpalatable vegetation (i.e., weeds), and increases in woody plant density. Studies with GRASP (McKeon et al., 2000) have addressed the issue of how to represent some of these degradation processes at a land-type scale (land types are usually defined as a combination of tree species and soil type). Degradation processes are generally specific to land-types rather than to broad scale pasture communities. Often land-types and fires occur as a mosaic in the landscape and hence, cannot be fully represented at a coarse 5 km-grid scale, as used in AussieGRASS.

Nonetheless, with AussieGRASS, the impact of increased grazing pressure on pasture and resource attributes can be evaluated for purposes of risk assessment, by indicators such as pasture utilization (amount of growth eaten by livestock), pasture or bare ground cover expressed relative to safe threshold values, and simulated soil loss. The results show that a 30% increase in livestock numbers substantially increases the risk of degradation (Table 21.2), with savanna systems appearing to be more sensitive to change in livestock than grasslands. For example, soil loss increased by 11% and 18% for grasslands and savannas, respectively, at the higher stocking rates. Similarly, the risk of damage to the pasture resource (average area where annual utilization exceeded 30%) substantially increased (+17% and 34% for grassland and savanna, respectively). Hence, this increase in stocking rate is unlikely to be sustainable at a continental level. If the feedbacks of soil loss and other degradation processes were included in the simulations, then the simulated impact on system components is likely to be greater than represented in Table 21.2. However, the trajectories of degradation and recovery are difficult to describe and parameterize. The combination of measurement and simulation of ground cover will provide the basis for modeling the land-type specific process.

Conclusion

We reviewed a range of issues by using the operational AussieGRASS spatial model, which simulates soil water, balance, and pasture dry matter dynamics across Australia's rangelands and native pastures. We identified major areas with potential for improvement: the assimilation of climate data from reanalysis; fire scar data from the early and pre-remote sensing era; improved simulation of the dynamics of feral and native herbivores; and better representation of the dynamics of woody plant cover.

A major emerging application for the AussieGRASS and other erosion or runoff models is the simulation of surface cover. An algorithm to estimate the contribution of tree litter to surface cover was developed to enhance remote sensed estimates. In terms of pasture cover, algorithms describing the relationship between cover or pasture mass and climate variables, notably vapor pressure deficit, have also been developed. These improvements enhance the continental simulation of surface cover and ability to parameterize the model.

We described a major simulation experiment to estimate risks associated with changing livestock numbers across the continent on attributes of the natural resource system, including hydrology, soil loss, and dry matter flow. Although the AussieGRASS model is not designed to fully represent degradation processes for individual land types, it nevertheless allows assessment of the general risks of land and pasture degradation. These simulations were also used to assess likely changes in carbon and nitrogen fluxes with changing grazing pressure. The simulations can be used to help understand the relationship between managing livestock production systems and the potential for greenhouse mitigation in Australia's savannas.

References

Barnston, A. G., S. J. Mason, L. Goddard, D. G. Dewitt, and S. E. Zebiak. 2003. Multimodel ensembling in seasonal climate forecasting at IRI. *Bulletin of the American Meteorological Society* 84, 1783–1796.

Burrows, W. H., B. K. Henry, P. V. Back, et al. 2002. Growth and Carbon stock change in eucalypt woodlands in north-east Australia: Ecological and greenhouse sink implications. *Global Change Biology.* 8, 769–784.

Carter, J. O., W. B. Hall, K. D. Brook, G. M. McKeon, K. A. Day, and C. J. Paull. 2000. Aussie GRASS: Australian grassland and rangeland assessment by spatial simulation. In *Applications of Seasonal Climate Forecasting in Agricultural and Natural Ecosystems—the Australian Experience*, eds. G. Hammer, N. Nicholls, and C. Mitchell. 329–350. Netherlands, Kluwer Academic Press.

Compo, G. P., J. S. Whitaker, and P. D. Sardeshmukh. 2006. Feasibility of a 100-year reanalysis using only surface pressure data. *Bulletin of the American Meteorological Society* 87, 175–190. DOI: 10.1175/BAMS-87-2-175.

Corfield, J. P. and B. S. Nelson. 2008. Scaling up from Ecograze—the Virginia Park experience. In *Proceedings of the 15th Biennial Conference of the Australian Rangeland Society*, Charters Towers, Queensland.

Craig, R., B. Heath, N. Raisbeck-Brown, M. Steber, J. Marsden, and R. Smith. 2002. The distribution, extent and seasonality of large fires in Australia, April 1998–March 2000, as mapped from NOAA-AVHRR imagery. In *Australian Fire Regimes: Contemporary Patterns (April 1998–March 2000) and Changes Since European Settlement*, eds. J. Russell-Smith, R. Craig, A. M. Gill, R. Smith, and J. Williams. Australia State of the Environment Second Technical Paper Series (Biodiversity), Department of the Environment and Heritage, Canberra. http://www.ea.gov.au/soe/techpapers/index.html.

Day, K. A., K. G. Rickert, and G. M. McKeon. 2005. Monitoring agricultural drought in Australia. In *Monitoring and Predicting Agricultural Drought: A Global Study*, eds. V. K. Boken, A. P. Cracknell and R. L. Heathcote, 369–385. Oxford, Oxford University Press.

DNR. 1999. Land cover change in Queensland 1995–1997: A Statewide Landcover and Trees Study (SLATS) Report, Aug, 1999. Department of Natural Resources, Brisbane.

Donohue, R. J., T. R. McVicar, and M. L. Roderick. 2008. Climate-related trends in Australian vegetation cover as inferred from satellite observations. *Global Change Biology* 15, 1025–1039.

Fensham, R. J. and J. E. Holman. 1999. Temporal and spatial patterns in drought related tree dieback in Australian savannah. *Journal of Applied Ecology* 36, 1035–1050.

Hall, W. B., G. M. McKeon, J. O. Carter, et al. 1998. Climate change and Queensland's grazing lands: II. An assessment of the impact on animal production from native pastures. *Rangeland Journal* 20, 177–205.

Hassett, R. C., H. L. Wood, J. O. Carter, and T. J. Danaher. 2000. A field method for statewide ground-truthing of a spatial model. *Australian Journal of Experimental Agriculture* 40, 1069–1079.

Henry, B. K. 2006. Improved climate forecasts for wool producers in Australia's pastoral zone. Final Report QNR30, Land, Water and Wool Climate sub-program. Available at: http://products.lwa.gov.au/files/PK071415.pdf

Hill, M. J., S. H. Roxburgh, J. O. Carter, G. M. McKeon, and D. J. Barrett. 2005. Vegetation state change and carbon dynamics in savanna woodlands of Australia in response to grazing, drought and fire: A scenario approach using 113 years of synthetic annual fire and grassland growth. *Australian Journal of Botany* 53, 715–739.

Hill, M. J., S. H. Roxburgh, J. O. Carter, and D. J. Barrett. 2006. Development of a synthetic record of fire probability and proportion of late fires from simulated growth of ground stratum and annual rainfall in the Australian tropical savanna zone. *Environmental Modelling and Software* 21, 1214–1229.

Jeffrey, S. J., J. O. Carter, K. B. Moodie, and A. R. Beswick. 2001. Using spatial interpolation to construct a comprehensive archive of Australian climate data. *Environmental Modelling and Software* 16, 309–330.

Kokic, P., R. Nelson, A. Potgieter, and J. Carter. 2004. *An Enhanced ABARE System for Predicting Farm Performance*. ABARE eReport 04.6, Canberra.

Lehmann, J. J. Skjemstad, S. Sohi, et al. 2008. Australian climate-carbon cycle feedback reduced by soil black carbon. *Nature Geoscience* 1, 832–835.

Lewis, D. 2002. *Slower Than the Eye Can See*. Tropical Savannas CRC, Darwin. N.T.

Lu, H, I. P. Prosser, C. J. Moran, J. C. Gallant, G. Priestley, and J. G. Stevenson. 2003. Predicting sheetwash and rill erosion over the Australian continent. *Australian Journal of Soil Research* 41, 1037–1062.

McKeon, G. M., A. J. Ash, W. B. Hall, and M. Stafford Smith. 2000. Simulation of grazing strategies for beef production in north-east Queensland. In *Applications of Seasonal Climate Forecasting in Agricultural and Natural ecosystems. The Australian Experience*, ed. G.L. Hammer, N. Nicholls and C. Mitchell, Dordrecht, The Netherlands, Kluwer Academic Publishers, pp. 227–252.

McKeon, G. M., W. B. Hall, B. K. Henry, G. S. Stone, and I. W. Watson. 2004. *Pasture Degradation and Recovery in Australia's Rangelands: Learning from History*. Queensland Department of Natural Resources, Mines and Energy. Queensland Department of Natural Resources Mines and Energy, Brisbane, 256pp.

McKeon, G. M., G. S. Stone, J. I. Syktus, et al. 2009. Climate change impacts on northern Australian rangeland livestock carrying capacity: A review of issues. *The Rangeland Journal* 31, 1–29.

Murphy, S. R. and G. M. Lodge. 2002. Ground cover in temperate native perennial grass pastures. I. A comparison of four estimation methods. *The Rangeland Journal* 24, 288–300.

NAMS. 2009. National Agricultural Monitoring System, www.nams.gov.au (accessed 01/04/09).

NRW. 2008. Land cover change in Queensland 2005–06: A Statewide Landcover and Trees Study (SLATS) Report, Feb, 2008. Department of Natural Resources and Water, Brisbane.

Pople, A. R., S. R. Phinn, N. Menke, G. C. Grigg, H. P. Possingham, and C. McAlpine. 2007. Spatial patterns of kangaroo density across the South Australian pastoral zone over 26 years: Aggregation during drought and suggestions of long distance movement. *Journal of Applied Ecology* 44, 1068–1079.

Rayner, D. P., K. B. Moodie, A. R. Beswick, N. M. Clarkson, and R. L. Hutchinson. 2004. New Australian daily historical climate surfaces using CLIMARC. Queensland Department of Natural Resources, Mines and Energy.

Rickert, K. G., J. W. Stuth, and G. M. McKeon. 2000. Modelling pasture and animal production. In *Field and Laboratory Methods for Grassland and Animal Production Research*, eds. L. T. Mannetje and R. M. Jones. New York, CABI Publishing, pp. 29–66.

Scanlan, J. C., A. J. Pressland, and D. J. Myles. 1996. Run-off and soil movement on mid slopes in north-east Queensland grazed woodlands. *Rangeland Journal* 18, 33–46.

Scanlan, J. C. and D. Berman. 1999. Determining the impact of the rabbit as a grazing animal in Queensland. In *People and Rangelands, Building the Future. Proceedings of the 6th International Rangeland Congress*, eds D. Eldridge and D. Freudenberger. Aitkenvale, Qld, 6th International Rangeland Congress Inc., pp. 520–521.

Stone, R. C. and A. Auliciems. 1992. SOI phase relationships with rainfall in eastern Australia. *International Journal of Climatology* 12, 625–636.

Stone, G., D. Bruget, J. Carter, R. Hassett, R. McKeon, and D. Rayner. 2008. Pasture production and condition issue. Chapter 4 Land. In State of the Environment 2007. Environmental Protection Agency, pp. 127–139. http://www.epa.qld.gov.au/register/p02256ay.pdf

Syktus, J., G. McKeon, N. Flood, I. Smith, and L. Goddard. 2003. Evaluation of a dynamical seasonal climate forecast system for Queensland. In *National Drought Forum 2003: Science for Drought*, eds. R. Stone and I. Partridge. Brisbane, Queensland Department of Primary Industries, pp. 160–172.

Tanner, C. B. and T. R. Sinclair. 1983. Efficient water use in crop production: Research or re-search? In *Limitation to Efficient Water Use in Crop Production*, eds. H. M. Taylor, W.R. Jordan and T.R. Sinclair. Madison, ASA, CSSA and SSA, pp. 1–27.

Wilson, J. R., B. Deinum, and F. M. Engels. 1991. Temperature effects on anatomy and digestibility of leaf and stem of tropical and temperate forage species. *Netherlands Journal of Agricultural Science* 39, 31–48.

Zlylstra, P. 2003. *Fire History of the Australian Alps, Prehistory to 2003*. Australian Alps Liaison Committee, May 2006. http://www.australianalps.deh.gov.au/publications/fire/pubs/alps-fire-history-chapter1.pdf [accessed 010409].

Section VI

Continental and Global Scale Modeling of Savannas

22

Carbon Cycles and Vegetation Dynamics of Savannas Based on Global Satellite Products

Christopher Potter

CONTENTS

Introduction

Savannas have been broadly defined as grassland-dominated ecosystems with sparse (<25%) tree cover (Eiten, 1986). Savannas in the tropical zones have been described as any ecosystem containing significant open grass cover (Grace et al., 2006). In seasonally warm biomes, savannas are undergoing rapid conversion by humans to other land uses; and while this process of savanna conversion has been less recognized than deforestation in the humid tropics, it could have just as great an impact on terrestrial carbon cycling as any other land cover change worldwide. Regional conversions rates of savannas may exceed 1% per year (San Jose and Montes, 2003), approximately twice as fast as that of rainforest conversion rates. On a global scale, this loss (often by burning and land use change) of native savanna cover in favor of agricultural uses is likely to constitute a significant flux of greenhouse gases (GHGs), mainly as carbon dioxide (CO_2), to the atmosphere (Seiler and Crutzen, 1980; Andreae et al., 1998).

This chapter presents a compilation of results from satellite remote sensing and global ecosystem modeling of savanna biomes over the past 25 years. The two main objectives of these studies were to (a) review global modeling of savanna carbon cycles since 1980 and (b) understand the current worldwide connections of climate to savanna vegetation dynamics. Both these objectives have implications for predicting how savanna biomes will be impacted by climate change in the decades to come and how the loss or gain of CO_2 from savannas may feedback to affect atmospheric warming patterns.

Background on Carbon Modeling and Remote Sensing Methods

The launch of NASA's Terra satellite platform in 1999 with the moderate resolution imaging spectroradiometer (MODIS) instrument on-board initiated a new era in remote sensing of the Earth system, with promising implications for carbon cycle research. Direct input of satellite vegetation index "greenness" data from the MODIS sensor into ecosystem simulation models is now used to estimate spatial variability in monthly NPP, biomass accumulation, and litter fall inputs to soil carbon pools. Global NPP of vegetation can be predicted using the relationship between leaf reflectance properties and the absorption of PAR, assuming that net conversion efficiencies of PAR to plant carbon can be approximated for different ecosystems or are nearly constant across all ecosystems (Running and Nemani, 1988; Goetz and Prince, 1998).

Operational MODIS algorithms generate the EVI (Huete et al., 2002) as global image coverages from 2000 till the present day. The EVI represents an optimized vegetation index, whereby the vegetation index from red and near infra-red spectral bands are designed to approximate vegetation biophysical properties derived from canopy radiative transfer theory and/or measured biophysical-optical relationships. The EVI was developed to optimize the greenness signal, or area-averaged canopy photosynthetic capacity, with improved sensitivity in high biomass regions. The EVI has been found useful in estimating absorbed PAR related to chlorophyll contents in vegetated canopies, and it has been shown to be highly correlated with processes that depend on absorbed light, such as GPP (Xiao et al., 2004; Rahman, 2005).

As documented in Potter (1999), the monthly NPP flux, defined as net fixation of CO_2 by vegetation, is computed in the CASA model on the basis of light-use efficiency (Monteith, 1972). Monthly production of plant biomass is estimated as a product of time-varying surface solar irradiance, S_r, and EVI from the MODIS satellite, and a constant light utilization efficiency term (e_{max}) that is modified by time-varying stress scalar terms for temperature (T) and moisture (W) effects (Equation 22.1).

$$\text{NPP} = S_r \, \text{EVI} \, e_{max} \, T \, W \qquad (22.1)$$

The e_{max} term is set uniformly at 0.39 g C M J^{-1} PAR, a value that derives from calibration of predicted annual NPP to previous field estimates (Potter et al., 1993). This model calibration has been validated globally by comparing predicted annual NPP with more than 1900 field measurements of NPP (Potter et al., 2003; Zheng et al., 2003). Interannual NPP fluxes from the CASA model have been reported (Behrenfeld et al., 2001) and validated against multiyear estimates of NPP from field stations and tree rings (Malmström et al., 1997). Our NASA-CASA model has been validated against field-based measurements of NEP fluxes and carbon pool sizes at multiple forest sites (Amthor et al., 2001; Hicke et al., 2002) and against atmospheric inverse model estimates of global NEP (Potter et al., 2003).

The T stress scalar is computed with reference to derivation of optimal temperatures (T_{opt}) for plant production. The T_{opt} setting will vary by latitude and longitude and reaches the middle 30s in low latitude zones. The W stress scalar is estimated from monthly water deficits, based on a comparison of moisture supply (precipitation and stored soil water) to potential ET demand using the method of Priestly and Taylor (1972). Monthly mean surface air temperature, solar radiation flux, and precipitation grids for global model simulations came from NCEP reanalysis products (Kistler et al., 2001).

After the year 2001, the MODIS 1-km land cover map (Friedl et al., 2002; Figure 22.1) was aggregated to 0.5 degree latitude or longitude resolution and was used to specify the predominant land cover class for the W term in each pixel as either savanna, forest, crop, pasture, or other classes such as water or urban area. These biomes span variations along horizontal (dense vs. sparse vegetation cover) and vertical (single vs. multistory vegetation structure) dimensions, canopy height, leaf type, and soil brightness. This MODIS land cover product uses a supervised classification approach in

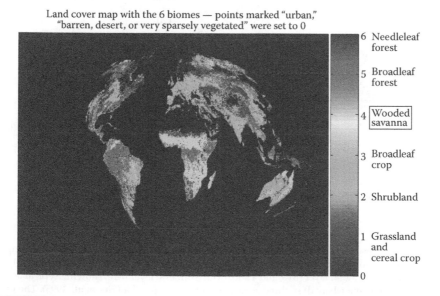

Land cover map with the 6 biomes — points marked "urban,"
"barren, desert, or very sparsely vegetated" were set to 0

6 Needleleaf forest

5 Broadleaf forest

4 Wooded savanna

3 Broadleaf crop

2 Shrubland

1 Grassland and cereal crop

0

FIGURE 22.1
(**See color insert following page 320.**) Global terrestrial biome map from the MODIS satellite instrument. Numbered biome codes are also listed in Table 22.2. Areas classified as urban or desert were labeled as zero.

which training sites are provided to a decision-tree classifier. The classifier then generates a decision tree that is exercised on the global data field, thus making a global map (Friedl et al., 2002). Before 2001, the global land cover map based on AVHRR satellite observations (DeFries et al., 1995) was used to define the same set of land cover types for the CASA model.

Carbon accumulation rates in forest biomass at the stand level are a function of both growth and mortality of trees. For the NASA-CASA model, Potter (1999) reported that these processes could be expressed in terms of the mean residence time (τ, in years) of carbon in the aboveground wood tissue pools. Tissue allocation ratios (α, percent of NPP to woody pools) were expressed in a similar manner, based on estimates from the global ecosystem literature. These default savanna values for τ (25 years) and α (45%) together determine the model's estimation of potential accumulation rates of biomass. These potential accumulation rates of woody biomass are based on the assumption of growth to mature plant status, but they are subject to validation and readjustment based on comparisons to field-based inventory measurements.

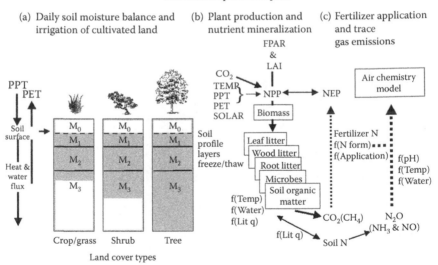

FIGURE 22.2

Schematic representation of components in the NASA-CASA model. The soil profile component (a) is layered with depth into a surface ponded layer (M_0), a surface organic layer (M_1), a surface organic-mineral layer (M_2), and a subsurface mineral layer (M_3), showing typical levels of soil water content (shaded) in three general vegetation types (DeFries et al., 1995). The production and decomposition component. (b) Shows separate pools for carbon cycling among pools of leaf litter, root litter, woody detritus, microbes, and soil organic matter, with dependence on litter quality (q). (c) Fertilizer application and trace gas emissions component estimates fluxes of CO_2 and nitrogen trace gases from soil microbial activity.

Evapotranspiration in NASA-CASA is connected to water content in the soil profile layers (Figure 22.2), as estimated using the NASA-CASA algorithms described by Potter (1999). The soil model design includes three-layer (M_1–M_3) heat and moisture content computations: surface organic matter, topsoil (0.3 m), and subsoil to rooting depth (1–10 m). These layers can differ in soil texture, moisture holding capacity, and carbon-nitrogen dynamics. Water balance in the soil is modeled as the difference between precipitation or volumetric percolation inputs, monthly estimates of PET, and the drainage output for each layer. Inputs from rainfall can recharge the soil layers to field capacity. Excess water percolates through to lower layers and may eventually leave the system as seepage and runoff.

Based on plant production as the primary carbon and nitrogen cycling source, the NASA-CASA model is designed to couple daily and seasonal patterns in soil nutrient mineralization and soil heterotropic respiration (R_h) of CO_2 from soils worldwide. The NEP can be computed as NPP minus R_h fluxes, excluding the effects of small-scale fires and other localized disturbances or vegetation regrowth patterns on carbon fluxes. The soil model uses a set of compartmentalized difference equations with a structure comparable to the CENTURY ecosystem model (Parton et al., 1992). First-order decay equations simulate exchanges of decomposing plant residue (metabolic and structural fractions) at the soil surface. The model also simulates surface SOM fractions that presumably vary in age and chemical composition. Turnover of active (microbial biomass and labile substrates), slow (chemically protected), and passive (physically protected) fractions of the SOM are represented. Along with moisture availability and litter quality, the predicted soil temperature in the M_1 layer controls SOM decomposition.

The soil carbon pools were initialized to represent storage and flux conditions in near steady state (i.e., an annual NEP flux less than 0.5% of annual NPP flux) with regard to mean land surface climate recorded for the period 1999–2000. This initialization protocol was found to be necessary to eliminate any notable discontinuities in predicted NEP fluxes during the transition to our model simulation years of interest before MODIS EVI availability. Initializing to near steady state does not, however, address the issue that some ecosystems are not in equilibrium with regard to net annual carbon fluxes, especially when they are recovering from past disturbances. For instance, it is openly acknowledged that the CASA modeling approach using MODIS satellite data inputs cannot capture all the effects of cutting and burning of vegetation and regrowth from recent wood harvest activities, although impacts of major wildfires are detectable (Potter et al., 2005).

Although previous versions of the CASA model (Potter et al., 1993) used an NDVI to estimate FPAR, the current model version (Potter et al., 2009), instead, has been calibrated to use MODIS EVI data sets as direct inputs to Equation 22.1, described above. In long-term (1982–2004) simulations, continuity between AVHRR and MODIS sensor data for inputs to NASA-CASA is an issue that must be addressed by recalibration of annual NPP results after

2000. NASA-CASA model predictions with 2001 monthly MODIS EVI inputs have been adjusted using the same set of field measurements of NPP (Olson et al., 1997; Potter et al., 2003; Zheng et al., 2003). To best match predictions with previously measured NPP estimates at the global scale ($R^2 = 0.91$), the model e_{max} term for 2001 MODIS EVI inputs was reset to 0.55 g C M J^{-1} PAR.

Predictions of Savanna NPP and Biomass Carbon Pools

Based on the pre–2001 land cover map of DeFries et al. (1995), the distribution of savanna vegetation carbon fluxes predicted by the CASA model across the continents showed that annual NPP was highest (greater than 800 g C m^{-2} year^{-1}) along the margins of the Amazon and Congo river basin (moist tropical forest) regions (Figure 22.3). The areas of lowest savanna NPP (less than 400 g C m^{-2} year^{-1}) were predicted to border the semiarid climate zones of Central and South America, Sahelian Africa, southern Africa, South Asia, and Australia.

A comparison of savanna carbon fluxes and aboveground biomass pools to other vegetation biomes from the CASA model was made by Potter (1999), who reported that total annual NPP for the biome generally called *wooded*

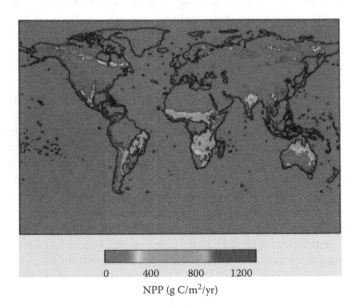

NPP (g C/m^2/yr)

FIGURE 22.3
(See color insert following page 320.) Net primary production of savanna biomes estimate at 0.5 degrees latitude and longitude from the CASA model. (Adapted from Potter, C. S. 1999. *Bioscience* 49, 769–778.)

TABLE 22.1

Ecosystem Carbon Estimates from the CASA Model for Global Biomes Classes

Class	Area Total (10⁶ km²)	AGB Olson (g C m²)	AGB Olson (Pg C)	AGB Average (g C m²)	AGB Total (Pg C)	NPP Average (g C m⁻² yr⁻¹)	NPP Total (Pg C yr⁻¹)
Broadleaf evergreen forest	15.4	8159	126.1	12,622	195.0	1075.4	16.6
Coniferous evergreen forest and woodland	13.5	11,381	153.5	6014	84.9	233.3	3.3
High latitude deciduous forest and woodland	5.9	11,918	70.1	5603	34.0	244.5	1.5
Tundra	8.6	428	3.8	1878	17.7	72.6	0.7
Mixed coniferous forest and woodland	7.5	5321	40.2	7386	56.9	386.2	3.0
Wooded grassland and shrubland	24.3	4540	111.2	8193	202.4	782.4	19.3
Temperate grassland	22.1	476	10.6	154	3.5	172.5	4.0
Desert	17.7	190	4.3	525	9.4	21.4	0.4
Cultivated	14.6	747	10.9	182	2.7	365.0	5.4
Broadleaf deciduous forest and woodland	3.6	7390	26.9	6715	24.7	397.4	1.5
Semiarid shrubland	11.5	971	11.1	1721	20.0	69.0	0.8
Total	145.4		567.8		651.7		56.4

Aboveground biomass (AGB) is estimated from extrapolation of site data reported by Olson et al. (1983) and from the NASA-CASA model calculated for both average values and totals by vegetation class.

grassland and shrubland was the highest of any in the world (Table 22.1), at around 19 Pg C year⁻¹ (1 Pg = 10¹⁵ g). Aboveground biomass (AGB) in savanna ecosystems worldwide was estimated by the CASA model at an average of slightly more than 8000 g C m⁻², which was second only to tropical evergreen forests at an average of 12,600 g C m⁻². Comparisons of these CASA model predictions were presented in this study to those based on measurement data sets compiled by Olson et al. (1983), whose estimates of AGB in savanna biomes was approximately 50% of the CASA model predictions. The most logical explanation for this discrepancy between AGB estimates in Table 22.1 for the savanna biome is a small sample size of measurements available at the time for compilation by Olson et al. (1983).

For a more up-to-date comparison of CASA model results to measurements, Grace et al. (2006) compiled data from a review of the literature to report that tropical savanna NPP rates can range from 100 to 1200 g C m⁻² year⁻¹. As also predicted in the CASA model results (Figure 22.3), lower-end NPP values were reported by Grace et al. (2006) for the arid and semiarid savannas occurring in extensive regions of Africa, Australia, and South

America. The global average NPP of the savanna studies reviewed was 720 g C m^{-2} year^{-1} (Grace et al., 2006), which is a close match to our CASA model prediction of the global average NPP of 782 g C m^{-2} year^{-1} for the savanna biome (Table 22.1; Potter, 1999).

Based largely on an earlier data compilation study by Saugier et al. (2001) that assigned 27.6 million km^2 global land coverage to tropical savannas and grasslands, Grace et al. (2006) reported that total carbon stored (aboveground and belowground) in this biome was 326 Pg C and savanna NPP averaged 19.9 Pg C year^{-1}. The CASA model prediction reported by Potter (1999) for aboveground biomass in savannas and grasslands was 202 Pg C globally, whereas NPP averaged 19.3 Pg C year^{-1} over 24.3 million km^2 global savanna coverage area (Table 22.1). Since mean annual NPP was nearly identical in the two predictions from Potter (1999) and Saugier et al. (2001), the comparison of living biomass estimates suggests that roughly 52% of savanna carbon pools is stored aboveground and 48% is stored belowground. Nonetheless, due to a paucity of soil measurements, this may be a conservative estimate of belowground carbon storage capacity of tropical savanna ecosystems.

In the absence of burning or other land cover change, the carbon sequestration rate (NEP sink) of savannas globally has been reported to average +14 g C m^{-2} year^{-1} (Grace et al., 2006). This is, however, an oversimplified generalization of annual NEP fluxes in any terrestrial ecosystem, because net yearly carbon sink and source fluxes can be highly dependent on seasonal climate anomalies. For example, Potter et al. (2009) predicted with the CASA model that savanna (*Cerrado*) NEP fluxes in the Brazilian states of Mato Grosso and Goiás switched from a net sink flux on the order of +30 g C m^{-2} year^{-1} during 2001 and 2003 to a net source flux on the order of −25 g C m^{-2} year^{-1} during the 2002–2003 El Niño period. According to the climate data inputs to these CASA runs, total precipitation across the states of Brazil was about 3% above the regional 5-year (2000–2004) mean in 2000 and 2001 and was about 6% below the regional 5-year mean in 2002.

Global Climate Controls on Savanna Vegetation Dynamics

New understanding of how future climate change may impact the productivity of savannas around the world will depend upon analysis of historical patterns of precipitation and temperature controls on savanna vegetation phenology. Over relatively short-time periods, the influence of atmosphere-ocean surface patterns, such as those associated with the El Niño-Southern Oscillation (ENSO) and the Arctic Oscillation (AO), have been noted as significant global teleconnections for atmospheric circulation and land surface climate (Glantz et al., 1991). *Teleconnection* is a term used in meteorological studies to describe simultaneous variation in climate and related processes over widely separated points on Earth. There are different phases in climate patterns such as the ENSO, which is simply called El Niño in the warm phase and La Niña in the cold phase. The ENSO warming at the sea surface, which is driven by changes in winds and ocean-atmosphere heat exchange, typically extends to about 30°N and 30°S latitude, with lags into continental land areas of several months (Trenberth and Caron, 2000). Previous studies have identified global connections between interannual SST and land surface processes (Dai et al., 1998; Myneni et al., 1998; Los et al., 2001; Potter et al., 2003).

In a recent study of global climate–vegetation dynamics, Potter et al. (2007) analyzed the MODIS Collection 4 from February 18, 2000 to December 18, 2005 (which represented the latest available version of MODIS products at that time) and reported strong ENSO climate controls over savanna ecosystems worldwide. The MODIS product for fraction absorbed of photosynthetically active radiation (FPAR) by vegetation canopies was used to track global changes in savanna cover (Myneni et al., 1997, 2001). The FPAR is a common measure of canopy greenness phenology in that it measures the proportion of available radiation in the photosynthetically active wavelengths (400–700 nm) that a vegetation canopy can absorb (Zhang et al., 2003). Land surface climate data sets used in this study were obtained from the NOAA National Center for Environmental Prediction (NCEP) reanalysis products (Kistler et al., 2001; Qian et al. 2006). Surface temperature and pressure indices were obtained from the NOAA NCEP Climate Prediction Center (Higgins et al., 2001; Zhou et al., 2001).

For the southern hemisphere latitude zones dominated by savanna cover, Potter et al. (2007) reported that MODIS monthly FPAR levels started at above average levels in 2001 and declined to below average levels as the 2002 El Niño event weakened in 2003–2005. The moderately strong El Niño event in 2002–2003 followed a more La Niña-like period in 1998–2001 (Lagerloef et al., 2003). During October–December 2002, El Niño anomalies peaked for most oceanic and atmospheric variables (e.g., equatorial SST, zonal wind, eastern Pacific thermocline depth) (McPhaden, 2004). The NINO34 anomalies (which cover SST over the area 5°S–5°N, 120°–170°W) reached 1.8°C in November 2002, approximately equal to peak values during the moderate-strength 1986–1987 and 1991–1992 El Niños.

In both the Northern and Southern Hemispheres, monthly FPAR in biomes dominated by savannas (grasslands and shrublands) were most strongly correlated with the NINO34 index variations from 2000 to 2005 (Table 22.2 and Figure 22.4a). The MODIS monthly FPAR anomalies were most closely correlated (positively or negatively) with NCEP reanalysis surface temperature variations within the savanna biome in the southwestern United States, central Asia, eastern Russia and China, southern Africa, and Australia. The MODIS monthly FPAR anomalies were most closely correlated with NCEP reanalysis monthly precipitation variations (Figure 22.4b) within the savanna biome in central Asia, southern Africa, and Australia. In the majority of regions, Pottter et al. (2007) found that seasonal FPAR cover in savannas increased most

TABLE 22.2a

Pearson's Correlation Values of MODIS FPAR (by Global Vegetation Biome) to Climate Index Anomalies over the Time Period 2000–2005 for the Northern Hemisphere Shift Periods in Months are Shown that Provided for the Highest Correlation Values

Biome	Description	AAO	AO	NAO	PNA	NINO12	NINO3	NINO34	Threshold Correlation from Randomization Test
1	Grasses and cereal crops	−0.35	−0.30	−0.25	0.31	0.26	0.48	0.59**	0.49
Best shift (mo)		14	11	7	13	14	14	14	
2	Shrubland	−0.35	−0.30	−0.25	0.27	0.27	0.47	0.63**	0.52
Best shift (mo)		13	8	6	0	14	14	14	
3	Broadleaf crops	0.36	−0.29	−0.25	−0.34	0.24	0.32	0.24	NA
Best shift (mo)		0	2	7	8	12	11	10	
4	Savanna	0.32	−0.23	−0.28	0.38	−0.17	0.19	−0.18	NA
Best shift (mo)		0	4	6	0	4	11	1	
5	Broadleaf forest	−0.33	−0.24	0.23	0.47	0.39	0.54*	0.53*	0.37
Best shift (mo)		8	5	12	4	12	12	11	
6	Needleleaf forest	−0.26	−0.19	0.30	0.37	0.44	0.50	0.38	NA
Best shift (mo)		9	7	4	5	13	12	11	

** $p < 0.05$; * $p < 0.10$.

TABLE 22.2b

Pearson's Correlation Values of MODIS FPAR (by Global Vegetation Biome) to Climate Index Anomalies over the Time Period 2000–2005 for the Southern Hemisphere Shift Periods in Months are Shown that Provided for the Highest Correlation Values

Biome	Description	AAO	AO	NAO	PNA	NINO12	NINO3	NINO34	Threshold Correlation from Randomization Test
1	Grasses and cereal crops	0.24	0.13	-0.21	-0.21	-0.35	-0.60**	0.76**	0.59
Best shift (mo)		0	14	0	12	0	13	13	
2	Shrubland	0.28	-0.17	0.11	-0.10	-0.39	-0.67**	-0.84**	0.66
Best shift (mo)		0	7	14	14	0	0	0	
3	Broadleaf crops	0.35	-0.23	0.25	-0.30	-0.55*	-0.57*	-0.53*	0.52
Best shift (mo)		0	5	10	14	1	1	13	
4	Savanna	0.45	-0.31	-0.25	-0.18	-0.52*	-0.61**	-0.64**	0.54
Best shift (mo)		0	5	5	12	0	0	0	
5	Broadleaf forest	-0.28	0.32	0.35	0.28	0.37	-0.29	-0.29	NA
Best shift (mo)		3	12	12	7	6	1	1	
6	Needleleaf forest	NA	NA	NA	NA	NA	NA	NA	NA
Best shift (mo)		NA	NA	NA	NA	NA	NA	NA	NA

** $p < 0.05$; * $p < 0.10$.

Climate index key: AAO is the Antarctic Oscillation, AO is the Arctic Oscillation, NAO is the North Atlantic Oscillation, PNA is the Pacific-North American Oscillation (Higgins et al., 2001).

NINO indices are sea surface temperature (SST) anomalies within the following regions:

NINO12: 0–10° S, 80° W–90° W.

NINO3: 5° N–5° S, 90° W–150° W.

NINO34: 5° N–5° S, 120° W–170° W.

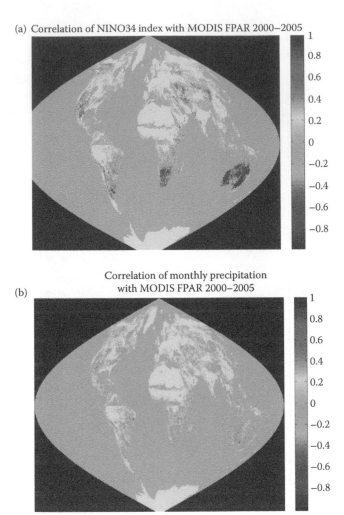

(a) Correlation of NINO34 index with MODIS FPAR 2000–2005

Correlation of monthly precipitation
with MODIS FPAR 2000–2005

(b)

FIGURE 22.4
(See color insert following page 320.) Correlations of monthly vegetation FPAR with climate
variables anomalies in 2000–2005. (a) NINO34 index and (b) NCEP precipitation. White areas
of the maps are those with more than one missing monthly FPAR value per year and are, therefore, not considered significant at the $p < 0.10$ level. The break line in correlation results across
the equatorial regions is an artifact of separating the Northern and Southern Hemisphere
biomes, which are 6 months out of phase.

strongly when rainfall amounts increased and FPAR cover decreased when
seasonal rainfall amounts decreased. These short-term climate-vegetation
dynamics imply a strong dependence of savanna zones upon future patterns
of rainfall (or net moisture deficits) over regions of the globe most likely to be
impacted by climate change.

Discussion of Possible Futures for Savanna Ecosystems

There are a variety of studies that offer projections for savanna ecosystem and global climate change over the next century. Some coupled Global Circulation Models (GCMs) with dynamic vegetation features have projected that ever-green forests will be succeeded by mixed forest, savanna, and grassland eco-system in eastern Amazonia (White et al., 1999; Cramer et al., 2001; Cramer et al., 2004) and accompanied by large emissions of carbon from soil and veg-etation (Cox et al., 2000; Jones et al., 2003; Cox et al., 2004). In contrast to these projections, Malhi et al. (2008) suggested that forest dieback due to drought and deforestation in the eastern Amazon would result mainly in replacement by agricultural ecosystems instead of expansion of natural savanna cover into former forest cover areas. This study by Malhi et al. (2008) further implied that future changes in precipitation, particularly in the dry season, are probably the most critical determinant of the climatic fate of the forest and savanna cover in the eastern and southern extents of the Amazon region.

In a continental scale analysis for African savannas, Sankaran et al. (2005) showed that water availability was the main limitation on the maximum cover of woody species in most African savanna systems, but that distur-bance dynamics control savanna structure below the maximum potential cover levels. In southern Africa, increases in air temperature predicted for 2100 range from 2° to 6°C (Reid et al., 2007). Overall rainfall is projected to become even more variable than it is now. Even if rainfall changes little from present-day levels, increases in temperature will elevate evaporation rates from soils, leading to frequent water deficits. The intensity of extreme events such as droughts is likely to increase. In such cases, grassy savanna may become less dominant (Woodward and Lomas, 2004), as desertification occurs in some areas and shrubs and trees benefit from higher levels of CO_2 in others (Bond and Midgley, 2000). Desanker et al. (1997) added that increases in drying of the African savanna biome in the next century could be com-pounded by other pressures such as conversion to agriculture.

In closing, the uncertain future for many human population centers living in or near savanna biomes around the world makes remote sensing and eco-system modeling of these areas a high priority for international research organizations. In addition, the potential losses of CO_2 to the atmosphere from savanna soils and diebacks of their woody plant communities are among the highest of any global biome. Fortunately, a new level of geographic detail has been added by satellite observations over the past decade to stud-ies of large-scale climate impacts at the biome scale. By extending this satel-lite historical data set, individual municipalities in savanna biomes can distinguish their unique contributions to global greenhouse gas exchange, which should influence local management policies and adaptation to climate change in an unprecedented manner.

References

Amthor, J. S., J. M. Chen, J. S. Clein, et al. 2001. Boreal forest CO_2 exchange and evapotranspiration predicted by nine ecosystem process models: Inter-model comparisons and relations to field measurements. *Journal of Geophysical Research—Atmospheres* 106, 33,623–33,648.

Andreae, M. O., T. W. Andreae, H. Annegarn, et al. 1998. Airborne studies of aerosol emissions from savanna fires in southern Africa, 2. Aerosol chemical composition. *Journal of Geophysical Research—Atmospheres* 103, 32119–32128.

Behrenfeld, M. J., J. T. Randerson, C. R. McClain, et al. 2001. Biospheric primary production during an ENSO transition. *Science* 291, 2594–2597.

Bond, W. J. and G. F. Midgley. 2000. A proposed CO2-controlled mechanism of woody plant invasion in grasslands and savannas. *Global Change Biology* 6, 865–869.

Cox, M. P., R. A. Betts, M. Collins, P. P. Harris, C. Huntingford, and C. D. Jones. 2004 Amazonian forest dieback under climate-carbon cycle projections for the 21st century, *Theoretical and Applied Climatology* 78, 137–156.

Cox, P. M., R. A. Betts, C. D. Jones, S. A. Spall, and I. J. Totterdell. 2000. Acceleration of global warming due to carbon-cycle feedbacks in a coupled climate model. *Nature* 408, 184–187.

Cramer W., A. Bondeau, S. Schaphoff, W. Lucht, B. Smith, and S. Sitch. 2004. Tropical forests and the global carbon cycle: impacts of atmospheric carbon dioxide, climate change and rate of deforestation. *Philosophical Transactions: Biological Sciences* 359, 331–343.

Cramer, W., A. Bondeau, F. I. Woodward, et al. 2001. Global response of terrestrial ecosystem structure and function to CO2 and climate change: Results from six dynamic global vegetation models. *Global Change Biology* 7, 357–373.

Dai, A., K. E. Trenberth, and T. R. Karl. 1998. Global variations in droughts and wet spells: 1900–1995. *Geophysical Research Letters* 25, 3367– 3370.

DeFries, R., M. Hansen, and J. Townshend. 1995. Global discrimination of land cover types from metrics derived from AVHRR Pathfinder data. *Remote Sensing of Environment* 54, 209–222.

Desanker, P. V., P. G. H. Frost, C. O. Justice and R. J. Scholes. (eds) 1997. The Miombo Network: Framework for a Terrestrial Transect Study of Land-Use and Land-Cover Change in the Miombo Ecosystems of Central Africa. International Geosphere-Biosphere Programme (IGBP), Stockholm, Sweden.

Eiten, G. 1986. The use of the term "savanna." *Tropical Ecology* 27, 10–23.

Friedl, M. A., D. K. McIver, J. C. F. Hodges, et al. 2002. Global land cover mapping from MODIS: Algorithms and early results. *Remote Sensing of Environment* 83, 287–302.

Glantz, M. H., R. W. Katz, and N. Nicholls. (eds) 1991. *Teleconnections Linking Worldwide Climate Anomalies.* Cambridge University Press, New York, 527pp.

Goetz, S. J and S. D. Prince. 1998. Variability in light utilization and net primary production in boreal forest stands. *Canadian Journal of Forest Research* 28, 375–389.

Grace, J., J. San José, P. Meir, H. S. Miranda, and R. Montes. 2006. Productivity and carbon fluxes of tropical savannas. *Journal of Biogeography* 33, 387–401.

Hicke, J. A., G. P. Asner, J. T. Randerson, et al. 2002. Satellite-derived increases in net primary productivity across North America, 1982–1998. *Geophysical Research Letters* 29, 1427, doi:10.1029/2001GL013578.

Higgins, R. W., Y. Zhou, and H.-K. Kim. 2001. Relationships between El Niño-Southern Oscillation and the Arctic Oscillation: A Climate-Weather Link. NCEP/Climate Prediction Center ATLAS 8 (http://www.cpc.noaa.gov/research_papers/ncep_cpc_atlas).

Huete, A., K. Didan, T. Miura, E. P. Rodriguez, X. Gao, and L. G. Ferreira. 2002. Overview of the radiometric and biophysical performance of the MODIS vegetation indices. *Remote Sensing of Environment* 83, 195–213.

Jones, C. D., P. M. Cox, R. L. H. Essery, D. L. Roberts, and M. J. Woodage. 2003. Strong carbon cycle feedbacks in a climate model with interactive CO2 and sulphate aerosols. *Geophysical Research Letters* 30, 1479.

Kistler, R., E. Kalnay, W. Collins, et al. 2001. The NCEP-NCAR 50-year reanalysis: Monthly means CD-ROM and documentation. *Bulletin of the American Meteorological Society* 82, 247–268.

Lagerloef, G., R. Lukas, F. Bonjean, et al. 2003. El Niño Tropical Pacific Ocean surface current and temperature evolution in 2002 and outlook for early 2003. *Geophysical Research Letters* 30, 1514, doi:10.1029/2003GL017096.

Los, S. O., G. J. Collatz, L. Bounoua, P. J. Sellers, and C. J. Tucker. 2001. Global interannual variations in sea surface temperature and land surface vegetation, air temperature, and precipitation. *Journal of Climate* 14, 1535–1549.

Malhi, Y., J. T. Roberts, R. A. Betts, T. J. Killeen, W. Li, and C. A. Nobre. 2008. Climate change, deforestation, and the fate of the Amazon. *Science* 319, 169–172. DOI: 10.1126.

Malmström, C. M., M. V. Thompson. G. P. Juday, S. O. Los, J. T. Randerson, and C. B. Field. 1997. Interannual variation in global scale net primary production: Testing model estimates. *Global Biogeochemical Cycles* 11, 367–392.

McPhaden, M. J. 2004. Evolution of the 2002/03 El Niño. *Bulletin of the American Meteorological Society* 677–695. DOI: 10.1175/BAMS-85-5-677.

Monteith, J. L. 1972. Solar radiation and productivity in tropical ecosystems. *Journal of Applied Ecology* 9, 747–766.

Myneni, R. B., C. J. Tucker, G. Asrar, and C. D. Keeling. 1998. Interannual variations in satellite-sensed vegetation index data from 1981 to 1991. *Journal of Geophysical Research* 103, 6145–6160.

Myneni, R. B., J. Dong, C. J. Tucker, et al. 2001. A large carbon sink in the woody biomass of Northern forests. *Proceedings of the National Academy of Sciences, USA* 98, 14784–14789.

Myneni, R. B., C. D. Keeling, and R. R. Nemani. 1997. Increased plant growth in the northern high latitudes trends during 1979–90. *Nature* 386, 698–701.

Olson, R. J., J. M. O. Scurlock, W. Cramer, W. J. Parton, and S. D. Prince. 1997. *From Sparse Field Observations to a Consistent Global Dataset on Net Primary Production.* IGBP-DIS Working Paper No. 16, IGBP-DIS, Toulouse, France.

Olson J. S., J. A. Watts, and L. J. Allison. 1983. *Carbon in Live Vegetation of Major World Ecosystems.* Report ORNL-5862. Oak Ridge National Laboratory, Oak Ridge, TN, 164pp.

Parton, W. J., B. McKeown, V. Kirchner, and D. Ojima. 1992. *CENTURY Users Manual.* Natural Resource Ecology Laboratory. Colorado State University, Fort Collins.

Potter, C. S., J. T. Randerson, C. B. Field, et al. 1993. Terrestrial ecosystem production: A process model based on global satellite and surface data. *Global Biogeochemical Cycles* 7, 811–841.

Potter, C. S. 1999. Terrestrial biomass and the effects of deforestation on the global carbon cycle. *BioScience* 49, 769–778.

Potter, C., S. Klooster, R. Myneni, V. Genovese, P. Tan, and V. Kumar. 2003. Continental scale comparisons of terrestrial carbon sinks estimated from satellite data and ecosystem modeling 1982–98. *Global and Planetary Change* 39, 201–213.

Potter, C., P. Tan, V. Kumar, et al. 2005. Recent history of large-scale ecosystem disturbances in North America derived from the AVHRR satellite record. *Ecosystems* 8, 808–827.

Potter, C., S. Boriah, M. Steinbach, V. Kumar, and S. Klooster. 2008. Terrestrial vegetation dynamics and global climate controls. *Climate Dynamics* 31, 67–78.

Potter, C., S. Klooster, A. Huete, et al. 2009. Terrestrial carbon sinks in the Brazilian Amazon and Cerrado region predicted from MODIS satellite data and ecosystem modeling. *Biogeosciences Discussions* 6, 1–23.

Priestly, C. H. B. and R. J. Taylor. 1972. On the assessment of surface heat flux and evaporation using large-scale parameters. *Monthly Weather Review* 100, 81–92.

Qian, T. A. Dai, K. E. Trenberth, and K. W. Oleson. 2006. Simulation of global land surface conditions from 1948 to 2004. Part I: Forcing data and evaluations. *Journal of Hydrometeorology* 7, 953–975.

Rahman, A. F., D. A. Sims, V. D., Cordova, and B. Z. El-Masri. 2005. Potential of MODIS EVI and surface temperature for directly estimating per-pixel ecosystem C fluxes. *Geophysical Research Letters* 32, L19404, doi:10.1029/2005GL024127.

Reid, H., L. Sahlén, J. Stage, and J. MacGregor. 2007. The economic impact of climate change in Namibia: How climate change will affect the contribution of Namibia's natural resources to its economy. Environmental Economics Programme Discussion Paper 07–02. International Institute for Environment and Development, London, 54pp.

Running, S. W and R. R. Nemani. 1998. Relating seasonal patterns of the AVHRR vegetation index to simulated photosynthesis and transpiration of forests in different climates. *Remote Sensing of Environment* 24, 347–367.

San José, J. J. and R. A. Montes. 2003. Neotropical savanna converted to food cropping and cattle feeding systems: Soil carbon and nitrogen changes over 30 years. *Forest Ecology and Management* 184, 17–32.

Sankaran, M., N. P. Hanan, R. J. Scholes, et al. 2005. Determinants of woody cover in African savannas. *Nature* 438, doi:10.1038.

Saugier, B., J. Roy, and H. A. Mooney. 2001. Estimations of global terrestrial productivity, converging toward a single number? In *Global Terrestrial Productivity*, eds. J. Roy, B. Saugier, and H. A. Mooney. San Diego CA, Academic Press, pp. 541–555.

Seiler, W. and P. C. Crutzen. 1980. Estimates of gross and net fluxes of carbon between the biosphere and the atmosphere from biomass burning. *Climatic Change* 2, 207–247.

Trenberth, K. E. and J. M. Caron. 2000. The Southern Oscillation revisited: Sea level pressures, surface temperatures and precipitation, *J. Climate* 13, 4358–4365.

White, A., M. G. R. Cannel, and A. D. Friend. 1999. Climate change impacts on ecosystems and the terrestrial carbon sink: A new assessment. *Global Environmental Change* 9, S21–S30.

Woodward, F. I. and M. R. Lomas. 2004. Simulating vegetation processes along the Kalahari transect. *Global Change Biology* 10, 383–392.

Xiao, X., Q. Zhang, B. Braswell, et al. 2004. Modeling gross primary production of temperate deciduous broadleaf forest using satellite images and climate data. *Remote Sensing of Environment* 91, 256–270.

Zhang, X., M. A. Friedl, C. B. Schaaf, et al. 2003. Monitoring vegetation phenology using MODIS. *Remote Sensing of Environment* 84, 471–475.

Zheng, D., S. Prince, and R. Wright. 2003. Terrestrial net primary production estimates for 0.5° grid cells from field observations—A contribution to global biogeochemical modeling. *Global Change Biology* 9, 46–64.

Zhou, S., A. J. Miller, J. Wang, and J. K. Angell. 2001. Trends of NAO and AO and their associations with stratospheric processes. *Geophysical Research Letters* 28, 4107–4110.

23

CENTURY-SAVANNA Model for Tree–Grass Ecosystems

William J. Parton, Robert J. Scholes, Ken Day, John O. Carter, and Robin Kelly

CONTENTS

Introduction

At least a sixth of the global land surface is covered by ecosystems in which both woody plants and grasses contribute to the NPP (Scholes and Hall, 1996); these ecosystems are broadly referred to as *savannas*. Classical theory suggests that trees have an advantage over grasses as a result of the tree's ability to root throughout the soil profile, whereas grasses predominantly root in the surface horizons (Walker et al., 1981). Although a vertical separation between tree and grass roots and their exploitation of sources of soil water can be observed (Brown and Archer, 1990; Dodd et al., 1998; Midwood et al., 1998), savanna soils are often shallow (<1 m); and in semiarid regions, soil moisture seldom penetrates to the depth where trees have an advantage. The primary niche difference between trees and grasses may instead be primarily temporal rather than a result of rooting patterns (Scholes and Walker, 1993; Sala et al., 1997). Savanna climates have a distinct seasonality: a hot, wet growing season and a warm, dry nongrowing season. Trees in savannas reach their full leaf area in the first weeks of the wet season, sometimes earlier if soil moisture is sufficient; whereas grasses reach peak leaf area later in the growing season with a decline before trees begin to shed their leaves (Archibald and Scholes, 2007). Therefore, trees have almost exclusive access to resources at both the beginning and the end of the growing season; whereas grasses may be more competitive during the middle.

The main factors that control plant production in savanna systems include precipitation, nutrient availability, and solar radiation. Aboveground yearly grass production in the savannas of the semiarid and sub-humid tropics is linearly related to annual rainfall (Scholes, 1993; DelGrosso et al., 2008). Precipitation provides an overall tree biomass constraint (Sankaran et al., 2005), but at annual scales, tree production is only weakly dependent upon rainfall during that year. Grassland plant production responds quickly to the addition of nitrogen (Owensby et al., 1970; Dodd and Laurenroth, 1978), with the largest response occurring when nitrogen and phosphorus are applied together (Donaldson et al., 1984). Tree production responses to the addition of water or fertilizer are much slower than the responses of grasses. Tree and grass production are positively correlated to solar radiation during time periods when soil water is available for plant growth. Savanna tree canopies are generally sparse and interrupted. Therefore, light availability is generally not considered a key factor in the competitive interaction, except in the case of near-closed woodlands (i.e., tree cover > 50%).

It is during tree establishment that the competitive dynamics may favor grasses (Bond, 2008). A well-established grass sward has a strong suppressive effect on tree recruitment, both through competition for light, water, and nutrients and also through the provision of fuel for frequent fires. Fires in savannas kill tree buds, leaves, and small branches within the flame zone (typically up to 3 m) but seldom kill larger trees. Therefore, sub-mature trees may be retained in the grass layer for decades before a combination of poor grass growth (due to drought or heavy grazing) and a sufficiently long fire-free period allows maturation.

Savanna Models

Although generalizations regarding savanna structure and function are difficult, several models have been developed to integrate the current body of knowledge about savanna systems (House et al., 2003). The most detailed of these models, MUSE-TREEGRASS (Simioni et al., 2000), mechanistically simulates the two-dimensional spatial distribution of tree and grass plants and represents competition for water and nutrients as a function of tree and grass live root biomass in each grid point (Simioni et al., 2000). Time steps that range from hourly to daily are used to simulate water flow, plant demography, light and energy balance, and plant growth for trees and grasses. Individual root and foliage crowns are represented in space with the aim of studying the effect of vegetation structure (density, crown shape, and spatial distribution) and fire on ecosystem functioning.

The SAVANNA ecosystem model (Coughenour, 1992, 1995) is a semi-spatially explicit mechanistic model that simulates tree population dynamics and growth, in addition to grass growth, without representing the exact spatial location of the trees and grasses. The model is capable of simulating, at a weekly-monthly time step, the role of landscape heterogeneity on ecosystem function over time. SAVANNA keeps track of patches in the landscape, the extent of which is a dynamic outcome of vegetation growth, recruitment, and mortality. The model simulates plant growth as a function of water stress, solar radiation, and rooting distributions, with competition for water and light and disturbances such as grazing, fire, and harvest as the major factors controlling tree or grass competition.

The highly parameterized daily time-step pasture growth model, GRASP (McKeon et al., 1990), which was developed in Australia and also applied in Zimbabwe, simulates aboveground grass yield as a function of a site's nitrogen index, tree density, and plant transpiration. Tree growth is not explicitly modeled; but tree-based transpiration, proportional to tree basal area (TBA), is represented, which reduces the amount of water available for grass growth. Pasture growth is a function of grass transpiration, radiation interception, temperature, air humidity, nitrogen availability, and pasture management (grazing, fire, irrigation, and mowing). Nitrogen uptake by grass is constant per unit of transpiration and stops when a user-specified maximum annual N uptake has been reached.

The objective of this chapter is to describe the theoretical basis and testing of a savanna model based on the well-known CENTURY ecosystem biogeochemistry model (Metherell et al., 1993; Parton et al., 1994). The CENTURY model has been extensively used to simulate ecosystem dynamics (plant and soil carbon, nitrogen, and phosphorus) for grasslands, forest, and crops around the world (Parton et al., 1993; Kelly et al., 1997; Paustian et al., 1997). The SAVANNA-CENTURY model uses the ideas first presented by Scanlan and Burrows (1990) to represent competition between trees and grasses in a simplified way: trees having an inverse-concave competitive effect on grass as the TBA increases and grasses being relatively more competitive in sites with higher productivity (more nutrients and soil water). This chapter documents the rules used in SAVANNA-CENTURY to predict competition between trees and grasses. Observed grass plant production data and soil carbon data from sites in Africa and Australia will be used to show that the SAVANNA-CENTURY model correctly simulates the impact of tree biomass on grass production, soil carbon levels for different soil textures, and the impact of burning treatments on tree–grass competition and soil carbon levels.

SAVANNA-CENTURY Model Description

CENTURY is a general purpose biogeochemically based terrestrial ecosystem model, widely used for the simulation of carbon, nitrogen, and phosphorus cycling. It keeps account of the main ecosystem organic matter pools (three in the soil and several in the plants), each of which has C:N:P:S ratios within defined ranges. The transfers between the pools are controlled by moisture, temperature, soil texture, and organic matter quality (the lignin : nitrogen ratio). Transfer of carbon to pools with a progressively slower turnover rate is associated with respiration of carbon and mineralization of the nutrients to maintain the specified C : nutrient ratio. Nitrogen (other nutrients behave similarly) is taken up by plants after losses due to leaching and N gas fluxes (N_2O, N_2, NH_3, and NOx) and additions due to deposition and fixation. Carbon assimilation by plants is a simple function of soil moisture and air temperature. In general, it is the availability of nitrogen that limits NPP. CENTURY 4.5 requires data for monthly rainfall, average maximum and minimum daily temperatures, and soil texture (top 20 cm), in addition to information regarding the maximum and minimum nutrient and lignin content of plant parts.

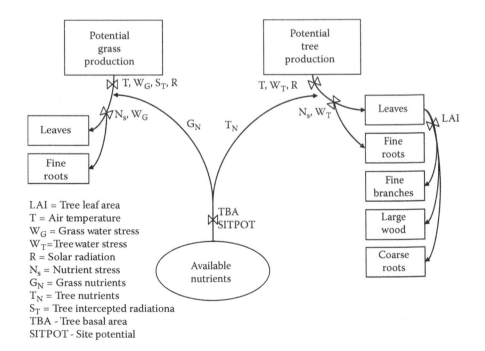

LAI = Tree leaf area
T = Air temperature
W_G = Grass water stress
W_T = Tree water stress
R = Solar radiation
N_s = Nutrient stress
G_N = Grass nutrients
T_N = Tree nutrients
S_T = Tree intercepted radiationa
TBA - Tree basal area
SITPOT - Site potential

FIGURE 23.1
Flow diagram for the SAVANNA-CENTURY model highlighting the factors that control competition between trees and grasses for soil nutrients and water.

The CENTURY-SAVANNA model (Metherell et al., 1993) simulates tree and grass growth simultaneously and includes competition for light, nutrients, and water (Figure 23.1). It is capable of simulating the impact of fire and animal disturbance on tree and grass biomass that results in altered competition between trees and grasses. The major factors that control competition for nutrients include TBA (increased nutrient uptake with a higher basal area) and production potential of the site (grasses are more competitive in high production sites). Grass production is reduced by light interception from the tree canopy, and rooting depth controls the amount of water available for plant growth (trees generally have deeper roots and greater water availability). The major factors that control potential plant growth for trees and grasses include water stress, temperature, and solar radiation. The equations used to describe potential plant growth for trees and grasses are presented below (Metherell et al., 1993). The water stress term is calculated as a function of rainfall and the available soil water within the tree and grass rooting depth (potentially more water is available for tree growth). Potential grass growth is reduced as a function of the fraction of light intercepted by the trees.

The major factor controlling competition between trees and grasses in SAVANNA-CENTURY is the competition for available nutrients. This relationship is derived from the Scanlan and Burrows (1990) equations relating grass production to TBA and site potential using the following equations:

$$Nt = 1 - c1^* \exp(-K * TBA),\qquad(23.1)$$

$$SITPOT = 1000.0 + 25.0 * ANP,\qquad(23.2)$$

$$K = 1.664 \exp(-0.00102 + SITPOT * Tmn),\qquad(23.3)$$

where Nt is the fraction of available nutrients for tree growth, SITPOT is the site potential for grass growth (kg ha^{-1}), TBA is the tree basal area (m^2 ha^{-1}), K is a parameter controlling the curvature of the TBA effect on Nt, Tmn is the nitrogen available for plant growth (gN m^{-2}), and ANP is the annual net primary production (function of annual precipitation).

Grass nutrient uptake is equal to 1.0 minus Nt. Figure 23.2 shows that the fraction of nutrients allocated to tree growth increases with increasing TBA and that sites with higher SITPOT values have fewer nutrients for enhancing tree growth. Scanlan and Burrows (1990) have shown that K (the curvature term) is a function of SITPOT, which linearly increases as a function of annual aboveground grass production in the absence of trees (Del Grosso et al., 2008). The TBA is calculated as a function of aboveground wood carbon using the following equations:

$$BAS = (W_c/(0.80 * ((W_c * 0.01)^{**}0.635))),\qquad(23.4)$$

$$TBA = (W_c/BAS) * K_2,\qquad(23.5)$$

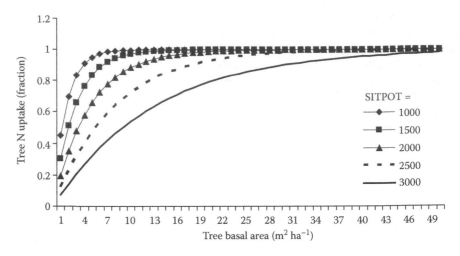

FIGURE 23.2
The impact of tree basal area (TBA) and site potential on the fraction of nitrogen allocated for tree growth.

where K_2 is a tree-specific coefficient for calculating TBA and W_c is the wood carbon in trees (large wood and fine branches).

Another important change to the SAVANNA-CENTURY model is its capability to represent the impact of soil texture on nitrogen gas fluxes (NH_3, N_2, NOx, and N_2O). Equation 23.6 shows the impact of clay content on soil NH_3 gas fluxes from urine and feces deposited by grazing animals, whereas Equation 23.7 shows the impact of clay content on the combined nitrogen gas losses from soil nitrogen mineralization.

$$Na = 0.70 + 0.50 * Cs, \tag{23.6}$$

$$Ng = 0.03 - 0.0677 * Cs, \tag{23.7}$$

where Cs is soil clay content (fraction), Na is the fraction of nitrogen lost from animal urine and feces deposition as NH_3, and Ng is the fraction of nitrogen lost from nitrogen mineralized during decomposition of soil organic matter (N_2, NOx, and N_2O). The impact of clay content on animal NH_3 loss is based on observed data (Schimel et al., 1986; Delgado et al., 1996) showing that NH_3 losses increase as the clay content decreases. NOx trace gas fluxes are one of the largest gas fluxes from savanna and grassland ecosystems (Martin et al., 1998) with NOx fluxes decreasing with increasing soil clay content. The parameters in Equations 23.6 and 23.7 were based on model tuning where the parameter values were adjusted so that the model results matched observed grass production data showing higher production in high clay sites (Dye and Spear, 1982).

Disturbances, such as fire, grazing, and tree death, greatly impact the competition between trees and grasses. The SAVANNA-CENTURY model

simulates these disturbances by using surface fire, grazing, and tree disturbance sub-routines. The surface fire sub-routine simulates differential
removal of live and dead grass biomass (standing dead and surface litter),
return of carbon (fine ash and wood charcoal) and nutrients from the fire,
and burning of surface dead wood pools (large wood and fine branches). The
grazing sub-routines simulate different grazing practices by removing different amounts of live and dead grass biomass. The tree disturbance sub-
routines simulate the impact of fire as it kills different parts of live trees
(leaves, fine branches, and large wood), the return of nutrients and charcoal
from fire events, animal grazing of leaves and fine branches (e.g., goat and
elephant grazing of savanna trees), and the physical removal of trees by other
forest management practices. The model is capable of simulating the differential impact of cold and hot fires on biomass removal, nutrient loss, and
charcoal returns; fires that burn the surface material but that have minimal
impacts on the trees; and hot fires that kill most of the trees and remove large
amounts of surface biomass. Fire and grazing disturbances are the major
cause of reductions in TBA, and they result in enhanced grass growth.

Model Testing and Calibration

Observed plant production and soil carbon data sets from Africa and Australia
(Beale, 1973; Dye and Spear, 1982; Scholes, 1987, Jones et al., 1990; Scanlan and
Burrows, 1990; Bird et al., 2000) were used to test the ability of the SAVANNA-
CENTURY model to simulate the impacts of (1) soil texture on plant production
and soil carbon levels, (2) fire frequency on soil carbon levels, and (3) TBA on
grass production. The observed plant production data sets from Zimbabwe
sand and clay Matopos sites (Dye and Spear, 1982) and the sand and clay
Klaserie Private Nature Reserve in South Africa (Scholes, 1987) were used to
help parameterize Equations 23.6 and 23.7 (showing nitrogen gas loss). The
Beale (1973) and the Scanlan and Burrows (1990) data sets were used to parameterize coefficients in Equations 23.1 through 23.3 (impact of TBA and site
potential on allocation of nutrients). The observed soil carbon data were not
used to calibrate model parameters and, thus, can be considered an independent test of the impact of soil texture and fire frequency on soil carbon levels.
The observed soil Δ ^{13}C data sets from Bird et al. (2000) is an independent test
(not used for model development) of the ability of the model to simulate the
impact of fire frequency on tree and grass soil carbon inputs.

The *Combretum* and *Acacia* sites located in the Klaserie Private Nature
Reserve in South Africa (Scholes, 1987) represent low and high clay content
sites with soil carbon and plant production data from tree removal and
control treatments. Grass production data were collected by sequential
bi-monthly clipping of the sward protected by herbivore exclusion cages for
1984 and 1985. Observed climate data in 1982–1985 were measured at the site;

whereas the long-term rainfall and temperature record from Phalaborwa, 20 km to the north, was used for the period 1924–1981. The SAVANNA-CENTURY model was set up to simulate the two sites by running the model for the 1900 year spin-up runs to estimate the model initial conditions and then represent the management of the treatments from 1982 to 1986 (tree clearing and control). Tree disturbance rates for the spin-up runs were estimated so that the initial tree biomass in 1983 matched the observed data from the sites.

Dye and Spear (1982) present annual grass plant production data from several sites in Zimbabwe (Matopos clay, Matopos sand, Naymandhlovu sand, and Tuli sand) with tree removal and control experiments maintained from 1960 to 1980. Wet season grass production was determined by annual clipping of all the plot grass biomass at the end of the growing season (May–July) on the cleared and uncleared sites from 1960 to 1980. The observed monthly climate data from 1960 to 1980 were used for all of the sites. Observed climate data from Bulawayo during 1941 to 1960 and 1980 to 2000 were used for the Matopos and Naymandhlovu (NYAM) sites, and observed weather data from 1901 to 1980 were used for the Tuli site. Observed soil carbon data for the Matopos clay and sand sites came from Bird et al. (2000). The model was set up to simulate the impact of fire frequency (annual, 3-year, and 5-year fire frequencies and protected) on soil carbon and Δ ^{13}C values for the sandy and clay sites at Matopos. Observed data from Bird et al. (2000) show the impact in 1998, after 50 years of fire treatments, on the soil carbon and Δ ^{13}C values. The model was set up to simulate these experiments by a spin-up run to estimate initial model values followed by the observed tree clearing experiments and the fire frequency treatments. The tree disturbance rates for the spin-up runs were estimated so that the observed initial tree cover values matched the observed fire treatment values in 1948 and 1960 for the tree clearing treatments.

The Nwanetsi site (24°27'S 31°53'E) fire trials in Kruger National Park, South Africa (van Wyk, 1971), used fire exclusion, annual burn, and triennial burn since 1959. We set up the model to simulate these different burning treatments and used the observed soil texture and monthly weather data from Satara (10 km from the site). We conducted a spin-up run to estimate initial soil conditions in 1959. The observed soil carbon data in 1988 (Jones et al., 1990) were used to test the simulated soil carbon levels.

The simulated impact of TBA on reducing grass plant production was tested using grass growth data from the Boatman site in Australia (Beale, 1973) and a Eucalyptus forest site in central Queensland, Australia (Scanlan and Burrows, 1990). The observed data sets show how grass growth decreases as TBA increases. The model was set up to simulate these experiments by running the model to equilibrium conditions before the beginning of the experiments. The model runs were then set up to simulate where the trees were permanently cleared from the site and were subsequently allowed to regrow. The observed soil texture and climate data from the two sites were used as inputs to the model.

Model Results

A comparison of the observed and simulated grass production (Figure 23.3) from tree clearing experiments in South Africa (Klaserie) and Zimbabwe (Tuli, Matopos, and Nyamandhlovu sites) shows a comparison of simulated and observed regressions of aboveground production to growing season

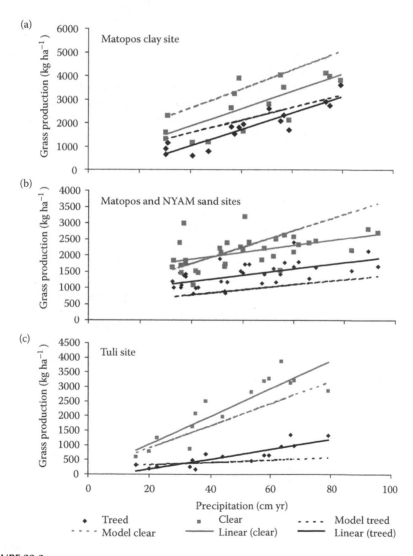

FIGURE 23.3

Comparison of simulated and observed grass production for cleared and controlled sites with (a) clay soils, (b) sandy soils, and (c) results from the Tuli site.

annual precipitation for the Matopos clay site (Figure 23.3a), Matopos sand site (Figure 23.3b, combined data from the Matopos and NYAM sand sites), and the Tuli sand site (Figure 23.3c). As expected, the observed data showed that the clearing of trees resulted in increased grass production (>50%) for all of the sites, with the model results generally similar to observed data. A comparison of the model results and observed data for the Matopos sand and clay sites (Figure 23.3a vs. Figure 23.3b) shows that plant production is higher for the cleared and uncleared clay sites when compared with the sandy sites. Comparison of the observed slope of the rainfall versus production curves for the cleared and uncleared sites suggests that the slopes were similar for the cleared and uncleared Matapos sand and clay sites; whereas the slope of the cleared sites was higher than that of the uncleared Tuli site. The model results are consistent with the observed data for the Tuli site (higher slope for the cleared sites); however, the model results showed that the cleared sites have higher slopes when compared with the uncleared sites at the Matopos clay and sandy sites. A comparison of observed and simulated plant production shows that the model tended to overestimate plant production for the cleared Matopos clay site and underestimate plant production for the Matopos sand sites. The observed and simulated plant production data for the cleared and uncleared Klaserie sites (Table 23.1) show higher plant production for the cleared sites (>50) and higher plant production for the clay sites when compared with the sandy sites. The overall comparison of observed versus model simulated plant production (Figure 23.4a) for all of the cleared and uncleared sites showed that the model did a good job of matching year-to-year changes in plant production and the impact of clearing and soil texture on plant production $(r^2 = 0.61 - y = 0.98 * x + 49.4)$. A general comparison of the observed and simulated soil C levels for all of the sites and treatments (Figure 23.4b) shows that the model correctly simulated $(r^2 = 0.98 - y = 226 + 0.89 * x)$ the impact of soil texture and burning treatments on soil C levels.

Table 23.2 shows a comparison of the observed and simulated soil C and Δ ^{13}C ratios for all of the sites. The impact of fire frequency on soil carbon was evaluated for the Matopos sand and clay sites and the Nwanetsi burning

TABLE 23.1

Comparison of Observed and Simulated Aboveground Grass Biomass (g/m^2) for Combretum (sand) and Acacia (clay) Sites in the Klaserie Private Nature Reserve with and without Tree Clearing

		Simulated		Observed	
Treatment		Jan 1984	March 1985	Jan 1984	March 1985
Acacia (clay)	Controlled	136	306	59	334
	Tress Cleared	170	418	88	615
Combretum (sand)	Controlled	68	140	49	253
	Trees Cleared	125	291	81	356

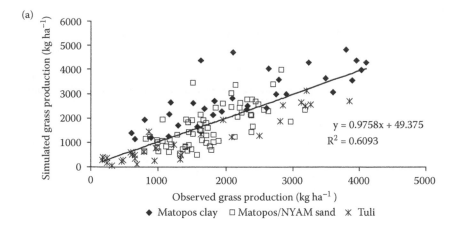

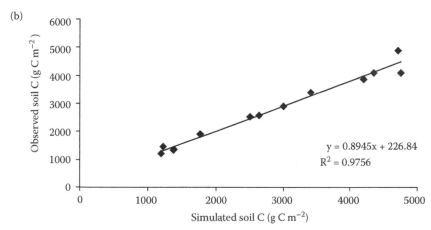

FIGURE 23.4
Combined comparisons of (a) observed and simulated grass production for the Matopos clay and sand sites and the Tuli site and (b) observed and simulated soil carbon levels for all of the clay and sand tree removal sites and fire frequency sites.

trials, whereas the impact of soil texture on soil carbon is demonstrated by a combination of all of the sites. The observed data and simulated model results for the three fire trials show that the highest soil C levels occur for the no-fire treatments and are lowest for the annual burning treatment (Table 23.2, Figure 23.5). The model and observed data also show that soil carbon levels increase from a minimum of 1200 gC m^{-2} in sandy soils to over 4000 gC m^{-2} for the highest clay soils (Nwanetsi sites), with a general pattern of increased carbon occurring with increased clay content (see Table 23.2).

Both the observed data and the model results for the soil Δ ^{13}C levels for the sand and clay Matopos burning trials (Table 23.2, Figure 23.5) show that Δ ^{13}C values are lowest (more negative) for fire removal treatment and highest (less

TABLE 23.2

Comparison of Observed and Simulated Soil C and A $\Delta^{13}C$ Values
for the Different Sites

Sites		Soil C		Soil A $\Delta^{13}C$		Clay Content
		Simulated	Observed	Simulated	Observed	
Matopos (sand)	No fire	1770	1908	−17.01	−20.80	0.07
	5-year fire	1390	1375	−15.95	−15.90	0.07
	Annual fire	1196	1216	−14.15	−15.50	0.07
Matopos (clay)	No fire	3416	3390	−16.12	−17.10	0.26
	5-year fire	3000	2894	−14.13	−14.27	0.26
	Annual fire	2630	2590	−13.13	−13.14	0.26
Nioanetsi (clay)	No burn	4720	4896	−14.17		0.60
	3-year fire	4760	4080	−13.73		0.60
	2-year fire	4355	4080	−13.64		0.60
	Annual	4200	3876	−13.42		0.60
Combretum (sand)		1228	1467	−17.78		0.08
Acacia (clay)		2500	2523	−14.39		0.25
Boatman (sand)		1380	1343			0.10

negative) for the annual fire treatments. The soil $\Delta^{13}C$ values for the clay soils are generally higher (less negative) by one to two Δ units. The model results show that as the fire frequency increases, there is an increase in grass cover and plant production (data not shown) and a relatively higher tree plant production for the sandy sites as compared with clay sites. The observed data from the Matopos clay and sand sites (Bird et al., 2000) show a clear pattern of increased tree cover with reduced fire frequency consistent with model results. The observed patterns of lower $\Delta^{13}C$ values for sandy soils suggest that tree carbon soil inputs are higher in sandy soil (tree $\Delta^{13}C$ values equal −26 to −28 vs. −12 to −14 for grass), whereas the increased soil $\Delta^{13}C$ with increased fire frequency is a result of the observed and simulated increased soil carbon inputs from C_4 gasses.

The simulated impact of increased TBA on reducing the grass aboveground plant production for the Boatman site (low production sites) and the Eucalyptus site (high production sites) is compared with the observed site patterns of decreased grass production with increased TBA (Figure 23.6a, b). The model and observed data show an exponential decrease in grass production with increased TBA, with the low production site having a much more rapid decrease in grass production with increased TBA. The model results suggest that trees are more competitive for available nutrients in low production sites when compared with high production sites (see Figure 23.2). The comparison of the observed versus simulated grass production in Figures 23.3 and 23.5 shows that the SAVANNA-CENTURY model correctly simulates grass production for sites with different climates, soil textures, and TBAs.

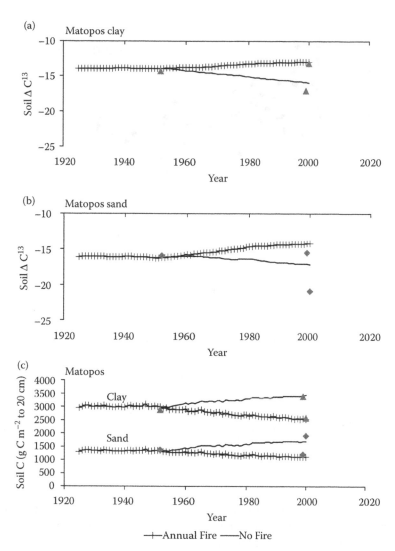

FIGURE 23.5
Comparison of observed vs. simulated soil Δ ¹³C values for the no burn and annual burn: (a) Matopos clay sites, (b) sandy sites, and (c) the observed vs. simulated soil carbon levels for the annual fire and no-fire treatments for the Matopos clay and sand sites.

Discussion

The results from this study show that the ultimate factor that controls grass production in grassland and savanna systems is precipitation but that the effect can be modeled via a mechanism involving the proximal factor of

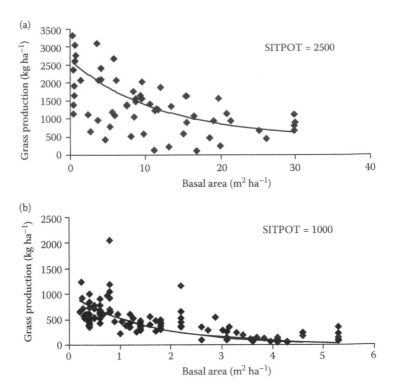

FIGURE 23.6
Comparison of the observed vs. simulated impact of tree basal area (TBA) on grass production at the (a) Boatman site (low sitpot, Beale, 1973) and the (b) Eucalypt site (high sitpot, Scanlan and Burrows, 1990—E5 site). The solid line on both figures represents the observed mean grass production changes as a function of TBA.

nitrogen mineralization and competition for the available nitrogen. Soil texture can modify grass production, with sites consisting of finer texture soils (higher clay content) generally experiencing higher grass production as a result of higher nutrient availability. Trees in the savanna systems tend to increase in cover over time if fire and animal disturbances are eliminated. This results in a reduction in grass production as the trees dominate in the competition for water and nutrient resources. The competitive advantage of grass tends to be higher for sites with higher potential to grow grasses as a result of higher annual rainfall or higher availability of nutrients.

The model versus observed data comparison (Figures 23.3 and 23.4) shows that the SAVANNA-CENTURY model can correctly simulate the impact of annual precipitation, soil texture, and tree cover on grass production in savanna ecosystems. The comparison of observed and simulated soil C and Δ ^{13}C values (Table 23.2, Figure 23.4b) shows that the model can also correctly simulate the impact of soil texture on soil carbon and soil fertility along with the impact of soil texture on relative carbon soil inputs for trees and grasses (more grass production for high fertility sites). The ability of the model to simulate the observed impact of increased precipitation on increased grass production (Figure 23.3) results from model assumptions regarding the impact of rainfall on plant water stress and nitrogen inputs to the systems. Model equations predict that as rainfall increases, water stress decreases and nitrogen inputs increase. A simple model that predicts decreased water stress with increased precipitation would not be able to predict the observations that grass production at both arid and humid sites respond to an increase in nitrogen fertilizer, with a 50% increase for arid sites and over a 100% increase at humid grassland sites (Owensby, 1970; Dodd and Lauenroth, 1978). The importance of nitrogen for plant production has been shown by Burke et al. (1997), where grassland soil nitrogen mineralization rates were positively correlated to annual rainfall. A comparison of simulated ANPP with observed ANPP for all of the sites and treatments (Table 23.1, Figures 23.3 and 23.4) shows that model results compared well with observed data for most of the sites and treatments; however, the model tended to overestimate ANPP for the cleared Matopos clay site and underestimate ANPP for the uncleared Matopos sandy sites. It is not clear why the model results differ from the observed data for these sites and treatments.

The model results and observed data from the fire frequency treatments in Zimbabwe and South Africa show a consistent trend for decreasing soil carbon levels with increasing fire frequency. Model results suggest that increased fire frequency causes less carbon to be returned to the soil system and, thus, decreases soil carbon levels. The model results also show that increasing the fire frequency increases the amount of nitrogen lost from the system, which results in a long-term decrease in total plant production. Both model results and observed data show that increased fire frequency caused an increase in the fraction of total plant production generated by grasses. The observed decrease in tree cover and increase in soil Δ ^{13}C levels, with increased fire frequency, support the simulated model results. The simulated increase in tree production for sandy sites as compared with clay sites for the Zimbabwe fire treatments is consistent with the observed data showing lower soil Δ ^{13}C values for the sandy sites (tree Δ ^{13}C of −26 vs. grass Δ ^{13}C of −13). The model predicted an increase in grass plant production with increased fire frequency, which results from the disturbance and removal of tree biomass associated with each fire. Higher grass production on clay soils results from model assumptions that sites with higher plant production (increasing with clay content) have a higher fraction of the nitrogen available for grass growth that is conceptually based on the Scanlan and Burrows (1990) data from Australia.

The ability of the SAVANNA-CENTURY model to simulate the observed increase in soil carbon levels and soil fertility levels with increased clay content results from the assumptions regarding the impact of soil texture on soil carbon stabilization and N gas losses from the soil (NH_3, N_2O, and NOx). The core model assumptions that control soil carbon dynamics are (1) stabilization of microbial products into slow organic matter decreases with increasing soil sand content and (2) stabilization of organic matter into passive pools (very slow turnover pools) increases with increased clay content. The observed pattern of higher soil fertility with increased clay content (Figure 23.3) results from the model assumptions that loss of NH_3 from animal deposition of urine and feces decreases with increased clay content (Schimel et al., 1986) and that soil NOx loss decreases with increased clay content (Martin et al., 1998). Model simulated reductions in gaseous N loss with increased clay content results in higher soil fertility due to the reduced gaseous loss of nitrogen from the soil. Sandy soils also have a higher NO_3 leaching loss.

The important theoretical assumptions in the model are that tree cover, fine root biomass, and lateral extension increase as the TBA increase, thus increasing the ability of the tree to extract soil water and nutrients from the soil system. The other core concept ruling tree–grass competition is that trees have a longer time to access water and nutrients, as live tree canopies exist several weeks longer than live grass canopies (Scholes and Walker, 1983; Archibald and Scholes, 2007). The model results and observed data showing that grasses are more competitive on higher plant production sites are based on the observation that grasses respond very quickly to increases in nutrient availability and rainfall (a 200% to 300% increase in grass production from 300 mm to 800 mm precipitation, see Figure 23.3); whereas tree cover and leaf area experience a minimal increase from annual changes in precipitation. This is also consistent with the observation that the removal of grasses will only result in a slow increase in tree cover (Knoop and Walker, 1985), whereas the removal of trees will result in a rapid increase in grass plant production (see Figure 23.3). The SAVANNA-CENTURY model assumption that tree roots are deeper than grass roots has minimal impact on the model results, as most of the available water and nutrients are within the rooting zone of both trees and grasses.

The ability of the model to simulate fire and animal disturbances on tree death is the major mechanism for promoting grass growth in a system that would otherwise be dominated by trees. The model simulates the impact of different tree disturbance events on tree biomass and growth; however, the model user must specify when the disturbance occurs and the extent to which the disturbance damages the TBA. The major improvement needs for the SAVANNA-CENTURY model include (1) adding a dynamic tree growth model that includes the differential impact of fire on small and large trees and the impact of drought stress on tree mortality, (2) adding a dynamic fire disturbance model that simulates the impact of fire weather, fire fuel load, and the probability of fire ignition, and (3) adding a soil runoff and erosion model.

References

Archibald, S. and R. J. Scholes. 2007. Leaf green-up in a semi-arid African savanna—separating tree and grass responses to environmental cues. *Journal of Vegetation Science* 18, 583–594.

Beale, I. F. 1973. Tree density effects on yields of herbage and tree components in south west Queensland mulga (*Acacia aneura* F. Muell.) scrub. *Tropical Grasslands* 7, 135–142.

Bird, M. I., E. M. Veenendaal, C. Moyo, J. Lloyd, and P. Frost. 2000. Effect of fire and soil texture on soil carbon in a sub-humid savanna Matopos, Zimbabwe. *Geoderma* 94, 71–90.

Bond, W. 2008. What limits trees in C_4 grasslands and *savannas*? *Annual Review of Ecology, Evolution, and Systematics* 39, 641–659.

Brown, J. R. and S. Archer. 1990. Water relations of a perennial grass and seedlings vs. adult woody plants in a subtropical savanna, Texas. *Oikos* 57, 366–374.

Burke, I. C., W. K. Lauenroth, and W. J. Parton. 1997. Regional and temporal variation in net primary production and nitrogen mineralization in grasslands. *Ecology* 78, 1330–1340.

Coughenour, M. B. 1992. Spatial modeling and landscape characterization of an African pastoral ecosystem: A prototype model and its potential use for monitoring drought. In *Ecological Indicators*, Vol. I, eds. D. H. McKenzie, D. E. Hyatt, and V. J. McDonald. London: Elsevier Applied Science, pp. 787–810.

Coughenour, M. B. 1995. The use of GIS/RS data in landscape-scale simulation models: The SAVANNA model as an example. In *Natural Vegetation as a Resource: A Remote Sensing Workbook for East and Southern Africa*, eds. J. M. O. Scurlock, M. J. Wooster, and G. D'Souza. Chatham, UK: NRI.

Del Grosso, S. J., J. Wirth, S. M. Ogle, and W. J. Parton. 2008. Estimating agricultural nitrous oxide emissions. *Transactions of the American Geophysical Union* 89, 529–530.

Delgado, J. A., A. R. Mosier, D. S. Schimel, W. J. Paton, and D. W. Valentine. 1996. Long-term ^{15}N studies in a catena of the shortgrass steppe. *Biogeochemistry* 32, 41–52.

Dodd, J. L. and W. K. Lauenroth. 1978. Analysis of the response of a grassland ecosystem to stress. In *Perspectives in Grassland Ecology*, ed. N. French. New York: Springer-Verlag, pp. 43–58.

Dodd, M. B., W. K. Lauenroth, and J. M. Welker. 1998. Differential water resource use by herbaceous and woody plant life forms in a shortgrass steppe community. *Oecologia* 117, 504–512.

Donaldson, C., G. Rootman, and D. Grossman. 1984. Long term nitrogen and phosphorus application to veld. *African Journal of Range and Forage Science* 1, 27–32.

Dye, P. J. and P. T. Spear. 1982. The effects of bush clearing and rainfall variability on grass yield and composition in southwest Zimbabwe. *Zimbabwe Journal of Agricultural Research* 20, 103–118.

House, J. I., S. R. Archer, D. Breshears, and R. J. Scholes. 2003. Issues in savanna theory and modelling. *Journal of Biogeography* 30, 1763–1777.

Jones, C. L., N. L. Smithers, M. C. Scholesn, and R. J. Scholes. 1990. The effect of fire frequency on the organic components of a basaltic soil in the Kruger National Park. *South African Journal of Plant and Soil* 7, 236–238.

Kelly, R. H., W. J. Parton, G. J. Crocker, et al. 1997. Simulating trends in soil organic carbon in long-term experiments using the century model. *Geoderma* 81, 75–90.

Knoop, W. T. and B. H. Walker. 1985. Interactions of woody and herbaceous vegetation in a southern African savanna. *Journal of Ecology* 73, 235–253.

Martin, R. E., M. C. Scholes, A. R. Mosier, D. S. Ojima, E. A. Holland, and W. J. Parton. 1998. Controls on annual emissions of nitric oxide from soils of the Colorado shortgrass steppe. *Global Biogeochemical Cycles* 12, 81–91.

McKeon, G. M., K. A. Day, S. M. Howden, et al. 1990. Northern Australian savannas: Management for pastoral production. *Journal of Biogeography* 17, 355–372.

Metherell, A. K., L. S. Harding, C. V. Cole, and W. J. Parton. 1993. *CENTURY soil organic matter model environment. Technical Documentation, Agroecosystem version 4.0.* Great Plains System Research Unit technical Report 4, USDA-ARS, Fort Collins, Colorado.

Midwood, A. J., T. W. Boutton, S. R. Archer, and S. E. Watts. 1998. Water use by woody plants on contrasting soils in a savanna parkland: Assessment with d^{2}H and d^{18}O. *Plant and Soil* 205, 13–24.

Owensby, C. E., R. M. Hyde, and K. L. Anderson. 1970. Effects of clipping and supplemental nitrogen and water on loamy upland bluestem range. *Journal of Range Management* 23, 341–346.

Parton, W. J., D. S. Schimel, D. S. Ojima, and C. V. Cole. 1994. A general model for soil organic matter dynamics: Sensitivity to litter chemistry, texture and management. In *Quantitative Modeling of Soil Forming Processes*, eds. R. B. Bryant and R. W. Arnold. SSSA Spec. Publ. 39. Madison, WI: ASA, CSSA and SSA, pp. 137–167.

Parton, W. J., M. O. Scurlock, D. S. Ojima, et al. 1993. Observations and modeling of biomass and soil organic matter dynamics for the grassland biome worldwide. *Global Biogeochemical Cycles* 7, 785–809.

Paustian, K., E.T. Elliott, and K. Killian. 1997. Modeling soil carbon in relation to management and climate change in some agroecosystems in central North America. In *Soil Processes and the Carbon Cycle*, eds. R. Lal, J. M. Kimble, R. F. Follett, and B. A. Stewart. Boca Raton, FL: CRC Press, pp. 459–471.

Sala, O. E., W. K. Lauenroth, and R. A. Golluscio. 1997. Plant functional types in temperate semi-arid regions. In *Plant Functional Types: Their Relevance to Ecosystem Properties and Global Change*, eds. T. M. Smith, H. H. Shugart, and F. I. Woodward. Cambridge: Cambridge University Press, pp. 217–233.

Sankaran, M., N. P. Hanan, R. J. Scholes, et al. 2005. Determinants of woody cover in African savannas. *Nature* 438, 846–849.

Scanlan, J. C. and W. H. Burrows. 1990. Woody overstorey impact on herbaceous understorey in *Eucalyptus* spp. communities in Central Queensland. *Austral Ecology* 15, 191–197.

Schimel, D. S., W. J. Parton, F. J. Adamsen, R. G. Woodmansee, R. L. Senft, and M. A. Stillwell. 1986. The rule of cattle in the volatile loss of nitrogen from a shortgrass steppe. *Biogeochemistry* 2, 39–52.

Scholes, R. J. 1987. Response of three semi-arid savannas on contrasting soils to the removal of the woody component. PhD Thesis, University of the Witwatersrand, Johannesburg.

Scholes, R. J. and D. O. Hall. 1996. The carbon budget of tropical savannas, woodlands and grasslands. In *Global Change: Effects on Coniferous Forests and Grasslands*, eds. A. I. Breymeyer, D. O. Hall, J. M. Mellilo, and G. I. Agren. SCOPE Vol. 56. Chichester: Wiley, pp. 69–100.

Scholes, R.J. and B.H. Walker. 1993. *An African Savanna: Synthesis of the Nylsvley Study*. Cambridge: Cambridge University Press.

Simioni, G., X. Le Roux, J. Gignoux, and H. Sinoquet. 2000. TREEGRASS: A 3D, process-based model to simulate plan interactions in tree–grass ecosystems. *Ecological Modelling* 131, 47–63.

van Wyk, P. 1971. Veld burning in the Kruger National Park, an interim report on some aspects of research. *Proceedings of the 12th Tall Timbers Fire Ecology Conference* 11, 9–31.

Walker, B. H., D. Ludwig, C. S. Holling, and R. S. Peterman. 1981. Stability of semi-arid savanna grazing systems. *Journal of Ecology* 69, 473–498.

24

Climate–Fire Interactions and Savanna Ecosystems: A Dynamic Vegetation Modeling Study for the African Continent

Almut Arneth, Veiko Lehsten, Kirsten Thonicke, and Allan Spessa

CONTENTS

Introduction

Savannas are inherently "disturbed" ecosystems, but the regularly recurring disruptions play such a fundamental ecological role (Scholes and Archer, 1997) that "episodic events" rather than "disturbance" may the more apt terminology. From an atmospheric perspective, fire is the most significant of these episodic events. Fires shape community species composition; tree to grass ratio and nutrient redistribution; and biosphere-atmosphere exchange of trace gases, aerosols, momentum, and energy. Savannas' estimated mean NPP of 7.2 ± 2.0 t C ha^{-1} year^{-1} amounts to nearly two thirds of tropical forest NPP (Grace et al., 2006); but remarkably little is known about savanna net carbon balance, especially for the African continent (Williams et al., 2007). In the absence of transient changes in the fire regime, such as could be introduced by climate change or fire-driven changes in land cover, savanna fires do not affect average annual net carbon uptake much, as the carbon released is rapidly recaptured by regrowing vegetation. However, pyrogenic emissions include particles, ozone, methane, and other volatile hydrocarbons, all of which affect tropospheric chemistry and climate (Scholes and Andreae, 2000). Moreover, fires affect convective uplift via the interplay of surface albedo, Bowen ratio, and available soil moisture. In that way, fire intensity, timing, and total area burnt may also affect atmospheric circulation patterns, as was discussed for the Australian Monsoon (Lynch et al., 2007).

Scientific consensus currently rates Africa as the continent with the largest fire activity globally, yet estimates of fire frequency, area burnt, total pyrogenic emissions, and seasonal and interannual variation are highly uncertain (van der Werf et al., 2006, 2008). Continental burnt area estimates from satellite remote sensing information are of the order $200-300 \times 10^4 \, km^2$ (Lehsten et al., 2009a; van der Werf et al., 2006), a likely underestimate (Roy and Boschetti, 2009). Spatial resolution and detection failures caused by clouds or dust, lags between recurring satellite overpass, and algorithms used to analyze the measured spectra are still problematic.

In the northern hemispheric part of the continent, the fire season lasts from approximately November to March with peak burnt area detected around December. In the southern hemisphere Africa, most of the fires take place between March and October, peaking around July (van der Werf et al., 2006). Pronounced interannual variation with regard to fire peak season or total area burnt is possible through two rainfall-driven mechanisms: the amount of grass fuel available after the wet season and the length of the dry season (Spessa et al., 2005).

Separating the effects of climate from those of episodic events on savanna structure and functioning is difficult and perhaps not even useful, as these are mutually dependent. Precipitation or, more precisely, plant available soil moisture is a key environmental variable that not only determines fuel moisture but also provides an important limit to African savanna carbon uptake (Arneth et al., 2006; Merbold et al., 2008) and, hence, fuel availability (Bucini and Hanan, 2007; Hély et al., 2007; Sankaran et al., 2005). Annual precipitation is projected to decrease in many regions of Africa, but its future seasonal or spatial distribution is uncertain (Giannini et al., 2008).

Here we investigate current climate–vegetation–fire interactions in Africa, using simulation experiments for a number of processes that operate from local to continental scales. We compare model results with previously published theory and observational and modeling analyses. Our chief objectives are twofold:

1. To present an integrative, mechanistic approach linking climate, fire regimes, and vegetation productivity and to discuss the main uncertainties relating to the modelling of fire.

2. To highlight the need to improve the capability of terrestrial models to simulate vegetation patterns in savannas. We specifically focus on factors determining tree to grass interactions due to their prominence

in biogeochemical cycling, atmospheric interactions, and ecosystem services for humans.

Dynamic Vegetation Models for Use in Savannas

Dynamic vegetation models (DVMs) simulate the global or regional dynamics and composition of vegetation in response to changes in climate and atmospheric CO_2 concentration (Cramer et al., 1999; Prentice et al., 2007). Vegetation is typically represented by a number of plant functional types (PFTs) for efficient computation of phenology and carbon and hydrological cycles over centuries. Bioclimatic limits provide constraints on the PFT range. Processes such as plant photosynthesis and autotrophic and heterotrophic respiration are explicitly calculated, and a set of carbon allocation rules determines plant growth. Plant establishment, growth, mortality, and decomposition and their response to resource availability (light, water, and, in some cases, nitrogen) modulate population dynamics. Crucially for seasonally dry ecosystems, DVMs contain a representation of the soil water balance that limits stomatal conductance and photosynthesis in periods of soil moisture deficit.

The sole episodic event, if at all explicitly represented, is fire; herbivory has so far not been incorporated (Cramer et al., 1999; Prentice et al., 2007). The DVMs have been successfully evaluated against a number of observations of the global terrestrial carbon or water cycles; but at the ecosystem scale, evaluations have often been confined to observations in forests, mostly from temperate or boreal environments (Friend et al., 2007; Prentice et al., 2007; Sitch et al., 2003). Important features of the savanna biome, such as forest–savanna boundaries or the variable ratio of trees to grasses found along environmental gradients (Bucini and Hanan, 2007), have so far not been successfully reproduced (Cramer et al., 2001, their Figure 24.2), most likely due to the incomplete representation of resource competition and/or fire–vegetation interactions.

Forest-gap models have been proposed as the appropriate tools for a more realistic linking of biogeochemical cycles to terrestrial ecology (Friend et al., 1997; Prentice et al., 2007; Smith et al., 2001). The DVM LPJ-GUESS (Smith et al., 2001), which is used in this chapter, is one example. Mortality and establishment are accounted for explicitly, which leads to the representation of each PFT by different-aged cohorts. The growth and mortality functions in LPJ-GUESS result in a competition between light-demanding PFTs that allocate carbon to fast growth and high maximum recruitment and PFTs that allocate carbon toward survival at low growth rates resulting from intense shading by neighbors. The allocation scheme is flexible and allows relatively more carbon to be allocated to root growth in water-limited environments. The overall canopy structure is determined from stochastic patch-scale dynamics, with a number of replicate patches representing the range of possible patch developments. Vegetation coverage of a grid cell equals the average over the replicated patches simulated for the location (Hickler et al., 2009; Smith et al., 2001). Processes related to canopy CO_2 and water exchange, litter decomposition, and soil carbon balance are as in the more traditional DVM LPJ (Sitch et al., 2003).

The process-based fire model SPITFIRE (Thonicke et al., 2010) is used for fire calculations in LPJ-GUESS (Lehsten et al., 2009a). SPITFIRE calculates fuel combustion, fire intensity, fire spread, and pyrogenic emissions based on a parameterization for ignition and simulated fuel moisture, amount, and type. Fuel moisture is linked to the soil moisture balance calculations in LPJ-GUESS in the case of live grasses and to a climatically driven fire-danger index in the case of dead grass and tree litter (Thonicke et al., 2009). Fine fuel with large surface-area-to-volume ratio such as tree or grass leaf litter equilibrates quickly to atmospheric moisture conditions, supporting intense fires of short duration. Area burnt per successful ignition may be calculated from fuel characteristics and in the model is assumed to circumscribe an elliptical shape of the burned area with the length of the major axis proportional to wind speed. Burnt area may also be prescribed, that is, the model is initialized with a burned area derived from a remote sensing product (Lehsten et al., 2009). Pyrogenic emissions are subsequently calculated by multiplying the area burnt with the combusted biomass, using a PFT-specific emission factor. Fuel moisture governs combustion completeness.

Tree to Grass Ratio as Affected by Mean Annual Precipitation, Fire Return Interval, and Variable Rooting Depths

Plant available soil water (θ) in arid and semiarid environments is fundamental for plant physiological activity and species survival. Since at the continental level θ has to be derived from water balance calculations, MAP is frequently taken as a surrogate for moisture supply to plants. Across Africa, woody cover is observed to increase with increasing precipitation over a relatively broad semiarid to mesic range between *c.* 400 and 1600 mm MAP; more arid regions sustain grass-dominated systems, whereas in regions with relatively high and more regular rainfall savannas make way for dense woodlands and forests with nearly closed canopies (Bucini and Hanan, 2007; Sankaran et al., 2005).

Root niche separation is often discussed as an important determinant of tree and grass coexistence, with grasses being competitive for water in the upper soil layer, whereas tree roots can also access soil moisture from deeper soil layers (for a summary, see Bucini and Hanan, 2007). Experimental evidence, however, is not conclusive with regard to observed root distribution and water sources accessed by trees and grasses (Daly et al., 2000, and references therein). An alternative hypothesis is that the frequency of episodic events such as fire or grazing keeps woody cover in savannas below a climatically possible equilibrium state (Bucini and Hanan, 2007; Sankaran et al., 2005; Scheiter and Higgins, 2009). Support comes from experimental fire exclusion, leading to an increased proportion of shrubs and trees and overall taller woody vegetation (Govender et al., 2006). However, causal relationships to provide evidence for either theory being more conclusive than the other cannot be easily determined from compilations of observations (Bucini and Hanan, 2007). As the environment becomes more humid, increasing MAP and θ clearly support trees, but a number of additional factors produce a divergence from simple climate-vegetation relationship. Most likely, no single process determines the coexistence of trees and grasses in African savannas (Bucini and Hanan, 2007; House et al., 2003; Sankaran et al., 2005).

Lehsten et al. (2010) analyzed the effects of fire disturbance versus variable rooting depths of trees and grasses in a simulation experiment along a precipitation gradient from northern equatorial Africa to the Sahel (Figure 24.1). Mean annual precipitation increases from *c.* 100 mm a^{-1} at the northern end to 1400 mm a^{-1} close to the equator. To test the effects of fire-related disturbance on tree to grass ratio, prescribed fire return intervals were varied between annual and infinite (no fire); ignition was set to always occur during the peak fire season (mid-March). The fire return interval was expressed as the probability of a location to burn, thus some variability in terms of the actual time between one burn event and the next was allowed with an imposed limit to the fire-free period not exceeding twice the prescribed return interval. In a second experiment, the relative proportion of tree and grass roots in the upper of the model's two soil layers (upper 500 mm) was varied linearly in six steps between 1 (tree) : 0.5 (grass) to 0.5 (tree) : 1 (grass). The remaining proportion tapped into the lower soil layer (1500 mm depth). Fire was excluded in this second experiment.

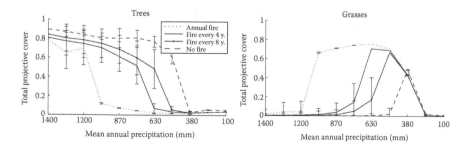

FIGURE 24.1

Simulated woody and grass fractional total projective cover (TPC) along a precipitation gradient. Average prescribed fire return interval was between 1 (annual fire) and infinity (no fire). Standard deviations relate to the stochastic representation of gap-forest dynamics.

In the high precipitation equatorial regions, African natural vegetation is characterized by dense forests that were reproduced in the model experiment (expressed as fractional total projective cover, TPC; Figure 24.1). In these regions, sizeable burnt areas cannot be detected from satellites (Roy et al., 2008; Tansey et al., 2008). In areas of MAP > 1200 mm, simulated woody cover was reduced only slightly (to *c.* 0.8) even when an artificially high fire frequency was applied. Fire intensity was strongly reduced by the permanently moist conditions. Even though the dominant tropical evergreen trees are much more susceptible to fire damage than raingreen trees dominating the more arid regions, fire had little effect on mortality. By contrast, at the arid end of the simulated gradient, an artificially prolonged fire return interval did not affect the simulated dominance of grassland vegetation (Figure 24.1), as soil moisture levels were inadequate for the survival of tree saplings.

The calculated tree to grass ratios were not only highly variable but also strongly modulated by the prescribed fire return interval between the evidently climatically determined dominance of trees when MAP > 1200 mm and that of grasses when MAP < 500 mm. Grasses became very competitive across a wide range of the simulation domain when fire frequency was set to annual. With lower fire frequency, however, trees and grass began to coexist in variable proportions. Complete exclusion of fire in the mesic regions resulted in clear dominance of woody vegetation until a climate threshold of around 400 mm MAP was reached.

Fire return intervals from 2 to 8 years are typical for the region covered by the simulated gradient (Tansey et al., 2008). LPJ-GUESS-SPITFIRE, hence, reproduces a correlation of woody cover to MAP that closely resembles the sigmoidal relationship shown in the analysis of Bucini and Hanan (2007). However, in contrast to the simulations with variable fire intervals, prescribing variable tree to grass root ratios in the upper and lower soil levels did not affect relative vegetation cover greatly. Grasses only became less competitive in comparatively arid regions (around MAP 400 mm) and when tree roots were nearly fully in the upper layer (Lehsten et al., 2010). The role of fire as a central determinant of canopy structure has also been claimed in fire-exclusion simulations across the entire African continent, which resulted in expansion of woody vegetation types, especially in regions around 400 to 800 mm MAP (Bond et al., 2005; Lehsten et al., 2009; Scheiter and Higgins, 2009). Despite large remaining uncertainties, there appears to be broad agreement between models of fire being a chief control over tree–grass competition on regional to continental scales. As with all model studies, we cannot completely exclude the possibility that the reproduction of observed patterns by fire, rather than root access to water, may be due to the wrong reasons, especially as the models do not account for a number of additional processes that have been debated as determinants of woody cover in savannas. However, considering the difficulty that first-generation vegetation models had in reproducing accurate savanna vegetation, this is an important step toward realistic process-based simulation studies in these globally important ecosystems. Further, these developments give us confidence that the model may also be applied to study the interactive effects of climate change, fire, vegetation dynamics, and atmospheric composition.

Vegetation–Fire Interactions: Intensity and Fuel Accumulation

Fire scorch height depends on fire intensity and, hence, on the amount and type of litter; whereas cambial damage is affected by fire residence time, bark thickness, and the PFT specific ability to survive trunk overheating. Small trees from young age cohorts are more likely to be susceptible to the effects of crown scorching and cambial damage than taller (older) trees. The probability of tree seedlings growing above the typical flame height depends not only on the number of recurring years without fire but also on the intensity of fires, which may be larger when return intervals are infrequent.

The competitiveness of grasses under a high burn frequency (annual; Figure 24.2) becomes visible in an example simulation located in the mesic region of the studied climatic gradient (870 mm MAP). Tree seedlings and young age cohorts are killed before they escape the flame height. At low fire frequency (e.g., 8 years), trees dominate and slowly decomposing woody litter can accumulate in the absence of fires. The higher average fuel accumulation eventually leads to intense fires and higher damage but not sufficient to completely kill the canopy. Nonlinear interactions such as these can result in grass coexisting with trees over a substantial number of years. Although the combusted biomass in some fires was up to four times larger in the woody- than in the grass-dominated systems, the higher frequency in the latter more than compensated for this effect. Over the 50-year simulation period, a total of 4.3 versus 1.1 kg C was combusted in the annual versus the 8-year experiment. These results are, however, in the absence of additional processes not represented in the model, such as herbivory by termites, which consume some of the dead material and, thus, reduce available litter in regions with infrequent fires.

Interactive Effects of Precipitation and NPP

Having determined that the model is in principle able to reproduce the dynamics of savanna ecosystems, including the effects of fire, the next logical step is to turn toward continental burn patterns, their drivers, and effects on carbon and trace gas emissions. As a starting point, annual area burnt may be prescribed from remote sensing products. This allows us to concentrate on the effects of climate on productivity of vegetation and climate–fire–vegetation

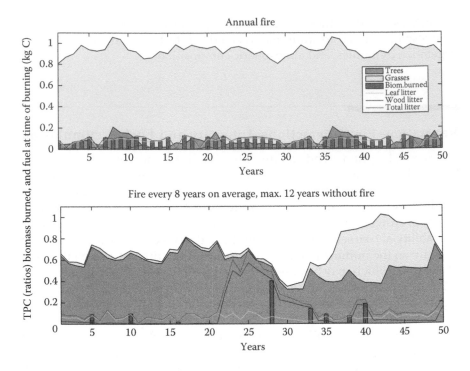

FIGURE 24.2

(See color insert following page 320.) Simulated tree and grass total projective cover, burnt biomass, and litter at the time of burning for a grid cell with 870 mm MAP over a 50 year period. Simulations used a repeated 27-year climatology calculated from daily NCEP data as in Weber et al. (2009). CO_2 concentration was set to 341 ppm. Fire return intervals were annual in the upper panel and eight years (on average) in the lower panel. Note that the climatic and fuel conditions modify the combustion completeness, hence some fires may result in very low burned biomass and, therefore, may not be visible in the figure. Grasses reach a maximum TPC < 1 to allow for a certain amount of radiation reaching the ground necessary for subsequent establishment; grasses vanish once trees occupy a higher value than the maximum possible TPC for grasses.

interactions, whereas additional factors such as ignition and fire spread as well as suppression are not required. A recent such estimate of pyrogenic carbon release was 723 ± 70 Tg C year^{-1}, based on a remotely sensed area burnt of 195.5 ± 24.3 10^4 km^2 (2001–06; L3JRC, Lehsten et al., 2009; Tansey et al., 2008). Using the MODIS burnt area product (Roy et al., 2008), this value increased to 790 ± 18 Tg C year^{-1} for the years 2000–2007 (Figure 24.3). Considering the large uncertainties associated with detecting fires from space (Roy and Boschetti, 2009), these totals are remarkably close and probably fortuitously so. They fall well within the much larger published range of pyrogenic carbon release of 300–1800 Tg C year^{-1} (Williams et al., 2007).

Typical NPP in fire-prone savannas and woodlands ranges from around 0.2 to 0.5 kg C m^{-2} year^{-1} between dry and humid regions (Williams et al., 2007). In the majority of fire-affected regions, not more than 40% of annual

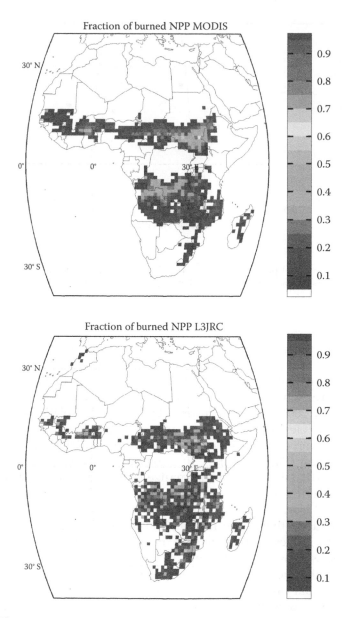

FIGURE 24.3
(**See color insert following page 320.**) Fraction of combusted NPP, burnt area prescribed from remote sensing products MODIS MCD45 (top panel; applied was an average burn frequency for each day and grid cell derived from the available data), and L3JRC (lower panel). NPP was calculated from a daily climatology. Displayed are averages for the years 2001–06. (Adapted from Lehsten, V. et al. 2009. *Biogeosciences* 6, 349–360; Weber, U. et al. 2009. *Biogeosciences* 6, 285–295.)

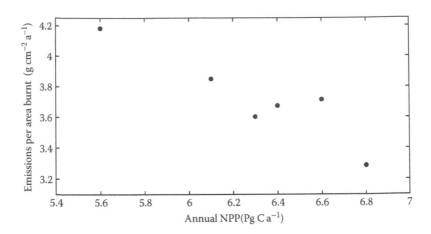

FIGURE 24.4
Simulated terrestrial net primary productivity (NPP) in savannas and woodlands and pyrogenic emissions per area burnt. (Data from Lehsten, V. et al. 2009a. *Biogeosciences* 6, 349–360.)

NPP was combusted (Figure 24.3). The fraction of NPP burnt rarely exceeded 50% of productivity in areas that appear to be located to the SW and NE of the equatorial rainforest belt. Annual NPP was strongly and positively correlated with precipitation (*P*; Lehsten et al., 2009; their Table 1). *P* also may have affected area burnt, the lowest value coinciding with the driest year; but over the entire 2001–06 period covered by L3JRC, the relationship between these two variables was poor. When normalized by area burnt, the highest pyrogenic carbon emissions were simulated for the driest year when total NPP was lowest (Figure 24.4). A 20% increase in NPP corresponded to a nearly proportional decline in normalized emissions. Though this result may seem counter-intuitive, since an increase in NPP also results in an increased fuel load and, hence, an increase rather than a decrease may be expected, it shows that in our model simulation (at least averaged over the continent) the influence of an increase in fuel moisture on combustion completeness outweighs the positive effect of higher fuel loads. Simple correlations such as the one observed here could, in principle, be used as indicators of the interannual variation in fire emissions in response to climate variation under current atmospheric (climate and CO_2) conditions, for example, in response to strong El-Niño patterns. However, additional variability introduced, for instance, by land management (van der Werf et al., 2008) will reduce their applicability for extrapolations into the future.

A more mechanistic approach links fire patterns not to NPP *per se* but to the amount and type of plant litter (Hély et al., 2007; Van der Werf et al., 2008). Across the African continent, strongly nonlinear and uni-modal relationships between precipitation and area burnt, fire emissions, and litter available for burning emerged when values for all grid cells were plotted, with peaks around 1000–1200 mm (Lehsten et al., 2009). Similar relationships had been demonstrated to exist in a regional burnt area analysis for northern Australia, albeit with slightly higher peak values (Spessa et al., 2005). Whether such patterns are uniform for the global savanna biome is still open to question. Differences in the seasonal distribution of precipitation, soil water storage, and vegetation together with different human land use patterns and population densities could be expected to introduce regional modifications. However, due to the overall strong impact of precipitation on productivity in seasonally dry ecosystems, and following the strong link of litter production with fire, the general global applicability of such a relationship seems likely.

Dynamic Simulation of Wildfires

Simulation experiments with burnt area prescribed are only applicable for periods covered by the remote sensing products. However, for studies of fire response to climate or human land use change, area burnt must be dynamically calculated. To do so, the spread of an average fire (under given climate conditions) is multiplied by the number of ignitions (n; Thonicke et al., 2010). An unresolved challenge in this respect is to separate effects of human-caused ignition from those of lightning. For present conditions, ignition sources in most regions of Africa are thought to be nearly exclusively anthropogenic, for instance, for rangeland management. However, humans are as important for ignition as they are for limiting its spread (Archibald et al., 2008). Arguably, in the absence of human presence, savannas might burn "in any case," albeit during a different time in the year. A clear distinction between human and natural fire regimes is, therefore, difficult to obtain. Based on an analysis of a sediment charcoal record, Bird and Cali (1998) argued that on millennial time scales the Holocene may be the first period in the last 400,000 years when large fire activity on the southern African continent can be detected during an interglacial period rather than during an interglacial-glacial climate transition.

For global applications, the anthropogenic ignitions have been related to population density (P_D) as (Thonicke et al., 2009)

$$n_{h,ig} = P1 * e^{-0.5 * \sqrt{P_D}} * a(N_d) * A \qquad (24.1)$$

with $a(N_d)$ being the number of ignitions per human per fire season day in a grid cell of area A. $P1$ is a fitting parameter derived from present burnt area products.

We applied the dynamic fire simulation to a larger test region around the precipitation gradient studied in Figure 24.1 (Figure 24.5). Using the standard

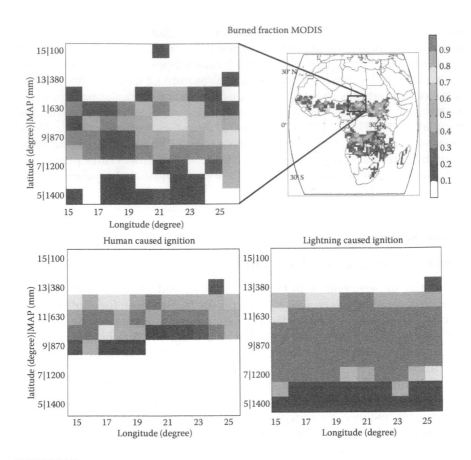

FIGURE 24.5
(See color insert following page 320.) Influence of ignition on average fraction of burned area over the period 2001–2007; upper right: continental burned fraction remotely sensed by MODIS MCD45; all other panels display an area defined by an MAP gradient from 1400 to 100 mm, located around 5°N to 15°N and 15°E 25°E. Upper left: Fractional burned area remotely sensed by MODIS. Lower left panel: Simulated, human ignition only. Lower right: Simulated, lightning ignition only (maximum assumption).

parameter value $P1 = 30$ used by Thonicke et al. (2010) for their global simulation of burned area resulted in an average burned area fraction of 0.137, less than the fraction of 0.211 estimated from the MODIS burned area product for that region. Adjustment of $P1$ to a value of 50 resulted in a simulated burned area fraction of 0.215.

With dynamic simulation of burnt area and adjusted $P1$, the general pattern is met, in terms of the largest proportion of burned area residing in the mesic regions of the gradient where the climate is still dry enough to allow fires not only to ignite but also to spread and litter production is sufficient to support sizeable fires. However, the model could not reproduce an apparent gradient of increasing burnt area from west to east and shows a strong cutoff toward the dry regions in the north, where simulated vegetation productivity rapidly declines to very low values (Figure 24.5).

The separation of human- and lightning-caused ignitions is a non-trivial problem that has not yet been solved. A simple starting point may be to ask whether lightning ignition is a potentially limiting factor at all. To do so, lightning strikes were prescribed from a monthly climatology (http://thunder. msfc.nasa.gov/data/index.html# GRIDDED_DATA), using a cloud-to-ground strike ratio of 0.3 adopted from Price and Rind (1993). Posing a hypothetical upper limit of lightning ignition, we allowed each single ground-lighting strike to cause an ignition under favorable burn conditions.

With this assumption, area burnt substantially increased across the model domain (Figure 24.5). Thus, at least for a maximum lightning scenario, lightning alone could be more than adequate as an ignition source. However, the assumption of a 100% ignition success rate of lighting strikes in this simulation must clearly be viewed as an artificial upper limit, in reality the ignition to lightning strike ratio is substantially lower. In a recent statistical analysis for southern Africa, climate emerged as one of the chief determinants of total burnt area (Archibald et al., 2008), with rainfall (amount and dry season length) and tree cover dominating as driving factors. Human effects have to be viewed as the balance between ignition and extinction, as anthropogenic landscape fragmentation reduces fire spread. Fire barriers are not yet included in LPJ-GUESS-SPITFIRE and the "lightning-only" scenario, therefore, will also overestimate total area burnt in their absence. What is more, a higher temporal resolution lightning climatology (daily rather monthly) would be needed to further investigate lightning-fire effects.

Summary

Process-based fire routines embedded in mechanistic vegetation models provide a powerful tool to investigate the dominant and competing hypotheses that underlie fire ecology in African savanna ecosystems. Although a number of possible important processes are not yet accounted for, the results in this chapter indicate realistically simulated climate–vegetation–fire interactions in terms of their effects on tree and grass coexistence on regional to continental scales. Analysis of the robustness of these results at finer spatial resolutions is hampered by the absence of suitable climate data.

The importance of correctly representing tree–grass dynamics based on process understanding cannot be overstressed; it goes well beyond a "mere" ecological challenge. Trees and grasses differ in a number of important physiological and growth characteristics (among others, photosynthetic type, canopy roughness, growth form, litter type, and litter chemical composition) and interact rather differently with climate. The ability to simulate the tree to grass ratios is *sine qua non* for studies of global change effects and feedbacks in these ecosystems due to the different biophysical and bio-geochemical exchanges of woody versus herbaceous vegetation with the atmosphere.

Fully dynamic modelling of wildfires is still at a very early stage, and it is particularly limited by quantitative understanding of climate versus human-driven fire regimes. We have implicitly accounted for anthropogenic effects here by (i) prescribing burnt area from a satellite product (which does not distinguish between lightning and human-caused fire) and (ii) discussing sensitivities of the simulations to ignition patterns. An estimated 90% of Africa's population depends on rain-fed crop production and pastoralism to meet its basic food supplies (Patt and Winkler, 2007), and these areas are located within the savanna biome. The increasing pressure that human land use or land cover change will have on fire regimes and emissions must be incorporated in future climate change and climate change feedback studies.

Acknowledgments

The work in this chapter was supported by the European Union CarboAfrica project (GOCE, 037132) and by grants from the Swedish Research Council and Formas to AA. AA acknowledges a Visiting Professorship at Helsinki University's Physics Department. Paul Miller provided a helpful and thorough editing of the draft.

References

Archibald, S., D. P. Roy, B. W. van Wilgen, and R. J. Scholes. 2008. What limits fire?: An examination of drivers of burnt area in Southern Africa. *Global Change Biology*, doi: 10.1111/j.1365-2486.2008.01754.x.

Arneth, A., E. M. Veenendaal, C. Best, et al. 2006. Water use strategies and ecosystem-atmosphere exchange of CO_2 in two highly seasonal environments. *Biogeosciences* 3, 421–437.

Bird, M. I. and J. A. Cali 1998. A million-year record of fire in sub-Saharan Africa. *Nature* 394, 767.

Bond, W. J., F. I. Woodward, and G. F. Midgley 2005. The global distribution of ecosystems in a world without fire. *New Phytologist* 165, 525–538.

Bucini, G. and N. P. Hanan. 2007. A continental-scale analysis of tree cover in African savannas. *Global Ecology and Biogeography* 16, 593–605.

Cramer, W., A. Bondeau, F. I. Woodward, et al. 2001. Global response of terrestrial ecosystem structure and function to CO$_2$ and climate change: Results from six dynamic global vegetation models. *Global Change Biology* 7, 357–373.

Cramer, W., D. W. Kicklighter, A. Bondeau, et al. 1999. Comparing global models of terrestrial net primary productivity (NPP): Overview and key results. *Global Change Biology* 5, 1–15.

Daly, C., D. Bachelet, J. M. Lenihan, R. P. Neilson, W. Parton, and D. Ojima. 2000. Dynamic simulation of tree–grass interactions for global change studies. *Ecological Applications* 10, 449–469.

Friend, A. D., A. Arneth, N. Y. Kiang, et al. 2007. FLUXNET and modelling the global carbon cycle. *Global Change Biology* 13, 610–633, doi: 10.1111/j.1365-2486.2006. 01223.x.

Friend, A. D., A. K. Stevens, R. G. Knox, and M. G. R. Cannell. 1997. A process-based, terrestrial biosphere model of ecosystem dynamics (Hybrid v3.0). *Ecological Modelling* 95, 249–287.

Giannini, A., M. Biasutti, I. M. Held, and A. H. Sobel. 2008. A global perspective on African climate. *Climatic Change* 90, 359–383.

Govender, N., W. S. W. Trollope, and B. W. Van Wilgen. 2006. The effect of fire season, fire frequency, rainfall and management on fire intensity in savanna vegetation in South Africa. *Journal of Applied Ecology* 43, 748–758.

Grace, J., J. San Jose, P. Meir, H. S. Miranda, and R. A. Montes. 2006. Productivity and carbon fluxes of tropical savannas. *Journal of Biogeography* 33, 387–400.

Hély, C., K. Caylor, P. Dowty, et al. 2007. A temporally explicit production efficiency model for fuel load allocation in Southern Africa. *Ecosystems* 10, 1116–1132.

Hickler, T., S. Fronzel, M. Araújo, O. Schwieger, W. Thuiller, and M.T. Sykes. 2009. An ecosystem-model-based estimate of changes in water availability differs from water proxies that are commonly used in species distribution models. *Global Ecology and Biogeography* 18, 304–313.

House, J. I., S. Archer, D. D. Breshears, and R. J. Scholes. 2003. Conundrums in mixed woody-herbaceous plant systems. *Journal of Biogeography* 30, 1763–1777.

Lehsten, V., K. J. Tansey, H. Balzter, et al. 2009a. Estimating carbon emissions from African wildfires. *Biogeosciences* 6, 349–360.

Lehsten, V., A. Arneth, K. Thonicke, A. Spessa, and I.-C. Prentice. 2010. Tree-grass coexistence in savannas—root distribution versus disturbances: a simulation study. *Ecography*, submitted.

Lynch, A. H., D. Abramson, K. Gorgen, J. Beringer, and P. Uotila. 2007. Influence of savanna fire on Australian monsoon season precipitation and circulation as simulated using a distributed computing environment. *Geophysical Research Letters* 34, L20801, doi:10.1029/2007GL030879.

Merbold, L., J. Ardö, A. Arneth, et al. 2008. Precipitation as driver of carbon fluxes in 11 African ecosystems. *Biogeosciences Discussion* 5, 4071–5105.

Patt, A. G. and J. Winkler. 2007. *Applying Climate Information in Africa: An Assessment of Current Knowledge.* Boston University, Source: http://www.iiasa.ac.at/Admin/ PUB/Documents/XO-07-006.pdf

Prentice, I. C. A. Bondeau, W. Cramer, et al. 2007. Dynamic global vegetation modelling: Quantifying terrestrial ecosystem responses to large-scale environmental change. In *Terrestrial Ecosystems in a Changing World*, eds. J. G. Canadell, D. E. Pataki, and L. F. Pitelka. 175–192. Berlin, Springer.

Price, C. and D. Rind. 1993. What determines the cloud-to-ground lightning fraction in thunderstorms. *Geophysical Research Letters* 20, 463–466.

Roy, D. and L. Boschetti. 2009. Southern Africa validation of the MODIS, L3JRC and GlobCarbon burned area products. *IEEE Transactions on Geoscience and Remote Sensing* 47, 132–1044.

Roy, D. P., L. Boschetti, C. O. Justice, and J. Ju. 2008. The collection 5 MODIS burned area product—Global evaluation by comparison with the MODIS active fire product. *Remote Sensing of Environment* 112, 3690–3707.

Sankaran, M., N. P. Hanan, R. J. Scholes, et al. 2005. Determinants of woody cover in African savannas. *Nature* 438, 846–869.

Scheiter, S., and S. I. Higgins. 2009. Impacts of climate change on the vegetation of Africa: An adaptive dynamic vegetation modelling approach (aDGVM). *Global Change Biology* 15, 2224–2246, doi: 10.1111/j.1365–2486.2008.01838.x.

Scholes, M. and M. O. Andreae. 2000. Biogenic and pyrogenic emissions from Africa and their impact on the global atmosphere. *AMBIO: A Journal of the Human Environment* 29, 23–29.

Scholes, R. J. and S. R. Archer. 1997. Tree–grass interactions in savannas. *Annual Review of Ecology & Systematics* 28, 517–544.

Sitch, S., B. Smith, I. C. Prentice, et al. 2003. Evaluation of ecosystem dynamics, plant geography and terrestrial carbon cycling in the LPJ dynamic global vegetation model. *Global Change Biology* 9, 161–185.

Smith, B., I. C. Prentice, and M. T. Sykes. 2001. Representation of vegetation dynamics in the modelling of terrestrial ecosystems: Comparing two contrasting approaches within European climate space. *Global Ecology & Biogeography* 10, 621–637.

Spessa, A., B. McBeth, and I. C. Prentice. 2005. Relationships among fire frequency, rainfall and vegetation patterns in the wet–dry tropics of northern Australia: an analysis based on NOAA-AVHRR data. *Global Ecology and Biogeography* 14, 439–454.

Tansey, K., J. M. Gregoire, P. Defourny, et al. 2008. A new, global, multi-annual (2000–2007) burnt area product at 1 km resolution. *Geophysical Research Letters* 35 L01401, doi:10.1029/2007GL031567.

Thonicke, K., A. Spessa, I.C. Prentice, S. P. Harrison, L. Dong, and C. Carmona-Moreno. 2009. The influence of vegetation, fire spread and fire behaviour on global biomass burning and trace gas emissions. *Biogeosciences* in press.

van der Werf, G. R., J. T. Randerson, L. Giglio, G. J. Collatz, P. S. Kasibhatla, and A. F. Arellano. 2006. Interannual variability in global biomass burning emissions from 1997 to 2004. *Atmospheric Chemistry and Physics* 6, 3423–3441.

van der Werf, G. R., J. T. Randerson, L. Giglio, N. Gobron, and A. J. Dolman. 2008. Climate controls on the variability of fires in the tropics and subtropics. *Global Biogeochemical Cycles* 22, GB3028, doi:10.1029/2007GB003122.

Weber, U., M. Jung, M. Reichstein, et al. 2009. The inter-annual variability of Africa's ecosystem productivity: A multi-model analysis. *Biogeosciences* 6, 285–295.

Williams, C. A., N. P. Hanan, J. Neff, R. J. Scholes, J. Berry, A. S. Denning, and D. Baker. 2007. Africa and the global carbon cycle. *Carbon Balance and Management* 2, doi: 10.1186/1750-0680-2-3.

Section VII

Understanding Savannas as Coupled Human–Natural Systems

25

People in Savanna Ecosystems: Land Use, Change, and Sustainability

Kathleen A. Galvin and Robin S. Reid

CONTENTS

Introduction

Pastoralism is one of the most sustainable ways that humans use savannas (Fritz et al., 2003; Hansen et al., 2005), if livestock are managed sustainably. Pastoralism is an economic activity suitable to savannas where, generally, crop farming is not possible or is highly uncertain. In savannas, herders keep cattle in sub-Saharan Africa, North America, South America, and Australia; sheep in the eastern Mediterranean; camels in North Africa and the Arabian Peninsula; and horses along with multiple stock from North Africa east across Asia (Sutton and Anderson, 2004). In the savannas of South America, such as pampas (Argentina) or the cerrado (Brazil), ranchers raise cattle (Wilcox, 1999).

Pastoral land tenure has traditionally been communal and resources are common pool (Behnke and Scoones, 1993). Pastoralists, especially in the arid and semiarid savannas, tend to be mobile and flexible in their management strategies (Fratkin and Mearns, 2003). This is because they live in environments that are inherently uncertain with high climatic variability, so the primary management strategy, *mobility*, depends on the ability of people to make quick decisions to track rain and resources. Pastoralists rely on mutual help for survival or for negotiating access to pasture (Galvin, 2009). Movement of livestock herds is conducted through reciprocal rights to common pool resources, sometimes belonging to other people. These rights to use another group's property is the basis for the nonexclusive tenure and land use systems common to pastoralism (Turner, 1999a, b).

In addition to a variable natural environment, the sociopolitical environment, such as state structures, markets, and wars, constantly changes and affects pastoral livelihoods and management (Salzman, 2004). The legacy of the European colonial model of economic growth, for example in Africa, focuses on agriculture, leaving herders with little political power (Smith, 2005).

However, changing climate, land tenure and land use policies, increasing human population growth, variable markets, and changing expectations are having immense effects on these systems. The land is rapidly moving toward exclusive systems of property ownership based on the notion that individuals invest and steward the land better than groups of individuals (Behnke, 2008). Thus, pastoralists are losing access to high value resources, often dry season refuge areas to agriculture and to conservation. Further, pastoralists often support privatization to keep agriculturalists, non-governmental organizations, and other powerful groups from gaining title to their lands (BurnSilver, 2007). Pastoralists are also diversifying their own economic activities into agriculture, wage labor, and businesses (Barrett et al., 2001; Little et al., 2001). All of these trends tend to lead to a loss of mobility that jeopardizes both rangeland resources and pastoral livelihoods (Fratkin and Mearns, 2003).

This chapter first presents a conceptual framework that embraces change and sustainability as fundamental components of pastoral human–ecological systems. We then describe the processes and consequences of change among pastoralists. Finally, we address the implications for sustainability of the linked human–environmental system. This review is not meant to be inclusive of all pastoral social–environmental systems globally, as will be seen; it represents a large number of pastoral systems.

Savannas as Coupled Human–Environment Systems

One of the fundamental challenges in science today is to understand how changes in land use and ecosystems affect human well-being, ecosystem structure and function, and how they are linked (Turner et al., 2007). Global savannas are transforming in the face of rapid changes in human populations, land use, and climate that challenge resilience and sustainability of the linked human–environmental system (Galvin et al., 2008). We present a linked human–environmental framework here to understand the interactions of the pastoral socio-environmental system. People are considered part of the system and not a disturbance of the system. This approach is useful, because it assumes that change is a natural state of a system and that resilience and vulnerability are inherent characteristics of the system (Reynolds et al., 2007).

It focuses on the relationships between system components. This means that for savanna systems, for example, climate affects forage availability, which, in turn, affects livestock productivity and herder management strategies. It incorporates the concepts of adaptation, which are the decision-making processes and actions that populations (individuals, groups, or organizations) take in response to current or predicted change and social learning (Brooks et al., 2005; Smit and Wandel, 2006). Finally, adaptive capacity encompasses the mechanisms that allow actions in response to changes, be they climatic or social (Nelson et al., 2007). Adaptive capacity rises with access to greater managerial ability; financial, technological, and informational resources; infrastructure; knowledge; institutions; political influence; kin and other networks; and learning ability. Adaptive capacity, thus, may be public or private.

Social resources including networks, institutions, and other forms of social capital provide important elements of adaptive capacity. Adaptive capacity is context specific and varies from place to place. Pastoralists, for example, use social capital to gain access to water and forage, connecting livestock to patchy resources in these environments (Hobbs et al., 2008a, b; Galvin et al., 2008). Local forms of social capital need to provide for flexible management in uncertain and changing environments. Land tenure change that promotes expansion of farming, conservation areas, or private property may reduce flexibility in the system by reducing movement. A system's adaptive capacity is not static. Population pressures or resource overuse, for example, may eventually reduce the coping ability of the system and narrow the range of potential response strategies (Adger, 2003). A system is resilient if it can adapt to stresses or perturbations without undergoing significant change (Berkes et al., 2003). Sometimes systems transform, creating new social–ecological systems (Nelson et al., 2007). For example, a changing climate may impact an ecosystem so that it can no longer support a traditional livelihood system such as farming or herding. Finally, what is adaptive today may not be adaptive in the future, so we can only talk about adaptive pastoral systems at particular points in time.

Land Use, Drivers, and Processes of Change in Savannas

Although livestock grazing is a predominant land use, people currently also use savannas for conservation (protected areas), mining, tourism, settlement, hunting, farming, tourism, recreation, and other purposes. Much of what was originally savanna are now used for settlements, cropland, and other purposes and are no longer recognizable as savanna. This is especially true in wetter areas (semiarid and dry sub-humid); where a full 35–50% of the land is now plowed for irrigated or rain-fed crops, and about 2–4% of the land is used for towns and cities (MEA, 2005). Savannas that have not been fully converted to other uses can be divided into intensive and extensive grazing lands. We define grazing lands here as those land cover categories (GLC, 2003) without significant tree cover (thus we exclude forest/woodland), cropland, or urban areas; these lands are dominated by shrubs and herbaceous plants. Intensive grazing lands, where human population is >20 people/km^2, support 76% of the people in grazing lands (Reid et al., 2008a); they can be highly fragmented and are well advanced in the fragmentation transition. Extensive rangelands, where human populations are less than 20 people/km^2, covering 91% of grazing lands, only support 24% of all people in grazing lands. Here, landscapes are relatively open, less fragmented, and the impacts of humans on savanna ecosystems is relatively light. About 11% of extensive and 4% of intensive grazing lands are now inside protected areas, where human use is limited (Reid et al., 2008a), a change that principally occurred in the last two centuries.

All savannas are in the process of change, in different directions in different places around the world, depending on their natural resource endowments, history, culture and geography, the drivers (or causes) of change, and how people respond and cope with (and modify) these drivers (Reid et al., 2008b). Behnke (2008, p. 306) argues that underlying these drivers of change are the "growing power of centralized, bureaucratic states and the spread of capitalism" (Figure 25.1). "It is this combination of factors that allows the

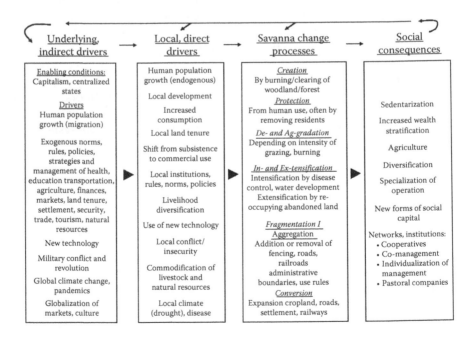

FIGURE 25.1
Schematic of the causal chain from underlying and local drivers of change, through savanna change processes to consequences for people.

political dimensions of resource control to recede and permits indigenous land users to view natural resources in modern economic terms as a commodity and as a source of personal profit" (Behnke, 2008, p. 306). The spread of capitalism has also caused a shift from subsistence, nomadic cultures to commercial, settled cultures. The rise of central governments created policies that dramatically changed land use in savannas and modified their extent and their function. Some of these changes are still happening today.

This context underlies a causal chain similar to that in Figure 25.1, where *underlying, indirect* drivers (some endogenous to local populations like culture, some exogenous like national government policy) influence drivers that are *local* and *direct*. Underlying drivers, often arising outside savannas, include global climate change, national government policy, globalization, and expansion of market economies (Behnke, 2008). These drivers affect large areas and people, and savannas respond to these drivers, often in unique ways. Government policy on land tenure is particularly important, from laws that allow herders who hold land in common to gain private title in Africa to rules about the size and arrangement of land parcels in Australian rangelands.

One important underlying driver is climate change, which may strongly affect savanna productivity and pastoral livelihoods. Most savannas, and in particular the arid and semiarid lands, are, by definition, highly variable in intra- and interannual precipitation (Ellis and Galvin, 1994); and pastoralists respond to these patterns. Climatic variability on the seasonal-to-interannual time scale is changing, and the El Niño/Southern Oscillation (ENSO) often causes increased variability (Wolter and Timlin, 1998). For example, in Africa, data trends "point to a probable increase in the frequency of extreme weather events in the short to medium time horizons, a conclusion also arrived at by the IPCC" (Magadza, 2003, p. 265). Temperatures in the Northern Hemisphere are now warmer than at any time in the past 100 years based on tree ring and ice core data (Hurrell and Stenseth, 2005). As CO_2 doubles, grasses will become less digestible (Bond et al., 2003), and this will support fewer livestock and wildlife in savannas in the future.

Local, direct drivers include local human population growth, shifts from subsistence to commercial ranching, use of new technology, local land tenure, and, more generally, local institutions (Figure 25.1). For example, some local, east African populations allocated land by customary law and enclosed (and fragmented) wetter savannas, long before exogenous, underlying policy made this legal at a national level (Behnke, 2008). Sometimes there is not a clear distinction among these classifications, with population growth and technology as prominent examples. One example is the invention of barbed wire technology in the late 1800s, when settlers suddenly had an economical way to control land and erected fences that cut up or fragmented rangeland, eventually ending free movement of wildlife and livestock and the people who depended on them (Krell, 2002). Recent subdivision of Kenyan group ranches, long after private titling was legal at the national level, not only secured ownership but also created new conflicts as herders found parcels were too small to support their livestock through drought.

These indirect and direct drivers cause six main change processes in savannas: (1) creation, (2) protection, (3) degradation or aggradation, (4) intensification or extensification, (5) fragmentation or re-aggregation, and (6) conversion (Figure 25.1). Indigenous people created some of the savannas around the world today by burning forests and woodlands like the open grassland created from former woodland in Maasailand (Dublin, 1995). People often protect savannas by removing local residents who have lived there for years, even millennia, assuming savannas are natural without people. The most ancient example of this is in Africa, where humans evolved in savanna woodlands for millions of years and governments carved all savanna parks from land previously used by herders and hunter gatherers (Brockington, 2002).

Many experts suggest that herders degrade savannas through overgrazing (Lamprey, 1983), but this may be less widespread than originally thought (Vetter, 2005). However, when grazing is heavy, long-lived grasses can decline and short-lived grasses can increase, as in Niger (Hiernaux, 1998) and South Africa (Fynn and O'Connor, 2000). Grass production can also decline in response to grazing, with negative effects on cattle arising from poor range-land condition during drought (Fynn and O'Connor, 2000). On the other hand, grazing aggrades some European pastures, by creating and maintaining structural heterogeneity, which, in turn, affects the number of plants and animals in pastures (Rook et al., 2004).

If farmers or ranchers use land more intensively, they produce more meat or milk for each amount of savanna resource used. Intensification highly modifies the structure and function of savannas. This increasing efficiency slows the need to clear additional land for pasture or feed crops; however, it may act to maintain savannas. For example, Swedish farmers rarely use low-yielding pastures for livestock grazing today, so the extent of livestock grazing has contracted significantly since the 1960s (Deutsch and Folke, 2005). An opposing process is that of extensification, where ranchers use each hectare of land less intensively, moving across the land to access heterogeneous resources. This happened initially after state support of livestock production vanished when the Soviet Union broke up (Alimaev and Behnke, 2008).

Many ranchers build fences and water points, which fragment savannas into smaller areas (Hobbs et al., 2008b). Here, fragmentation can cut migration routes of wildlife entirely (Boone and Hobbs, 2004), but more subtly it can reduce the diversity of grazing-sensitive, native plant species, even far from waterpoints (Landsberg et al., 2003). However, once fragmentation occurs and the best land is excised for intensive uses, ranchers can start to re-aggregate or un-fragment the remaining drier savanna landscapes (Behnke, 2008). In the northern Great Plains, this re-aggregation means un-fragmented ownership and use but not necessarily the removal of fences to create less fragmented landscapes (Lackett and Galvin, 2008). In Australia, ranchers often manage more than one property or lease paddocks owned by others and move stock between paddocks (Stokes et al., 2008); but they still remove fences and un-fragment the land physically.

Savanna structure and function are usually disrupted when people convert land to other uses. In the bio-diverse Brazilian cerrado, farmers expanded soybean production, partially to provide feed for livestock in Asia (Viglizzo and Frank, 2006). In the rangelands at the base of Mt. Kilimanjaro, agro-pastoralists and agriculturalists (both small-scale and large-scale farming enterprises) are converting precious swamplands and riverbanks into crops for local and national markets (Campbell et al., 2000). In Machakos, Kenya, although farmers converted open savanna pastures into horticultural and other cash farms, soil degradation declined with increased investments in soil conservation (Tiffen et al., 1994).

Transitions: Consequences of Change for People

Some generalizations emerge from an examination of changes in savannas across the world (Figure 25.1). First, these ecosystems are characterized by variability in climate, with frequent droughts. Climate change with increased warming may lead to increases in primary plant production initially; but subsequent increases in CO_2 may lead to declines in pasture productivity. At the same time, rangelands are fragmenting, driven by population pressure, expansion of farming, land tenure change, and policy such as protected area conservation and privatization of land. As markets expand and land use changes, pastoral families often face a need for cash to pay school fees, buy new technologies, and service other needs associated with more complex societal and economic structures.

However, it seems that pastoralists have been adapting to these changes by continuing to gain access to water and forage for their livestock as the examples below illustrate. Since many live in dry areas with low productivity, mobility and flexibility is essential for pastoral adaptation. Inputs like food supplements and trucking of livestock could compensate for the loss of mobility; but since most pastoralists live in developing countries with limited capacity to invest in pastoralism, this does not often occur.

East African pastoral societies have undergone much change, particularly in the last two to three decades. From having had livelihoods focused on highly mobile livestock subsistence herding, many pastoralists are now more sedentary with market-oriented livestock production, sometimes with a change in livestock emphasis. Pastoralists are increasingly focusing on agricultural production (e.g., Little et al., 2001). In Kajiado District, Kenya, communally owned land in group ranches is rapidly changing into private land. Group ranches, established in the 1960s through the 1980s, were supported by the Maasai to secure the land against non-Maasai who were moving in and expropriating the better watered land (Kimani and Pickard, 1998). The group ranches are today subdividing into private parcels, causing sedentarization (Thornton et al., 2006). Subdivided and sedentarized households tend to move their herds less. Further, Maasai are diversifying economic activities including agriculture, wage labor, or business as well as intensifying livestock production through changing breeds of livestock. Nevertheless, livestock continue to embody a significant element of livelihoods. In the extreme drought of 2000, for instance, households in the group ranches were able to move long distances toward the unfragmented areas. Social networks, based on kinship and clan relations, age-set membership, and stock associations formed the institutions on which grazing access within and across boundaries was negotiated. However, as Maasai savannas continue to be subdivided into private parcels, more formal institutions for movement are expected to emerge. It remains to be seen whether the social institutions needed for mobility can remain flexible enough to respond to quickly changing events.

China has made enormous changes to its land tenure system in the last 30 years. China's commune system was dismantled in the 1980s and the rangelands, though technically owned by the state, were allocated to villages or groups of households. More recent policy changes have resulted in grassland use rights being granted to individuals. Despite this, Banks et al. (2003) found the collective tenure arrangements persisting in western China. Small groups of often, kin-related households are effectively excluding outsiders. Banks et al. (2003, p. 134) write, "in all case study areas, village boundaries in pasture are monitored and enforced by the local households using the grassland." However, community-based management is not supported by the current Chinese legal framework, which emphasizes household tenure. Banks (2003) suggests that groups of communities can solve the collective action problems. "Group tenure arrangements found in Xinjiang (China) are largely based on pre-socialist social formations. Being based on pre-existing kinship structures, the group members also have repeated and multiplex relations. All of these characteristics are conducive to the accumulation of social capital ..." (Banks, 2003, p. 2137). He suggests that communities align loosely with government's pasture rules in a sort of co-management arrangement. In the end, herders would benefit by remaining flexible in their herd management and at low cost to the government.

Australia's rangelands occupy more than 70% of the land area, 60% of which is used for commercial livestock production (Pickup, 1998). Australian rangelands are key areas for meat and wool production and have a long history of fragmentation based on European land laws. Rangelands with a longer history of settlement generally have had stronger policy pressure for closer settlement, more periods of demand for land from immigrants or soldiers returning from the two World Wars, and more boom periods of exaggerated expectations for pastoralism (Stokes et al., 2008, pp. 94–95).

At the same time, there is an increasing consolidation of properties in parts of Australia to offset rising production costs through efficiencies of scale.

Pastoralists combine properties for other reasons: to combine several properties to specialize operations between properties, to plan for inheritance in large families, speculative property trading, and expansion of successful businesses (Stokes et al., 2004). In each case, fragmentation and consolidation, pastoralists often need to access resources not available on a single property. Some pastoralists use a system called *agistment*, in which livestock are transferred between farms to areas with better forage and water (McAllister et al., 2006). Agistment networks are groups of kin, friends, or business partners who work together to provide adequate forage and water for their livestock. Livestock, through prior agreements, are moved between private parcels of land when resources become low in their current locations. This system of trust and social networking allows ranchers to manage for spatiotemporal variability in resources (McAllister et al., 2006).

Future trends and impacts of global warming and socioeconomic changes are very difficult to predict. Likewise, social capital is tenuous in its form, structure, and duration, so that it comes and goes depending on needs. For example, in drought, many networks of human relationships may come into play and then dissolve until the next crisis. The examples described here suggest that for pastoralists who continue to need to access savannas, a combination of old and new social, political, financial, and human capital enables them to manage for flexibility. This trend toward co-management seems promising, at least in the short run. However, it remains to be seen whether linking to outside and to higher level institutions is going to provide the long-term flexibility to deal with the changes and surprises that will certainly occur in the future.

Sustainable Savannas

Although it is impossible to suggest ways to ensure sustainability of savannas and the peoples who depend upon them around the globe, we can point to some general principles. We assume here that it is not possible to sustain savanna ecosystem structure and function unless we also sustain human society. We assume that if we maintain or promote the adaptive capacity and resilience of human societies and savanna ecosystems, their structure and function will be sustained over the long term. The first issue is the need to define what the sustainability would look like in a particular place for different types of human users. As we described above, people convert savannas to other uses like wheat fields, granite quarries, cities, and industrial areas; we will not concern ourselves with sustainability of these uses. Instead, we focus our attention on sustaining savannas for livestock grazing, partly because this is one of the more sustainable ways to produce food in savannas. The same principles here may also apply to human uses that promote ecosystem protection, such as biodiversity conservation, tourism, recreation, and others.

For livestock grazing and pastoralism, six key principles (synthesized from Sandford, 1983; Fernández-Giménez and Swift, 2003) seem to sustain the resilience of both household economies and savanna ecosystems: maintaining mobility (and opportunistic management more generally), promoting diversity (both income and vegetation cover), creating water and forage reserves, improving profitability, maintaining exchange (of social and natural capital), and promoting restoration (of a functional coupled human–environment system).

As mentioned earlier, mobility is a keystone concept that promotes the sustainability of savannas, by allowing herders and livestock to access ephemeral flushes of water and forage and heterogeneous vegetation types. This also allows some parts of the savanna to rest from grazing as herders move to other parts. Without access to larger areas through movement, herders in Kenya can support fewer livestock, as much as 25% fewer as land sizes shrink (Boone et al., 2005). There is still a lively debate as to whether this rotational grazing in commercial grazing systems conserves the productivity of savannas compared with continuous grazing, with a recent review strongly suggesting that there is no difference in many experimental studies (Briske et al., 2008). What confers greater productivity in these commercial systems is better overall management of herds, regardless of grazing system. Interestingly, it may be that rotational grazing, which logically demands closer observation of animals and the savanna that supports them (Briske et al., 2008), may be the reason that rotational grazing is still supported by some proponents and practitioners alike (Teague et al., 2004). Mobility is widely supported as a key strategy in many common property pastoral systems (Hobbs et al., 2008a).

Access to a diversity of sources of household income and a diversity of vegetation patches is also a key to sustainability and resilience of pastoral households and savanna ecosystems (McCabe, 2003; Homewood et al., 2009). Income diversity allows pastoral families to survive through difficult droughts, prolonged dry seasons, and bouts of disease, when livestock production is poor. For wildlife and livestock, access to heterogeneous vegetative resources sustains populations by allowing herbivores to access forage of different quality and quantity at different times of the year (Hobbs et al., 2008b).

For both livestock and wildlife, it is critically important to have access to reserves of forage and water during dry periods in savannas. At a local scale, pastoralists in east Africa often reserve forage for vulnerable animals near the homestead to maintain their weight, especially in the dry season. On a larger scale, areas of the landscape are often saved for grazing of larger herds of livestock in the dry season. This is the principle behind grazing leases on public land in the western United States, for example. Similarly, the great migrations of wildlife globally often move from dry to wet season ranges or from winter to summer ranges.

One of the biggest challenges of maintaining intact savannas is that livestock grazing is often less profitable than other uses, including crop production in Africa (Norton-Griffiths et al., 2008) and sale for ex-urban development in the United States (Hansen et al., 2002). In the case of crop farming, this clearly depends on rainfall, with livestock better competing as a profitable land use with crops as rainfall declines (Norton-Griffiths et al., 2008). This matters, because ranching and pastoralism are often more compatible with higher biodiversity and other ecosystem services than either crop farming (Fritz et al., 2003) or housing development (Hansen et al., 2005). Improving the profitability of ranching, whether through intensification (Coppock et al., 2009), payments for ecosystem services (Daily et al., 2009) or other methods, are ways to keep working ranches viable and sustain savanna ecosystems.

A fifth principle is the idea of exchange of social and natural capital. Social and landscape networking and exchange of social and natural capital can be a foundation of hedging risk in pastoral systems (Galvin, 2008). Mutual assistance networks allow families to survive hard winters, droughts, and long dry seasons. Australian agistment arrangements allow pastoralists who do not have enough forage to link with those who do, which provides assistance and uses pastures more effectively. This serves to reconnect fragmented landscapes as well (Stokes et al., 2008).

A final principle, which is rarely applied, is the restoration of a functional coupled human–environment system using all the above principles, to promote resilience and sustainability of savannas landscapes and pastoral society. Socially, adoption of agistment arrangements or creating broader grazing associations can restore mobility, access to diversity, exchange, and may improve profitability. Co-management of state-owned resources with local communities to restore some access to historical resources may be useful (Brockington et al., 2006). Ecologically, restoration of the function of savanna landscapes is expensive but is sometimes needed.

In conclusion, we suggest that the sustainability and resilience of linked human–natural systems in savannas depends on thinking out of the box concerning our common assumptions about how best to manage and use landscapes. Conventional wisdom and much economic theory suggest that intensification of use, specialization of products, and privatization of land and other resources is the route to greater human development and profitability of enterprises. We suggest here that we also need to consider extensification, mobility, connectivity, and diversity as concepts and practices that will promote sustainability for savannas themselves and the people who depend upon them.

Acknowledgment

The writing of this review was supported, in part, by the National Science Foundation project, Decision-Making in Rangeland Systems: An Integrated Ecosystem-Agent-Based Modeling Approach to Resilience and Change (DREAMER), Grant No. SES-0527481.

References

Adger, W. N. 2003. Social aspects of adaptive capacity. In *Climate Change, Adaptive Capacity and Development*, ed. J. B. Smith, R. J. T. Klein, S. Huq, London, Imperial College Press.

Banks, T. 2003. Property rights reform in rangeland China: On the road to the household ranch. *World Development* 31, 2129–2142.

Banks, T., C. Richard, L. Ping, and Y. Zhaoli. 2003. Community-based grassland management in Western China. *Mountain Research and Development* 23, 132–140.

Barrett, C. B., T. Reardon, and P. Webb. 2001. Nonfarm income diversification and household livelihood strategies in rural Africa: Concepts, dynamics and policy implications. *Food Policy* 26, 315–331.

Behnke, R. H. 2008. The drivers of fragmentation in arid and semi-arid landscapes. In *Fragmentation in Semi-arid and Arid Landscapes*, ed. K. A. Galvin, R. S. Reid, R. H. Behnke, and N. T. Hobbs, 305–340. Dordrecht, Springer.

Behnke, R. H., I. Scoones, and C. Kerven. 1993. *Range ecology at disequilibrium. new models of natural variability and pastoral adaptation in African Savannas*. Overseas Development Institute. London. International Institute for Environment and Development, 248pp.

Berkes, F., J. Colding, and C. Folke, eds. 2003. *Navigating Social-Ecological Systems: Building Resilience for Complexity and Change*. Cambridge University Press, Cambridge, UK.

Bond, W. J., G. F. Midgley, and F. I. Woodward. 2003. The importance of low atmospheric CO_2 and fire in promoting the spread of grasslands and savannas. *Global Change Biology* 9, 973–982.

Boone, R. B., S. B. BurnSilver, P. K. Thornton, J. S. Worden, and K. A. Galvin. 2005. Quantifying declines in livestock due to land subdivision. *Rangeland Ecology & Management* 58, 523–532.

Boone, R. B. and N. T. Hobbs. 2004. Lines around fragments: Effects of fencing on large herbivores. *African Journal of Range & Forage Science* 21, 79–90.

Briske, D. D., Derner, J. D., Brown, et al. 2008. Rotational grazing on rangelands: Reconciliation of perception and experimental evidence. *Rangeland Ecology & Management* 61, 3–17.

Brockington, D. 2002. Fortress conservation: The preservation of the Mkomazi Game Reserve, Tanzania. International African Institute, Oxford.

Brockington, D., J. Igoe, and K. Schmidt-Soltau. 2006. Conservation, human rights, and poverty reduction. *Conservation Biology* 20, 250–252.

BurnSilver, S. 2007. Critical Factors Affecting Maasai Pastoralism: The Amboseli Region, Kajiado District, Kenya. Ph.D. Colorado State University, Fort Collins, Colorado.

Campbell, D. J., H. Gichohi, A. Mwangi, and L. Chege. 2000. Land use conflict in Kajiado District, Kenya. *Land Use Policy* 17, 337–348.

Coppock, D. L., D. L. Snyder, L. D. Sainsbury, M. Amin, and T. D. McNiven. 2009. Intensifying beef production on Utah Private Land: Productivity, profitability, and risk. *Rangeland Ecology & Management* 62, 253–267.

Daily, G. C., S. Polasky, J. Goldstein, et al. 2009. Ecosystem services in decision making: Time to deliver. *Frontiers in Ecology and the Environment* 7, 21–28.

Deutsch, L. and C. Folke. 2005. Ecosystem subsidies to Swedish food consumption from 1962 to 1994. *Ecosystems* 8, 512–528.

Dublin, H. T. 1995. Vegetation dynamics in the Serengeti-Mara ecosystem: The role of elephants, fire and other factors. In, *Serengeti II: Dynamics, Management and Conservation of an Ecosystem*, ed. A. R. E. Sinclair and P. Arcese, 71–90. Chicago, Illinois, USA, University of Chicago Press.

Ellis, J. and K. A. Galvin. 1994. Climate patterns and land use practices in the dry zones of east and west Africa. *BioScience* 44, 340–349.

Fernández-Giménez, M. E. and D. M. Swift. 2003. Strategies for sustainable grazing management in the developing world. In *Proceedings of the VIIth International Rangelands Congress*, Durban, South Africa, pp. 821–831.

Fratkin, E. M. and R. Mearns. 2003. Sustainability and Pastoral Livelihoods: Lessons from East African Maasai and Mongolia. *Human Organization* 62, 112–122.

Fritz, H., S. Said, P. C. Renaud, S. Mutake, C. Coid, and F. Monicat. 2003. The effects of agricultural fields and human settlements on the use of rivers by wildlife in the mid-Zambezi valley, Zimbabwe. *Landscape Ecology* 18, 293–302.

Fynn, R. W. S. and T. G. O'Connor. 2000. Effect of stocking rate and rainfall on rangeland dynamics and cattle performance in a semi-arid savanna, South Africa. *Journal of Applied Ecology* 37, 491–507.

Galvin, K. A. 2008. Responses of pastoralists to land fragmentation: Social capital, connectivity and resilience. In *Fragmentation in Semi-arid and Arid Landscapes*, ed. K. A. Galvin, R. S. Reid, R. H. Behnke, and N. T. Hobbs, 369–389. Dordrecht: Springer.

Galvin, K. A. 2009. Transitions: Pastoralists living with change. *Annual Review of Anthropology* 38, 185–198.

Galvin, K. A., N. Hobbs, Thompson, R. S. Reid, and R. H. Behnke, eds. 2008. *Fragmentation of Pastoral Rangelands. Consequences for Human and Natural Systems.* Dordrecht, The Netherlands: Springer.

GLC. 2003. Global Land Cover 2000 Database. Joint Research Centre, European Commission, http://www-gem.jrc.it/glc2000.

Hansen, A. J., R. L. Knight, J. M. Marzluff, et al. 2005. Effects of exurban development on biodiversity: Patterns, mechanisms, and research needs. *Ecological Applications* 15, 1893–1905.

Hansen, A. J., R. Rasker, B. Maxwell, et al. 2002. Ecological causes and consequences of demographic change in the new west. *BioScience* 52, 151–162.

Hiernaux, P. 1998. Effects of grazing on plant species composition and spatial distribution in rangelands of the Sahel. *Plant Ecology* 138, 191–202.

Hobbs, N. T., K. A. Galvin, C. J. Stokes, et al. 2008a. Fragmentation of rangelands: Implications for humans, animals, and landscapes. *Global Environmental Change* 18, 776–785.

Hobbs, N. T., R. S. Reid, K. A. Galvin, and J. E. Ellis. 2008b. Fragmentation of arid and semi-arid ecosystems: implications for people and animals. In *Fragmentation in Semi-arid and Arid Landscapes*, ed. K. A. Galvin, R. S. Reid, R. H. Behnke, and N. T. Hobbs, pp. 25–44. Dordrecht: Springer.

Hurrell, J. W. and N. C. Stenseth. 2005. Global climate change: Building links between the climate and ecosystem impact research communities. *Climate Research* 29, 181–182.

Homewood, K., P. Trench, and P. Kristjanson. 2009. *Staying Maasai? Livelihoods, Conservation and Development in East African Rangelands.* London: Springer.

Kimani, K. and J. Pickard. 1998 Recent trends and implications of Group Ranch sub-division and fragmentation in Kajiado district, Kenya. *The Geographical Journal* 164, 202–213.

Krell, A. 2002. *The Devil's Rope: A Cultural History of Barbed Wire*. London: Reaktion Books.

Lackett, J. M. and K. A. Galvin. 2008. From fragmentation to reaggregation of range-land in the Northern Great Plains. In *Fragmentation in Semi-arid and Arid Landscapes*, ed. K. A. Galvin, R. S. Reid, R. H. Behnke, and N. T. Hobbs, 111–134. Dordrecht: Springer.

Lamprey, H. F. 1983. Pastoralism yesterday and today: The overgrazing controversy. In *Tropical Savannas. Ecosystems of the World*, ed. F. Bourlière, Amsterdam: Elsevier.

Landsberg, J., C. D. James, S. R. Morton, W. J. Muller, and J. Stol. 2003. Abundance and composition of plant species along grazing gradients in Australian rangelands. *Journal of Applied Ecology* 40, 1008–1024.

Little, P. D., K. Smith, B. A. Cellarius, D. L. Coppock, and C. Barrett. 2001. Avoiding disaster: Diversification and risk management among East African Herders. *Development and Change* 32, 401–433.

Magadza, C. H. D. 2003. Engaging Africa in adaptation to climate change. In *Climate Change. Adaptive Capacity and Development*, ed. S. Huq, R. J. T. Klein, J. B. Smith, London: Imperial College Press.

McAllister, R. R. J., I. J. Gordon, M. A. Janssen, and N. Abel. 2006. Pastoralists responses to variation of rangeland resources in time and space. *Ecological Applications* 16, 572–583.

McCabe, J. T. 2003. Sustainability and livelihood diversification among the Maasai of northern Tanzania. *Human Organisation* 62, 100–111.

MEA (Millennium Ecosystem Assessment). 2005. *Ecosystems and Human Well-Being: Synthesis*. Washington, DC: Island Press.

Nelson, D. R., W. N. Adger, and K. Brown. 2007. Adaptation to environmental change: Contributions of a resilience framework. *Annual Review of Environment and Resources* 32, 11.1–11.25.

Norton-Griffiths, M., S. Serneels, M. Y. Said, et al. 2008. Land use economics in the Mara area of the Serengeti ecosystem. In *Serengeti III: Human Impacts on Eco-system Dynamics*. ed. A. R. E. Sinclair, C. Packer, S. A. R. Mduma, and J. M. Fryxell, pp. 379–416. Chicago: Chicago University Press.

Pickup, G. 1998. Desertification and climate change—The Australian perspective. *Climate Research* 11, 51–63.

Reid, R. S., K. A. Galvin, and R. L. Kruska. 2008a. Global significance of extensive graz-ing lands and pastoral societies: An introduction. In *Fragmentation in Semi-arid and Arid Landscapes*, ed. K. A. Galvin, R. S. Reid, R. H. Behnke, and N. T. Hobbs, pp. 1–24. Dordrecht: Springer.

Reid, R. S., H. Gichohi, M. Y. Said, et al. 2008b. Fragmentation of an peri-urban savanna, Athi-Kaputiei Plains, Kenya. In *Fragmentation in Semi-arid and Arid Landscapes*, ed. K. A. Galvin, R. S. Reid, R. H. Behnke, and N. T. Hobbs, pp. 195–224. Dordrecht: Springer.

Reynolds, J. F., D. M. Stafford Smith, E. F. Lambin, et al. 2007. Global desertification: Building a science for dryland development. *Science* 316, 847–851.

Sandford, S. 1983. *Management of Pastoral Development in the Third World*. Chichester, UK: John Wiley & Sons.

Salzman, P. C. 2004. *Pastoralists: Equality, Hierarchy, and the State.* Boulder, CO: Westview Press.

Smit, B. and J. Wandel. 2006. Adaptation, adaptive capacity and vulnerability. *Global Environmental Change* 16, 282–292.

Smith, A. B. 2005. *African Herders: Emergence of Pastoral Traditions.* Walnut Creek, CA: Altamira Press.

Stokes, C. J., A. J. Ash, and R. R. J. McAllister. 2004. Fragmentation of Australian rangelands: Risks and trade-offs for land management. In *Living in the Outback: Conference Papers,* ed. G. Bastin, D. Walsh and S Nicolson, pp. 39–48. Alice Springs, Australia: Australian Rangeland Society.

Stokes, C. J., R. R. J. McAllister, A. J. Ash, and J. E. Gross. 2008. Changing patterns of land use and land tenure in the Dalrymple Shire, Australia. In *Fragmentation in Semi-arid and Arid Landscapes,* ed. K. A. Galvin, R. S. Reid, R. H. Behnke, and N. T. Hobbs, pp. 93–121. Dordrecht: Springer.

Sutton, M. Q. and E. N. Anderson. 2004. *Introduction to Cultural Ecology.* Walnut Creek, CA: Altamira Press.

Teague, W. R., S. L. Dowhower, and J. A. Waggoner. 2004. Drought and grazing patch dynamics under different grazing management. *Journal of Arid Environments* 58, 97–117.

Thornton, P. K., S. B. Burnsilver, R. B. Boone, and K. A. Galvin. 2006. Modelling the impacts of Group Ranch Subdivision on Agro-Pastoral Households in Kajiado, Kenya. *Agricultural Systems* 87, 331–356.

Tiffen, M., M. Mortimore, and F. Gichuki. 1994. *More People, Less Erosion.* Nairobi, Kenya: ACTS Press.

Turner, B. L., E. F. Lambin, and A. Reenberg. 2007. The emergence of land change science for global environmental change and sustainability. *Proceedings of the National Academy of Sciences* 104, 20666–20671.

Turner, M. D. 1999a. The role of social networks, indefinite boundaries and political bargaining in maintaining the ecological and economic resilience of the transhumance systems of Sudan-Sahelian West Africa. In *Managing Mobility in African Rangelands* ed. M. Niamir-Fuller, pp. 97–123. London: Beijer International Institute of Ecological Economics.

Turner, M. D. 1999b. Spatial and temporal scaling of grazing impact on the species composition and productivity of Sahelian annual grasslands. *Journal of Arid Environments* 41, 277–297.

Vetter, S. 2005. Rangelands at equilibrium and non-equilibrium: Recent developments in the debate. *Journal of Arid Environments* 62, 321–341.

Viglizzo, E. F. and F. C. Frank. 2006. Land-use options for Del Plata Basin in South America: Tradeoffs analysis based on ecosystem service provision. *Ecological Economics* 57, 140–151.

Wilcox, R. W. 1999. "The law of the least effort"—Cattle ranching and the environment in the savanna of Mato Grosso, Brazil, 1900–1980. *Environmental History* 4, 338–368.

Wolter, K. and M. S. Timlin. 1998. Measuring the strength of ENSO events—how does 1997/98 rank? *Weather* 53, 315–324.

26

Social and Environmental Changes in Global Savannas: Linking Social Variables to Biophysical Dynamics

Matthew D. Turner

CONTENTS

Introduction

Savanna environments have a long history of the human occupation. Given this history, savanna structure and function are shaped by past and present human land uses (Bird and Cali, 1998). In many areas of the world, it is difficult to understand savanna dynamics without understanding human institutions and land-use practices. Fire, grazing, farming, nontimber product extraction, and logging are all the activities that have significant effects on savannas over extensive areas of the world (Wilson, 1989; Fairhead and Leach, 1996; Bassett and Zueli, 2000). Moreover, these interactions and their impacts are highly variable across space and time.

This chapter is concerned with ways to conceptualize and therefore model important relationships among social and biophysical factors in a spatially explicit fashion. To develop an understanding of these relationships, modeling necessarily simplifies the complex systems that are savanna ecology and human society. Successful modeling efforts are those that lead to increased understanding—a goal that is more likely when these efforts are built from strong empirical understandings rather than data-driven exercises. Still, our understanding of the complex web of interactions that constitute the society–environment interface in the world's savannas can only occur through simplifying assumptions that reduce complexity to make causal relationships tractable.

This chapter is organized into four parts. First, a brief overview of the relationship between human-managed activities and savanna dynamics is provided. This section is followed by a discussion of models that explicitly seek to explore causal relationships between social and ecological dynamics in savanna environments. These models are based on various conceptualizations of the nature of these relationships. In the third section, I discuss these conceptualizations and in so doing, will argue that a consideration of savanna ecology reveals that how land-use practices are performed is as important ecologically as their magnitude (or spatial extent). Moreover, more rigorous social analysis requires greater engagement at the level at which land-use decisions are made. This may require greater attention to factors affecting decisions made by land managers within households. The fourth section presents a case of household-level analysis of the social factors influencing how land-use practices (grazing) are performed.

Society–Nature Relationships in Savanna Environments

Savanna environments are farmed, grazed, mined, and built upon by humans in a multitude of different ways. The savanna biome covers a significant portion of the sociocultural variation of the world. Savanna environments are found in some of the poorest countries of the world and some of the richest. Savanna environments also represent some of the least and most densely populated areas of the world. There are two major implications of this simple set of observations. First, it is extremely difficult to make general statements about socioenvironmental relationships in savanna regions. The importance of various "factors" or "drivers" of change will differ depending on the local context (Geist et al., 2006). A second and related point is that the wide variation in the attributes of human societies that reside in savanna environments means that economy–nature, culture–nature, population–nature couplings are not tight. While savanna environments influence and are influenced by human land-use practices, neither determines the other. It is therefore best to think of the society and environment interface in savanna environments as that between open, loosely coupled systems.

Savannas are generally defined by their structure. Human land-use practices such as farming, grazing, fire setting, and logging have important effects on the prevalence of trees. Therefore the prevalence and type ("open," "dense," "closed") of savannas are strongly affected by human land-use practices at different spatial and temporal scales. Savannas are selectively cleared for farming and for the harvesting of timber. These effects are direct and sudden, generally affecting areas in proximity to human settlements and/or transportation infrastructure. Fire is a major anthropogenic and "natural" factor affecting the savanna structure over much broader areas primarily through its effect on the recruitment of trees (Bird and Cali, 1998). The importance of fire in shaping savanna structure in any location is influenced not only by the setting of fires by humans but by how land uses influence fire spread. Farming and roads can act as fire breaks limiting fire spread. Grazing by livestock, through the reduction of fuel load, can reduce the spread of fire (e.g., Bassett and Zueli, 2000). Grazing may also influence savanna structure by influencing either the competition between herbaceous and lignaceous vegetation (Belsky, 1994) or tree recruitment (Reid and Ellis, 1995).

The functioning of savannas can also be influenced by human land-use practices. Analogous to their impacts on savanna structure, human land-use practices can significantly affect species composition due to species-specific extraction (logging, grazing) and differential sensitivities of species to human-mediated disturbances. Changes in species composition, through changes in ecology, will affect savanna functioning. Human land uses will also affect biogeochemical cycling not only through their effects on structure and species composition but through increases in nutrient fluxes to and from the system and/or changes in the speed of nutrient cycling (Detling, 1989; Powell et al., 1995). While harvests stemming from agricultural activities (livestock and crops) can be seen as nutrient exports from the system, these same activities, if supplemented by fertilizer and feed supplements (livestock), also are sources of nutrients to the system (Swift et al., 1989; Manlay et al., 2001). A general feature however of land-use practices has been to transform nutrient forms to those that are more available to plants thereby increasing nutrient leakage (volatilization, leaching) from the system while accelerating the rate of nutrient cycling. For example both fire and grazing work to mobilize nutrients (make more available) while at the same time increase losses from the savanna ecosystem.

Socioecological Models and Modes of Causal Inference

Simply put, socioecological modeling approaches range across the spectrum of system modeling and the empirical analyses of relationships among social and ecological data sets. Given that simulation modeling efforts require adoption of system boundaries, system structure (unchanging parameterizations), a common socioecological currency and time steps among other

elements, they vary significantly in the assumptions made and sophistication across the human society–savanna ecology divide (Taylor, 2005). For
example, biophysical modelers may define these parameters based on an
understanding of the major biophysical processes that shape savanna ecology which may not be appropriate for modeling human behavior. As a result,
changes in human land uses are simply brought in as boundary conditions.
For example, ecosystem models of savannas may use nutrient flows as a common currency tying system components together. Human behavior may be
more appropriately modeled using a different currency (money, information,
etc.) but by being incorporated into the model, human behavioral dynamics
is conceptualized as being shaped by issues of nutrient utilization efficiency
(diet satisfying or maximizing) *or* is treated as a boundary condition.
Similarly, economic approaches to land-use decision making (setting money
as currency) will often ignore the biophysical constraints to different cropping options, and therefore treat savanna ecology as not part of the model
(influenced by human decisions) but as setting broad boundary constraints.

Recent advances have been made toward more integrated socioeconomic
modeling. Agent-based modeling shows promise by allowing the analyst to
explore the spatial land-use patterns produced by basic behavioral rules
(Janssen et al., 2000; Walker and Janssen, 2002; Parker and Manson, 2003;
Manson, 2005; Gross et al., 2006). Work tying a household decision model
(PHEWS—Pastoral Household and Economic Welfare Simulator) to the
SAVANNA ecosystem model (Coughenour, 1993) has been used successfully
to explore the food security implications of changes in household resource
access due to changes in rainfall, subdivision of pastures, livestock mobility,
and wildlife–livestock competition (Thornton et al., 2003, 2006; Galvin et al.,
2006). The dynamic of this coupled model approach comes from the biophysical model tracing how changes in rainfall and vegetative production influence households' subsistence budgets. Distribution of productive resources
among households, livelihood strategies, relative prices, and institutional
rules are based on local surveys and fixed within the model. Therefore, the
ability of these models to address social dynamics beyond that strictly resulting from simulated vegetative changes is extremely limited. These limitations
of the SAVANNA-PHEWS model, one of the most advanced efforts for socioenvironmental modeling in savanna regions, point to the continued difficulties of simulating socioenvironmental relations with mathematical models.

More empirical approaches to the study of socioecological relations involve
a range of model building and statistical analytical approaches. Despite the
promise of agent-based modeling, a very large fraction of empirical analyses
tying social and biophysical changes in savanna regions are best described
as spatial correlative approaches. Of these, the most common approach has
been to tie remotely sensed land cover changes to readily available data such
as land use, transportation infrastructure, market locations, or population
distribution data (Reenberg, 1999; Lambin et al., 2000; Serneels and Lambin,
2001). Implicitly or explicitly, human presence or activity is measured by the

data that is available at such scales, with causation inferred through the spatial correlation of these features and the dependent variable (land cover, etc.). The spatial analytical revolution has increased the tools and sophistication of these approaches shedding new light on the patterns and processes that underlie land cover change (Liverman et al., 1998).

While these spatial correlative approaches are useful starting points for further exploration, a number of concerns can been raised about their use for characterizing socioenvironmental relations. First, spatial correlation is not a necessary or sufficient condition to infer a causal relationship among several variables (Sayer, 1992). Second, reliance on readily available social data raises the specter of correlations in fact being spurious—being driven by the relationships of independent variables with those variables that actually have causal connections to dependent variables. For example, population density is a variable that is strongly correlated with a number of different social parameters associated with land use (Netting, 1993; Turner et al., 1993; Tiffen et al., 1994) such as "market access" which is often correlated with population density (Ensminger, 1992; Tiffen et al., 1994). So, are market-stimulated forms of commercial agriculture really caused by higher population densities? Third, a subset of these analyses displays a circularity in logic due to their limited engagement with savanna ecology or human society. If ecological and social changes are measured by land cover change and land use or population changes respectively, relationships between these sets of variables are either uninteresting or simple truisms. For example, identifying a relationship between agricultural expansion and decline in forest cover does not provide new insights with respect to understanding socioecological relationships beyond providing more data on patterns and rates of deforestation since the increase in one landcover type (farming) necessarily occurs at the expense of the latter (forest cover).

Studies that better address these issues generally seek to work at spatiotemporal scales where social and ecological data are more available and for which there are deeper understandings about the causal relationships among biophysical and social variables. Such work tends to move the social unit of analysis from the human population within a region to particular households managing particular parts of a landscape. In this way, the modeled decision-making process includes a range of different social and ecological factors. More discussion of such approaches will continue below.

Conceptualizations of Socioecological Relations

All modeling approaches, whatever the spatial or social organizational scales they adopt, are based on conceptualizations of the causal relations that tie human decisions to savanna ecology. In this regard, models vary significantly in the complexity and depth of the causal connections made between the ecological and social realms. This variation and its effects on the products of socioecological modeling will be described below.

Savanna Land-Use Ecology

A major starting point for conceptualizing the society–environment interface in savanna ecosystems is to consider the "land-use ecology" of the savanna ecosystem of concern (Figure 26.1). Land-use ecology involves the relationship between human land uses prevalent in the area and the ecological parameters of most concern to the analyst. Land-use ecology is the bidirectional causal interface between human land-use practices and savanna ecology. Serious engagement with land-use ecology leaves open the possibility of a myriad of different ecological responses to human activities. Such responses may often be nonlinear and contingent and therefore run counter to simple conceptual models that dominate particularly the social scientific literature linking social and environmental change. Two common models along these lines are the "stock depletion" and "tipping bucket" models. The

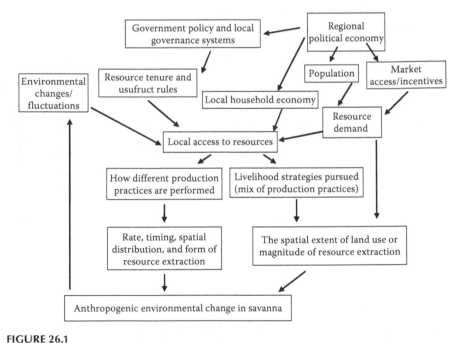

FIGURE 26.1
Relations between broader social and environmental change as mediated by land managers' access to productive resources.

"stock depletion" model treats ecological response as linearly related to human resource use. Ecological relations are a depletable stock—an expansion of population, cash cropping, timber extraction will necessarily lead to commensurate levels of environmental transformation. Dependence in this model is often implicit—analyses that trace changes only to the point of documenting "greater pressures on the environment" rely on this model. Without clearly stated caveats, references to "greater pressures on the environment" are viewed as instances of "environmental change" by most audiences. Somewhat more sophisticated treatments incorporate a threshold concept, above which increases in human pressures will have a disproportionate impact on the environment (the bucket tips). More recent nonengagements with land-use ecology are the widespread use of spatial analytics (connectivity, fragmentation, fractal dimension, etc.) as dependent variables with little attempt to move beyond these highly abstracted concepts to what they mean for local ecosystems.

The prevalence of such conceptualizations cannot be explained simply by analyst ignorance. As contributions to this volume demonstrate, scientists have developed sophisticated understandings of biophysical dynamics in savanna ecosystems. One major reason is the analytical goal of many modeling efforts. Land uses can affect different ecological parameters in different ways. Grazing may cause the increase in the prevalence of one species while reducing the other. It may have limited effects at certain pasture sites while having severe effects at other pasture sites. It may stimulate growth in the short term but reduce productive potential in the long term. It may redistribute the availability of one nutrient (e.g., phosphorus) while reducing the overall availability of another (e.g., nitrogen). In short, the impacts of land-use practices on savanna ecosystems are never simple and always contingent on biophysical features, social geography, and history. Focusing on specific ecological parameters as dependent variables reveals these contingencies but could be seen as overly narrow and unsatisfactorily complex. Abstracting from these messy realities is facilitated by choosing aggregate dependent variables, such as biodiversity, vegetative cover, ecosystem services, sustainability, and using spatial analytics as proxy measures for them. Over broad spatial scales, the lack of available data on parameters of concern (species composition, soil fertility, nutrient fluxes, etc.) has led many studies to rely on spatial analytics of land cover data as dependent variables.

The degree of analytical engagement with land-use ecology will have a significant effect on how social and environmental relationships are modeled. By engaging with land-use ecology, different land-use parameters are identified as important. Rather than simply focusing on the aggregate magnitude of resource extraction, the manner in which resources are extracted may have just as, if not more important effects on the ecological parameter of concern. By "manner of resource extraction," I refer to the following: the spatial pattern of extraction; the seasonality of extraction; the portion of ecological population that is extracted (e.g., demographic cohort or development stage of ecological population); or the morphological parts of the organism extracted (e.g., roots, leaves, stem, fruits, seeds of plants; horn, hair, wool, body, etc. of animals).

An important example of the ecological importance of how a resource is extracted in savanna systems is that of the ecological effects of livestock grazing. On the annual grasslands of the dry savannas of West Africa, where livestock husbandry retains certain levels of mobility, how animals are grazed is more important than the aggregate stocking rate at spatial scales ranging from a village territory to the whole region. This results from: (1) these grasslands are more sensitive (in terms of productivity and species composition) to the timing (duration and seasonality) of grazing than to the aggregate level of animals stocked over a yearly cycle; and (2) grazing management results in a wide variation in the magnitude and timing of grazing pressures experienced from one grazing site to another. Similarly, the effects of fire on savanna structure and species composition are strongly determined by the timing of fire during the dry season.

In sum, how resource extraction affects the ecological parameters of concern will shape the parameterization of society–environment models. If the manner in which resources are extracted matters ecologically, then social factors affecting *how* (rather than simply how much) resources are extracted need to be incorporated into the model. As shown in Figure 26.1, this opens up a range of different causal relationships tying social and environmental processes.

Social Depth of Causal Chains

Socioecological modeling varies significantly not only with respect to the level of engagement with land-use ecology but also with respect to the "depth" of the causal chains tying social factors to variation in land-use practices. What does bringing the "social" into savanna ecosystem modeling mean? In many cases, it may simply involve using spatially registered data that are readily available as proxies for resource demand. Such data include population density, transportation infrastructure, settlement locations, cleared land, or cropped area. Spatial analyses tying these factors to measures of environmental change have proven to be both popular and useful. However, a number of concerns have been raised about such approaches. As mentioned earlier, the use of readily available social data may lead to problems with any causal analysis (spurious correlations, circularities in logic, ignoring "how" resources are extracted … etc.). An additional set of concerns are raised by social scientists who see the chains of social causation in such analyses as insufficiently elaborated to identify social factors that may be most important and policy-relevant. Without such elaboration, finding that a factor (population density) is correlated with some environmental change variable is unsatisfactory since the relationship could be driven by other factors that happen to be correlated with population density (government activities, market access, land tenure) or are the underlying causes for population density variation (government investments, resettlement programs, market incentives, etc.). Deeper chains of social change have a better chance of revealing the underlying webs of causation that are not only more accurate depictions of causation but reveal more nodes of possible policy intervention.

Given the inherent spatial heterogeneity of savanna environments, characterization or modeling of environmental change needs to be spatially explicit. Social factors such as land tenure status, labor availability, wealth, or livelihood strategy can be tied to particular units of land through their linkage to their managing households (or other social units). These social variables can become tractable by reducing analytical spatial scales from regional to those approaching village territories. At these scales, the social landscape is more tractable as the households managing particular parcels of land are revealed. Such analyses (i.e., household level) have two major advantages over analyses that, due to approach or spatial scale, obscure the social identities of resource managers. The first is that a greater range of social variables influencing land use and resource management can be measured and evaluated. The resources (land, labor, capital, knowledge, social networks, markets, etc.) to which land managers have access can be included in the analysis which represents an improvement over analyses that rely simply on population density and infrastructure to estimate land uses. The second is that managers can be directly questioned about the reasons behind their decisions and in so doing, other factors behind the magnitude and manner of land-use change are revealed.

Despite the promise of greater incorporation of social variables into the modeling of savanna socioecosystems (Liverman et al., 1998; Turner et al., 2007), there remain limits to the degree to which social change can be adequately modeled in a spatially explicit manner. Barriers include:

1. The multiple scales of organizational change through which social change occurs. Bringing the analyses down to the unit making land-use decisions (usually at the household level) may lead one to ignore decisions made at the community (common property, etc.), district (allocations/restrictions), and national (policy, infrastructure investments) scales that not only strongly influences households' access to resources (e.g., Figure 26.1) but directly affect savanna ecologies (management of protected areas, provision of land to outside commercial interests, etc.).

2. A related point is that the networks (social relations, markets, information, etc.) within which human societies are entangled, while not having clear boundaries, often transcend those that must be drawn in socioecological modeling.

3. Contemporary decisions about land use are influenced by social as well as ecological histories. Therefore, simply exploring causal relationships between contemporary social and ecological change may lead to insufficient explanations.

4. Some important social "variables," particularly those that describe relations among people, are likely to be inadequately captured by a single number or category. Social power, trust, social capital, spirituality, leadership, openness, tension, ... are all "variables" that may play important roles in affecting the relationship between human communities and savanna ecosystems that are difficult to capture quantitatively.

These and other considerations point to the inherent limits to modeling efforts—particularly in analyzing the dynamics of the social realms that surround local socioecological relations. These issues point to the need for modeling efforts that are informed by: the prior social and ecological work in an area; engagements with local people about the meanings they attach to their past and present land-use practices; and a general understanding of broader social changes occurring in the region. If positioned well, such models can help reveal the ecological implications of broader socioenvironmental change (market trends, policy changes, immigration, recurrent drought, changes in wealth distribution, etc.) by analyzing how changes in specific social variables linked to such broader changes influence decisions made at the community and household level. This is not a particularly ambitious modeling approach. Broader changes are treated effectively as external changes that influence internal social variables that shape managers' access to resources which, in turn, changes how they manage local resources. Using the case of grazing management in dry savanna regions, the following section provides an example of such a modeling approach.

Case: Social Factors Influencing Grazing Patterns in Tropical Savannas

As described above, humans manage fire, grazing, and farming in savanna environments to meet their subsistence needs and respond to commercial opportunities. Utilizing remotely sensed imagery, spatial modeling work has been performed on fire spread (Menault et al., 1991; Eva and Lambin, 1998; Mbow et al., 2000; Berjak and Hearne, 2002; Laris, 2002) and cropped area dynamics (Lambin, 1988; Reenberg and Lund, 1998; Reenberg, 1999; Lambin et al., 2000; Serneels and Lambin, 2001; Favier et al., 2004). Livestock grazing is a major form human-managed land use in savanna regions with direct (selective grazing, nutrient distribution) and indirect (spread of fire) effects on ecological dynamics. In many of the world's savanna regions, livestock grazing is managed by humans in open rangeland situations. Decisions made by the herders coupled with livestock behavior determine the locations where livestock graze, drink, rest, and ruminate (Coppolillo, 2000; Turner et al., 2005; Butt et al., 2009). Differences in the prevalence of these activities across savanna landscapes, may lead not only to differences in grazing pressure (Turner et al., 2005) but also to the potential of livestock-mediated nutrient transfers (Tolsma et al., 1987; Georgiadis, 1989; Weltzin and Coughenour, 1990; Powell et al., 1995; Augustine, 2003).

Compared to cropping and anthropogenic fire, grazing across open rangelands is less tractable through remotely sensed data. Efforts to model the distribution of these activities by herder-managed livestock in open rangeland situations have generally relied on simple diffusion models or optimum foraging approaches (e.g., Andrew, 1988; de Boer and Prins, 1989; Senft, 1989; Coughenour, 1991; Bailey et al., 1996). As a result, the role of manager decisions have been grossly simplified or ignored. What is the human role in determining the spatiotemporal distribution of livestock activities in open rangeland situations? More recent work utilizing map-based herder accounts of grazing itineraries (Guerin et al., 1986; Turner and Hiernaux, 2002) or GPS herd tracking data (Coppolillo, 2000; Burnsilver et al., 2003; Baker and Hoffman, 2006; Butt et al., 2009) have provided more household-level herd movement information that can be used to tie social factors to variations in grazing geographies. Such work has demonstrated that the spatiotemporal distributions of livestock activity diverge strongly from that predicted from diffusion models and show significant inter-household variation (Turner et al., 2005; Turner and Hiernaux, 2008; Butt et al., 2009).

Such work could be combined with ethnographic and household survey studies focused on:

1. Linking broader social changes to households' access to productive resources (land, labor, livestock, cash, social networks).
2. The distribution of these resources among households. Such mixed methods approaches promise to provide more thorough socioenvironmental depictions by tying social change to distributions of grazing pressure and on to ecological change.

For example, broader social changes affecting livestock rearing in the Sudano-Sahelian region of West Africa include:

1. Shifts in livestock ownership from herding professionals to those with other occupations.
2. Increased reliance on contracted labor to provide herding services as owners have limited interest or ability to herd their own livestock.

3. Diversification of the labor resources of herding families into cropping agriculture and migrant labor in order to meet subsistence needs (Turner and Hiernaux, 2008).

These trends reduce the availability of herding labor, self-ownership rates in managed herds, and the prevalence of professional management. Analysis of approximately 4000 grazing itineraries within three agropastoral territories in western Niger found that grazing management plays a major role in affecting livestock distribution in relation to vegetation density, consistent with statements by local informants (Turner et al., 2005). The magnitude of labor investment into herding (herded > herd-release > free pasture) was found to increase the dispersion of livestock leading to greater presence of the livestock where palatable vegetation is more plentiful—thus, reducing the potential for localized overgrazing. Labor availability in the household, self-ownership rates of herds, and existence of managers from herding ethnicities, all household-level factors that are shown to have declined with current social trends, were both identified by local informants and found through quantitative analysis to be positively associated with the herd management mode, the age of herder, and the distance covered by the itinerary (Turner and Hiernaux, 2008). These factors in turn are positively associated with greater grazing dispersion and adjustment of grazing pressures to changing forage availabilities across dry savanna landscapes—the two major factors affecting the ecological impacts of a livestock population (Hiernaux and Turner, 1996; Hiernaux, 1998; Turner, 1998). In short, such modeling work helps trace the ecological implications of broader social changes that have occurred in the region over the past 20 years. By changing the distribution of resource access among households, the social changes listed earlier have led to grazing patterns that are more likely to lead to localized declines in vegetative productivity.

Conclusions

The dynamics of the world's savannas have and continue to be shaped by human society. Our understanding of and ability to model the ecology of savannas has increased significantly over the past twenty years. Still there are real limitations of quantitative modeling to predict or effectively simulate the complexity of the socioenvironmental change. For these loosely coupled open systems, we must work to collect longer-term social, land use, and ecological data to inform and supplement our simulation modeling efforts. Our ability to monitor both human land uses and ecological change in savannas has increased rapidly. The collection of longer-term social data is lagging behind. As demonstrated in the case above, understandings of broader social changes influencing resource management is critical for policy-relevant modeling work. Greater resources allocated to longer-term studies tying social and land-use change (Guyer et al., 2007) would greatly improve our abilities to understand society–environment relations in savanna environments.

Given the conceptual complexities of socioenvironmental relations in savanna environments, modeling of these relations will be difficult and context-dependent. Still, greater attention to issues of social complexity and causal depth; how land use is practiced; and the spatiotemporal heterogeneity of ecological response will lead to promising lines of inquiry and modeling efforts. This should help us build spatially explicit models that are informed by understandings of social change; trace the implications of such change on the distribution of productive resources; characterize how changing resource access shapes households' land-use practices and explicitly ties ecological response to how land uses are performed. Given the interlinked futures of the savanna biomes and human populations, expansions of such efforts are much needed to understand the change we are witnessing and to develop conservation and development solutions.

References

Andrew, M. H. 1988. Grazing impact in relation to livestock watering points. *Trends in Ecology and Evolution* 3, 336–340.

Augustine, D. 2003. Long-term, livestock-mediated redistribution of nitrogen and phosphorus in an East African savanna. *Journal of Applied Ecology* 40, 137–149.

Bailey, D. W., J. E. Gross, E. A. Laca, et al. 1996. Mechanisms that result in larger herbivore grazing distribution patterns. *Journal of Range Management* 49, 386–400.

Baker, L. E. and M. T. Hoffman. 2006. Managing variability: Herding strategies in communal rangelands of semiarid Namaqualand, South Africa. *Human Ecology* 34, 765–784.

Bassett, T. J. and K. B. Zuel. 2000. Environmental discourses and the Ivorian Savanna. *Annals of the Association of American Geographers* 90, 67–95.

Belsky, A. J. 1994. Influences of trees on savanna productivity: Tests of shade, nutrients and tree–grass competition. *Ecology* 75, 922–932.

Berjak, S. G. and J. W. Hearne. 2002. An improved cellular automaton model for simulating fire in a spatially heterogeneous savanna system. *Ecological Modeling* 148, 133–151.

Bird, M. I. and J. A. Cali 1998. A million year record of fire in sub-Saharan Africa. *Nature* 394, 767–769.

Burnsilver, S., R. B. Boone, and K. A. Galvin 2003. Linking pastoralists to a heterogeneous landscape: The case of four Maasai group ranches in Kajiado District, Kenya. In *People and the Environment: Approaches for Linking Household and Community Surveys to Remote Sensing and GIS*, eds. J. Fox, C. Mishra, R. R. Rindfuss, and S. J. Walsh. Kluwer Academic Publishers, Boston, pp. 173–199.

Butt, B., A. Shortridge, and A. WinklerPrins 2009. Pastoral herd management, drought coping strategies, and cattle mobility in southern Kenya. *Annals of the Association of American Geographers* 99, 309–334.

Coppolillo, P. B. 2000. The landscape ecology of pastoral herding: Spatial analysis of land use and livestock production in East Africa. *Human Ecology* 28, 527–560.

Coughenour, M. B. 1991. Spatial components of plant–herbivore interactions in pastoral, ranching, and native ungulate ecosystems. *Journal of Range Management* 44, 530–542.

Coughenour, M. B. 1993. *Savanna—A Spatial Ecosystem Model. Model Description and User Guide*. NREL, Colorado State University, Fort Collins, CO.

de Boer, W. F. and H. H. T. Prins 1989. Decisions of cattle herdsmen in Burkina Faso and optimal foraging models. *Human Ecology* 17, 445–464.

Detling, J. K. 1989. Grassland and savannas: Regulation of energy flow and nutrient cycling by herbivores. In *Concepts of Ecosystem Ecology*, eds. L. R. Pomeroy and J. J. Alberts. Springer-Verlag, New York, pp. 131–148.

Ensminger, J. 1992. *Making a Market: The Institutional Transformation of an African Society*. Cambridge University Press, Cambridge.

Eva, H. and E. Lambin 1998. Burnt area mapping in Central Africa using ATSR data. *International Journal of Remote Sensing* 19, 473–497.

Fairhead, J. and M. Leach 1996. *Misreading the African Landscape*. Cambridge University Press, Cambridge.

Favier, C., J. Chave, A. Fabing, D. Schwartz, and M. A. Dubois 2004. Modelling forest–savanna mosaic dynamics in man-influenced environments: Effects of fire, climate and soil heterogeneity. *Ecological Modeling* 171, 85–102.

Galvin, K. A., P. K. Thorton, J. R. de Pinho, J. Sunderland, and R. B. Boone 2006. Integrated modeling and its potential for resolving conflicts between conservation and people in rangelands of East Africa. *Human Ecology* 34, 155–183.

Geist, H. J., W. McConnell, E. Lambin, E. Moran, D. Alves, and T. Rudel 2006. Causes and trajectories of land-use/cover change. In *Land-Use and Land-Cover Change: Local Processes and Global Impacts*, eds. E. Lambin, and H. J. Geist. Springer, New York, pp. 41–70.

Georgiadis, N. J. 1989. Microhabitat variation in an African savanna: Effects of woody cover and herbivores in Kenya. *Journal of Tropical Ecology* 5, 93–108.

Gross, J. E., R. R. J. McAllister, N. Abel, D. M. Stafford Smith, and Y. Mauru 2006. Australian rangelands as complex adaptive systems: A conceptual model and preliminary results. *Environmental Modelling & Software* 21, 1264–1272.

Guerin, H., C. Sall, D. Friot, B. Ahokpe, and A. Ndoye 1986. Ébauche d'une méthodologie de diagnostic de l'alimentation des ruminants domestiques dans un système agropastoral: L'example de Thyssé-Kaymor - Sonkorong au Sénégal. *Cahiers de la Recherche et Développement* 9–10, 60–69.

Guyer, J., E. Lambin, L. Cligget, et al. 2007. Temporal heterogeneity in the study of African land use. *Human Ecology* 35, 3–17.

Hiernaux, P. 1998. Effects of grazing on plant species composition and spatial distribution in rangelands of the Sahel. *Plant Ecology* 138, 191–202.

Hiernaux, P. and M. D. Turner 1996. The effect of the timing and frequency of clipping on nutrient uptake and production of Sahelian annual rangelands. *Journal of Applied Ecology* 33, 387–399.

Janssen, M. A., B. H. Walker, J. Langridge, and N. Abel. 2000. An adaptive agent model for analysing co-evolution of management and policies in a complex rangeland system. *Ecological Modelling* 131, 249–268.

Lambin, E. 1988. L'apport de la télédetection dans l'étude des systèmes agraires d'Afrique: L'exemple du Burkina Faso. *Africa* 58, 337–352.

Lambin, E. F., M. D. A. Rounsevell, and H. J. Geist 2000. Are agricultural land-use models able to predict changes in land-use intensity? *Agriculture, Ecosystems and Environment* 82, 321–331.

Laris, P. 2002. Burning the seasonal mosaic: Preventative burning strategies in the wooded savanna of southern Mali. *Human Ecology* 30, 155–186.

Liverman, D., E. F. Moran, R. R. Rindfuss, and P. C. Stern, eds. 1998. *People and Pixels: Linking Remote Sensing and Social Science.* National Academy Press, Washington, DC.

Manlay, R. J., D. Masse, J. L. Chotte, et al. 2001. Carbon, nitrogen, and phosphorus allocation in agro-ecosystems of a West African savannah. II. The soil component under semi-permanent cultivation. *Agriculture, Ecosystems and Environment* 88, 233–248.

Manson, S. M. 2005. Agent-based modeling and genetic programming for modeling land change in the Southern Yucatán Peninsular Region of Mexico. *Agriculture, Ecosystems & Environment* 111, 47–62.

Mbow, C., T. T. Nielson, and K. Rasmussen 2000. Savanna fires in east-central Senegal: Distribution patterns, resource management and perceptions. *Human Ecology* 28, 561–583.

Menault, J.-C., L. Abbadie, F. Lavenu, P. Loudjani, and A. Poudaire 1991. Biomass burning in West African savannas. In *Global Biomass Burning*, ed. J. S. Levine. MIT Press, Cambridge, MA, pp. 133–142.

Netting, R. M. 1993. *Smallholders, Householders. Farm Families and the Ecology of Intensive, Sustainable Agriculture.* Stanford University Press, Stanford.

Parker, D. C. and S. M. Manson 2003. Multi-agent systems for the simulation of land-use and land-cover change: A review. *Annals of the Association of American Geographers* 93, 314–337.

Powell, J. M., S. Fernandez-Rivera, T. O. Williams, and C. Renard, eds. 1995. *Livestock and Sustainable Nutrient Cycling in Mixed Farming Systems of Sub-Saharan Africa.* International Livestock Centre for Africa, Addis Ababa, Ethiopia.

Reenberg, A. 1999. Agricultural systems in space and time—The dynamic mosaic of land use. *Geografisk Tidsskrift/Danish Journal of Geography, Special Issue* 26, 181–190.

Reenberg, A. and C. Lund 1998. Land use and land right dynamics—Determinants for resource management options in eastern Burkina Faso. *Human Ecology* 26, 599–620.

Reid, R. S. and J. E. Ellis 1995. Impacts of pastoralists on woodlands in South Turkana, Kenya: Livestock-mediated tree recruitment. *Ecological Applications* 5, 978–992.

Sayer, A. 1992. Chapter 6: Theory and method I: Abstraction, structure and cause. *Method in Social Science: A Realist Approach.* Routledge, London, pp. 175–203.

Senft, R. L. 1989. Hierarchical foraging models: Effects of stocking and landscape composition on simulated resource use by cattle. *Ecological Modeling* 46, 283–303.

Serneels, S. and E. F. Lambin 2001. Proximate causes of land-use change in Narok district, Kenya: A spatial statistical model. *Agriculture Ecosystems & Environment* 85, 65–81.

Swift, M. J., P. G. H. Frost, B. M. Campbell, J. C. Hatton, and K. B. Wilson 1989. Nitrogen cycling in farming systems derived from savanna: Perspectives and challenges. In *Ecology of Arid Lands*, eds. M. Clarholm and L. Bergström. Kluwer Academic Publishers, Dordrecht, pp. 63–76.

Taylor, P. J. 2005. *Unruly Complexity: Ecology, Interpretation, Engagement.* University of Chicago Press, Chicago.

Thornton, P. K., S. B. Burnsilver, R. B. Boone, and K. A. Galvin 2006. Modelling the impacts of group ranch subdivision on agro-pastoral households in Kajiado, Kenya. *Agricultural Systems* 87, 331–356.

Thornton, P. K., K. A. Galvin, and R. B. Boone 2003. An agro-pastoral household model for the rangelands of East Africa. *Agricultural Systems* 76, 601–622.

Tiffen, M., M. Mortimore, and F. Gichuki 1994. *More People: Less Erosion. Environmental Recovery in Kenya.* John Wiley, New York.

Tolsma, D. J., W. H. O. Ernst, and R. A. Verwey 1987. Nutrients in soil and vegetation around two artificial waterpoints in eastern Botswana. *Journal of Applied Ecology* 24, 991–1000.

Turner, M. D. 1998. Long-term effects of daily grazing orbits on nutrient availability in Sahelian West Africa: 1. Gradients in the chemical composition of rangeland soils and vegetation. *Journal of Biogeography* 25, 669–682.

Turner, M. D. and P. Hiernaux 2002. The use of herders' accounts to map livestock activities across agropastoral landscapes in semi-Arid Africa. *Landscape Ecology* 17, 367–385.

Turner, M. D. and P. Hiernaux 2008. Changing access to labor, pastures, and knowledge: The extensification of grazing management in Sudano-Sahelian West Africa. *Human Ecology* 26, 59–80.

Turner, M. D., P. Hiernaux, and E. Schlecht 2005. The distribution of grazing pressure in relation to vegetation resources in semi-arid West Africa: The role of herding. *Ecosystems* 8, 668–681.

Turner, B. L., G. Hyden, and R. Kates, eds. 1993. *Population Growth and Agricultural Change in Africa.* University Press of Florida, Gainesville.

Turner, B. L., E. F. Lambin, and A. Reenberg 2007. The emergence of land change science for global environmental change and sustainability. *Progress of the National Academy of Sciences* 104, 20666–20671.

Walker, B. H. and M. A. Janssen 2002. Rangelands, pastoralists and governments: Interlinked systems of people and nature. *Philosophical Transactions: Biological Sciences* 357, 719–725.

Weltzin, J. F. and M. B. Coughenour 1990. Savanna tree influence on understory vegetation and soil nutrients in northwestern Kenya. *Journal of Vegetation Science* 1, 325–334.

Wilson, K. B. 1989. Trees in fields in southern Zimbabwe. *Journal of Southern African Studies* 15, 369–383.

Section VIII

Synthesis

27

Current Approaches to Measurement, Remote Sensing, and Modeling in Savannas: A Synthesis

Michael J. Hill and Niall P. Hanan

CONTENTS

Introduction

This book provides a comprehensive review of the current state of knowledge and methodology associated with measurement and modeling of ecosystem function in savanna systems. This chapter develops a broad synthesis that draws the measurement, remote sensing and modeling strands together, and summarizes the key issues and deficiencies in data and analysis identified in the individual chapters. As a preamble to this discussion, it is important to recognize the pivotal role that analysis of National Oceanic and Atmospheric Administration Advanced Very High Resolution Radiometer (NOAA AVHRR) Normalised Difference Vegetation Index (NDVI) time series played in our current understanding of the seasonal and interannual dynamics of the West-African savanna. The Sahelian region of West Africa was the crucible wherein the appreciation of the importance of spectral information from remote sensing in understanding land and vegetation dynamics was born. Therefore the first part of this chapter provides a concise review of this research legacy. This legacy research links directly with much of the global scale modeling discussed in Section 6, and provides a background context for the current impetus to retrieve quantitative land surface properties from the new generation of sensors.

Following this review, a summation of the chapters, section by section, is provided along with identification of key issues raised and potential connections with data, methods, and concepts highlighted in other chapters and sections. Finally, we outline a set of issues that should be addressed in order to achieve a comprehensive measurement, modeling, and decision support approach for savannas.

Legacy Research: Coarse Resolution, Time-Series Remote Sensing, and Savannas

Savannas are intimately connected with the development of land systems remote sensing. Much of the excitement about remote sensing of land surface dynamics arose from initial analysis of vegetation change (Tucker et al., 1986) and monitoring of net primary productivity (Prince and Astle, 1986; Prince and Tucker, 1986; Tucker and Sellers, 1986; Prince, 1991a) in Sahelian Africa. This research focused on the surface dynamics evident from time series of NDVI from the NOAA AVHRR sensor. This led to the application of regression analysis to estimate biomass (e.g., Diallo et al., 1991; Prince, 1991a; Wylie et al., 1991), and examination of relationships with perceived desertification and climatic drivers (Hanan et al., 1991; Farrar et al., 1994). This also led to development of light/radiation use efficiency models (Prince, 1991b) and coupling of remote sensing time series data to more sophisticated models (e.g., Mougin et al., 1995; Lo Seen et al., 1995). The HAPEX SAHEL experiment (Gourtorbe et al., 1997) acted as a further impetus for remote sensing of the Sahelian savanna.

A long lineage of research can now be traced from this initial work including assessment of desertification (e.g., Nicholson et al., 1998), surface phenology (e.g., Potter and Klooster, 1999; Chidumayo, 2001; Heumann et al., 2007), burn scar retrieval (e.g., Roy et al., 1999), land degradation (e.g., Wessels et al., 2004), rainfall relationships (e.g., Du Plessis, 1999; Chamaille-Jammes et al., 2006), fire dynamics (Anyamba et al., 2003), and long-term examination of Sahelian dynamics (e.g., Anyamba and Tucker, 2005). The development of wide-swath radar scatterometers led to further examination of biomass and land surface properties (Frison et al., 1998, 2000; Jarlan et al., 2002, 2003; Zine et al., 2005). This broad scale analysis of land surface dynamics with AVHRR NDVI was enhanced by availability of data from the Moderate Resolution Imaging

Spectroradiometer (MODIS) and other global scale sensors, and spread to other global savannas including the cerrado (Franca and Setzer, 1998; 2001; Ferreira et al., 2003), Colombian llanos (Anaya et al., 2009), and the Australian tropical savanna and rangelands (e.g., Roderick et al., 1999; Donohue et al., 2008). A major campaign was conducted in southern Africa (SAFARI 2000—Southern African regional Science Initiative; Swap et al., 2003) that included substantial research connecting MODIS and other EOS products with ground measurements in savannas.

As analytical techniques, and application of multiscale analysis advanced, methods for retrieval of quantitative surface properties such as tree cover fraction (Hansen et al., 2003; Gill et al., 2009) and fractional cover of photosynthetic and nonphotosynthetic vegetation have been developed (Scanlon et al., 2002; Guerschman et al., 2009). In addition, parameters derived from models such as Ross-Thick-Li-Sparse Roujean (RTLSR; Roujean et al., 1992; Li et al., 1995) applied to MODIS reflectance data (Schaaf et al., 2002), and Rahamn-Pinty-Verstraete (RPV; Rahman et al., 1993; Pinty et al., 2002) and GO/SGO models (Simple Geometric Optical) applied to Muti-angle Imaging SpectroRadiometer reflectance data (MISR; Chopping et al., 2006) provide simplified angular absorption and reflectance information in the Bidirectional Reflectance Distribution Function (BRDF) that could be used to retrieve surface properties in savannas (Armston et al., 2007; Liesenberg et al., 2007) and shrublands (Chopping et al., 2008; Su et al., 2009). This BRDF information may also be used to calculate a clumping index that could help to provide better regional to global scale estimates of canopy cover in savannas, although variation with tree structure, canopy shape, tree density, Leaf Area Index (LAI) and other factors can be complex (Chen et al., 2005).

This more than 25-year history of characterizing the dynamics of savanna systems from broad scale, coarse resolution time series from AVHRR and MODIS has lead to wider understanding of the coupling of climate and land surface behavior (see Williams, Chapter 17), carbon dynamics (see Barrett, Chapter 18), grassland interactions with herbivory and fire (see Carter et al., Chapter 21), and the maturation of fire hotspot and burn scar mapping (see Roy et al., Chapter 12). It has also provided the means to drive global and regional models of vegetation and carbon dynamics (see Potter, Chapter 22). This history has to some extent molded the current remote sensing applications and models that are applied to savannas.

Summation of the Chapters

Biogeographical and Ecological Perspectives

In this first section, the chapters set out to give a concise overview of the geography of savannas, some of their broad characteristics and major disturbances, and the state of current ecological knowledge and understanding of tree–grass systems. In Chapter 1, Hill et al., constrain the treatment to the major large-scale continental concentrations of tropical and subtropical savannas in Africa, Australia, and South America. This does underrepresent the diversity and breadth of distribution of tree–grass systems (specifically in North America and Asia), but was necessitated by the global scale of the assessment and the lack of fully coherent and comprehensive land cover, or ecoregion definition of tree–grass systems, which are inherently difficult to classify due to their heterogeneity. The analysis highlights many issues that are expanded upon in subsequent chapters in the book:

a. The huge area involved (e.g., global scale modeling Chapter 22, Potter et al.).

b. The diversity of tree cover (e.g., remote sensing and modeling tree biomass and demography, Chapters 8 through 11, 13 through 16).

c. The variability in climate and seasonality of growth (e.g., the different flux sites in Section 2).

d. The significance of the savanna ecosystems for biodiversity, wildland fires (e.g., remote sensing, Chapter 12, and modeling, Chapter 16, fire in savannas), cattle and other livestock production (e.g., grazing influence on fractional cover, Chapter 10, Asner et al., fire, Chapter 15, Liedloff and Cook, and plant demography, Chapter 16, Jeltsch et al.), and human populations (Chapter 25, Galvin et al., and Chapter 26, Turner).

It also highlights the contrast between continents, and the diversity and gradients at regional scale within continents, for example, for human population densities, influence of seasonal flooding, and physiognomy of vegetation. The single most significant impression left by this analysis is the extraordinary diversity and complexity of savannas. This is illustrated by a selection of photos from three continents showing some of the diversity in physiognomy across the tropical savannas (Figure 27.1).

In Chapter 2, Hanan and Lehmann provided a detailed review of tree–grass interactions. They discussed the different conceptual approaches that have led to models: namely, the Walter hypothesis of differential rooting depth and access to water favoring trees; and bottleneck approaches that assume trees will out-compete grasses in the absence of disturbances like fire and herbivory; and a balanced concept, where intertree competition can limit tree density. Crucially, they concluded that none of these conceptual models, on their own, can explain tree density in tropical savannas, and that a merging of conceptual models is needed. This reinforces the general impression gained from Chapter 1, that the enormous diversity of tree–grass patterns, climate and seasonal responses, and timing, spatial distribution and intensity of fire, herbivory and human disturbances probably requires

FIGURE 27.1
(See color insert following page 320.) Photos showing variation in global savannas. (a) Australia: (1) sorghum woodland (photo: Marks); (2) *Aristida* spp. woodland, Larimah sandy site, NATT; (3) *Triodia* spp. woodland, Larimah loam site, NATT; and (4) Dichanthium woodland, Larimah clay site, NATT (photos: Hill).

detailed understanding at multiple, hierarchical levels of process and explicitly scale-dependent and coupled models. It also links to the detailed treatment of environmental factors, spatially explicit processes and upscaling of grid models covered in Section 4. Hanan and Lehmann went on to summarize the current state of knowledge concerning factors controlling savanna vegetation dynamics (Chapter 2, Figure 27.2), and provide a qualitative analysis of savanna dynamics with varying climate, fire, and large herbivore disturbance patterns (Chapter 2, Table 27.1).

Carbon, Water, and Trace Gas Fluxes

The chapters in this section covered carbon, water, and trace gas fluxes for savannas on four continents: northern Australia, the Brazilian cerrado, southern Africa (Kruger National Park), and Texas and Californian oak savannas. Vourlitis and da Rocha (Chapter 5), Hutley and Beringer (Chapter 3), Litvak et al. (Chapter 6), and Baldocchi et al. (Chapter 7) provide summary data, including biomass and NEE for their respective systems, whilst a summary of the southern-African site has been published by Archibald et al. (2009). The carbon stocks and fluxes for the five sites are compared in

(b)

FIGURE 27.1 Continued
(See color insert following page 320.) (b) South America: (1) cerrado (photo: Bustamante);
(2) Colombian llanos—gallery forest (photo: Devia); (3) Colombian llanos—rolling savanna
(photo: Etter); and (4) Colombian llanos—grassland with sparse shrubs (photo: Repizzo).

Table 27.1. These savannas are highly variable due to a number of factors. At
these sites biomass carbon is highest for the woodlands and lower where
more open physiognomy prevails. Whilst wet season maximum NEE can be
as high as 4 gC m^{-2} d^{-1}, annual average NEE is quite small—1 gC m^{-2} d^{-1} or
less and the interannual variability is very high (e.g., Archibald et al., 2009).
The variability is emphasized by the standard deviations for the NEE values
in Table 27.1. At all of the tropical sites, the systems exhibit pulsed responses
to rainfall, hence intra-annual and intraseasonal variation is also high. In
Chapter 4, Hanan et al. develop an elegant relationship between fPAR, soil
water potential, and maximum daytime NEE. In Chapter 5, Vourlitis and da
Rocha provide revealing relationships between range, maximum and minimum NEE, and annual rainfall and length of dry season for a gradient of
rainforest, cerradão and cerrado sites.

These savannas all harbor substantial stocks of carbon in biomass and soil.
In general, they are weak sinks, due to a combination of climatic factors
including: the length of the dry seasons; the long-term cycles of good and
poor monsoons (e.g., see the interannual variations in modeled NEE in
Northern Australia, Barrett, Chapter 18); the low fertility of the soils and low
productivity of native grasslands (see Bustamante et al., Chapter 19;
Etter et al., Chapter 20); the range in MAP (see Hill et al., Chapter 1) and its

FIGURE 27.1 Continued
(See color insert following page 320.) (c) Africa: (1) seasonal settlement in Sahelian savanna near Hombori, Mali (photo: Prihodko); (2) south Sahelian savanna at Lakamane, Mali (photo: Hanan); (3) agriculturally modified Sudanian savanna near Segou, Mali (photo: Prihodko); (4) Sudanian savanna, Foret Classe Faya, Mali (photo: Hanan); (5) Sudanian savanna, Monte Mandinge, Mali (photo: Hanan); and (6) knobthorn savanna, Kruger Park, South Africa (photo: Prihodko).

interaction with temperature influencing C turnover; and the diversity of woody/shrubby vegetation, in terms of floristics, phenology, ecophysiology, and density (e.g., Vourlitis et al., Chapter 5). The savannas also have very substantial and delicate water balances intimately interlinked with climate, physiognomy, and edaphology (e.g., Hanan et al., Chapter 4; Litvak et al., Chapter 6; Baldocchi et al., Chapter 7). These linkages beg for more caution and understanding of the consequences, before wholesale conversion to alternate land cover and land use occurs.

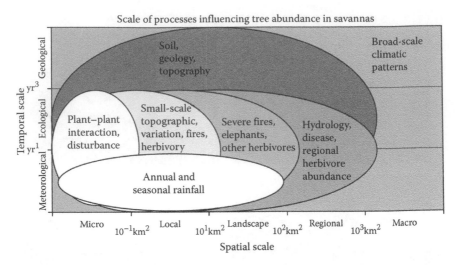

Scale of processes influencing tree abundance in savannas

FIGURE 27.2

A spatio-temporal framework for processes which influence savanna tree density through their effects on dispersal, germination, recruitment and mortality of savanna trees. (Reprinted from Gillson, L. 2004. *Landcape Ecology* 19, 883–894. With permission from Springer.)

The cerrado savanna is being cleared at a rapid rate (Bustamante et al., Chapter 19). The Guinean savanna in West Africa has a similar rainfall regime to the cerrado (Hill et al., Chapter 1), and potential for greater clearing and fragmentation must be substantial. Australia provides a case study of potential impacts arising from conversion of savanna (a temperate savanna) to agriculture, without clear understanding of the hydrological and nutrient cycling consequences; this resulted in substantial land degradation due to dryland salinity (e.g., Lambers, 2003) and soil acidification from fertilizer applications to exotic nitrogen-fixing pasture plants (Bolan et al., 1991). With tropical savanna systems potentially subject to greatly increased clearing in the twenty-first century (e.g., Tilman et al., 2001), in systems subject to high intensity, seasonal rainfall patterns, and highly acid soils (e.g., see Hill et al., Chapter 1), there is an urgent need for more flux measurements, more hydrological studies (see Gordon et al., 2005 for human modification of global water vapor flow), and more coupling of these data with models that can simulate long-term incremental changes that might ensue from removal of the natural savanna vegetation.

In Chapters 19 and 20, Bustamante and Ferreira, and Etter et al., indicate that grasslands in the cerrado and llanos respectively, are not significant carbon sinks due to low fertility. Analysis in Australian savannas has shown that whilst the soil is a major carbon stock, the main flux potential is for carbon loss with grassland degradation (Hill et al., 2006). However, Bustamante et al. (Chapter 19) also report variable outcomes for carbon fluxes after conversion of savanna, depending upon tillage techniques in cropping systems, and management of exotic pastures. This suggests that nutrient cycling is

TABLE 27.1

Comparison of Biomass, Soil Carbon, and NEE for the Savanna Ecosystems Described in Section 2

Parameter	Howard Springs[a] (Eucalypt.)	Cerrado[b] (campo sujo, sensu stricto, and campo limpo)	Cerradão[b] (Transitional Forests and Woodlands)	Edwards Plateau[c] (Oak-Juniper)	California[d] (Blue Oak)	South Africa[e] (Skukuza–Acacia and Combretum)
Biomass AG t ha^{-1}	34	0.8–16	46	114 (dense woodland site)	37	~17.1[f]
Biomass BG t ha^{-1}	17	7.6–33	23	NA	12	NA
Soil OC t ha^{-1}	140	17–67 (50 cm)	26–61 (50 cm)	NA	NA	0.81 (*Acacia*); 0.66 (*Combretum*) mg kg^{-1}
Annual average NEE gC m^{-2} d^{-1}	–1.17	–0.19 ±0.42	0.19 ± 0.24	–0.82	–0.25 ±0.12	0.069 ± 0.365[g]
Wet season maximum NEE gC m^{-2} d^{-1}	~ –3.0	~ –2.0	~ 0.0	~ –4 (in pulses)	~ –3.0	~ –2.7

[a] Hutley et al., Chapter 3.
[b] Vourlitis et al., Chapter 5.
[c] Litvak et al., Chapter 6.
[d] Baldocchi et al., Chapter 7.
[e] Hanan et al., Chapter 4.
[f] Hanan unpublished.
[g] Archibald et al., 2009 — calculated by the authors for a 365 day year from 5 years of annual values.

also a key area for further research. When Barrett (2001) established a continental database of soil and biomass carbon measurements to support the development of his VAST model, it provided a stark indication of how poor the field data coverage was for Australia, a continent with both huge undisturbed tropical savannas, and huge areas of agriculturalized temperate savanna.

Remote Sensing

The list of key surface properties potentially retrievable using remote sensing data and methods documented in this volume is impressive (Table 27.2), however calibration, validation, spatial and temporal coverage, and emergence of agreed methodology continue to be limitations. The potential for quantitative retrievals places particular emphasis on planned future satellite sensor combinations (the DESDYNI and HyspIRI Decadal Survey missions; NRC, 2007) which will supply global coverage with LiDAR, L band radar, monthly 60 m resolution hyperspectral imagery, and a high frequency of multichannel thermal imagery.

In Chapter 8, Lucas et al., describe the current state of the art in synergistic application of LiDAR and Radar for height, crown structure, demography, above-ground biomass, and component biomass of woody vegetation in savannas. This review is supplemented by examples of canopy and woody fraction retrieval from LiDAR in African savannas (Asner et al., Chapter 10). This highly quantitative approach is contrasted with more applied approaches to woody cover using Landsat TM with substantial field calibration and validation (Danaher et al., Chapter 9), and using a regression tree approach with Landsat TM and JERS-1 imagery validated with airborne LiDAR (Bucini et al., Chapter 11).

In Chapter 12, Roy et al., indicate that there are still issues with accuracy and consistency of fire products including incomplete active fire retrievals due to obscuration, overpass times, pixel size and detector sensitivity, and considerable differences in spatial and temporal properties of burn scar retrievals. The burn scars, in particular, are crucial elements in understanding and modeling dynamics of savannas. The current and future geostationary sensor systems such as (GOES-R) Geostationary Operational Environmental Satellite-R Series are potentially very valuable in so far as they provide improved diurnal coverage and detection of active fires.

These chapters make it clear that multiple data sources are essential for better quantitative retrievals. Multispectral data at various resolutions can be used to retrieve woody cover and fractional cover of NPV, PV and bare ground, and time series or unmixing methods can be used to add extra dimensionality such that subpixel cover proportions can be estimated (e.g., Asner et al., 2005), but this is the limit of quantitative retrievals from multispectral optical data. Fusion of multispectral data across scales (e.g., Landsat and MODIS), fusion of active and optical data products (e.g., LiDAR

TABLE 27.2

Key Savanna Vegetation Attributes; Available or Potential Retrieval Methods

Attribute	Spatial Extent	Method	Chapter
Vegetation height	Airborne transect (LiDAR) and satellite (InSAR)	Mix of LiDAR (points for calibration) and Radar (mapping)	8, 10
Stem location, number, and density	Airborne transect (LiDAR); Kruger Park—Landsat/JERS1	Tree crown delineation, counting, and indices based on penetration; regression tree on backscatter and reflectance (green band)	8
Crown and canopy properties	Airborne transect (LiDAR)	Crown delineation, LiDAR reflection; spatial indices	8
Full structural characterization	Transect/regional	Coupled LiDAR/optical/radar data; radar backscatter models	8
Above-ground biomasss	Airborne transect	Allometric equations using LiDAR derived height	8, 10
Component biomass	Airborne swaths (LiDAR, SAR, hyperspectral)	Semiempirical backscatter models; LiDAR allometrics with hyperspectral canopy description	8, 10
Foliage projected cover (FPC)	Regional–Australian States (Landsat)	Time series statistical analysis using thresholds and standard errors to decouple woody and herbaceous signals	9
Woody cover	Regional–Australian States (Landsat); Africa	Time series of FPC and standard error of fitted line; regression tree with landsat and radar	9, 11
Fractional cover of PV, NPV, and B	Site to continental (airborne or wide swath global imaging)	Spectral unmixing, spatial scaling, and hyperspectral to multispectral band association	10

continued

TABLE 27.2 (continued)

Key Savanna Vegetation Attributes; Available or Potential Retrieval Methods

Attribute	Spatial Extent	Method	Chapter
Canopy chemistry (Water, N, lignin, cellulose, tannins, minerals)	10 km (Airborne or narrow swath satellite)	Imaging spectroscopy with spectral transformations, multiple regressions, direct/ indirect estimation using absorption features and key wavelengths	10
Fire hotspots	Global	Thresholds for middle infrared brightness temperature plus contextual (neighborhood) analysis	12
Burn scars	Global, continental, and regional	Threshold changes in reflectance; hand digitized from NDVI; increased reflectance in short wave infrared; and angular properties	12
Fire frequency and seasonality	Global	From temporal distributions of burn scars and hot spots	12
Fuel load and moisture	Airborne transect, narrow satellite swath	From NPV biomass and vegetation moisture content from imaging spectroscopy plus models	10

tree structure and hyperspectral canopy chemistry), and incorporation of multiple remote sensing data types into assimilation frameworks promises substantial progress. To date, there are only a few examples of the data assimilation approach in savannas (e.g., Barrett, Chapter 19; Renzullo et al., 2008), but this is likely to be increasingly valuable in future.

Landscape Modeling

The chapters in the section on "Patch- to Landscape-Scale Processes" raise a number of very important issues regarding scale dependence, environmental and spatial processes, and upscaling from the patch to landscape to region (Table 27.3). The two-layer, spatially diverse arrangements of trees, shrubs, palms, and grassy understorey in savanna systems introduce a fundamental requirement for understanding the way spatial and nonspatial processes are influenced by demographics and arrangement (Table 27.3), and what the hierarchies of importance, and the thresholds of significance are for upscaling to landscapes and regions in order to capture aggregate ecosystem outcomes. A summary of the relationship between spatial and temporal scales and processes which affect tree demography in savannas was developed by Gillson (2004; Figure 27.2). This simple schema is useful in visualizing the possible connections between landscape and regional scale, wherein drivers of change might influence whole ecosystems as a result of processes that actually operate at relatively small spatial scale, but incrementally change the ecosystem over a human lifetime (e.g., Archer et al., 2001).

This section provides a very complete overview of the patch to landscape issues. In Chapter 13, Van Langevelde et al., describe the range of explanations for tree–grass coexistence (Table 27.3) and whether or not these explanations have been incorporated into models. They then go on to concentrate on environmental processes and draw attention to some key areas of deficiency in understanding and model integration, namely, nutrient cycling, tree demography, and fundamental understanding of plant traits. Yet, there is a very well developed model, Savanna-CENTURY (Parton et al., Chapter 24) that explicitly simulates pool dynamics of C, N, P, and S on a site/point basis. The point about plant traits is especially interesting, since savannas such as the cerrado are known hotspots for plant biodiversity (see Hill et al., Chapter 1), and deciduous behavior is significant to varying degrees across global savannas. For example, in West Africa, trees can be classified into two types of survival strategy: deep rooting to access soil water or scleromorphism to limit water loss; or avoidance through shedding of leaves in the dry season (de Bie et al., 1998). These strategies represent different ecophysiological paths, but also manifest different canopy phenologies. This diversity in plant strategies is also manifested in the cerrado (Damascos et al., 2005) and the Australian tropical savanna (Williams et al., 1995), but is only one aspect of a range of other plant traits that may significantly differentiate plant functional behaviors in these ecosystems (Damascos et al., 2005).

TABLE 27.3

Key Elements for Modeling of Landscape Scale Savanna Processes

Domain	Issues	Features
Environmental (van Langevelde, Chapter 13)	Coexistence explanations	Niche separation; disturbance; root–shoot partitioning; facilitation
	Essential environmental components	Water, fire, herbivory, and soil nutrients
	Issues not yet well addressed	Role of nutrient availability Variation in plant traits Demography of trees
Spatial (Meyer et al., Chapter 14)	Importance of spatial properties	Density-dependent process versus density-independent environmental process
	Patch formation and dynamics	Rain and fire driven
	Spatial analysis and grid models	Processes linked to clumped versus scattered (random) trees
	Spatial modeling	Moment equations; matrix models; cellular automata; grid models (multicell interactions, e.g., agent-based models) Focus on processes with strong spatial structure
Fire (Liedloff and Cook, Chapter 15)	Interaction between rainfall and disturbance	Fire effects on tree demography and biomass
	Contrasts between Australian and African savannas	Rainfall variability and length of dry season
Upscaling ecological process (Jeltsch et al., Chapter 16)	Application of approaches to spatial scaling from local to large scales	State and transition models— transition matrices to translate small-scale process changes to regional scale outcomes based on thresholds
	Horizontal processes	Approach can accommodate interaction among spatial units at multiple scales

The tree–grass systems that we regard as savannas exhibit a wide range of spatial arrangements of trees (Figure 27.3). These include open woodland (e.g., northern Australia, Kruger), grassland with gallery forests along rivers and streams (e.g., Colombian llanos), forest–savanna mosaics (e.g., Guinean savanna), woodland–grassland mosaics (e.g., oak woodlands of United States), grassland with woody vegetation of mixed physiognomy (e.g., cerrado sensu stricto), grassland with sparse trees and shrubs (e.g., parts of most savannas in Australia, South America, and Africa), and "biomes" containing all or

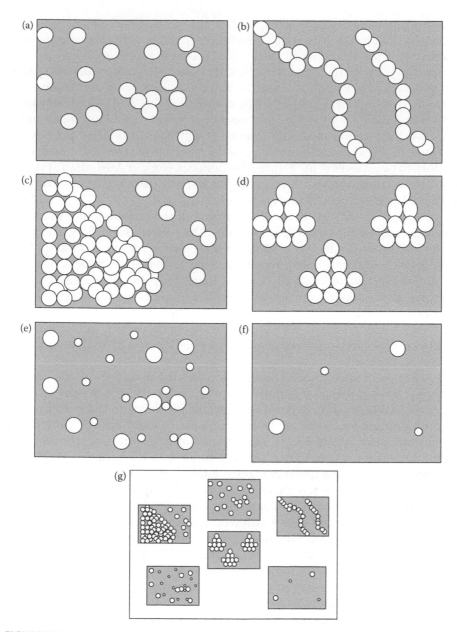

FIGURE 27.3
Spatial vegetation patterns in "savanna" systems. (a) open woodland; (b) grassland with gallery forests along rivers and streams; (c) forest–savanna mosaics; (d) woodland–grassland mosaics; (e) grassland with woody vegetation of mixed physiognomy; (f) grassland with sparse trees and shrubs; and (g) "biomes" containing all or many of the above landscapes. Grassland (shaded); trees (large circles); and shrubs/palms (small circles).

many of the above landscapes (Figure 27.3; most savannas at global land cover scale). In Chapter 14, Meyer et al., focus on spatially explicit processes and modeling. They draw a key distinction between processes where spatial properties are important and those that are essentially independent of spatial arrangement. They then discuss the importance of patch formation, and describe methods for spatially explicit modeling (Table 27.3). A key element of this is an inventory of spatial processes in savannas and how they can be addressed by spatial modeling (Meyer et al., Table 27.2). They provide an example of a grid model wherein spatial properties of trees and grasses are important (see Meyer et al., Figure 27.4). Seminal work in Australian arid systems has shown the importance of the patch structure of the woody and herbaceous layers for ecosystem function (Ludwig and Tongway, 1995; Ludwig et al., 2000). This has led to the development of the leakiness index concept (the extent to which landscapes are subject to surface water flow, and poor water infiltration; Ludwig et al., 2002), and exploration of its potential application across wide areas using remote sensing to define patchiness (Ludwig et al., 2006; Ludwig et al., 2007). Clearly, there is a point of intersection here between grid modeling, and spatially explicit quantitative surface properties that can be derived from remote sensing. An additional significant model, SAVANNA (Coughenour, 1992) is partially spatially explicit in that it tracks patches of trees and grass in the landscape (see Parton et al. in Chapter 23 for a brief description).

In Chapter 15, the FLAMES model simulates the dynamics of one hectare stands of trees and grass (Liedloff and Cook, Chapter, 15). In the case studies, trees die and biomass is reduced by high fire frequency in the wetter north, and by drought in the dryer areas. The authors highlight the differences in climate, particularly length of dry season, in contrasting the stability of the tree–grass system in northern Australia with the behavior of African savannas. This is not a spatially explicit model but does deal with demography, that is, density of woody and herbaceous components. This approach could be spatially explicit if operated on a 1 ha grid, but this would entail interaction between grid cells which would increase computation loads substantially. It is also possible to imagine how tree structure and biomass (from combined LidAR and Radar remote sensing), and herbaceous layer fractional cover and fuel properties (from a high frequency hyperspectral sensor such as HyspIRI) could be used to set initial conditions and dynamic burn characteristics (e.g., Andersen et al., 2005).

When the key environmental and spatial processes are captured in a spatially explicit modeling framework, particularly where there is dynamic interaction between spatial units, then the task becomes very complex, and computationally demanding. Hence there is a need for hierarchical approaches that take account of the target scale, and only apply the level of parameter, spatial grain, and process detail that is necessary. For this to work, an explicit understanding of the scale by process interaction is needed, and a method of upscaling is also needed. In Chapter 16, Jeltsch et al. provide a good example of a grid model (as introduced by Meyer et al., Chapter 14) that combines the interaction of climate, herbivory, and water and vegetation dynamics. However, importantly, they go on to discuss a strategy—involving state and transition modeling—for upscaling in these systems where processes change substantially between scales (Jeltsch et al., Figures 27.4 and 27.5). State and transition models (Westoby et al., 1989) have been widely used to capture vegetation dynamics (see Briske et al., 2005) in response to drivers of change such as climate, and disturbances such as grazing (e.g., Bestelmeyer et al., 2003) and clearing (e.g., Yates and Hobbs, 1997). They have also been used to assess potential carbon outcomes in rangelands subject to potential degradation from the interaction of drought and herbivory (e.g., Hill et al., 2005, 2006). Jeltsch et al. conduct simulations at $5 \times 5\ m^2$ scale within $1\ km^2$ grids, then aggregate these up to an entire area. By defining a series of critical process scales for savanna landscapes—and these may vary between systems based on topographic, edaphic, and disturbance patterns— this approach could enable an effective hierarchical, scale-dependent simulation of the complex dynamic processes in savanna systems. Depending upon the scale, these could be supported by surface description from remote sensing across a range of types, spectral coverage, spatial resolutions, and temporal frequencies. They may then provide a great deal of more detailed surface dynamics into global scale models that estimate gross exchanges between the surface and the atmosphere and inform future climate predictions.

Regional and Continental Scale Modeling and Applications

In the development of this book, the editors conceived of Section 5, regional remote sensing applications, and Section 6, global models, as relatively separate topics. However, it became clear from examining the work, that these chapters represent a continuum in terms of model applications, but just a couple of categories in terms of the use of remote sensing. The modeling continuum includes some of the models addressed in Section 4 on patch to landscape processes, and is illustrated in Table 27.4. This ranges from site and stand-based models such as Savanna-CENTURY and FLAMES, and grid models such as that of Jeltsch et al. (that do not use remote sensing), to regional "grid" approaches that focus on land cover/use change detection (based on moderate resolution remote sensing), to regional dynamical models, dependent on remote sensing canopy description, and burn scar mapping, to a range of "global" models such as SiB, LPJ, and CASA that represent the surface as a photosynthetic gradient, in some cases segmented by plant functional types or land cover types (Table 27.4). All of these models, except the land cover change grids utilize climate as a major driver. The use of remote sensing can be categorized into two types: (1) moderate resolution data such as from Landsat TM, sometimes supplemented with coarse resolution data, such as from MODIS, is used to define land cover and land cover change, and changes in land system processes are inferred from *in situ*

TABLE 27.4

Local/Regional/Global Approaches to Modeling/Estimating Carbon and Vegetation Dynamics in Savannas Covered in this Book

Region and Scale	Model	Properties	Remote Sensing Data	Product	Issues/Feature	Chapter
Site (Global)	SAVANNA-CENTURY	Climate, soil texture, soil nutrient, and plant fiber-driven	None	Cycling of C, N, P, and S in pools; vegetation dynamics	Addresses climate-nutrient interaction explicitly	23
Tree–grass stands (Australia)	FLAMES	Dynamic rainfall driven moderated by soils, vegetation, fire weather, and regime	None	Tree basal areas; C fluxes; fuel inputs and burn components	Predicts tree survival, demographic change at ha scale	15
Landscape (Africa)	Grid-based ecohydrological model	Dynamic rainfall-driven two layer soil, and vegetation model moderated by vegetation management	None	Soil moisture; shrub/grass demographics/proportions	Incorporates spatial dynamics of water-vegetation interaction	16
Regional (Colombian llanos)	GIS (IDRISI)	Logistic regression	Landsat TM and ETM; CBERS; MOD09A1 derived burned area	Land cover change; C emissions	Predictive model–future scenarios	20
Regional (Brazilian cerrado)	None [Image analysis and literature review]	Change detection	Landsat ETM and MODIS NDVI	Land cover change	Quantitative estimates from site measurements assigned to land use classes	19

Scale	Model	Description	Data inputs	Outputs	Notes	Ref
Regional (Australian tropical savanna)	Vegetation And Soil carbon Transfer (VAST)	C pools; 22 estimated parameters; model-data fusion algorithm minimizing objective cost function	$FPAR_{canopy}$ from AVHRR NDVI; ATSR hotspots; AVHRR burned area	C fluxes; NEP; NPP and NBP	Accommodates grid, point and other data with variable and sparse coverage	18
Regional (Australian savanna woodland)	AUSSIE GRASS	Dynamic, daily climate driven moderated by fire, herbivory, and management	Tree FPC and grassland FC from NDVI; AVHRR and MODIS burned area	Pasture production; soil erosion; vegetation C and N flux	Oriented towards climate-fire-herbivory dynamics; used in prediction of drought and climate change impacts	21
Continent (Africa)	SiB	"Big leaf" photosynthesis driven by stress modifiers	AVHRR NDVI	Photosynthesis		17
Continent (Africa)	LPJ-GUESS with SPTFIRE	Forest-gap model driven by competition for light among PFTs	L3JRC and MODIS burned area	Tree, grass, litter and burned fraction, and C emissions from burning	Oriented towards exploration of large scale climate–vegetation interactions	24
Global	NASA CASA	Light use efficiency model	MODIS Land Cover and MODIS EVI	NPP	Global dynamics at biome level	22

measurements or experiments and measurement reported in the literature; and (2) coarse resolution time series of vegetation indices (from MODIS or AVHRR) are used to retrieve canopy dynamics, growth may be parameterized for different remotely sensed land cover classes, and disturbance due to fire may be incorporated from burn scar mapping (Table 27.4).

Leaving aside the simple change detection-accounting approaches, it is evident that current modeling approaches make almost no use of advanced products from the MODIS sensor, multiscale approaches that combine Landsat TM and AVHRR or MODIS data, or combinations of optical, active, and passive remote sensing data. For example, Radarsat, Advanced Land Observation Satellite Phased Array L-band Synthetic Aperture Radar (ALOS PALSAR) and ENVISAT Advanced Synthetic Aperture Radar (ASAR) are widely available with global coverage at fine resolution, and land surface temperature from MODIS and soil moisture products from Advanced Microwave Scanning Radiometer (AMSRE-EOS) are available with high temporal frequency and global coverage at coarse resolution, but these data have not yet been incorporated into regional or global scale savanna model studies. By contrast, it is significant that the MODIS burned area product and the MODIS land cover product are important in the global scale modeling, since these products add information at a level of spatial fidelity previously unavailable. In addition, major calibration and validation of products derived from moderate and coarse resolution data (see Danaher et al., Chapter 9 and Carter et al., Chapter 21) specifically to define both woody and herbaceous cover for carbon accounting and drought management of pastoral feed resources have resulted in effective applied modeling systems that deliver information into the public domain. There are recent examples of coupling more complex biogeochemical modeling to land cover change time series developed from Landsat TM data for the United States (Liu et al., 2004). However, this approach has yet to be applied to savannas except in limited circumstances: detailed, fine scale mapping of land cover change from remote sensing and aerial photography for the cerrado is currently under way (Mercedes Bustamante, pers. comm.).

Among these chapters, there is a major focus on the carbon cycle and carbon dynamics, however there are interesting contrasts and complementarities. Sparse field data with very limited range of measurements represents a major barrier to the regional—continental scale modeling of savanna dynamics. Model-data assimilation (or fusion) described by Barrett in Chapter 18, provides a way forward to improve modeling of savanna dynamics since the framework is able to accommodate diverse data types at a range of spatial scales, temporal frequencies, and discontinuities. This framework requires a model built to match available data which can be explicitly tailored to ingest and utilize multiple remote sensing data types and products, along with the climate data and ground observations. A framework for this was outlined by Barrett et al. (2005) and some initial work was completed (Renzullo et al., 2008).

The capture of complex patch processes addressed with transition probabilities and upscaled via aggregation based on some scale by process hierarchy has potential to provide to the global models inputs more representative of the landscape-level environment/spatial dynamics. In Chapter 17, Williams indicates that interpretation of climate interactions in SiB simulations would be improved with data on vegetation traits and behavior. The LPJ model which uses plant functional types (PFT) might benefit from better upscaling of nutrient, water, and phenological traits, as well as better spatial representation of PFTs and traits. It is clear that there is a pressing need to build new modeling frameworks that can flexibly accommodate more remote sensing data and products from current and proposed sensing systems, and can couple modeling and scaling approaches from different scales in order to better capture the complex dynamic processes in savannas.

Interactions between Humans and Savanna Systems

This book is oriented towards the biophysical savanna system. However, we felt it important to try to gain some insights into how potential advances in measurement, remote sensing, and modeling of savannas might aid understanding and modeling of a truly coupled human–environment savanna system. The book focuses on dynamics of tree–grass systems; the only exception to this specifically deals with two South American savannas where large scale conversion of savanna to agriculture and silviculture is currently advanced or a potential near term phenomenon (see Chapters 19 and 20). Therefore in Section 7, the chapters by Galvin and Reid, and Turner are oriented substantially toward the socioecological dynamics of pastoralism, since this takes place within natural tree–grass systems.

In Chapter 25, Galvin and Reid present a framework for understanding the interactions in a pastoral socioenvironment system. This framework emphasizes adaptive capacity as the functional core—lack of adaptive capacity is a problem. They describe how savannas face pressures from different drivers, the major change processes these trigger, and the social consequences of these changes (Chapter 25, Table 27.1). These savanna change processes represent many elements that are the target of biophysical models seeking to capture vegetation and ecosystem functional change: burning and clearing (e.g., Chapter 15, Liedloff and Cook), exclusion of humans, grazing, and burning (e.g., Chapter 16, Jeltsch et al.), intensification or extensification through resource improvement or removal of limitations (such as water access), changes to boundaries and administrative rules that lead to changes in "disturbance factors" (e.g., management of Government land), and conversion to non-savanna (e.g., Chapters 19, 20). Finally, they identify key principles for pastoralism in sustaining resilience of both human economies and savanna ecosystems (Table 27.5) which returns strongly to the core of their framework-adaptive capacity.

These principles strongly link spatial properties of the ecosystems and human behavior, with economics and systems of social support. The

TABLE 27.5

Principles for Pastoral Sustainability in Savannas (Chapter 25, Galvin and Reid)

Principles for Pastoral Sustainability	Description
Mobility	Maintenance of livestock movement and access— "modern transhumance"
Diversity	Vegetation patches (resources in time and space) and sources of income (urban-agrarian mixes)
Reserves	Saving landscapes for dry season forage, and managing water resources
Profitability	Intensification of livestock systems or subsidy or pastoralism for ecosystem husbandry
Exchange	Human and resource exchanges to hedge risk
Restoration	Holistic application of first five principles partially adapted to modernizing societies

importance of spatial diversity, spatial access, spatially explicit control and management of resources, and spatial dynamics of grazing and fire provides further impetus to the development of spatially explicit biophysical models that include some of the deficient areas identified by van Langevelde (Chapter 13), and utilize greatly improved surface parameter retrievals from remote sensing (Section 3).

In Chapter 26, Turner examines how social variables can be linked to biophysical dynamics in savannas described as "open, loosely-coupled systems." The increased understanding that is a measure of success in modeling savannas is characterized by a trade-off between capture of "the complex web of interactions" and "simplifying assumptions ... to make causal relationships tractable." The socioecological modeling approaches seeking to address this issue are summarized in Table 27.6. They represent the different disciplinary channels through which modeling approaches have developed, as well as the major differences in paradigms between biophysical and social scientists. Agent-based models seek to bridge this divide. However, Turner offers two important approaches to conceptualizing socioecological relations: land-use ecology—to know how resources are extracted as well as how much; and increased social depth of causal chains—greater measurement of how social variables influence resource management, and how managers approach the decision-making process. This improvement requires spatial explicitness and Turner describes some major limitations to this: multiple scales; entangled social networks with fuzzy boundaries; importance of social histories; and difficulties in "capturing" social variables in a quantitative way.

In essence, there is a need to place modeling efforts within a detailed *a priori* social and ecological understanding, including human motivations and transformation of aspatial, multiscale, economic, administrative, and historical frameworks for decisions and behaviors down to the target level of

TABLE 27.6

Socioecological Models (Chapter 26, Turner)

Model Type	Characteristics	Limitation
Biophysio-centric	Changes in human land used as boundary conditions	Humans as disturbance
Econo-centric	Savanna ecology as a boundary condition	Static view of dynamic ecology
Integrated	Agent-based models	Complex and computationally demanding Require behavioral and decision-making data
Coupled biophysical and socioeconomic	For example, PHEWS-SAVANNA — dynamic rainfall and vegetation; fixed social, economic, and administrative rules	Hybrids that live within the restrictions imposed by established model structure
Spatial correlative	Link remotely sensed land cover changes to changes in land use, transportation, populations, etc.	Correlation is not causation Multiple, nonspecific correlations Circularity—poor linkage to ecology and society

the modeling approach. Here there is a parallel with the suggestions arising from van Langevelde, Meyer et al., and Jeltsch et al., in Section 4, involving methods for both upscaling and downscaling that deal with discontinuities in scale-dependent biophysical drivers, and drivers of human behavior. The case study in Chapter 26 looks at grazing and social patterns based on herd tracking and diffusion models; remote sensing can contribute information to these approaches through mapping of grazing-based piospheres around water points in arid lands, as well as to characterizing impacts from herd tracking (Sonneveld et al., 2009).

The Key Needs for an Ecology-Model-Remote Sensing Synthesis

The contributors to this book have documented an impressive array of field, technological, and simulation-based science focused on savannas. However, there is a noticeable lack of coherence between the "components"—site measurements, remote sensing retrievals and products, and models and modeling. The site network for both intensive and extensive measurements in savannas is fragmented and limited, and, with the exception of FLUXNET sites, has no consistent core set of observations that are routinely or widely taken (e.g., the datasets used by Sankaran et al., 2005; Barrett, 2001). The remote sensing retrievals and their value for surface description and/or as inputs to models are highly dependent upon platform and operational scale, and the methods and products are often only tested, or demonstrated for one ecosystem and one campaign. There has been substantial co-analysis of flux measurements with MODIS product 7×7 pixel tiles across the FLUXNET sites (Olson et al., 2004 and Heinsch et al., 2006), and Hanan et al. (Chapter 4) used MODIS EVI to provide FPAR estimates for their savannas. Sensors that could deliver a much better set of quantitative land surface measurements are planned, but launch dates are still several years away and subject to slippage. The coupled human–environment of savannas demands spatially explicit data collection and modeling at the scale of human activity (see Chapter 26, Turner), and vegetation demography (see Chapter 14, Meyer et al.) and frameworks for both up- and downscaling across biophysical and socioeconomic dataspaces. The regional and global scale models do not appear to accommodate upscaled process and land component data, and do not utilize the full breadth of new coarse resolution imagery, but most of these are "legacy" models with long histories that parallel the development of vegetation indices. Key elements of landscape process at savanna scale are captured in different models, such as nutrient cycling by Savanna-CENTURY (Chapter 24, Parton et al.), tree demography and patch dynamics by grid models (Chapter 16, Jeltsch et al.), and fire-induced patch dynamics by FLAMES (Chapter 15, Liedloff and Cook), but there is a lack of integration or a clear scaling path to regional and global dynamics.

However, if the contributions to this book are closely examined, it can be seen that all of the components needed to greatly improve the remote measurement and multiscale dynamic modeling of the natural vegetation systems of savannas are present. A global framework and team-based approach is needed to bring the elements together. Such a framework and team structure could be provided under a range of international structures such as Global Earth Observation System of Systems (GEOSS) or International Geosphere–Biosphere Program (IGBP) through a specific savanna working group which would form a cross-cutting activity under a number of themes, for example, Terrestrial Ecosystems, Biodiversity, Sustainable Agriculture and Desertification, and Human Health and Well-Being. However, at the science level there is a need for a coordinated measurement and modeling research program which could be sponsored by national research agencies such as NASA or NSF. This book has identified many areas where data, technology, methods, and models are currently advancing, but also where deficiencies and incoherencies remain. We provide the following as some of the areas where further work, and integration of approaches could be beneficial.

1. A more coherent approach to collection of field data in savannas would improve our fundamental understanding and ability to model and manage global savannas.

2. "Savanna" data needs to be staged on the internet via a system similar to that operated by the US Government through the Oak Ridge National Laboratory Data Archive and Access Center (ORNL DAAC) for long-term open access by scientific and other stakeholder communities.

3. Prior to launch of the new generation of land and vegetation observation sensors, coordinated calibration and validation of retrieval methods, and surface description products specific to two layer, tree–grass systems across a diversity of tree–grass landscapes is needed.

4. The deficiencies in landscape-scale data and modeling need to be addressed through collaboration across disciplines on savanna physiognomy, phenology, and physiology; capture of spatially explicit nutrient cycling and transfer; efficient computation of grid models; up- and downscaling approaches; and spatial and temporal aggregation frameworks for feeding regional and global scale models that capture aggregate effects of landscape processes.

5. In the process of "merging" the different landscape-scale concepts of tree–grass dynamics, the kind of conceptual model for interaction of climate and disturbance drivers outlined by Hanan and Lehmann should be developed by coupling and/or hybridizing existing landscape-scale models, but with an explicit focus on scale-dependent process and scale-induced process-redundancy.

6. At this landscape scale, there is a pressing need for a major collaboration between social and biophysical scientists, after the design of the LTER projects in Phoenix and Baltimore (e.g., Cadenasso et al., 2006), that bridges the paradigm differences and develops the approaches outlined by Turner, and Galvin and Reid, involving landscape ecology modelers.

7. The current suite of remote sensing data, for example, MODIS optical and thermal products, passive microwave and active radar data, and LiDAR data (such as ICESAT GLAS), should be combined with other spatial data in new models, designed for dedicated ingestion of remote sensing data, and applied along with available field data in a model-data assimilation framework to explore the full power of current data to reduce uncertainties and describe dynamics in global savannas (e.g., Barrett et al., 2005; Williams et al., 2009).

8. The measurement and modeling community should work with computer programmers, and web server technologists, to design multiscale computing and modeling architectures across which efficient transfer and aggregation of scale-dependent processes and effects can be propagated in order to make computation, and hierarchical application of simulations and products more effective and targeted.

High performance computing and distributed modeling initiatives are ongoing at the land–atmosphere interface for climate change modeling (e.g., Lokupitiya et al., 2009; Shukla et al., 2009).

9. The outcomes here are unpredictable, but almost certainly do not involve a model of everything, but a framework, a set of couplings, improved models at different scales, improved data networks and calibrated and validated spatially explicit remote sensing retrievals that collectively promote our ability to measure, manage, and predict future changes in the global savannas which are so important for the cultural and economic wellbeing of so many.

References

Anaya, J. A., E. Chuvieco, and A. Palacios-Orueta. 2009. Aboveground biomass assessment in Colombia: A remote sensing approach. *Forest Ecology and Management* 257, 1237–1246.

Andersen, H-E., R. J. McGaughey, and S. E. Reutebuch. 2005. Estimating forest canopy fuel parameters using LIDAR data. *Remote Sensing of Environment* 94, 441–449.

Anyamba, A., C. O. Justice, C. J. Tucker, and R. Mahoney. 2003. Seasonal to interannual variability of vegetation and fires at SAFARI 2000 sites inferred from advanced very high resolution radiometer time series data. *Journal of Geophysical Research Atmospheres* 108, D13, SAF 43–1, DOI 10.1029/2002JD002464.

Anyamba, A. and C. J. Tucker. 2005. Analysis of Sahelian vegetation dynamics using NOAA-AVHRR NDVI data from 1981–2003. *Journal of Arid Environments* 63, 596–614.

Archer, S., T. W. Boutton, and K. A. Hibbard. 2001. Trees in grasslands: Biogeochemical consequences of woody plant expansion. In *Global Biogeochemical Cycles in the Climate System*, eds. E-D. Schulze, S. P. Harrison, M. Heimann, et al., pp. 115–138. San Diego, Academic Press, 350pp.

Armston, J. D., Scarth, P. F. Phinn, S. R., and T. J. Danaher. 2007. Analysis of multi-date MISR measurements for forest and woodland communities, Queensland, Australia. *Remote Sensing of Environment*. 107, 287–298.

Asner, G. P., D. E. Knapp, E. N. Broadbent, P. J. C. Oliveira, M. Keller, and J. N. Silva. 2005. Selective logging in the Brazilian Amazon. *Science* 310, 480–482.

Barrett, D. J. 2001. NPP Multi-Biome: VAST Calibration Data, 1965–1998, report, Oak Ridge Natl. Lab. Distrib. Active Arch. Cent., Oak Ridge, Tenn., Available at http://www.daac.ornl.gov/).

Barrett, D. J., M. J. Hill, L. B. Hutley, et al. 2005. Prospects for improving savanna carbon models using multiple constraints model-data assimilation methods. *Australian Journal of Botany* 55, 689–714.

Bestelmeyer, B. T., J. R. Brown, K. M. Havstad, R. Alexander, G. Chavez, and J. E. Herrick. 2003. Development and use of state-and-transition models for rangelands. *Journal of Range Management* 56, 114–126.

de BSie, S., P. Ketner, M. Paasse, and C. Geerling. 1998. Woody plant phenology in the West Africa savanna. *Journal of Biogeography* 25, 883–900.

Bolan, N. S., M. J. Hedley, and R. E. White. 1991. Processes of soil acidification during nitrogen cycling with emphasis on legume-based pastures. *Plant and Soil* 134, 53–63.

Briske, D. D., S. D. Fuhlendorf, and F. E. Smeins. 2005. State-and-transition models, thresholds and rangeland health: A synthesis of ecological concepts and perspectives. *Rangeland Ecology and Management* 58, 1–10.

Cadenasso, M. L., S. T. A. Pickett, and M. J. Grove. 2006. Integrative approaches to investigating human–natural systems: The Baltimore ecosystem study. *Natures Sciences Sociétés* 14, 4–14.

Chamaille-Jammes, S., H. Fritz, and F. Murindagomos. 2006. Spatial patterns of the NDVI–rainfall relationship at the seasonal and interannual time scales in an African savanna. *International Journal of Remote Sensing* 27, 5185–5200.

Chen, J. M., C. H. Menges, and S. G. Leblanc. 2005. Global mapping of foliage clumping index using multi-angular satellite data. *Remote Sensing of Environment* 97, 447–457.

Chidumayo, E. N. 2001. Climate and phenology of savanna vegetation in southern Africa. *Journal of Vegetation Science* 12, 347–354.

Chopping, M., L. Su, A. Laliberte, A. Rango, D. P. C. Peters, and N. Kollikkathara. 2006. Mapping shrub abundance in desert grasslands using geometric-optical modeling and multiangle remote sensing with CHRIS/Proba. *Remote Sensing of Environment* 104, 62–73.

Chopping, M. J., G. G. Moisen, L. Su, A. Laliberte, A. Rango, J. V. Martonchik, and D. P. C. Peters. 2008. Large area mapping of southwestern forest crown cover, canopy height, and biomass using the NASA Multiangle Imaging Spectro-Radiometer. *Remote Sensing of Environment* 112, 2051–2063.

Damascos, M. A., C. H. B. A. Prado, and C. C. Ronquim. 2005. Bud composition, branching patterns and leaf phenology in cerrdao woody species. *Annals of Botany* 96, 1075–1084.

Donohue, R. J., M. L. Roderick, and T. R. McVicar. 2008. Deriving consistent long-term vegetation information from AVHRR reflectance data using a cover-triangle-based framework. *Remote Sensing of Environment* 112, 2938–2949.

du Plessis, W. P. 1999. Linear regression relationships between NDVI, vegetation and rainfall in Etosha National Park, Namibia. *Journal of Arid Environments* 42, 235–260.

Ferreira, L. G., H. Yoshida, A. Huete, and E. E. Sano. 2003. Seasonal landscape and spectral vegetation index dynamics in the Brazilian Cerrado: An analysis within the large-scale biosphere–atmosphere experiment in Amazonia (LBA). *Remote Sensing of Environment* 87, 534–550.

Franca, H. and A. W. Setzer. 1998. AVHRR temporal analysis of a savanna site in Brazil. *International Journal of Remote Sensing* 19, 3127–3140.

Franca, H. and A. W. Setzer. 2001. AVHRR analysis of a savanna site through a fire season in Brazil. *International Journal of Remote Sensing* 22, 2449–2461.

Frison, P.-L., E. Mougin, and P. Hiernaux. 1998. Observations and interpretation of seasonal ERS-1 wind scatterometer data of Northern Sahel (Mali). *Remote Sensing of Environment* 63, 233–242.

Frison, P.-L., E. Mougin, L. Jarlan, M. A. Karam, and P. Hiernaux. 2000. Comparison of ERS wind-scatterometer and SSMI data for Sahelian vegetation monitoring. *IEEE Transactions on Geoscience and Remote Sensing* 38, 1794–1803.

Gill, T. K., S. R. Phinn, J. D. Armston, and B. A. Pailthorpe. 2009. Estimating tree-cover change in Australia: Challenges of using the MODIS vegetation index product. *International Journal of Remote Sensing* 30, 1547–1565.

Gillson, L. 2004. Evidence of hierarchical patch dynamics in an east African savanna? *Landscape Ecology* 19, 883–894.

Gordon, L. J., W. Steffen, B. F. Jönsson, C. Folke, M. Falkenmark, and A. Johannessen. 2005. Human modification of global water vapor flows from the land surface. *Proceedings of the National Academy of Sciences* 102, 7612–7617.

Gourtorbe, J. P., T. Lebel, A. J. Dolman, J. H. C. Gash, P. Kabat, Y. H. Kerr, B. Monteny, S. D. Prince, J. N. M. Stricker, A. Tinga, and J. S. Wallace. 1997. An overview of HAPEX-Sahel: A study in climate and desertification. *Journal of Hydrology* 188–189, 4–17.

Guerschman, J., M. J. Hill, L. Renzullo, D. Barrett, A. Marks, and E. Botha. 2009. Estimating fractional cover of photosynthetic vegetation, nonphotosynthetic vegetation and bare soil in the Australian tropical savanna region upscaling the EO-1 Hyperion and MODIS sensors. *Remote Sensing of Environment* 113, 928–945.

Hanan, N. P., Y. Prevost, A. Diouf, and O. Diallo. 1991. Assessment of desertification around deep wells in the Sahel using satellite imagery. *Journal of Applied Ecology* 28, 173–186.

Hansen, M., R. S. DeFries, J. R. G. Townshend, M. Carroll, C. Dimiceli, and R. A. Sohlberg. 2003. Global percent tree cover at a spatial resolution of 500 meters: First results of the MODIS vegetation continuous fields algorithm. *Earth Interactions* 7, 1–15.

Heinsch, F. A., M. Zhao, S. R. Running, et al. 2006. Evaluation of remote sensing based terrestrial productivity from MODIS using regional tower eddy flux network observations. *IEEE Transactions on Geoscience and Remote Sensing* 44, 1908–1925.

Heumann, B. W., J. W. Seaquist, L. Eklundh, and P. Jönssön. 2007. AVHRR derived phonological change in the Sahel and Soudan, Africa, 1982–2005. *Remote Sensing of Environment* 108, 385–392.

Hill, M. J., S. H. Roxburgh, J. O. Carter, G. M. McKeon, and D. J. Barrett. 2005. Vegetation state change and carbon dynamics in savanna woodlands of Australia in response to grazing, drought and fire: A scenario approach using 113 years of synthetic annual fire and grassland growth. *Australian Journal of Botany* 53, 715–739.

Hill, M. J., S. J. Roxburgh, G. M. McKeon, J. O. Carter, and D. J. Barrett. 2006. Analysis of soil carbon outcomes from interaction between climate and grazing pressure in Australian rangelands using Range-ASSESS. *Environmental Modelling and Software* 21, 779–801.

Jarlan, L., P. Mazzega, and E. Mougin. 2002. Retrieval of land surface parameters in the Sahel from ERS wind scatterometer data: A "brute force" method. *IEEE Transactions on Geoscience and Remote Sensing* 40, 2056–2062.

Jarlan, L., P. Mazzega, E. Mougin, F. Lavenu, G. Marty, P. L. Frison, and P. Hiernaux. 2003. Mapping of Sahelian vegetation parameters from ERS scatterometer data with an evolution strategies algorithm. *Remote Sensing of Environment* 87, 72–84.

Lambers, H. 2003. Dryland salinity: A key environmental issue in southern Australia. *Plant and Soil* 257, v–vii.

Li, X., A. H. Strahler, and C. E. Woodcock. 1995. A hybrid geometric optical-radiative transfer approach for modeling albedo and directional reflectance of discontinuous canopies. *IEEE Transactions on Geoscience and Remote Sensing* 33, 466–480.

Liesenberg, V., L. Soares Galvao, and F. J. Ponzoni. 2007. Variations in reflectance with seasonality for classification of the Brazilian savanna physiognomies with MISR/Terra data. *Remote Sensing of Environment* 107, 276–286.

Liu, S., T. R. Loveland, and R. M. Kurtz. 2004. Contemporary carbon dynamics in terrestyrial ecosystems in the southeastern plains of the United States. *Environmental Management* 33, Suppl. 1, S442–S456.

Lokupitiya, E., S. Denning, K. Paustian, et al. 2009. Incorporation of crop phenology in Simple Biopshere Model (SiBcrop) to improve land-atmosphere carbon exchange from croplands. *Biogeosciences* 6, 969–986.

Lo Seen, D., E. Mougin, S. Rambal, A. Gaston, and P. Hiernaux. 1995. A regional Sahelian grassland model to be coupled with multispectral satellite data. II: Toward the control of its simulations by remotely sensed indices. *Remote Sensing of Environment* 52, 194–206.

Ludwig J. A. and D. J. Tongway 1995. Spatial organisation of landscapes and its function in semi-arid woodlands Australia. *Landscape Ecology* 10, 51–63.

Ludwig J. A., J. A. Wiens, and D. J. Tongway. 2000a. A scaling rule for landscape patches and how it applies to conserving soil resources in savannas. *Ecosystems* 3, 84–97.

Ludwig, J. A., R. W. Eager, G. N. Bastin, V. H. Chewings, and A. C. Liedloff. 2002. A leakiness index for assessing landscape function using remote-sensing. *Landscape Ecology* 17, 157–171.

Ludwig, J. A., R. W. Eager, A. C. Liedloff, G. N. Bastin, and V. H., Chewings. 2006. A new landscape leakiness index based on remotely sensed ground-cover data. *Ecological Indicators* 6, 327–336.

Ludwig, J. A., G. N. Bastin, V. H. Chewings, R. W. Eager, and A. C. Liedloff. 2007. Leakiness: A new index for monitoring the health of arid and semiarid landscapes using remotely sensed vegetation cover and elevation data. *Ecological Indicators* 7, 442–454.

Mougin, E., D. Lo Seen, S. Rambal, A. Gaston, and P. Hiernaux, 1995. A regional Sahelian grassland model to be coupled with multispectral satellite data. I. Model description and validation. *Remote Sensing of Environment* 52, 181–193.

Nicholson, S. E., C. J. Tucker, and M. B. Ba. 1998. Desertification, drought, and surface vegetation: An example from the West African Sahel. *Bulletin of the American Meteorological Society* 79, 815–829.

Olson, R. J., S. K. Holladay, R. B. Cook, et al. 2004. FLUXNET: Database of fluxes, site characteristics and flux-community information. Oak Ridge, Tennessee, Environmental Sciences Division, Oak Ridge National Laboratory, ORNL/TM – 2003/204 (available at: http://public.ornl.gov).

Potter, C. S. and S. A. Klooster. 1999. Dynamic global vegetation modeling for prediction of plant functional types and biogenic trace gas fluxes. *Global Ecology and Biogeography* 8, 473–488.

Prince, S. D., 1991a. Satellite remote sensing of primary production: Comparison of results for Sahelian grasslands 1981–1988. *International Journal of Remote Sensing* 12, 1301–1311.

Prince, S. D., 1991b. A model of regional primary production for use with coarse resolution satellite data. *International Journal of Remote Sensing*, 12, 1313–1330.

Prince, S. D. and W. L. Astle. 1986. Satellite remote sensing of rangelands in Botswana. I. Landsat MSS and herbaceous vegetation. *International Journal of Remote Sensing* 7, 1533–1553.

Prince, S. D. and C. J. Tucker. 1986. Satellite remote sensing of rangelands in Botswana. II. NOAA AVHRR and herbaceous vegetation. *International Journal of Remote Sensing* 7, 1555–1570.

Renzullo, L. J., D. J. Barrett, A. S. Marks, et al. 2008. Multi-sensor model-data fusion for estimation of hydrologic and energy flux parameters. *Remote Sensing of Environment* 112, 1306–1319.

Roderick, M. L., I. R. Noble, and S. W. Cridland. 1999. Estimating woody and herbaceous vegetation cover from time series satellite observations. *Global Ecology and Biogeography* 8, 501–508.

Roujean, J-L., Leroy, M., and Deschamps P-Y. 1992, A bidirectional reflectance model of the earth's surface for the correction of remotely sensed data. *Journal of Geophyical Research* 97, 20455–20468.

Roy, D. P., L. Giglio, J. D. Kendall, and C. O. Justice. 1999. Multi-temporal active-fire based burn scar detection algorithm. *International Journal of Remote Sensing* 20, 1031–1038.

Sankaran, M., N. P. Hanan, R. J. Scholes, et al. 2005. Determinants of woody cover in African savannas. *Nature* 438, 846–849.

Scanlon, T. M., J. D. Albertson, K. K. Caylor, and C. A. Williams. 2002. Determining land surface fractional cover from NDVI and rainfall time series for a savanna ecosystem. *Remote Sensing of Environment* 82, 376–388.

Seaquist, J. W., T. Hickler, L. Eklundh, J. Ardo, and B. W. Heumann. 2008. Disentangling the effects of climate and people on Sahel vegetation dynamics. *Biogeosciences Discussions* 5, 3045–3067.

Shukla, J., R. Hagedorn, B. Hoskins, et al. 2009. Revolution in climate prediction is both necessary and possible. A declaration at the world modelling summit for climate prediction. *Bulletin of the American Meteorological Society* DOI: 10.1175/2008BAMS2759.1, 175–178.

Sjöström, M., J. Ardö, L. Eklundh, et al. 2009. Evaluation of satellite based indices for gross primary production estimates in a sparse savanna in the Sudan. *Biogeosciences* 6, 129–138.

Swap, R. J., H. J. Annegarn, J. T. Suttles, et al. 2003. Africa burning: A thematic analysis of the Southern African Regional Science Initiative (SAFARI 2000). *Journal of Geophysical Research* 108, NO. D13, 8465, doi: 10.1029/2003JD003747.

Sonneveld, B. G. J. S., M. A. Keyzer, K. Georgis, S. Pande, A. Seid Ali, and A. Takele. 2009. Following the Afar: Using remote tracking systems to analyze pastoralists' trekking routes. *Journal of Arid Environments* 73, 1046–1050.

Su, L.,Y. Huang, M. J. Chopping, A. Rango, and J. V. Martonchik. 2009. An empirical study on the utility of BRDF model parameters and topographic parameters for mapping vegetation in a semi-arid region with MISR imagery. *International Journal of Remote Sensing* 30, 3463–3483.

Tilman, D., J. Fargione, B. Wolff, et al. 2001. Forecasting agriculturally driven global environmental change. *Science* 292, 281–284.

Tucker, C. J., C. O. Justice, and S. D. Prince. 1986. Monitoring the grasslands of the Sahel 1984–1985. *International Journal of Remote Sensing* 7, 1571–1581.

Tucker, C. J. and P. J. Sellers, 1986. Satellite remote sensing of primary production. *International Journal of Remote Sensing* 11, 1395–1416.

Wessels, K. J., S. D. Prince, P. E. Frost, and D. van Zyl. 2004. Assessing the effects of human-induced land degradation in the former homelands of northern South Africa with a 1 km AVHRR NDVI time-series. *Remote Sensing of Environment* 91, 47–67.

Westoby, M., B. Walker, and I. Noy-Meir. 1989. Opportunistic management for range-lands not at equilibrium. *Journal of Range Management* 42, 266–274.

Williams, R. J., B. A. Myers, W. J. Muller, G. A. Duff, and D. Eamus. 1997. Leaf phenology of woody species in a North Australian tropical savanna. *Ecology* 78, 2542–2558.

Williams, M., A. D. Richardson, M. Reichstein, et al. 2009. Improving land surface models with FLUXNET data. *Biogeosciences Discussions* 6, 2785–2835.

Yates, C. J. and R. J. Hobbs. 1997. Woodland restoration in the Western Australian wheatbelt: A conceptual framework using a state and transition model. *Restoration Ecology* 5, 28–35.

Zine, S., L. Jarlan, P.-L. Frison, E. Mougin, P. Hiernaux, and J.-P. Rudant. 2005. Land surface parameter monitoring with ERS scatterometer data over the Sahel: A comparison between agro-pastoral and pastoral areas. *Remote Sensing of Environment* 96, 438–452.

Index

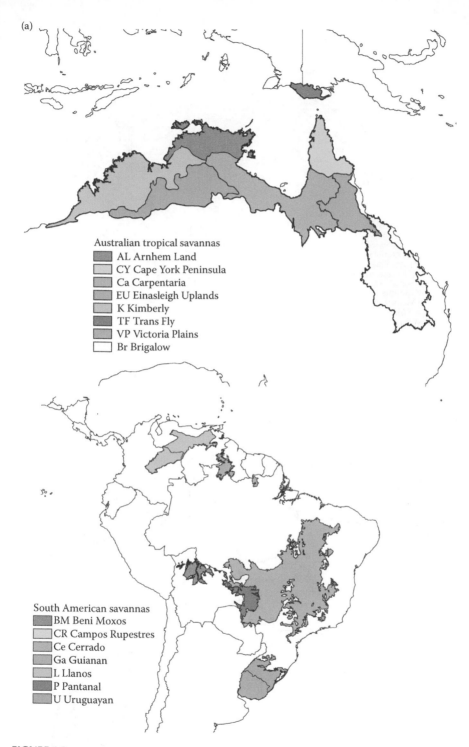

(a)

Australian tropical savannas
- **AL** Arnhem Land
- **CY** Cape York Peninsula
- **Ca** Carpentaria
- **EU** Einasleigh Uplands
- **K** Kimberly
- **TF** Trans Fly
- **VP** Victoria Plains
- **Br** Brigalow

South American savannas
- **BM** Beni Moxos
- **CR** Campos Rupestres
- **Ce** Cerrado
- **Ga** Guianan
- **L** Llanos
- **P** Pantanal
- **U** Uruguayan

FIGURE 1.2
(a) WWF ecoregions for Australian, South American, and African tropical and subtropical savannas. (Adapted from Olson, D. M. et al. 2001. *Bioscience* 51, 933–938.) (b) MODIS IGBP land cover maps for Australian, South American, and African tropical and subtropical savannas. (Adapted from Friedl, M. A. et al. 2002. *Remote Sensing of Environment* 83, 287–302.)

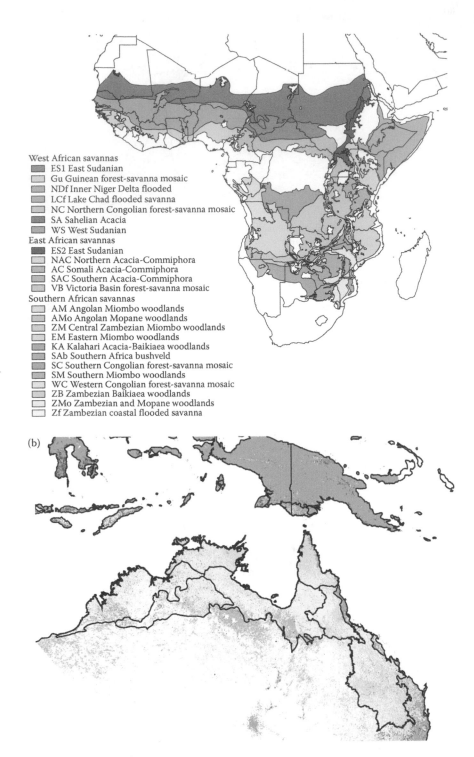

West African savannas
- ES1 East Sudanian
- Gu Guinean forest-savanna mosaic
- NDf Inner Niger Delta flooded
- LCf Lake Chad flooded savanna
- NC Northern Congolian forest-savanna mosaic
- SA Sahelian Acacia
- WS West Sudanian

East African savannas
- ES2 East Sudanian
- NAC Northern Acacia-Commiphora
- AC Somali Acacia-Commiphora
- SAC Southern Acacia-Commiphora
- VB Victoria Basin forest-savanna mosaic

Southern African savannas
- AM Angolan Miombo woodlands
- AMo Angolan Mopane woodlands
- ZM Central Zambezian Miombo woodlands
- EM Eastern Miombo woodlands
- KA Kalahari Acacia-Baikiaea woodlands
- SAb Southern Africa bushveld
- SC Southern Congolian forest-savanna mosaic
- SM Southern Miombo woodlands
- WC Western Congolian forest-savanna mosaic
- ZB Zambezian Baikiaea woodlands
- ZMo Zambezian and Mopane woodlands
- Zf Zambezian coastal flooded savanna

(b)

FIGURE 1.2 Continued.

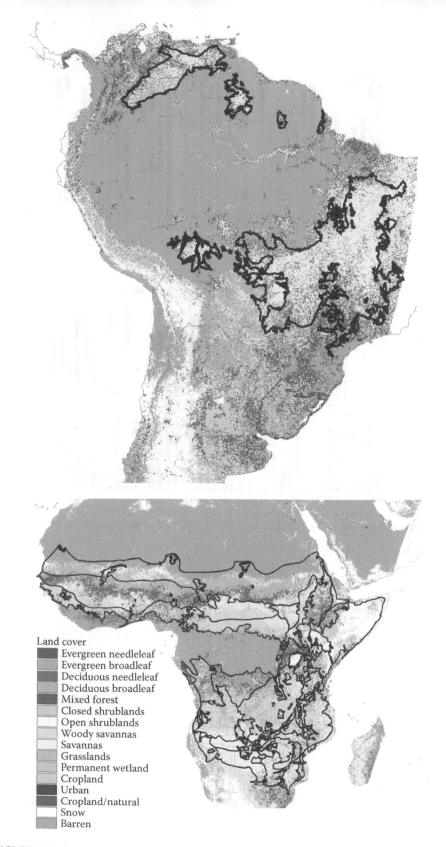

Land cover
- Evergreen needleleaf
- Evergreen broadleaf
- Deciduous needleleaf
- Deciduous broadleaf
- Mixed forest
- Closed shrublands
- Open shrublands
- Woody savannas
- Savannas
- Grasslands
- Permanent wetland
- Cropland
- Urban
- Cropland/natural
- Snow
- Barren

FIGURE 1.2 Continued.

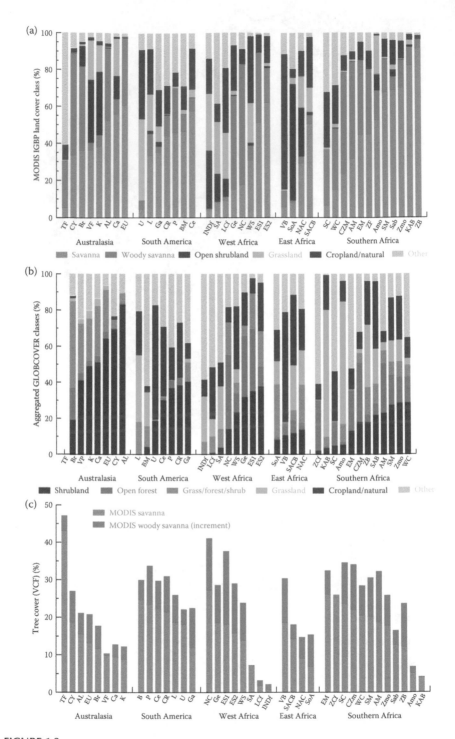

FIGURE 1.3
Proportion of savanna ecoregions occupied by (a) MODIS IGBP land cover classes; (b) aggregated classes from the ESA GLOBCOVER; and (c) percentage tree cover from MOD44 VCF data for MODIS IGBP savanna and woody savanna (full bar height) land cover classes.

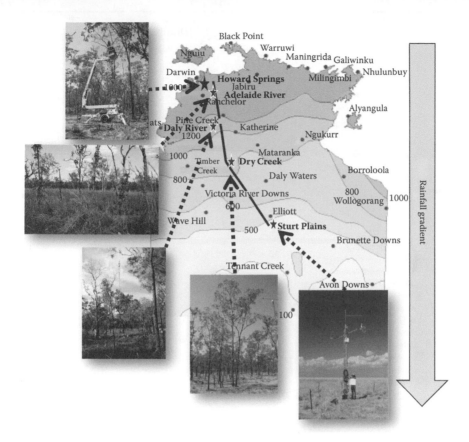

FIGURE 3.2
Locations and photographs of savanna flux tower sites, Northern Territory, Australia. This region features extensive open-forest and woodland savanna ecosystems with scattered seasonal swamps and monsoon vine forest patches. The network of sites provides data describing mass and energy exchange across this biogeographical gradient.

FIGURE 4.1
Photograph of the Skukuza flux tower showing associated vegetation.

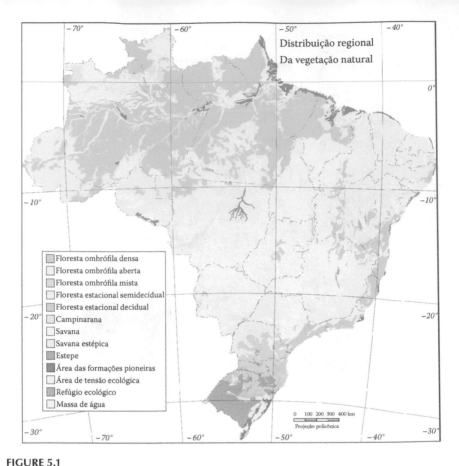

FIGURE 5.1
The distribution of vegetation of Brazil. Savanna (cerrado) is shown in tan, and Area de Tensao Ecologica (transitional forest) is shown in gray-blue. (Accessed from the Instituto Brasileiro de Geografia e Estatística [http://www.ibge.gov.br/servidor_arquivos_geo/], verified on March 30, 2009.)

FIGURE 6.2
(a) View from the top of the woodland flux tower. (b) View from the top of the savanna flux tower. (c) Grassland tower pasture.

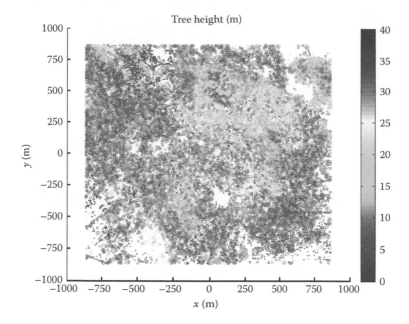

FIGURE 7.1
A map of tree location, height, and crown size of a blue oak savanna, near Ione, CA. The horizontal dimensions of the figure are 1000 m × 1000 m, and the flux measurement tower is at the center. Canopy structure was quantified using an airborne LIDAR during the summer of 2009.

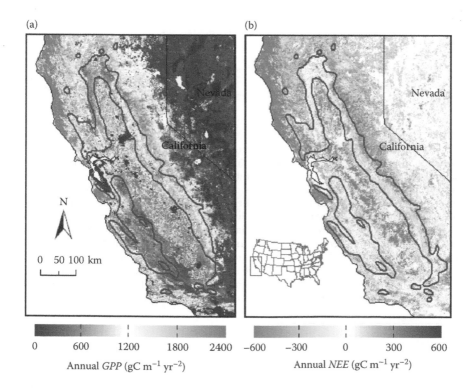

FIGURE 7.4
Spatial map of net ecosystem carbon exchange for the blue oak biome: (a) is for annual gross primary productivity (*GPP*); and (b) is for net annual ecosystem carbon exchange (*NEE*). (Adapted from Xiao, J. et al. 2008. *Agricultural and Forest Meteorology* 148, 1827–1847; and derived using data from the MODIS sensor and the AmeriFlux network.)

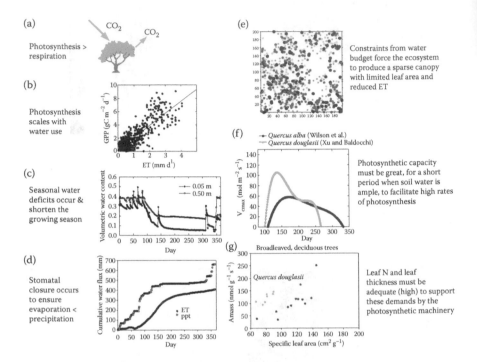

(a) CO_2 CO_2

Photosynthesis > respiration

(b)

Photosynthesis scales with water use

(c)

Seasonal water deficits occur & shorten the growing season

(d)

Stomatal closure occurs to ensure evaporation < precipitation

(e)

Constraints from water budget force the ecosystem to produce a sparse canopy with limited leaf area and reduced ET

(f) Quercus alba (Wilson et al.)
Quercus douglasii (Xu and Baldocchi)

Photosynthetic capacity must be great, for a short period when soil water is ample, to facilitate high rates of photosynthesis

(g) Broadleaved, deciduous trees

Quercus douglasii

Leaf N and leaf thickness must be adequate (high) to support these demands by the photosynthetic machinery

FIGURE 7.5

An eco-hydrological explanation for the structure and function of a blue oak savanna and its sustenance. For the oak savannas to survive in highly seasonal wet/dry, cool/hot climates, they must comply with several linked constraints. (a) First, annual photosynthesis must exceed ecosystem respiration; (b) the oaks experience difficulty in achieving high rates of photosynthesis during the growing season, because the lack of summer rains causes volumetric soil moisture to drop below 5% and induces physiological stress; (c) further, the act of photosynthesis comes with a cost—evaporation; (d) consequently, annual evaporation must be constrained such that it does not exceed annual precipitation, or the trees invest in deep roots that tap ground water; (e) the landscape can reduce its total evaporation by establishing a partial canopy; (f) it can assimilate enough carbon if it adjusts and maximizes its photosynthetic capacity during the short spring wet period; and (g) high rates of photosynthesis are not free and come with producing thick leaves, rich in nitrogen and Rubisco.

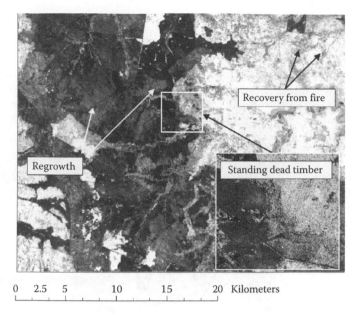

0   2.5   5   10   15   20   Kilometers

FIGURE 8.6
Composite of Landsat-derived FPC and ALOS L-band HH and HV data in RGB. Areas of dead standing timber (green; with high HH σ^0, low FPC, and HV σ^0; note that confusion exists with rough surfaces, particularly when surface and soil moisture is high), regrowth after clearance (red; low HH and HV σ^0 and high FPC), and fire (orange; moderate HH and HV and high FPC) are highlighted. Inset is an AIRSAR image showing areas of standing dead timber (blue). (Adapted from Lucas, R. M., et al. 2008b. *Patterns and Processes in Forest Landscapes: Multiple uses and sustainable management*, eds. R. Lafortezza, J. Chen, G. Sanesi and T. R. Crow, Springer, New York, pp. 213–240.)

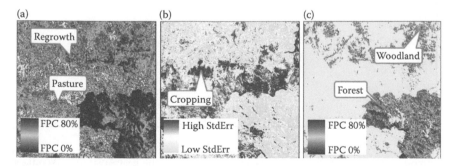

FIGURE 9.4
Example of time series approach to discriminate woody extent: (a) the FPC derived from a 2008 Landsat TM image showing high photosynthetic activity across the image; (b) the standard error of a linear fit to the time series with dark areas representing a low standard error; (c) indicative of stable perennial vegetation and used in conjunction with the minimum of the time series to estimate the 2008 woody extent product.

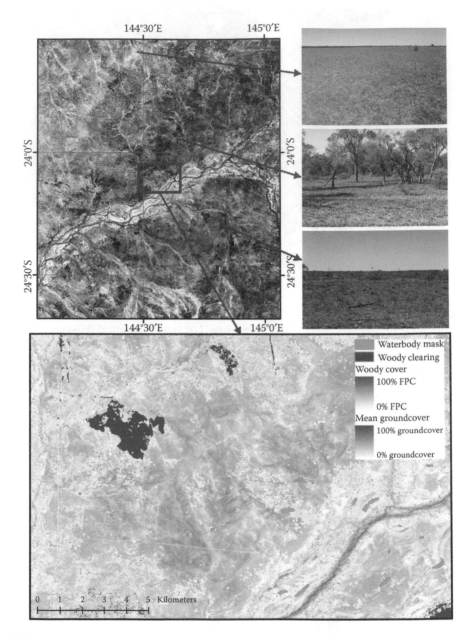

FIGURE 9.8
Clockwise from top left. Landsat 5 TM image of a portion of the East Australian Woodlands with bands 5,4,3 displayed as RGB. Photographs show examples of open pasture, open woodland communities, and recently cleared areas. Map shows information products developed from time series analysis of Landsat imagery including waterbody location, areas of woody clearing, mean ground cover, and woody foliage projective cover. © The State of Queensland (Department of Environment and Resource Management).

FIGURE 10.5
Detailed fusion of spectral and LiDAR data collected over the Kruger National Park, South Africa, by the Carnegie Airborne Observatory. (Adapted from Asner, G. P. et al. 2007. *Journal of Applied Remote Sensing* 1, DOI:10.1117/1111.2794018.) The imagery shows the 3-D partitioning of the landscape, including actual shadows cast by woody vegetation canopies onto the herbaceous and bare soil surface below. Co-alignment of imaging data opens new doors for quantitative 3-D fractional cover analysis of savannas.

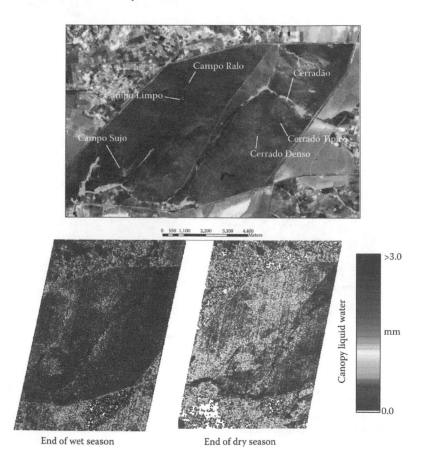

FIGURE 10.10
Canopy water content in the Brazilian Cerrado savannas using Earth Observing-1 Hyperion imaging spectrometer data. Notice how canopy water content decreased variably across this savanna region based on differences in plant physiognomy: Campo Limpo, Campo Sujo, and Campo Ralo ecosystems are dominated by herbaceous species; Cerradao, Cerrado Tipico, and Cerrado Denso are areas dominated by drought-tolerant woody species.

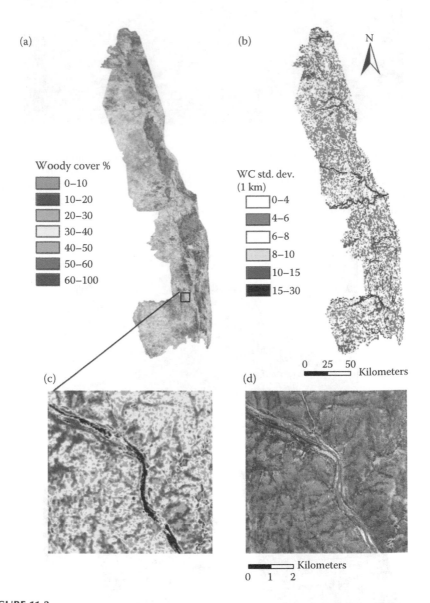

FIGURE 11.3
(a) KNP woody cover percent map (90-m nominal resolution). (b) Heterogeneity map for Kruger National Park with focal resolution of 1 km. (c) Woody cover map zoom on Sabie River shows false color infrared-composite of a 4-m resolution IKONOS image (4–28–2001) (photo-synthetically active vegetation in red shades), and (d) woody cover percent estimates.

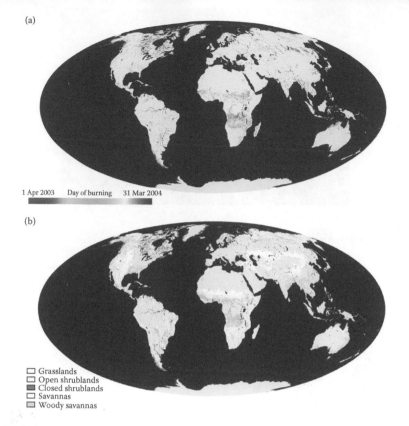

FIGURE 12.1

Satellite-derived global fire distribution: (a) global burned area distribution, derived from the MODIS burned area product for April 2003 to March 2004 shown superimposed over MODIS true color surface reflectance to provide geographic context. (Adapted from Roy, D. P. 2008. *Remote Sensing of Environment* 112, 3690–3707.) (b) Global distribution of savannas, grasslands, and shrublands defined by the MODIS land cover product (Adapted from Friedl, M. A., et al. 2002. *Remote Sensing of Environment* 83, 287–302.)

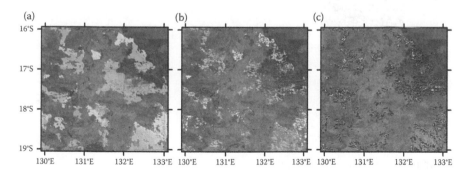

FIGURE 12.2

Comparison of burned areas in Northern Australia in October 2002, as detected by the MODIS burned area product (Adapted from Roy, D. P. et al. 2008. *Remote Sensing of Environment* 112, 3690–3707) and the MODIS active fire product (Adapted from Giglio, L. et al. 2003. *Remote Sensing of Environment* 87, 273–282), superimposed on MODIS surface reflectance to provide geographic context. An area of 4° of longitude by 4° of latitude is shown. (a) The approximate day of burning defined by the MODIS 500 m burned area product with day of burning shown in a rainbow color scale (first day blue, last day red); (b) the date of the first detection by the MODIS 1km active fire product (Terra and Aqua, day and night detections), in the same color scale; (c) the number of times each pixel is detected by the MODIS active fire product (red = 1, yellow = 2, green = 3, cyan = 4).

FIGURE 14.1
Spatial structure in a semiarid savanna.

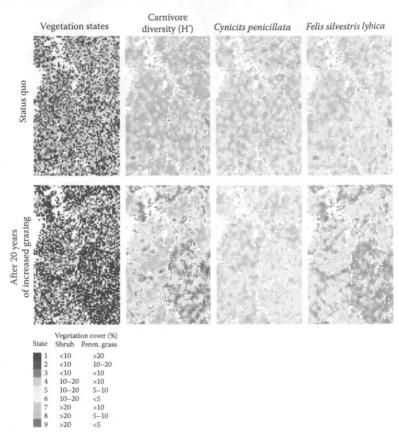

FIGURE 16.5
Sample application of the upscaled regional vegetation model (70 km × 100 km) for a region in the southern Kalahari (South Africa, between 26°48′46 S, 20°44′56 E and 27°09′51 S, 21°07′20 E): 1. column: vegetation states, 2. column: carnivore diversity (Shannon Index) from dark gray = high (up to 0.70) to light gray = low diversity (up to 0.46), 3. & 4. column: relative abundance of yellow mongoose (*Cynictis penicillata*) and African wildcat (*Felis silvestris lybica*) from dark gray = high to light gray = low. The upper row shows the status quo (distribution of vegetation states derived by remote sensing) and the lower row shows predicted changes after 20 years under the scenario of 20% increase of current grazing intensity for all farms. The proportion of shrub-dominated vegetation states increases under this scenario, leading to a decrease in carnivore diversity and abundance of single species. However, the intensity of decrease is different between species: *Cynictis penicillata* on average decreases by only 2%, whereas *Felis silvestris* decreases by 18%.

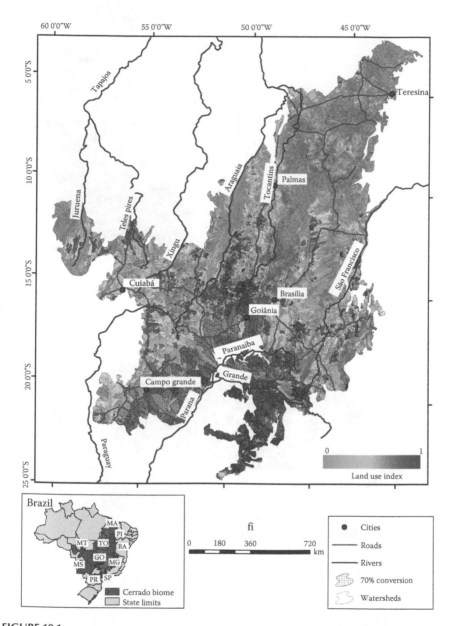

FIGURE 19.1
Land use in the cerrado biome relative to 15,612 watersheds (0–100% range). Black dots correspond to watersheds with conversion over 70%.

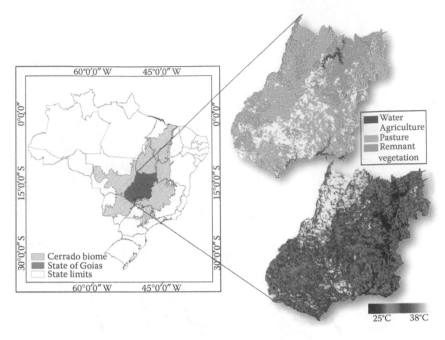

FIGURE 19.2
Land surface temperature for the State of Goiás (core cerrado region) (1 km spatial resolution MOD11 daytime product for day-of-year 161).

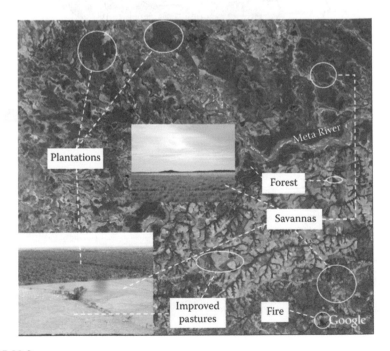

FIGURE 20.2
Satellite image showing the main land covers of the Colombian Llanos and the savanna transformation process (Insert photos by A. Etter).

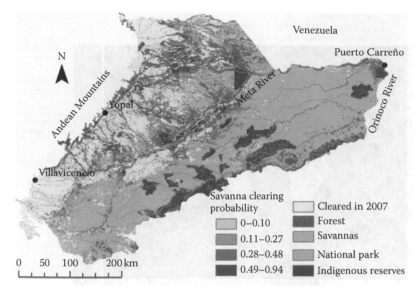

FIGURE 20.5
Predicted future agricultural frontier expansion into the savanna ecosystems of the Colombian Llanos for the period 2007–2020 obtained by applying the logistic regression model. The progression of the expansion is shown by the red shadings from darker to lighter ones.

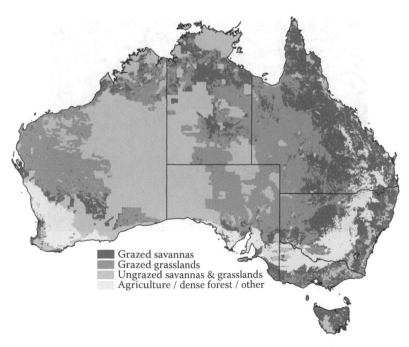

FIGURE 21.1
Map showing distribution of grazed and ungrazed savannas and grasslands in Australia.

Land cover map with the 6 biomes—points marked "urban,"
barren, desert, or very sparsely vegetated were set to 0

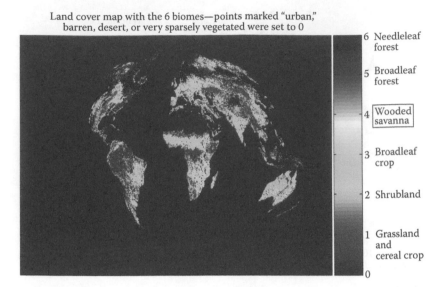

6 Needleleaf
forest

5 Broadleaf
forest

4 Wooded
savanna

3 Broadleaf
crop

2 Shrubland

1 Grassland
and
cereal crop

0

FIGURE 22.1
Global terrestrial biome map from the MODIS satellite instrument. Numbered biome codes are
also listed in Table 22.2. Areas classified as urban or desert were labeled as zero.

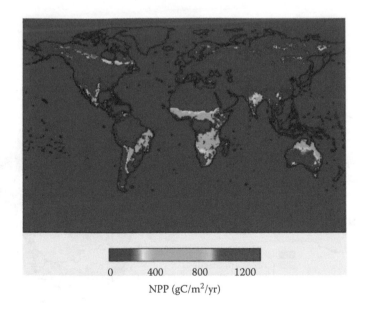

NPP (gC/m^2/yr)

FIGURE 22.3
Net primary production of savanna biomes estimate at 0.5 degrees latitude and longitude from
the CASA model. (Adapted from Potter, C. S. 1999. *Bioscience* 49, 769–778.)

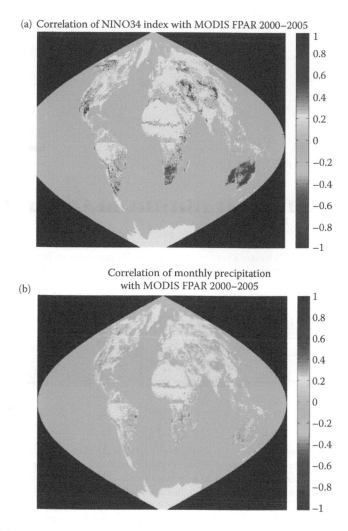

(a) Correlation of NINO34 index with MODIS FPAR 2000–2005

Correlation of monthly precipitation
with MODIS FPAR 2000–2005

(b)

FIGURE 22.4

Correlations of monthly vegetation FPAR with climate variables anomalies in 2000–2005. (a) NINO34 index and (b) NCEP precipitation. White areas of the maps are those with more than one missing monthly FPAR value per year and are, therefore, not considered significant at the $p < 0.10$ level. The break line in correlation results across the equatorial regions is an artifact of separating the Northern and Southern Hemisphere biomes, which are 6 months out of phase.

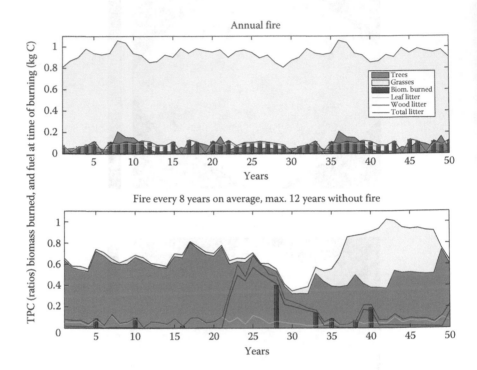

FIGURE 24.2

Simulated tree and grass total projective cover, burnt biomass, and litter at the time of burning for a grid cell with 870 mm MAP over a 50 year period. Simulations used a repeated 27-year climatology calculated from daily NCEP data as in Weber et al. (2009). CO_2 concentration was set to 341 ppm. Fire return intervals were annual in the upper panel and eight years (on average) in the lower panel. Note that the climatic and fuel conditions modify the combustion completeness, hence some fires may result in very low burned biomass and, therefore, may not be visible in the figure. Grasses reach a maximum TPC < 1 to allow for a certain amount of radiation reaching the ground necessary for subsequent establishment; grasses vanish once trees occupy a higher value than the maximum possible TPC for grasses.

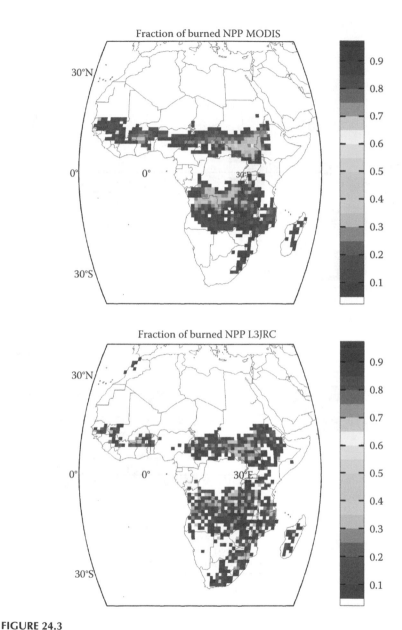

FIGURE 24.3

Fraction of combusted NPP, burnt area prescribed from remote sensing products MODIS MCD45 (top panel; applied was an average burn frequency for each day and grid cell derived from the available data), and L3JRC (lower panel). NPP was calculated from a daily climatology. Displayed are averages for the years 2001–06. (Adapted from Lehsten, V. et al. 2009. *Biogeosciences* 6, 349–360; Weber, U. et al. 2009. *Biogeosciences* 6, 285–295.)

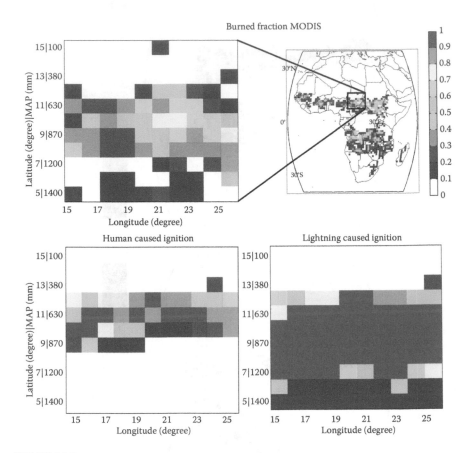

FIGURE 24.5
Influence of ignition on average fraction of burned area over the period 2001–2007; upper right: continental burned fraction remotely sensed by MODIS MCD45; all other panels display an area defined by an MAP gradient from 1400 to 100 mm, located around 5°N to 15°N and 15°E 25°E. Upper left: Fractional burned area remotely sensed by MODIS. Lower left panel: Simulated, human ignition only. Lower right: Simulated, lightning ignition only (maximum assumption).

FIGURE 27.1
Photos showing variation in global savannas. (a) Australia: (1) sorghum woodland (photo: Marks); (2) *Aristida* spp. woodland, Larimah sandy site, NATT; (3) *Triodia* spp. woodland, Larimah loam site, NATT; and (4) Dichanthium woodland, Larimah clay site, NATT (photos: Hill). (b) South America: (1) cerrado (photo: Bustamante); (2) Colombian llanos—gallery forest (photo: Devia); (3) Colombian llanos—rolling savanna (photo: Etter); and (4) Colombian lla-nos—grassland with sparse shrubs (photo: Repizzo).

FIGURE 27.1 Continued.
(c) Africa: (1) seasonal settlement in Sahelian savanna near Hombori, Mali (photo: Prihodko); (2) south Sahelian savanna at Lakamane, Mali (photo: Hanan); (3) agriculturally modified Sudanian savanna near Segou, Mali (photo: Prihodko); (4) Sudanian savanna, Foret Classe Faya, Mali (photo: Hanan); (5) Sudanian savanna, Monte Mandinge, Mali (photo: Hanan); and (6) knobthorn savanna, Kruger Park, South Africa (photo: Prihodko).

Printed and bound by CPI Group (UK) Ltd, Croydon, CR0 4YY

01/11/2024

01782618-0007